Orchids: Phytochemistry and Biology

Orchids: Phytochemistry and Biology

Editor: Finnley Johnson

Callisto Reference
www.callistoreference.com

Callisto Reference,
118-35 Queens Blvd., Suite 400,
Forest Hills, NY 11375, USA

Visit us on the World Wide Web at:
www.callistoreference.com

ISBN: 978-1-64116-820-5 (Hardback)

Cataloging-in-Publication Data

Orchids : phytochemistry and biology / edited by Finnley Johnson.
p. cm.
Includes bibliographical references and index.
ISBN 978-1-64116-820-5
1. Orchids. 2. Botanical chemistry. 3. Biology. I. Johnson, Finnley.
QK495.O64 O73 2023
584.15--dc23

Table of Contents

Preface

Orchid is a large family of flowering plants, which belong to genus Orchidaceae. Orchid plants produce complex flowers having a large specialized lip (labellum). Most of the orchids are perennial. They are found in terrestrial and epiphytic habitats. Orchid phytochemistry refers to the study of phytochemicals or the chemicals derived from orchid plants. Some of the common phytochemicals detected in orchids are alkaloids, phenols, flavonoids, tannins and proteins. Orchid research focused on the biology of orchids involves the cellular and molecular processes of orchid plants such as pigment and floral scent formation, pollination, bioactive compounds, and plant-microbe interactions in orchid species. This book includes some of the vital pieces of works being conducted across the world, on various topics related to orchids. Its extensive content provides the readers with a thorough understanding of the phytochemistry and biology of orchids. The book will prove to be immensely beneficial to students and researchers studying this family of flowering plants.

Various studies have approached the subject by analyzing it with a single perspective, but the present book provides diverse methodologies and techniques to address this field. This book contains theories and applications needed for understanding the subject from different perspectives. The aim is to keep the readers informed about the progresses in the field; therefore, the contributions were carefully examined to compile novel researches by specialists from across the globe.

Indeed, the job of the editor is the most crucial and challenging in compiling all chapters into a single book. In the end, I would extend my sincere thanks to the chapter authors for their profound work. I am also thankful for the support provided by my family and colleagues during the compilation of this book.

Editor

1

Functional Insights into the Roles of Hormones in the *Dendrobium officinale-Tulasnella* sp. Germinated Seed Symbiotic Association

Tao Wang [1,2], Zheng Song [1], Xiaojing Wang [1], Lijun Xu [1], Qiwu Sun [1,*] and Lubin Li [1,*]

1 State Key Laboratory of Forest Genetics and Tree Breeding, Key Laboratory of Silviculture of the State Forestry Administration, Research Institute of Forestry, Chinese Academy of Forestry, Beijing 100091, China; wangtao5757@126.com (T.W.); songzheng566@163.com (Z.S.); wangxjwork@163.com (X.W.); xulijun19935@126.com (L.X.)

2 Beijing Botanical Garden, Beijing 100093, China

* Correspondence: sqw@caf.ac.cn (Q.S.); lilubin@126.com (L.L.)

Abstract: *Dendrobium* is one of the largest genera in the Orchidaceae, and *D. officinale* is used in traditional medicine, particularly in China. *D. officinale* seeds are minute and contain limited energy reserves, and colonization by a compatible fungus is essential for germination under natural conditions. When the orchid mycorrhizal fungi (OMF) initiates symbiotic interactions with germination-driven orchid seeds, phytohormones from the orchid or the fungus play key roles, but the details of the possible biochemical pathways are still poorly understood. In the present study, we established a symbiotic system between *D. officinale* and *Tulasnella* sp. for seed germination. RNA-Seq was used to construct libraries of symbiotic-germinated seeds (DoTc), asymbiotic-germinated seeds (Do), and free-living OMF (Tc) to investigate the expression profiles of biosynthesis and metabolism pathway genes for three classes of endogenous hormones: JA (jasmonic acid), ABA (abscisic acid) and SLs (strigolactones), in *D. officinale* seeds and OMF under symbiotic and asymbiotic conditions. Low concentrations of endogenous JA, ABA, or SLs were detected in the *D. officinale-Tulasnella* symbiont compared with the asymbiotic tissues. Gene annotation results suggest that the expression of DEGs (differentially expressed genes) related to JA and ABA biosynthesis from *D. officinale* were down-regulated, while most of the key DEGs related to SL biosynthesis from *D. officinale* were up-regulated in the symbiotic germinated seeds compared with the asymbiotic germinated seeds. Moreover, in the OMF, we found a significantly up-regulated differential expression of the JA and ABA biosynthesis-related genes in the symbiotic interaction, with the opposite expression trends to those found in *Dendrobium*. This indicates that *Dendrobium* seed symbiotic germination may be stimulated by the apparent involvement of the OMF in the production of hormones, and relatively low concentrations of endogenous JA, ABA, or SLs might be maintained to promote the growth of the *D. officinale-Tulasnella* symbiotic protocorm-like body. These results will increase our understanding of the possible roles played by endogenous hormones in the regulation of the orchid-fungus symbiosis.

Keywords: *Dendrobium officinale*; *Tulasnella* sp.; symbiosis; endogenous hormones

---

1. Introduction

In the majority of orchid species, seed germination is largely dependent on mycorrhizal fungi under natural conditions because of the lack of nutritional reserves [1]. Symbiotic seed germination, which is essential for orchid propagation and reintroduction, can be used for the conservation of the *Dendrobium* species in their natural habitats [2]. The process of orchid seed germination under natural conditions is complex and unique, and involves various physiological and biochemical processes such as colonization, plant growth stimulation, and the plant response to the fungus [1]. Although successes in orchid seed germination are regularly reported in the literature, it is important to note that when comparing the asymbiotic and symbiotic germination of the same species, the rates of orchid seed germination, development of the protocorms, and growth is usually better in the latter [3,4]. Many studies have shown that orchid mycorrhizal fungi (OMF) promote metabolic processes in orchid seeds [4,5]. Although orchid seeds contain very low nutrient reserves, recent analyses of the unique properties of protocorms have shown that during symbiotic seed germination, the expression of putative genes involved in metabolism, transcriptional regulation, and signal transduction pathways is upregulated [5–7]. Miura et al. (2018) examined an orchid-mycorrhizal symbiosis by analyzing the transcriptome of *Bletilla striata* associated with *Tulasnella* sp. at an early developmental stage, and the essential genes required for the establishment of a mutualistic relationship with AM fungi and/or rhizobia in most terrestrial plants were identified from *B. striata* [8]. These studies indicate that the symbiotic association formed from the germinated seed is able to interact with mycorrhizal fungi.

When mycorrhizal fungi undergo symbiotic interactions with orchids, phytohormones produced by the orchid and/or the microbial partners could play key roles [9–11]. The symbiosis between *Cymbidum goeringii* and a *Rhizoctonia*-like mycorrhizal fungi causes the release of hormones, which can promote the growth of *C. goeringii* seedlings [9]. The results of comparative transcriptomic and proteomic analyses suggested that an ABA receptor protein, PYR1, showed a significantly higher expression during the early symbiotic germination stages (when the embryo swells, enlarges, and emerges from the seed coat) in *D. officinale* [10]; the developmental stages are defined according to Stewart et al. [12]. The role of JA signaling is well known in plant–pathogen interactions [11]. A recent study also detected higher levels of lipoxygenase in *Oncidium sphacelatum* at the green protocorm stage of seed development with *Ceratobasidium* sp., and this could possibly result in an enhanced JA biosynthesis in the green protocorms with the fungal symbiont [6]. These studies indicate that the functions of the growth-inhibiting hormones ABA and JA could be important for symbiotic seed germination and growth of *Dendrobium*.

In higher plants, ABA is derived from the oxidative cleavage of carotenoids. Isopentenyl diphosphate (IPP) formed in the 1-deoxy-D-xylulose-5-phosphate (DXP) pathway is converted to geranylgeranyl pyrophosphate (GGPP), from which carotenoids are synthesized [13,14]. The carotenoid β-carotene is hydroxylated to form the xanthophyll zeaxanthin, which is further converted into ABA by a series of enzymes, such as zeaxanthin epoxidase (ZXE) and short-chain dehydrogenase/reductase (SDR). It has been reported that the symbiotic fungi, including OMF, can also produce compounds that are similar to the plant hormome ABA, although studies have shown that the ABA biosynthetic pathways seem to differ between higher plants and fungi [15]. Fungi synthesize ABA via a direct pathway from isopentenyl diphosphate (IDP) and farnesyl diphosphate (FDP), which are derived from the mevalonic acid (MVA) pathway [16,17]. After a series of cyclization, isomerization, and oxidization reactions, ABA is synthesized from farnesyl pyrophosphate (FPP). Labeling experiments have been performed to characterize the ABA biosynthesis pathway in *Botrytis cinerea, Cercospora rosicolla*, and some other species of fungi [18–20]. Previous studies have shown that 1′,4′-*trans*-diol-ABA is the main precursor of ABA in *B. cinerea* and that Δ*BcABA4* transformants accumulate 1′,4′-*trans*-diol-ABA [17]. The *BcABA4* gene encodes a putative SDR (short-chain-type dehydrogenase/reductase) which is involved in the ABA biosynthesis in *Arabidopsis* [16].

The JA biosynthesis pathways of plants and OMF are more similar. JA synthesis is initiated in the plastid by the oxygenation of linoleic or α-linolenic acid by lipoxygenases (LOXs) to form (13*S*)-hydroperoxy linoleic or α-linolenic acid. This is then followed by the first committed step of JA biosynthesis catalyzed by allene oxide synthase (AOS) that converts 13-HPOT to a highly reactive allene oxide, which is converted to *cis*-(+)-12-oxophytodienoic acid by allene oxide cyclase (AOC). *cis*-(+)-OPDA is then transported to the peroxisome, where 12-oxophytodienoate reductase 3 (OPR3) and three rounds of β-oxidation convert it to JA [21].

Previous studies have shown that plants generally produce multiple strigolactone (SL) species which are released from the roots and induce germination in seeds of the *Striga* and *Phelipanche* species in the soil [22,23]. SLs or their biosynthetic precursors act as plant hormones to inhibit shoot branching in plants [24], which was discovered in studies using a series of enhanced shoot branching mutants [22–25]. SLs also function as host recognition signals for AM (arbuscular mycorrhizal) fungi, as well as branching factors in AM fungi [22,23]. A hypothetical SL-biosynthetic pathway was initially proposed, with β-carotene as a substrate for carotenoid-cleavage dioxygenase enzymes (CCD7 and CCD8, which act sequentially in the pathway). Further investigation of the *MAX1* gene, which encodes a cytochrome P450 in the CYP711A1 clade, acting downstream to CCD8 and of its orthologs in *Arabidopsis* and rice revealed that this gene is expressed in all vascular tissues, and functions only in the late steps of SL synthesis [26]. D14/DAD2 (a probable strigolactone esterase) may be a putative SL receptor involved in SL signaling pathway regulation. Although it has been hypothesized that SL phytohormones have important functions during plant-fungus interactions and symbioses [23], little is known about the role of SLs in the symbiotic associations between orchids and OMF.

These phytohormones are well known to control plant development and to trigger important signaling events [15]. Moreover, plant hormones produced by mycorrhizal fungi have been shown to improve the growth of medicinal orchids [4]. In addition, the possible biosynthesis pathways for plant and fungal hormones in the orchid–fungal symbiosis are poorly understood at present. The manner in which these molecules favor the invasion of plant tissues and the development of fungi inside plant tissues is also largely unknown. To this end, we performed comparative transcriptomic and hormone profiling between symbiotically and asymbiotically germinated *D. officinale* seeds in an effort to gain possible functional insights into the biosynthesis and metabolism of ABA, JA, and SLs in the regulatory module in the *Dendrobium–Tulasnella* symbiosis. We also predicted some unigenes involved in phytohormone production by the symbiotic fungus *Tulasnella* sp., and these genes had different expression profiles in symbiotic and asymbiotic germinated seeds.

2. Results

2.1. Detecting Mycorrhizal Fungus Colonization in Germinated Seeds of D. officinale

Both morphological observation and Trypan blue staining, which showed that the intracellular fungal hyphae stained blue in the symbiotic germinated seeds (DoTc; Figure 1), indicated that a symbiotic relationship had been established between *D. officinale* germinated seeds and *Tulasnella* sp., while in the group of seeds germinated without fungi with a low germination rate (0.1%), we did not find hyphae, and these were therefore defined as asymbiotic germinated seeds (Do).

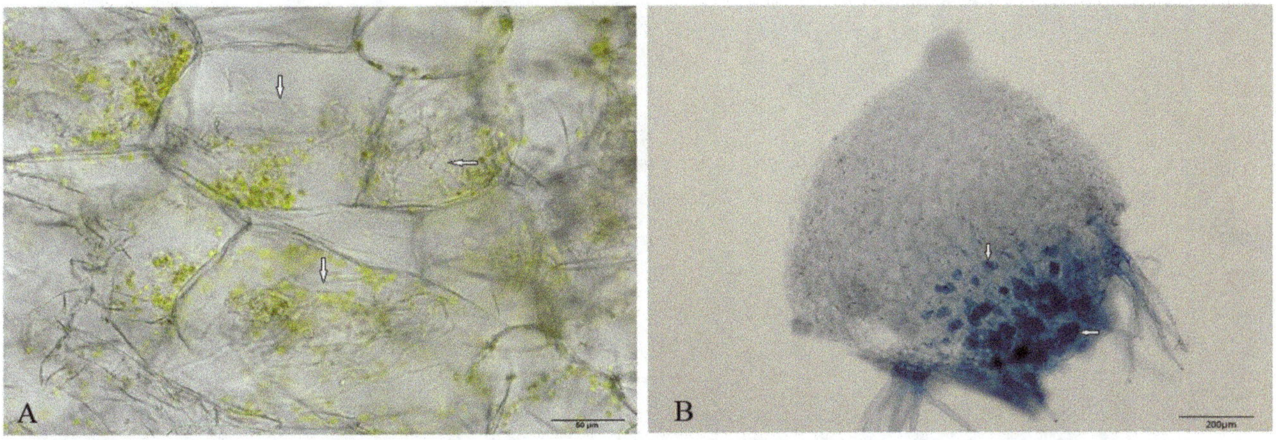

Figure 1. The morphological observation of intracellular hyphae in the symbiotic germinated seeds of *D. officinale*. The microscopic observation of intracellular hyphae in a free-hand section (**A**) of germinated seeds were stained blue following Trypan blue staining (**B**). The arrows indicate the symbiotic mycelia.

2.2. Transcriptome Sequence Assembly and Annotation.

To globally characterize the transcriptome of *D. officinale* and its *Tulasnella* endosymbiont with enhanced sequence coverage, nine cDNA libraries were constructed using RNA extracted from DoTc, Do, and free-living *Tulasnella* mycelium (Tc). In total, 667,270,886 Illumina paired-end raw reads were generated. The raw reads are available in the NCBI SRA database under the accession number SRP136098. After discarding the adaptor and low-quality sequences, we obtained 642,119,290 clean reads (accounting for 96 Gb of DNA sequence). The percentages of clean reads among the raw tags (Q20) ranged from 95.52% to 96.11% (Table S1). The cleaned reads of *D. officinale* from the DoTc and Do libraries were then mapped to the *D. officinale* genome database [27] with 45.7–50.27% and 60.62–61.04% of the reads in the DoTc and Do libraries, respectively, mapping to the genome (Table S2). The novel genes annotated in *Dendrobium officinale* are shown in Table S3.

The de novo assembly of the clean reads was performed and resulted in 50,597 unigenes that ranged from 201 to 12,137 bp in length, with an N50 length of 1488 bp in the *Tulasnella* transcriptome. The size distribution of the transcripts and unigenes in *Tulasnella* is shown in Table S4. The transcriptome assembly produced a substantial number of large unigenes: 15,254 unigenes were >1000-bp in length (Table S4). All assembled unigenes (Table S5) were annotated using BLASTX searches against the NR, NT, KO, SwissProt, PFAM, GO, and KOG databases (Table S6). In total, there were 1637 (3.23%) unigenes that were annotated in all databases, and 35,616 (79.39%) that were annotated in at least one database (Table S6), providing significant BLAST results. Among the annotated unigenes, 23,758 (46.95%) showed significant similarity to known proteins in the Nr database, 23,532 unigenes (46.50%) in the GO database, and 9105 unigenes (17.99%) in the KOG database (Table S6).

2.3. Differentially Expressed Genes in Symbiotic and Asymbiotic Germinated Seeds and Mycorrhizal Fungi

To compare the differences in gene expression between the symbiotic germinated seeds, asymbiotic germinated seeds, and mycorrhizal fungi, we compared the transcriptome profiles between the symbiotic and asymbiotic germinated seeds (DoTc vs. Do) and symbiotic and asymbiotic mycorrhizal fungi (DoTc vs. Tc). A total of 3367 DEGs (differentially expressed genes) were obtained that included 2230 up-regulated and 1137 down-regulated genes in the DoTc vs. Do comparison (Figure 2A and Table S7), and 7720 differentially expressed unigenes that included 7028 up-regulated and 692 down-regulated unigenes in the DoTc vs. Tc comparison (Figure 2B and Table S8).

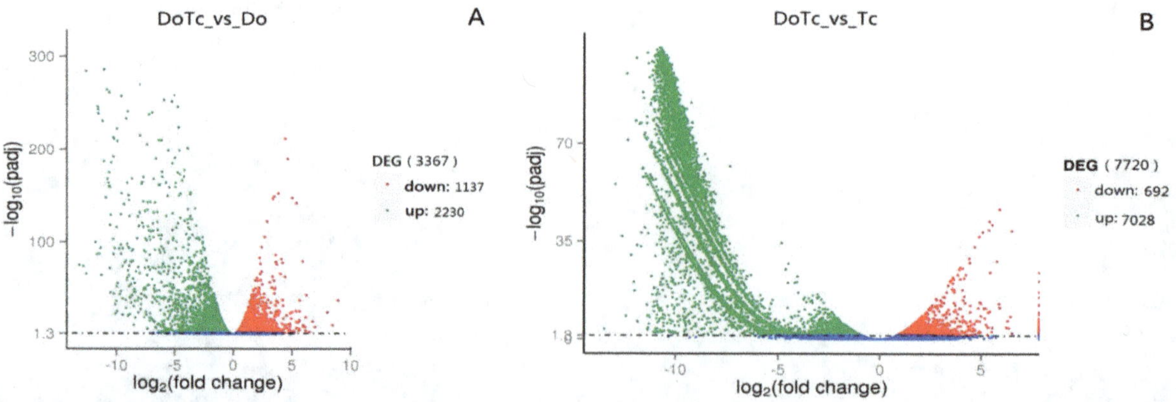

Figure 2. The distribution of differentially expressed genes (DEGs) in symbiotic and asymbiotic germinated seeds. (**A**) DEGs originating from *D. officinale* in the symbiotic germinated seeds (DoTc vs. Do), and (**B**) DEGs originating from OMF (orchid mycorrhizal fungi) in the symbiotic germinated seeds (DoTc vs. Tc). Genes in which the relative level of expression was down-regulated are shown in red, and those that were up-regulated are shown in green.

The annotation results suggest that orchid seed germination stimulated by the fungal association apparently involves plant hormone signal transduction, plant-pathogen interaction, and metabolic pathways from *D. officinale* that were obviously up-regulated (Figure S1) in the symbiotic germinated seeds (DoTc vs. Do). However, in the mycorrhizal fungus, the expression of genes for proteins implicated in the ribosome, proteasome biosynthesis, and protein processing in the endoplasmic reticulum were found in the symbiotic interaction (Figure S1, DoTc vs. Tc).

It is well known that plant hormones have important functions during plant-fungal interactions. In the present study, we found that DEGs related to hormone pathways show a significant differential expression (Figure S1A); we thus sought to gain possible functional insights into the biosynthesis and metabolism of three types of plant hormones—JA, ABA, and SL—in the regulatory module of the *Dendrobium–Tulasnella* symbiosis. In the ABA biosynthesis pathway, the expression of *ZXEs* and *SDRs* in *D. officinale* was significantly down-regulated in the symbiotic germinated seeds compared with the asymbiotic germinated seeds. However, the possible ABA biosynthesis-related short-chain type dehydrogenase/reductase-encoding gene *BcABA4* [16] and the P450 monooxygenase-encoding genes *BcABA1/2* [17] were significantly up-regulated in the symbiotic germinated seeds compared with free-living OMF (Table 1). Moreover, the JA biosynthesis pathway-related genes from the plant and OMF showed similar expression profiles in the symbiotic seeds, but with opposite trends. We have observed that some JA biosynthesis-related enzyme-encoding genes, such as LOX, AOS, OPR, and 3-ketoacyl-CoA thiolase (KAT) are down-regulated in the symbiotic germinated seeds relative to their expression in the asymbiotic germinated seeds, while the key genes in the *Tulasnella* JA biosynthesis pathway showed an up-regulated expression in the symbiotic germinated seeds. In the SL biosynthesis pathway, the isomerase-encoding gene *D27*, two carotenoid-cleavage dioxygenases *CCD7* and *CCD8*, and the *CYP450* monooxygenase gene *MAX1* were all up-regulated in the symbiotic germination seeds. In addition, five *D14/DAD2* genes were also found to be up-regulated in the *D. officinale-Tulasnella* symbiotic germination seeds.

Table 1. The DEGs in the DoTc vs. Do and DoTc vs. Tc comparisons involved in the ABA (abscisic acid), JA (jasmonic acid), and SL (strigolactone) biosynthesis pathways.

Gene Name	Gene ID		Description
ABA pathway		log2(DoTc/Do)	
DoZXE-1	Dendrobium_GLEAN_10076665	−2.11	Zeaxanthin epoxidase
DoZXE-2	Novel00252	−2.00	Zeaxanthin epoxidase
DoSDR-1	Dendrobium_GLEAN_10042288	−0.57	Short-chain type dehydrogenase/reductase
DoSDR-2	Dendrobium_GLEAN_10068100	−1.33	Short-chain type dehydrogenase/reductase
DoSDR-3	Dendrobium_GLEAN_10127951	−2.1	Short-chain type dehydrogenase/reductase
ABA pathway		log2(DoTc/Tc)	
BcABA4-1	Cluster-2641.12008	7.24	Short-chain type dehydrogenase/reductase dehydrogenase/reductase
BcABA4-2	Cluster-2152.0	2.42	Short-chain type dehydrogenase/reductase
BcABA1/2-1	Cluster-2641.19250	10.43	P450 monooxygenase protein CYP51
BcABA1/2-2	Cluster-2641.15199	10.01	P450 monooxygenase protein CYP51
JA pathway		log2(DoTc/Do)	
DoLOX	Dendrobium_GLEAN_10090503	−0.67	Linoleate 13S-lipoxygenase
DoAOS-1	Dendrobium_GLEAN_10010006	−0.98	Allene oxide synthase
DoAOS-2	Dendrobium_GLEAN_10012197	−0.40	Allene oxide synthase
DoOPR	Dendrobium_GLEAN_10113554	−1.65	12-oxophytodienoate reductase
DoKAT	Dendrobium_GLEAN_10097185	−2.50	3-ketoacyl-CoA thiolase

Table 1. *Cont.*

Gene Name	Gene ID		Description
JA pathway		log2(DoTc/Tc)	
TcLOX-1	Cluster-2641.17948	10.60	Linoleate 13S-lipoxygenase
TcLOX-2	Cluster-2641.19486	10.33	Linoleate 13S-lipoxygenase
TcLOX-3	Cluster-2641.18695	10.34	Linoleate 13S-lipoxygenase
TcLOX-4	Cluster-2641.17988	10.66	Linoleate 13S-lipoxygenase
TcLOX-5	Cluster-2641.16009	9.35	Linoleate 13S-lipoxygenase
TcLOX-6	Cluster-6210.0	6.67	Linoleate 13S-lipoxygenase
TcLOX-7	Cluster-1378.1	5.99	Linoleate 13S-lipoxygenase
TcAOS-1	Cluster-2641.19289	10.40	Allene oxide synthase
TcAOS-2	Cluster-2641.20726	10.07	Allene oxide synthase
TcAOS-3	Cluster-2641.16940	7.81	Allene oxide synthase
TcAOS-4	Cluster-2641.8055	6.54	Allene oxide synthase
TcAOS-5	Cluster-2641.30606	7.21	Allene oxide synthase
SL pathway		log2(DoTc/Do)	
DoHMGR-1	Dendrobium_GLEAN_10079864	0.87	Hydroxymethylglutaryl-CoA reductase
DoHMGR-2	Dendrobium_GLEAN_10079865	0.91	Hydroxymethylglutaryl-CoA reductase
DoD27	Dendrobium_GLEAN_10126990	5.22	β-carotene isomerase
DoCCD7-1	Dendrobium_GLEAN_10008301	4.21	Carotenoid cleavage dioxygenase 7
DoCCD7-2	Dendrobium_GLEAN_10009315	2.98	Carotenoid cleavage dioxygenase 7
DoCCD7-3	Dendrobium_GLEAN_10049685	9.92	Carotenoid cleavage dioxygenase 7
DoCCD8-1	Dendrobium_GLEAN_10070249	3.87	Carotenoid cleavage dioxygenase 8
DoCCD8-2	Dendrobium_GLEAN_10072088	4.08	Carotenoid cleavage dioxygenase 8
DoMAX1	Dendrobium_GLEAN_10109899	2.00	Cytochrome P450 711A1
DoD14-1	Dendrobium_GLEAN_10025168	2.45	Probable strigolactone esterase
DoD14-2	Dendrobium_GLEAN_10114182	1.52	Probable strigolactone esterase
DoD14-3	Dendrobium_GLEAN_10143398	1.86	Probable strigolactone esterase
DoDAD2-1	Dendrobium_GLEAN_10013491	2.01	Probable strigolactone esterase
DoDAD2-2	Dendrobium_GLEAN_10114181	1.34	Probable strigolactone esterase

2.4. Expression of DEGs Related to ABA, JA, and SL Hormone Biosynthesis Detected by qRT-PCR

In the symbiotic and asymbiotic germinated seed samples, we successfully detected the expression of key genes related to ABA, JA, and SL biosynthesis, and the results are shown in Figure 3. Two *CCD7* and *CCD8* genes and the *MAX1* gene, which are key genes in the *D. officinale* SL biosynthesis pathway, all showed up-regulated expression in the symbiotic germinated seed (DoTc) samples, based on expression profiles determined from transcriptome sequencing. In the symbiotic roots of *D. officinale* seedlings (SR), these genes were all down-regulated compared to the control asymbiotic root samples (ASR), which showed expression trends that were opposite to the SGS samples, with the exception of *CCD8-2*. In the ABA and JA biosynthesis pathways, the expression of the *ZXE*, *SDR-1* and *SDR-2*, *LOX*, *AOS-1* and *AOS-2*, and *OPR* genes from *D. officinale* was significantly down-regulated in the SGS and SR samples compared to one or both of the asymbiotic control samples (Do and ASR). Moreover, we also assayed the differential expression of four *BcABA* genes (*BcABA4-1*, *BcABA4-2*, *BcABA1/2-1*, and *BcABA1/2-2*) which could be involved in ABA biosynthesis in the OMF *Tulasnella* sp. in the SGS and SR samples and free-living *Tulasnella* mycelium (Tc). We found that the expression in the SGS and SR samples was significantly up-regulated compared with that in the Tc samples, and the expression of these four genes in the SGS samples was also based on the expression trends from the transcriptome sequencing and was more significantly upregulated than in the SR samples.

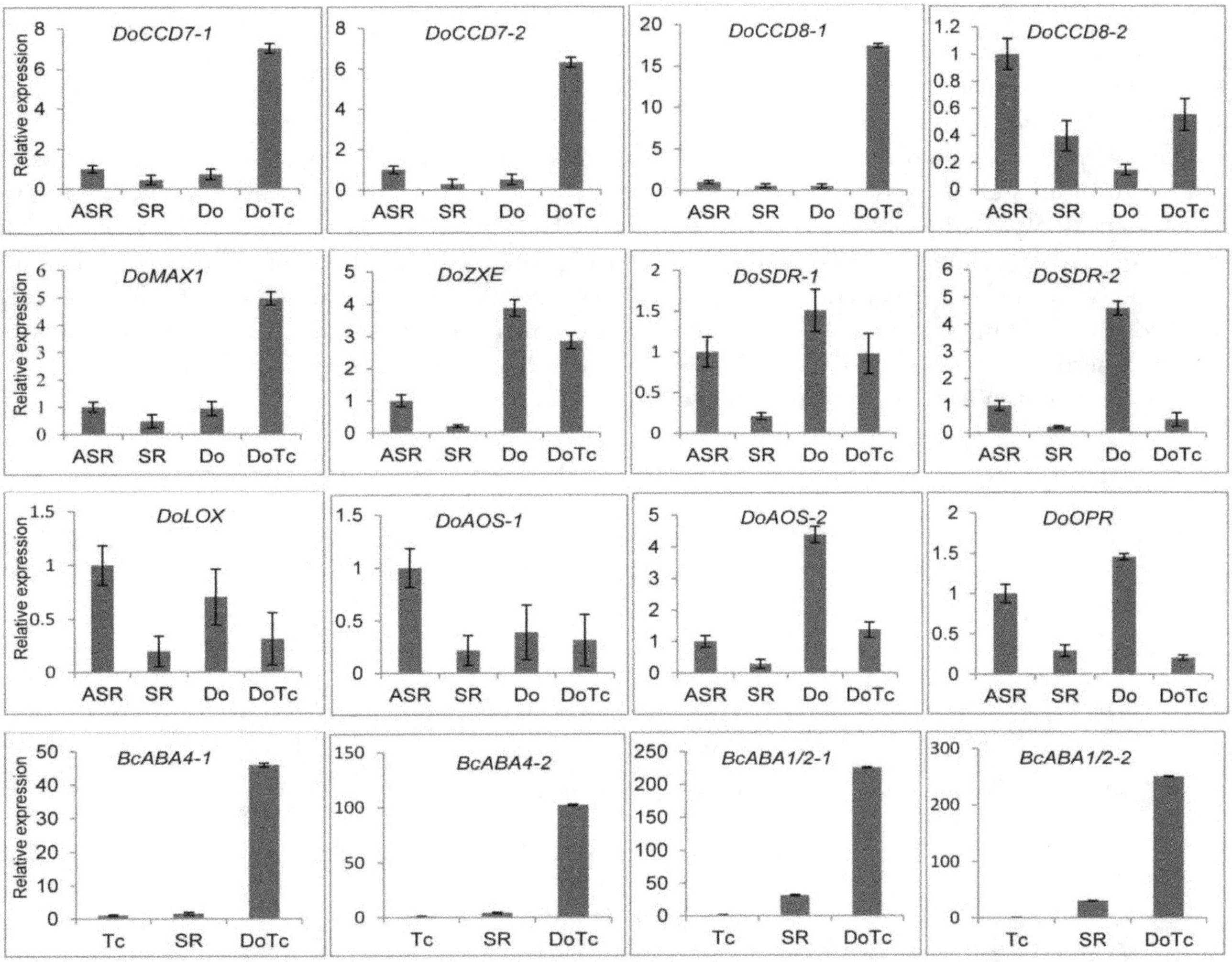

Figure 3. The expression levels of hormone biosynthesis-related genes determined by qRT-PCR. DoTc: symbiotic germinated seeds; Do: asymbiotic germinated seeds; SR: symbiotic roots of *D. officinale* seedlings; ASR: asymbiotic roots of *D. officinale* seedlings; Tc: free-living *Tulasnella* mycelium. Each data point represents the mean of three biological replicates and three experimental replicates. Bars represent the standard error of the mean, SE ($n = 9$).

2.5. Assays of ABA, JA, and SL Contents

In order to perform a deep analysis to determine their possible functions, we assayed the absolute content of ABA, JA, and 5-deoxystrigol (5-DS) in the symbiotic germinated seeds, asymbiotic germinated seeds, and free-living *Tulasnella* mycelium, corresponding to the RNA-Seq libraries DoTc, Do, and Tc, respectively. The results showed that in the symbiotic *D. officinale* germinated seeds, the JA concentration was slightly higher than in free-living *Tulasnella* mycelium and much lower than in the asymbiotic germinated seeds (Figure 4), and the ABA concentration was lower than in the two types of control samples (Figure 5). 5-DS, a precursor of SLs, is an essential index for quantifying SLs [28]. The 5-DS contents in the different tissues varied from 0.21 to 2.03 ng/g·FW); the lowest content was in the free-living *Tulasnella* mycelium, followed by the symbiotic germinated seeds, with the highest content found in the asymbiotic germinated seeds (Figure 5).

3. Discussion

The role of JA signaling is well known in plant–pathogen interactions, and it also possibly results in the enhanced biosynthesis of JA in the green protocorms of *Oncidium sphacelatum* in association with the fungal symbiont *Ceratobasidium* sp. [6]. We have observed that the JA concentration in the symbionts is lower than in asymbiotic germinated seeds, but it is higher than in free-living fungal hyphae of *Tulasnella* sp. (Figure 4). Moreover, the trend of changes in JA concentration seen in the symbionts and control samples is supported by the relative expression of JA biosynthesis pathway genes. We observed that some JA biosynthesis-related enzyme-encoding genes, such as *LOX*, *AOS*, *OPR*, and *KAT*, were down-regulated in the symbiotic germinated seeds relative to their expression in the asymbiotic germinated seeds, which suggested that JA, as a kind of defense-associated plant hormone, may also be involved in the regulation of *D. officinale* seed association with mycorrhizal fungi. We can speculate that the colonization of the protocorms by the *Tulasnella* fungus resulted in the downgrading of JA biosynthesis-related genes, and allowed the fungus to enter cortical cells unimpeded and set up an efficient mutualism. The key genes in the *Tulasnella* JA biosynthesis pathway showed a significantly up-regulated expression, indicating that JA produced by *Tulasnella* might accumulate to a high level in the symbiotic germinated seeds.

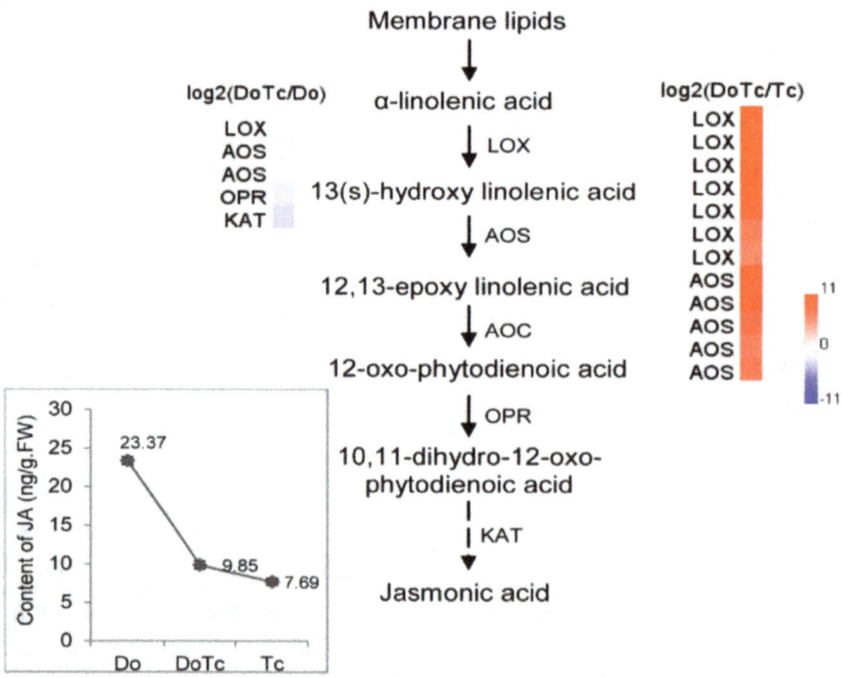

Figure 4. The genes involved in the biosynthesis of the oxylipin JA identified in the symbiotic germinated seeds of *D. officinale* in association with a *Tulasnella* sp. isolate. LOX: Linoleate 13S-lipoxygenase; AOS: Allene oxide synthase; AOC: Allene oxide cyclase; OPR: 12-oxophytodienoate reductase; KAT: 3-ketoacyl-CoA thiolase; JMT: Jasmonate O-methyltransferase. Red and blue indicate up- and down-regulated transcripts, respectively, from the DoTc vs. Do and DoTc vs. Tc comparisons. Each data point represents the mean of six biological replicates. Bars represent the standard error of the mean, SE ($n = 6$).

The putative biosynthesis network consisting of the ABA and SL biosynthesis pathways in the symbiotic germinated seeds of the *D. officinale-Tulasnella* symbiosis is shown in Figure 5. In this network, the *HMGR* gene that encodes 3-hydroxy-3-methylglutaryl-CoA reductase, which catalyzes the reduction of 3-hydroxy-3-methylglutaryl-CoA in the mevalonate pathway, showed an induced expression in the symbiotic germinated seeds. Geranylgeranyl-PP is then catalyzed to form β-carotene, which is the critical branching point for both SL and ABA, and many of the catalyzing enzyme-related genes showed different degrees of expression in the two types of germinated seeds.

Many studies have shown a negative correlation between the concentrations of endogenous ABA and rates of seed germination, which supports the notion that high concentrations of endogenous ABA in seeds may play a critical role in preventing germination in species such as wheat (*Triticum aestivum* L.) [29], cacao (*Theobroma cacao* L.) [30], and rapeseed (*Brassica rapa* L.) [31]. In orchids, it has been suggested that high concentrations of ABA occur in seeds that are difficult to germinate such as *Epipactis helleborine* [32], and *Calanthe tricarinata* [3]. Our results showed that in symbiotic germinated *D. officinale* seeds, the ABA concentration was lower than in the control, indicating that low endogenous ABA levels may have positive effects on *D. officinale* symbiotic seed germination. Previous studies have shown that in symbiotic *D. officinale* seedlings, the ABA concentration was lower than in the control [33], which is in accord with the lower ABA concentration trends observed in the symbiotic germinated seeds compared to the asymbiotic germinated seeds. In our study, the expression of *SDR* and *ZXE*, two genes in the ABA biosynthesis pathway, was significantly suppressed in the symbiotic germinated seeds. These results agree with those of a recent study in which high concentrations of exogenously applied ABA were also found to suppress seed germination in vitro [3].

In our study, the relative transcription of most of the key MVA pathway genes from *Tulasnella* was significantly up-regulated, as was the transcription of some critical ABA biosynthesis unigenes with homologs in *Botrytis*; examples are two putative *BcABA4* unigenes (Cluster-2641.12008 with 7.24-fold up-regulation and Cluster-2152.0 with 2.42-fold up-regulation) and two putative P450 monooxygenase protein CYP51 genes (*BcABA1* or *BcABA2*; Cluster-2641.19250 with 10.43-fold up-regulation and Cluster-2641.15199 with 10.01-fold up-regulation) in the symbiotic germinated seeds compared to the asymbiotic fungus *Tulasnella* sp. (Figure 5 and Table 1), indicating that symbiotic *Tulasnella* sp. might produce more activated ABA than the control. These results are consistent with previous reports showing that ABA affects mycelium growth [11]. In *Ceratocystis fimbriata*, the exogenous application of ABA increased fungal growth slightly. In *Magnaporthe oryzae*, ABA increased spore germination and the formation of appressoria, specialized infection structures for breaking down the plant cell wall and allowing invasion [11]. Moreover, the *Tulasnella* sp. ABA-INSENSITIVE-like protein (Cluster-2641.20009 with 8.94-fold up-regulation) and five ABA receptor PYL-related unigenes (6.54- to 10.06-fold up-regulation) were also found to be significantly up-regulated in the symbiotic germinated seeds compared with the asymbiotic *Tulasnella* sp. fungus. There is also emerging evidence to suggest that ABA plays an important role in compatible interactions in mutualistic plant-microbe interactions [18], and it can promote the infection and establishment of compatible interactions with arbuscular mycorrhizal fungi [16]. However, we found that the ABA concentration in the orchid symbionts is also lower than in the free-living fungal mycelium of *Tulasnella* sp. Based on these results, we can hypothesize that the symbiotic fungus *Tulasnella* produces more activated ABA to promote the recognition and establishment of compatible interactions with its mutualist orchid partner in early seed germination (stages 1–3); it is during these early stages that an abundance of pelotons is usually formed. After protocorm formation, the germinated seed develops to the fourth stage (emergence of leaf from shoot region), the reinvaded hyphae conspicuously decrease, and the hyphae are degraded in the symbiotic orchid cell [34]. Thus, there might be both a collapsed hyphae and healthy pelotons that can be observed. We have detected one putative abscisic acid 8′-hydroxylase-encoding gene, *CYP707A* (Cluster-3911.0), from *Tulasnella* that is highly expressed in symbiotic germinated seeds but not in asymbiotic *Tulasnella*, and this enzyme can degrade and deplete endogenous ABA levels to regulate intracellular ABA homeostasis in plants [16,35]. In line with its important role as a phytohormone, ABA concentrations in the plant are controlled by a tightly regulated balance between biosynthesis, inactivation, and degradation [36]. At this stage, low concentrations of ABA in the symbiont could promote *D. officinale* seed germination and inhibit the growth of *Tulasnella* mycelia.

In our study, we also detected a precursor of SLs (5-DS) in the germinated seeds of *D. officinale*, and in addition, we found that genes encoding key enzymes involved in the conversion of β-carotene to carlactone and mature SLs, *D27*, *CCD7*, *CCD8*, and *MAX1* are expressed in both types of germination seeds, indicating that SLs might also play important roles in regulating *D. officinale* seed germination. Moreover, the content of 5-DS in the asymbiotic germinated seeds (2.03 ng/g·FW) was much higher than in the symbiotic germinated seeds (0.81 ng/g·FW), and this result was supported by the slow growth rate of the asymbiotic germinated seeds compared with that of the symbiotic germinated seeds, indicating that high concentrations of SLs or their biosynthetic precursors might inhibit growth.

In other plants, studies have shown that SLs or their biosynthetic precursors act as plant hormones that inhibit shoot branch growth [22–25]. In addition, the expression of the SL biosynthesis and signal pathway genes increased significantly (>2-fold) in the symbiotic germinated seeds compared to the asymbiotic germinated seeds, which possibly have enhanced the branching of the *Tulasnella* fungus within orchid tissues to maximize the number of pelotons formed. This model might be supported by previous studies showing that SLs also promote branched growth in AM fungi [22,23]. Nevertheless, based on the lower level of 5-DS in the symbiotic germinated seeds and the higher level of 5-DS in the asymbiotic germinated seeds, we can speculate that the symbiotic fungus *Tulasnella* might suppress the biosynthesis of orchid SLs in order to maintain a low concentration of SLs to accelerate the germination rate in the *D. officinale-Tulasnella* symbiosis. Additionally, the germinated seed developed to the fourth stage (emergence of the leaf from the shoot region), the reinvaded hyphae conspicuously decreased, and the hyphae degraded in the symbiotic orchid cell [34]. At this stage, a low concentration of SLs in the symbiont might also slow the growth of the reinvaded *Tulasnella* mycelium.

Previous studies have suggested that ABA is one of the regulators of the strigolactone biosynthesis that acts through an as-yet-unknown mechanism [36]. Moreover, Toh et al. have proposed that host-exudate SLs reduce ABA levels to alleviate seed thermoinhibition in *Arabidopsis* [37,38]. This suggests that there could be complicated cross-talk among the hormones in the orchid-fungus symbionts that warrant future study.

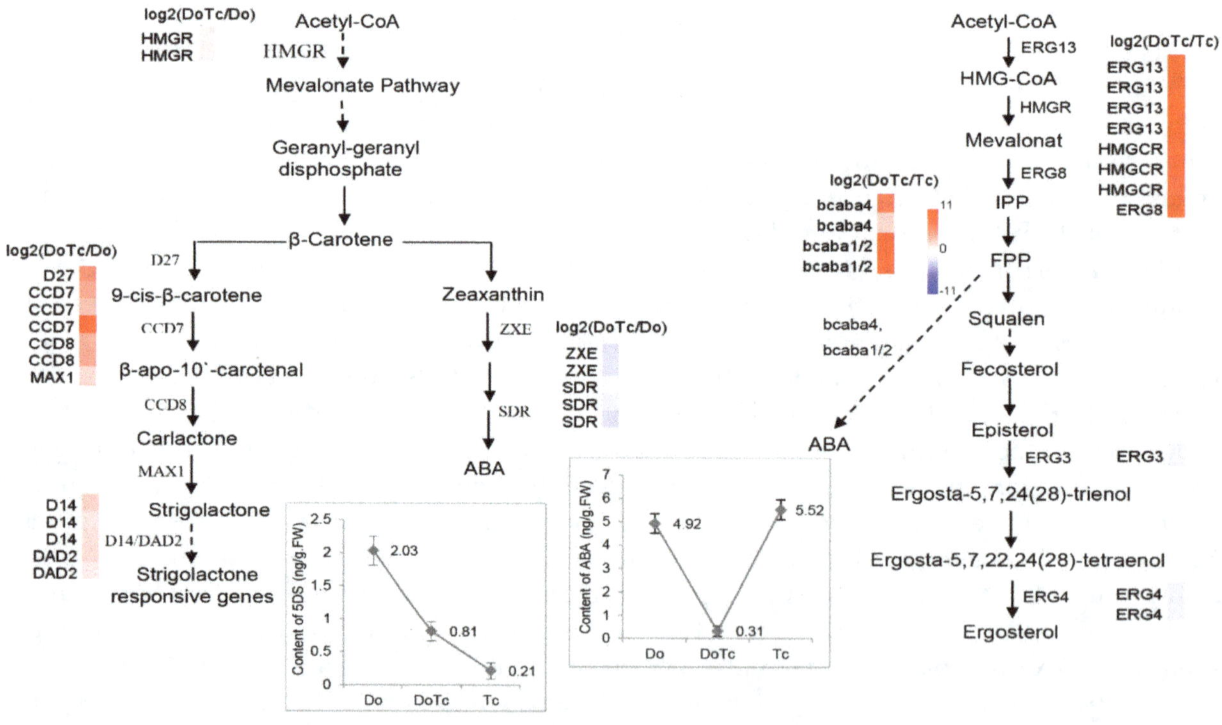

Figure 5. The diagram showing the proposed stigmasterol, ABA, and ergosterol biosynthesis pathways in the symbiotic germinated seeds of *D. officinale* in the symbiotic association with a *Tulasnella* sp. isolate. HMGR: Hydroxymethylglutaryl-CoA reductase; D27: β-carotene isomerase; CCD: carotenoid cleavage dioxygenase; MAX: more axillary growth; ZXE: zeaxanthin epoxidase, SDR: short-chain dehydrogenase/reductase; HMG: Hydroxymethylglutaryl; ERG13/HMGS: Hydroxymethylglutaryl-CoA synthase; ERG8: phosphomevalonate kinase; FPP: Farnesyl-Pyrophosphate; ERG4: sterol C-24 reductase. Red and blue indicate up- and down-regulated transcripts, respectively, from the DoTc vs. Do and DoTc vs. Tc comparisons. Each data point represents the mean of six biological replicates. Bars represent the standard error of the mean, SE ($n = 6$). The solid arrows represent direct relationships, and the dashed arrows represent indirect relationships.

4. Materials and Methods

4.1. Symbiotic and Asymbiotic Germination of D. officinale Seeds

D. officinale plants were grown in the greenhouse of the Chinese Academy of Forestry (Beijing, China) at day/night temperatures of 24 °C/18 °C, a relative air humidity of 65–75%, a 10-h day/14-h night photoperiod, and an irradiance of 700 mol $m^{-2} \cdot s^{-1}$. Six *D. officinale* seed capsules were collected from the greenhouse and divided into two groups after surface sterilization. Seeds from the first group were cultured with *Tulasnella* sp., which was isolated from *Cymbidium goeringii* roots (collected from Suizhou, China); a BLAST search showed that the ITS sequences from this isolate were 99% homologous to the ITS from *T. calospora* isolate Pv-PC-2-1 (NCBI accession No: GU166418.1) [39]. The seeds with *Tulasnella* sp. were cultured on 8 g/L oatmeal agar (OMA) medium; this group was defined as DoTc. The second group of seeds was cultured on the same OMA medium without fungi. Seeds were sown and incubated according to the procedures described by Xu et al. [39]. The DoTc seeds germinated and developed to the fourth stage (appearance of the first true leaf, germination rate around at 90%) [12,34], at which point they were cleared in KOH and stained using Trypan blue [39,40]. Microscopic observation showed that the intracellular hyphae stained blue in the DoTc germinated seeds, which indicated that a symbiotic system was established between *D. officinale* and *Tulasnella* sp. The group of seeds germinated without fungi had a very low germination rate (0.1%), and we did not find fungal hyphae; these were defined as asymbiotic germinated seeds (Do). Free-living fungal mycelium was maintained on an OMA medium at 25 °C for 10 days (Tc). DoTc, Do, and Tc (three biological replicates) were collected, immediately frozen in liquid nitrogen, and stored at −80 °C prior to RNA extraction.

4.2. cDNA Library Construction and Nucleotide Sequencing

The total RNA was extracted from each sample using the RNeasy Plant Mini Kit (Qiagen, Hilden, Germany) and treated with an RNase-free DNase I digestion kit (Aidlab, Beijing, China) to remove residual genomic DNA. RNA degradation and contamination was monitored by electrophoresis on 1% agarose gels. RNA purity was checked using the NanoPhotometer® spectrophotometer (Implen, Westlake Vilagge, CA, USA). RNA concentration was measured with the Qubit® RNA Assay Kit in the Qubit® 2.0 Fluorometer (Life Technologies, Carlsbad, CA, USA). RNA integrity was assessed using the RNA Nano 6000 Assay Kit on the Bioanalyzer 2100 system (Agilent Technologies, Santa Clara, CA, USA).

For each sample, 3 μg of the total RNA was used as input material for the RNA-Seq library preparations. Sequencing libraries were generated using the NEBNext® Ultra™ RNA Library Prep Kit for Illumina® (NEB, Ipswitch, MA, USA) following the manufacturer's recommendations, and index codes were added to attribute the sequencing reads to each sample. The PCR products were purified (AMPure XP system) and the library quality was assessed on the Agilent Bioanalyzer 2100. The RNA-Seq libraries were sequenced on an Illumina Hiseq 4000 instrument to generate 150 bp paired-end reads. All samples were sequenced with three biological replicates.

4.3. Transcriptome Data Analysis

4.3.1. Quality Control and cDNA Assembly

Raw reads in the FASTQ format were first processed using in-house custom Perl scripts. In this step, clean reads were obtained by removing reads containing adapters, poly-*N*-containing reads, and low quality reads from the raw data. Additionally, the Q20 and Q30 scores, and the GC content of the clean reads were calculated simultaneously. All downstream analyses were based on high-quality clean data.

Sequence read mapping to the *D. officinale* Reference Genome: The index of the *D. officinale* reference genome [27] was built using Bowtie v2.2.3, and the paired-end reads were aligned to the reference genome using TopHat v2.0.12 with mismatch set to 2 and the default values used for the other parameters. The OMF transcriptome de novo assembly was accomplished using Trinity [41] with the min_kmer_cov set to 2 by default and all other parameters set to the default values. The coding sequences were predicted using Estscan v3.0.3 with the default settings [42]. All assembled unigene sequences in this study are given in Table S5.

4.3.2. Quantification of Gene Expression Levels

HTSeq v0.6.1 was used to count the read numbers mapped to each *D. officinale* gene. The FPKM (expected number of Fragments Per Kilobase of transcript sequence per Million base pairs sequenced) was then calculated for each gene based on the length of the gene and the read count mapped to the gene [43].

Gene expression levels in the OMF were estimated using RSEM with the default settings [44]. The clean reads were mapped to the reference sequences which were assembled by Trinity [41] using the clean read data from the Tc library, and the transcript read counts were obtained and converted to FPKM for gene expression level analysis in the OMF [43].

4.3.3. Differential Gene Expression Analysis

A differential expression analysis of the two conditions/groups (two biological replicates per condition) was performed using the DESeq R package (1.18.0) with the default settings [45]. DESeq provides statistical routines for determining differential expression in digital gene expression data using a model based on the negative binomial distribution. The resulting *p*-values were adjusted using the Benjamini–Hochberg procedure for controlling the false discovery rate. Genes with a $\log_2^{(Fold\_Change)} > 1$ and adjusted p-value < 0.001 identified by DESeq were considered to be differentially expressed.

4.3.4. Unigene Functional Annotation and Metabolic Pathway Analysis

Unigene function was annotated based on the seven databases reported by Liu et al. [46]. Metabolic pathway analysis was performed using the Kyoto Encyclopedia of Genes and Genomes Pathway database (KEGG: http://www.genome.jp/kegg/). We used the KOBAS [47] software to test the statistical enrichment of DEGs in the KEGG pathways.

4.4. Hormone Extraction, Fractionation, and UPLC-ESI-qMS/MS Analysis

The concentrations of the ABA, JA, and SL hormones in the Do, DoTc, and Tc samples were measured using ultraperformance liquid chromatography coupled with a tandem quadrupole mass spectrometer equipped with an electrospray interface (UPLC-ESI-qMS/MS). Six biological replicates were used in the hormone quantification measurements.

1. 5-DS: To isolate 5-DS, 1 g (fresh weight) samples of symbiotic or asymbiotic seeds were frozen in liquid nitrogen, ground to a powder, transferred to 2 mL microcentrifuge tubes, and then extracted with 10 mL acetone with ultrasound in a water bath for 15 min, after which the tubes were kept at −20 °C for at least 12 h. After centrifugation at 8000 rpm for 5 min, the supernatants were transferred into new tubes and centrifuged at 13,000 rpm at 4 °C for 5 min. The lower organic phases were then transferred to new tubes and evaporated to dryness in a stream of dry N_2. The crude extracts were dissolved in 3 mL normal hexane and filtered through spin columns (500 mg, 6 mL, Agilent Technologies Corporation, Santa Clara, CA, USA) with 5 mL of normal hexane, and the elution products were then evaporated to dryness in a stream of N_2. The crude extracts were then dissolved in 200 μL of methanol and filtered through spin columns (Ultra-Free MC, 0.22 μm pore size; Millipore, Tokyo, Japan), and subjected to an HPLC-MS/MS analysis.

Identification of 5-DS by LC–MS/MS: HPLC separation was conducted with an Agilent 1290 HPLC instrument (Agilent Technologies Corporation) fitted with a poroshell 120 SB-C18 column (2.1 mm × 150 mm, 2.7 μm), and 2 μL was injected for the HPLC-MS/MS analysis. The column temperature was kept at 30 °C. The binary mobile phase system consisted of methanol (A) and water (B) in 0.1% formic acid under a gradient condition (0–0.5 and 5.1–8.00 min, 80% A, 3–5 min, 90% B) with a flow rate of 0.3 mL/min. Mass spectrometry was performed with a SCIEX-6500Qtrap (MS/MS) mass spectrometer (AB SCIEX LLC., Framingham, MA, USA) equipped with an electrospray source. The drying and nebulizing gas was nitrogen generated from pressurized air in an N2G nitrogen generator (MIU instruments, Hangzhou, China). The curtain gas was set to 25 psi, the nebulizer gas was set to 55 psi, the heater gas was set to 65 psi, and the ion spray voltage was set to 5000 V. The atomization temperature was set to 350 °C. Analytes were quantified in the multiple-reaction-monitoring (MRM) mode for the known 5-DS detection, at m/z 331.0/234.1 with a declustering potential (DP) of 45 V and collision energy (CE) of 14 V.

2. ABA and JA: To isolate the hormones, 1 g (fresh weight) samples of symbiotic or asymbiotic seeds were frozen in liquid nitrogen, ground to a powder, and transferred to 2 mL microcentrifuge tubes. The ground tissue was extracted in 10 mL isopropanol/hydrochloric acid buffer by shaking at 4 °C for 30 min, and 20 mL of dichloromethane was then added and the mixture was vibrated at 4 °C for a further 30 min. The tubes were centrifuged at 13,000 rpm at 4 °C for 5 min, and the lower organic phase was transferred to a new tube and was then evaporated to dryness in a stream of N_2. The extracts were dissolved in 400 μL of methanol and filtered through spin columns (Ultra-Free MC, 0.22 μm pore size; Millipore, Tokyo, Japan), and subjected to an HPLC-MS/MS analysis.

ABA and JA were measured with an HPLC-ESI-qMS/MS instrument (AQUITY UPLC™ System/Quattro Ultima Pt; Waters, Milford, MA, USA) fitted with an AQUITY UPLC BEH C18 column (1.7 μm, 2.1 × 100 mm, Waters, Milford, MA, USA). The column temperature was kept at 40 °C. A binary mobile phase system consisting of 2% (v/v) methanol, 0.05% formic acid, and 5 mmol/L ammonium acetate in water (A) and acetonitrile (B) under a gradient condition (0–0.25 and 5.01–6.00 min, 90% (A), 4–5 min, 100% (B)) with a flow rate of 0.3 mL/min. The sample chamber temperature was set to 10 °C and the injection volume was 10 μL. The capillary voltage was controlled at 20 V. The ion source temperature was set to 150 °C, the cone gas flow was 50 L/h, desolvation temperature was 400 °C, and the desolvation gas flow was 800 L/h. Analytes were also quantified in multiple-reaction-monitoring (MRM) mode for the known ABA and JA detections.

For ABA quantification, the declustering potential was −25 V, the collision energy −25 V, and the ion transition was at m/z 263.2/204.0; for JA quantification, the declustering potential was −32 V, the collision energy −12 V, and the ion transition was at m/z 59.0/109.0.

4.5. Differential Gene Expression Detected by qRT-PCR

The symbiotic and asymbiotic root samples from the *D. officinale* seedlings (10 cm in height) that were previously used to prepare the RNA-Seq libraries were used to perform qRT-PCR assays for gene expression quantification. The plant materials were flash-frozen in liquid nitrogen and stored at −80 °C prior to RNA extraction. RNA isolation, qRT-PCR, and calculations of relative mRNA levels were performed as previously described [48]. Each data point represents the mean of three biological replicates and three experimental replicates. *DoEF1α* and *DoRPL30* [49] from *D. officinale*, and *TcEF1α* [50] and *TcTUBB* (Cluster-2641.19171, Table S5) from *Tulasnella* were used as reference genes for the normalization of gene expression, respectively. The primer sequences are shown in Table S9, and their amplification efficiencies were within the range of 95–105%.

5. Conclusions

In this study, we generated transcriptome datasets of *D. officinale* germinated seeds to examine the expression profiles of genes related to the biosynthesis of three types of endogenous hormones—JA, ABA, and SLs—in the *D. officinale-Tulasnella* symbionts, asymbiotic germinated seeds, and the free-living OMF. We also performed comprehensive analyses of the contents of individual hormones, and assayed gene expression using qRT-PCR in the symbiotic and asymbiotic seed samples. The results allowed us to propose potential regulatory mechanisms for JA, ABA, and SLs from the orchid and

OMF in the *D. officinale-Tulasnella* symbionts. Our study will provide new insights into the molecular mechanisms that regulate the orchid-fungus symbiosis.

Author Contributions: T.W. designed the experiments, analyzed the data and wrote the paper; Z.S., X.W., and L.X. performed the experiments; L.L. conceived, led, and supervised the study. Q.S. and L.L. were responsible for this work.

References

1. Arditti, J. Factors affecting the germination of orchid seeds. *Bot. Rev.* **1967**, *33*, 1–97. [CrossRef]
2. Da Silva, T.J.A.; Tsavkelova, E.A.; Zeng, S.; Ng, T.B.; Parthibhan, S.; Dobranszki, J.; Cardoso, J.C.; Rao, M.V. Symbiotic in vitro seed propagation of *Dendrobium*: Fungal and bacterial partners and their influence on plant growth and development. *Planta* **2015**, *242*, 1–22. [CrossRef] [PubMed]
3. Lee, Y.; Lu, C.; Chung, M.; Yeung, E.C.; Lee, N. Developmental changes in endogenous abscisic acid concentrations and asymbiotic seed germination of a terrestrial orchid, *Calanthe tricarinata* Lindl. *J. Am. Soc. Hortc. Sci.* **2007**, *132*, 246–252.
4. Liu, H.; Luo, Y.; Liu, H. Studies of mycorrhizal fungi of Chinese orchids and their role in orchid conservation in China: A review. *Bot. Rev.* **2010**, *76*, 241–262. [CrossRef]
5. Zhao, M.; Zhang, G.; Zhang, D.; Hsiao, Y.; Guo, S. ESTs analysis reveals putative genes involved in symbiotic seed germination in *Dendrobium officinale*. *PLoS ONE* **2013**, *8*, e72705. [CrossRef] [PubMed]
6. Valadares, R.B.S.; Perotto, S.; Santos, E.C.; Lambais, M.R. Proteome changes in *Oncidium sphacelatum* (Orchidaceae) at different trophic stages of symbiotic germination. *Mycorrhiza* **2014**, *24*, 349–360. [CrossRef] [PubMed]
7. López-Chávez, M.Y.; Guillén-Navarro, K.; Bertolini, V.; Encarnación, S.; Hernández-Ortiz, M.; Sánchez-Moreno, I.; Damon, A. Proteomic and morphometric study of the in vitro interaction between *Oncidium sphacelatum* Lindl. (Orchidaceae) and *Thanatephorus* sp. RG26 (Ceratobasidiaceae). *Mycorrhiza* **2016**, *26*, 353–365. [CrossRef] [PubMed]
8. Miura, C.; Yamaguchi, K.; Miyahara, R.; Yamamoto, T.; Fuji, M.; Yagame, T.; Imaizumi-Anraku, H.; Yamato, M.; Shigenobu, S.; Kaminaka, H. The mycoheterotrophic symbiosis between orchids and mycorrhizal fungi possesses major components shared with mutualistic plant-mycorrhizal symbioses. *Mol. Plant Microbe Interact.* **2018**. [CrossRef] [PubMed]
9. Chen, Y.W.; Wu, J.R.; Duan, X.T.; Qin, H.X. Plant hormones produced from symbionts of *Cymbidium goeringii* and four Rhizoctonia-like strains. *Chin. Guizhou Agric. Sci.* **2017**, *45*, 84–86. [CrossRef]
10. Chen, J.; Liu, S.S.; Kohler, A.; Yan, B.; Luo, H.M.; Chen, X.M.; Guo, S.X. iTRAQ and RNA-seq analyses provide new insights into regulation mechanism of symbiotic germination of *Dendrobium officinale* seeds (Orchidaceae). *J. Proteome Res.* **2017**, *16*, 2174–2187. [CrossRef] [PubMed]
11. Chanclud, E.; Morel, J. Plant hormones: A fungal point of view. *Mol. Plant Pathol.* **2016**, *17*, 1289–1297. [CrossRef] [PubMed]
12. Stewart, S.L.; Zettler, L.W. Symbiotic germination of three semiaquatic rein orchids (*Habenaria repens*, *H. quinquiseta*, *H. macroceratitis*) from florida. *Aquatic Bot.* **2002**, *72*, 25–35. [CrossRef]
13. Nambara, E.; Marion-Poll, A. Abscisic acid biosynthesis and catabolism. *Ann. Rev. Plant Biol.* **2005**, *56*, 165–185. [CrossRef] [PubMed]
14. Oritani, T.; Kiyota, H. Biosynthesis and metabolism of abscisic acid and related compounds. *Nat. Prod. Rep.* **2003**, *20*, 414–425. [CrossRef] [PubMed]
15. Gong, T.; Shu, D.; Yang, J.; Ding, Z.T.; Tan, H. Sequencing and transcriptional analysis of the biosynthesis gene cluster of abscisic acid-producing *Botrytis cinerea*. *Int. J. Mol. Sci.* **2014**, *15*, 17396–17410. [CrossRef] [PubMed]
16. Lievens, L.; Pollier, J.; Goossens, A.; Beyaert, R.; Staal, J. Abscisic acid as pathogen effector and immune regulator. *Front. Plant Sci.* **2017**, *8*, 587. [CrossRef] [PubMed]
17. Siewers, V.; Smedsgaard, J.; Tudzynski, P. The P450 monooxygenase BcABA1 is essential for abscisic acid biosynthesis in *Botrytis cinerea*. *Appl. Environ. Microb.* **2004**, *70*, 3868–3876. [CrossRef] [PubMed]
18. Stec, N.; Banasiak, J.; Jasiński, M. Abscisic acid: An overlooked player in plant–microbe symbioses formation. *Acta Biochim. Pol.* **2016**, *63*, 53–58. [CrossRef] [PubMed]

19. Hirai, N.; Yoshida, R.; Todoroki, Y.; Ohigashi, H. Biosynthesis of abscisic acid by the non-mevalonate pathway in plants, and by the mevalonate pathway in fungi. *Biosci. Biotech. Biochem.* **2000**, *64*, 1448–1458. [CrossRef] [PubMed]
20. Ding, Z.; Zhang, Z.; Zhong, J.; Di, L.; Zhou, J.; Yang, J.; Xiao, L.; Shu, D.; Tan, H. Comparative transcriptome analysis between an evolved abscisic acid-overproducing mutant *Botrytis cinerea* TBC-A and its ancestral strain *Botrytis cinerea* TBC-6. *Sci. Rep.* **2016**, *6*, 37487–37499. [CrossRef] [PubMed]
21. Zdyb, A.; Salgado, M.G.; Demchenko, K.N.; Brenner, W.G.; Płaszczyca, M.; Stumpe, M.; Herrfurth, C.; Feussner, I.; Pawlowski, K. Allene oxide synthase, allene oxide cyclase and jasmonic acid levels in *Lotus japonicus* nodules. *PLoS ONE* **2018**, *13*, 1–16. [CrossRef] [PubMed]
22. Liao, D.; Liu, J.; Liu, J.; Yang, X.; Chen, X.; Gu, M.; Gu, A. Advances in the response and modulation of phytohormones on arbuscular mycorrhizal symbiosis. *J. Plant Nutr. Fert.* **2016**, *22*, 1679–1689. [CrossRef]
23. Prandi, C.; Cardinale, F. *Strigolactones: A New Class of Plant Hormones with Multifaceted Roles*; John Wiley & Sons: Chichester, UK, 2014. [CrossRef]
24. Umehara, M.; Hanada, A.; Yoshida, S.; Akiyama, K.; Arite, T.; Takeda-Kamiya, N.; Magome, H.; Kamiya, Y.; Shirasu, K.; Yoneyama, K.; et al. Inhibition of shoot branching by new terpenoid plant hormones. *Nature* **2008**, *455*, 195–200. [CrossRef] [PubMed]
25. Umehara, M.; Cao, M.; Akiyama, K.; Akatsu, T.; Seto, Y.; Hanada, A.; Li, W.; Takeda-Kamiya, N.; Morimoto, Y.; Yamaguchi, S. Structural requirements of strigolactones for shoot branching inhibition in rice and *Arabidopsis*. *Plant Cell Physiol.* **2015**, *56*, 1059–1072. [CrossRef] [PubMed]
26. Saeed, W.; Naseem, S.; Ali, Z. Strigolactones biosynthesis and their role in abiotic stress resilience in plants: A critical review. *Front. Plant Sci.* **2017**, *8*, 1487–1500. [CrossRef] [PubMed]
27. Yan, L.; Wang, X.; Liu, H.; Tian, Y.; Lian, J.; Yang, R.; Hao, S.; Wang, X.; Yang, S.; Li, Q.; et al. The genome of *Dendrobium officinale* illuminates the biology of the important traditional Chinese orchid herb. *Mol. Plant* **2015**, *8*, 922–934. [CrossRef] [PubMed]
28. Liu, Y.; Li, G.; Gui, R.; Zhang, H.; Hu, X. Determination of strigolactones extracted from root of *Phyllostachys edulis* by ultra performance liquid chromatography. *J. Zhejiang Forest. Coll.* **2013**, *30*, 607–610.
29. Walker-Simmons, M. ABA levels and sensitivity in developing wheat embryos of sprouting resistant and susceptible cultivars. *Plant Physiol.* **1987**, *84*, 61–66. [CrossRef] [PubMed]
30. Pence, V.C. Abscisic acid and the maturation of cacao embryos in vitro. *Plant Physiol.* **1992**, *98*, 1391–1395. [CrossRef] [PubMed]
31. Finkelstein, R.R.; Crouch, M.L. Rapeseed embryo development in culture on high osmoticum is similar to that in seeds. *Plant Physiol.* **1986**, *81*, 907–912. [CrossRef] [PubMed]
32. Van, K.G. Abscisic acid in terrestrial orchid seeds: A possible impact on their germination. *Lindleyana* **1987**, *2*, 84–87.
33. Wu, H.; Song, X.; Yang, S.; Wang, J. Effects of symbiotic fungi on the physiological characteristics of *Dendrobium catenatum* Lindley (Orchidaceae). *Chin. Plant Sci. J.* **2011**, *29*, 738–742. [CrossRef]
34. Chen, J.; Wang, H.; Liu, S.S.; Li, Y.Y.; Guo, S.X. Ultrastructure of symbiotic germination of the orchid *Dendrobium officinale* with its mycobiont, *Sebacina* sp. *Aust. J. Bot.* **2014**, *62*, 229–234. [CrossRef]
35. Umezawa, T.; Okamoto, M.; Kushiro, T.; Nambara, E.; Oono, Y.; Kobayashi, M.S.M.; Koshiba, T.; Kamiya, Y.; Shinozaki, K. CYP707A3, a major ABA 8′-hydroxylase involved in dehydration and rehydration response in *Arabidopsis thaliana*. *Plant J.* **2006**, *46*, 171–182. [CrossRef] [PubMed]
36. Zeevaart, J.A.D.; Creelman, R.A. Metabolism and physiology of abscisic acid. *Annu. Rev. Plant Biol.* **1988**, *39*, 439–473. [CrossRef]
37. Lopez-Raez, J.A.; Kohlen, W.; Charnikhova, T.; Mulder, P.; Undas, A.K.; Sergeant, M.J.; Verstappen, F.; Bugg, T.D.; Thompson, A.J.; Ruyter-Spira, C.; et al. Does abscisic acid affect strigolactone biosynthesis? *New Phytol.* **2010**, *187*, 343–354. [CrossRef] [PubMed]
38. Toh, S.; Kamiya, Y.; Kawakami, N.; Nambara, E.; McCourt, P.; Tsuchiya, Y. Thermoinhibition uncovers a role for strigolactones in *Arabidopsis* seed germination. *Plant Cell Physiol.* **2012**, *53*, 107–117. [CrossRef] [PubMed]
39. Xu, L.; Tian, J.; Wang, T.; Li, L. Symbiosis established between orchid and *Tulasnella* spp. fungi. *J. Nucl. Agric. Sci.* **2017**, *31*, 876–883. [CrossRef]
40. Zhao, X.; Yang, J.; Liu, S.; Chen, C.; Zhu, H.; Cao, J. The colonization patterns of different fungi on roots of *Cymbidium hybridum* plantlets and their respective inoculation effects on growth and nutrient uptake of orchid plantlets. *World J. Microbiol. Biotech.* **2014**, *30*, 1993–2003. [CrossRef] [PubMed]

41. Grabherr, M.G.; Haas, B.J.; Yassour, M.; Levin, J.Z.; Thompson, D.A.; Amit, I.; Xian, A.L.F.; Raychowdhury, R.; Zeng, Q.; Chen, Z.; et al. Full-length transcriptome assembly from RNA-Seq data without a reference genome. *Nat. Biotechnol.* **2011**, *29*, 644–652. [CrossRef] [PubMed]
42. Haas, B.J.; Papanicolaou, A.; Yassour, M.; Grabherr, M.; Blood, P.D.; Bowden, J.; Couger, M.B.; Eccles, D.; Li, B.; Lieber, M.; et al. De novo transcript sequence reconstruction from RNA-seq using the Trinity platform for reference generation and analysis. *Nat. Protoc.* **2013**, *8*, 1494–1512. [CrossRef] [PubMed]
43. Trapnell, C.; Williams, B.A.; Pertea, G.; Mortazavi, A.; Kwan, G.; van Baren, M.J.; Salzberg, S.L.; Wold, B.J.; Pachter, L. Transcript assembly and quantification by RNA-seq reveals unannotated transcripts and isoform switching during cell differentiation. *Nat. Biotechnol.* **2010**, *28*, 511–515. [CrossRef] [PubMed]
44. Li, B.; Dewey, C.N. RSEM: Accurate transcript quantification from RNA-Seq data with or without a reference genome. *BMC Bioinform.* **2011**, *12*, 323–339. [CrossRef] [PubMed]
45. Wang, L.; Feng, Z.; Wang, X.; Wang, X.; Zhang, X. DEGseq: An R package for identifying differentially expressed genes from RNA-seq data. *Bioinformatics* **2011**, *26*, 136–138. [CrossRef] [PubMed]
46. Liu, S.; Chen, J.; Li, S.; Zeng, X.; Meng, Z.; Guo, S. Comparative transcriptome analysis of genes involved in GA-GID1-DELLA regulatory module in symbiotic and asymbiotic seed germination of *Anoectochilus roxburghii* (Orchidaceae). *Int. J. Mol. Sci.* **2015**, *16*, 30190–30203. [CrossRef] [PubMed]
47. Mao, X.; Cai, T.; Olyarchuk, J.G.; Wei, L. Automated genome annotation and pathway identification using the KEGG Orthology (KO) as a controlled vocabulary. *Bioinformatics* **2005**, *21*, 3787–3793. [CrossRef] [PubMed]
48. Qian, X.; Sun, L.; Xu, X.; Zhu, B.; Xu, H. Differential expression of *VvLOXA* diversifies C6 volatile profiles in some *Vitis vinifera* table grape cultivars. *Int. J. Mol. Sci.* **2017**, *18*, 2705–2719. [CrossRef] [PubMed]
49. An, H.; Zhu, Q.; Pei, W.; Fan, J.; Liang, Y.; Cui, Y.; Lv, N.; Wang, W. Whole-transcriptome selection and evaluation of internal reference genes for expression analysis in protocorm development of *Dendrobium officinale* Kimura et Migo. *PLoS ONE* **2016**, *11*, e0163478. [CrossRef] [PubMed]
50. Fochi, V.; Chitarra, W.; Kohler, A.; Voyron, S.; Singan, V.R.; Lindquist, E.A.; Barry, K.W.; Girlanda, M.; Grigoriev, I.V.; Martin, F.; et al. Fungal and plant gene expression in the *Tulasnella calospora–Serapias vomeracea* symbiosis provides clues about nitrogen pathways in orchid mycorrhizas. *New Phytol.* **2017**, *213*, 365–379. [CrossRef] [PubMed]

2

Genomic Characterization and Expression Analysis of Basic Helix-Loop-Helix (bHLH) Family Genes in Traditional Chinese Herb *Dendrobium officinale*

Yue Wang [1,2] and Aizhong Liu [3,*]

1 Key Laboratory of Economic Plants and Biotechnology, Yunnan Key Laboratory for Wild Plant Resources, Kunming Institute of Botany, Chinese Academy of Sciences, Kunming, Yunnan 650201, China; wangyue@mail.kib.ac.cn

2 Bio-Innovation Center of DR PLANT, Kunming Institute of Botany, Chinese Academy of Sciences, Kunming, Yunnan 650201, China

3 Key Laboratory for Forest Resources Conservation and Utilization in the Southwest Mountains of China, Ministry of Education, Southwest Forestry University, Kunming, Yunnan 650224, China

* Correspondence: liuaizhong@mail.kib.ac.cn

Abstract: *Dendrobium officinale* Kimura et Migo is of great importance as a traditional Chinese herb due to its abundant metabolites. The family of basic helix-loop-helix (bHLH) transcription factors widely exists in plants and plays an essential role in plant growth and development, secondary metabolism as well as responses to environmental changes. However, there is limited information on *bHLH* genes in *D. officinale*. In the present study, a total of 98 putative *DobHLH* genes were identified at the genomic level, which could be classified into 18 clades. Gene structures and conserved motifs in *DobHLH* genes showed high conservation during their evolution. The conserved amino acids and DNA bindings of DobHLH proteins were predicted, both of which are pivotal for their function. Furthermore, gene expression from eight tissues showed that some *DobHLH* genes were ubiquitously expressed while other *DobHLH* genes were expressed in the specific tissues. Expressional changes of *DobHLH* genes under MeJA and ABA treatments were detected by qRT-PCR. The protein–protein interactions between DobHLHs were predicted and several interactions were confirmed by yeast two hybrid. Therefore, our results here contribute to the understanding of *bHLH* genes in *D. officinale* and lay a foundation for the further functional study of its biological processes.

Keywords: *Dendrobium officinale*; basic helix-loop-helix transcription factors; genome-wide identification; expression analysis; protein–protein interaction; skin care

---

1. Introduction

Dendrobium officinale Kimura et Migo (or known as *D. catenatum*) is a perennial herbal plant, belonging to the family Orchidaceae. *D. officinale* plants are widely distributed across southern provinces in China, with limited spread to southeast Asian countries [1]. *D. officinale* is of great importance due to its medicinal and ornamental uses. *D. officinale* ranks first among the nine Chinese herbs for longevity and has a documented use in folk medicine for over 1300 years [2]. In the past decades, about 190 compounds have been identified and isolated from *D. officinale* plants including phenanthrenes, bibenzyls, saccharides and glycosides, alkaloids, essential oils, and others such as phenols, acids, esters, and amides [3]. Among them, polysaccharides, bibenzyls, alkaloids, and flavones have been confirmed as the major and bioactive ingredients [3,4]. The abundance of polysaccharides in its stems and the specific alkaloids make *D. officinale* have anticancer [1], anti-oxidant [5], and anti-inflammatory effects [1,6] as well as support immune modulation [7,8], hepatop protection, and hypoglycemia [8,9]. Due to its multiple health benefits, demand for high-quality *D. officinale* is increasing across medicine, food, and skin care.

As an epiphytic plant, wild *D. officinale* prefers to grow in a warm and humid environment. In nature, *D. officinale* inevitably suffers environmental stresses such as high temperature and drought, slowing growth, and lowering yield [10]. Despite abundant polysaccharides, the contents of other active components in *D. officinale* are low [11–13]. At present, many potential genes involved in the biosynthesis of polysaccharides, alkaloids, and flavonoids have been identified in *D. officinale* through transcriptome and genomic sequencings [4,14]. In *D. officinale*, polysaccharides in the stem, leaves, and flowers are one of the major active ingredients, and sucrose synthase (Susy) and two gene families (β-galactosidase and galacturonosyltransferase) are related to the contents and richness, respectively [2]. In addition, 10 sucrose phosphate synthase and 15 Susy genes showed marked expansion [2]. The upstream pathways of alkaloid biosynthesis in plants are conserved and clear, which are responsible for the formation of strictosidine. The downstream steps of the alkaloid biosynthesis in *D. officinale* are modification of strictosidine including cytochrome P450s (CYP450s)-mediated oxidation and hydroxylation reactions [15], which needs further investigation. However, transcription regulation of the biosynthesis has been largely unknown. As an important switch of gene expression, transcription factor (TF) can activate or repress the expression of specific target genes through interaction with cis-elements in the gene promoter region, which moderates various biological processes such as growth, response to stresses, and the biosynthesis of secondary metabolites [16].

Basic helix-loop-helix (bHLH) TFs are widespread in the eukaryotes and also are the second largest family of TFs in plants [17–20]. bHLH TFs are named due to the presence of a highly conserved bHLH domain, and the bHLH domain of 50–60 amino acids contains one basic region and one HLH region [21,22]. The basic region of 15 amino acids is at the N-terminus of the bHLH domain, while the HLH region of 40–50 amino acids is at the C-terminus of the bHLH domain including two alpha helices and one less conserved loop with variable lengths. The bHLH domain can specifically recognize and bind a conserved cis-element CANNTG, named E-box [23,24]. The two helices from the same bHLH TF or different bHLH TFs can interact to form a homodimer or a heterodimer, which then binds different regions in the promoter to regulate the expression of their target genes [25]. The other sequences of bHLH TFs are variable and less conserved.

The bHLH TF was first found in murine muscle [21], and later in animals [23] and plants [22,26]. Based on differences in the sequences, DNA binding, bHLH domain and functions, the bHLH TF family in animals is divided into six lineages (group A–F) [23]. In plants, most bHLH TFs share a similar structure with that of group B from animals, which can bind the G-box (CACGTG), although they also bind E-box [16]. Unlike in animals, Arabidopsis bHLH TFs are divided into many more clades [22]. The classification of plant bHLHs is gradually being clarified and most plants contain 15–26 clades due to their identification in more plants including tomato [27], maize [28], grape [29], cotton [30], jujube [25], and *Brachypodium distachyon* [31].

bHLH TFs play important roles in plant secondary metabolites. The first bHLH protein identified in plant was *R* gene in maize, which can regulate the expression of at least two genes in the flavonoid/anthocyanin pathway [32]. In Arabidopsis, the three bHLH proteins Glabra3 (GL3), Enhancer of Glabra3 (EGL3), and Transparent Testa8 (TT8) can interact with the myeloblastosis (MYB) transcription factor and WD40 protein to form a MYB–bHLH–WD complex, activating multiple genes in the biosynthesis of anthocyanin and resulting in anthocyanin accumulation [33]. An increasing number of research is reporting that bHLH TFs also participate in the regulation of other secondary metabolites including flavonols [33], terpenoid indole alkaloids [34], and tanshinone [35]. Recently a bHLH TF was found to be responsible for the specific anthocyanin of lip tissues in *Dendrobium* hybrids [36]. Identification of bHLH TFs in *D. officinale* is the first step to understanding their roles in the accumulation of bioactive compounds.

In addition, bHLH TFs participate in plant growth and development including photomorphogenesis, light signal transduction, flowering time, and the development of various tissues such as flower and root [16]. For example, two Arabidopsis phytochrome interacting factors (AtPIF4 and PIF3) can interact with phytochrome to control the expression of the genes involved in the regulation of light response [22,37]. A large number of bHLH TFs play a vital role in the response to environmental stresses such as salt, drought, and high temperature. Inducer of CBF expression 1 (ICE1), belonging to the III b subclade of Arabidopsis bHLH family, can activate downstream genes through the mediation of C-repeat binding factor (CBF) in response to freezing stress [38].

Although there has been much research on bHLH TFs in various plants, a comprehensive investigation of the bHLH family in *D. officinale* has yet not been carried out. The competence of genome sequencing for *D. officinale* makes the search for potential genes associated with important traits possible [2,39]. In the present study, we first identified a total of 98 candidate DobHLH members. Their phylogenetic relationship, conserved bHLH domain and motifs, gene structures, and cis-elements were characterized. The expression patterns in eight tissues were analyzed from the RNA-Seq data. The expression levels of some *DobHLH* genes under hormone treatments were detected by quantitative real time-polymerase chain reaction (qRT-PCR). The possible protein–protein interactions were predicted, and some interactions were confirmed by yeast two hybrid. Accordingly, our results provide information on the bHLH family in *D. officinale* and lay the foundation for further investigation into its function in its biological process.

2. Results

2.1. Genome-Wide Identification of DobHLH Members in Dendrobium officinale

After the hidden markov model (HMMER) search and confirmation of the bHLH domain presences with online CD-search tool and SMART, a total of 98 members in *D. officinale* were considered DobHLH candidates. They were named from DobHLH1 to DobHLH98 according to their relationship with AtbHLH proteins. The basic information of DobHLH members is shown in Table 1 and Table S1. The DobHLH proteins contain 85–662 AAs with molecular weights of 9.53–74.54 kDa and isoelectric points of 4.41–10.78. The subcellular localization of 98 DobHLH proteins were predicted by WOLF PSORT. The results showed that most of the predicted proteins (89) were localized in the nucleus, and five and four DobHLH proteins were in the cytoplasmic and mitochondrial, respectively. The 98 DobHLH genes were distributed in 85 scaffolds. The CDS lengths of the *DobHLH* genes varied from 258 bp (*DobHLH95* and *DobHLH96*) to 1989 bp (*DobHLH24*) while the lengths of genomic DNA were from 496 bp (*DobHLH79*) to 26,581 bp (*DobHLH58*).

Compared with the numbers of bHLH members in other plant species, the number in *D. officinale* was similar to that found in grape [29] and Chinese jujube [25] and was less than that found in most other species with over 100 bHLH members such as Arabidopsis (166), rice [40], *Brachypodium distachyon* [31], and tomato [27]. Some important crops had up to 200 or even many more bHLH members including maize [28], poplar [41], wheat [42], cotton [30], and *Brassica napus* [43]. The density of bHLH members in the *D. officinale* genome was approximately 0.28, which was similar to peach, but higher than that found in the genomes of the two lower plants *Volvox carteri* (0.024) and *Chlorella vulgaris* (0.081) [44]. These results showed that the number of *bHLH* genes in different species could be connected with the genome size.

2.2. Phylogenetic Relationship of DobHLH Proteins

To reveal the evolution of bHLH members, an unrooted-tree was constructed using MEGA 7.0 with the neighbor-joining method including 98 DobHLHs and 166 AtbHLHs (Figure 1). The phylogenetic tree could be divided into 22 clades, and each clade included two to 31 members. Ninety-eight DobHLHs were distributed in 18 clades (clade 1 to clade 18) and four clades (clade 19, 20, 21, 22) contained bHLH proteins only from Arabidopsis, without proteins from *D. officinale*, which suggested that gene deletion could occur during the evolution of *D. officinale*. In most clades, there were fewer bHLH proteins from *D. officinale* than from Arabidopsis. Clade 4 and 6 had more DobHLHs than AtbHLHs and clade 9 and 16 had the same number of bHLH members (three and one).

Table 1. The list of 98 *DobHLH* genes in *D. officinale*.

Gene Name	Gene ID	NCBI References	Clade	Gene Name	Gene ID	NCBI References	Clade
DobHLH1	LOC110116721	XP_020706080.1	1	DobHLH50	LOC110116300	XP_020705489.1	10
DobHLH2	LOC110115756	XP_020704772.1	1	DobHLH51	LOC114579352	XP_028549432.1	10
DobHLH3	LOC110107102	XP_020692919.1	1	DobHLH52	LOC110095032	XP_020676060.1	11
DobHLH4	LOC110096422	XP_020678032.1	1	DobHLH53	LOC110113329	XP_020701532.1	11
DobHLH5	LOC110116474	XP_020705700.1	1	DobHLH54	LOC110113204	XP_020701328.1	11
DobHLH6	LOC110098508	XP_020681021.1	1	DobHLH55	LOC110114625	XP_020703217.1	12
DobHLH7	LOC110111619	XP_020699225.1	1	DobHLH56	LOC110106166	XP_020691604.1	12
DobHLH8	LOC110112576	XP_020700508.1	1	DobHLH57	LOC110112285	XP_020700113.1	12
DobHLH9	LOC110108487	XP_020694816.2	1	DobHLH58	LOC110108563	XP_020694908.1	12
DobHLH10	LOC110110298	XP_020697356.1	2	DobHLH59	LOC110101829	XP_020685553.1	13
DobHLH11	LOC110093619	XP_020674221.1	2	DobHLH60	LOC110095526	XP_020676768.1	13
DobHLH12	LOC110115484	XP_020704390.1	2	DobHLH61	LOC110111433	XP_028555769.1	13
DobHLH13	LOC110093741	XP_020674407.1	3	DobHLH62	LOC110093428	XP_020673967.1	13
DobHLH14	LOC110102219	XP_020686105.1	3	DobHLH63	LOC110104725	XP_020689612.1	14
DobHLH15	LOC110098909	XP_020681520.1	3	DobHLH64	LOC110102342	XP_028552167.1	14
DobHLH16	LOC110107046	XP_028551992.1	3	DobHLH65	LOC110107722	XP_020693735.1	14
DobHLH17	LOC110096269	XP_020677790.1	3	DobHLH66	LOC110100649	XP_020683910.1	14
DobHLH18	LOC110116147	XP_020705280.1	4	DobHLH67	LOC110114258	XP_028550306.1	14
DobHLH19	LOC110094813	XP_028551573.1	4	DobHLH68	LOC110114447	XP_028556766.1	14
DobHLH20	LOC110107826	XP_020693884.2	4	DobHLH69	LOC110109277	XP_028547820.1	14
DobHLH21	LOC110094726	XP_020675682.1	4	DobHLH70	LOC110096494	XP_020678137.1	14
DobHLH22	LOC110114654	XP_020703261.1	5	DobHLH71	LOC110111863	XP_028551171.1	14
DobHLH23	LOC110111891	XP_020699606.1	5	DobHLH72	LOC110107433	XP_028552737.1	14
DobHLH24	LOC110097687	XP_020679864.1	5	DobHLH73	LOC110114740	XP_020703378.2	14
DobHLH25	LOC110103241	XP_020687528.1	6	DobHLH74	LOC110095482	XP_020676694.1	14
DobHLH26	LOC110094094	XP_020674906.1	6	DobHLH75	LOC110099461	XP_020682274.1	14
DobHLH27	LOC110116682	XP_020706025.1	6	DobHLH76	LOC110099214	XP_028548138.1	14
DobHLH28	LOC110099101	XP_020681804.1	6	DobHLH77	LOC110098199	XP_020680601.1	15
DobHLH29	LOC110113808	XP_020702166.2	6	DobHLH78	LOC110098270	XP_020680696.1	15
DobHLH30	LOC110094435	XP_020675326.1	6	DobHLH79	LOC110094287	XP_020675140.1	15
DobHLH31	LOC110116479	XP_020705705.1	6	DobHLH80	LOC110113891	XP_020702260.1	15
DobHLH32	LOC110114462	XP_020703009.1	6	DobHLH81	LOC110112441	XP_020700324.1	15
DobHLH33	LOC110092865	XP_020673218.1	6	DobHLH82	LOC110101026	XP_020684454.1	15
DobHLH34	LOC110109085	XP_020695652.1	7	DobHLH83	LOC110111081	XP_020698439.1	15
DobHLH35	LOC110096203	XP_020677673.1	7	DobHLH84	LOC110104845	XP_020689772.1	15
DobHLH36	LOC110112097	XP_020699850.1	7	DobHLH85	LOC110095998	XP_020677402.1	15
DobHLH37	LOC110107930	XP_020694038.1	8	DobHLH86	LOC110110399	XP_020697513.2	15
DobHLH38	LOC110115493	XP_020704404.1	8	DobHLH87	LOC110109507	XP_028547425.1	15
DobHLH39	LOC110094861	XP_020675853.1	8	DobHLH88	LOC114578594	XP_028547764.1	15
DobHLH40	LOC110112072	XP_020699818.1	8	DobHLH89	LOC110105526	XP_020690722.2	15
DobHLH41	LOC110114469	XP_028556744.1	9	DobHLH90	LOC110100961	XP_020684346.1	15
DobHLH42	LOC110107963	XP_020694091.1	9	DobHLH91	LOC110112336	XP_020700191.1	16
DobHLH43	LOC110106259	XP_028550041.1	9	DobHLH92	LOC110110400	XP_020697514.1	17
DobHLH44	LOC110107318	XP_028547365.1	10	DobHLH93	LOC110112399	XP_028555091.1	17
DobHLH45	LOC110103817	XP_028551326.1	10	DobHLH94	LOC110108630	XP_020695011.1	17
DobHLH46	LOC110114754	XP_020703396.1	10	DobHLH95	LOC110110529	XP_020697710.1	17
DobHLH47	LOC110107832	XP_020693899.1	10	DobHLH96	LOC110108826	XP_020695307.1	17
DobHLH48	LOC110107031	XP_020692824.1	10	DobHLH97	LOC110105593	XP_020690812.1	17
DobHLH49	LOC110098511	XP_020681024.1	10	DobHLH98	LOC110100228	XP_020683310.1	18

Moreover, an unrooted tree was constructed with bHLH proteins from *D. officinale* and rice (Figure S1). For as many as 15 clades, there were fewer bHLH proteins from *D. officinale* than from rice. Clades 6 and 17 contained more bHLH proteins from *D. officinale* than from rice. There were more bHLH proteins in the rice genome than in *D. officinale*, which could result from genomic or random duplication after differentiation with *D. officinale* [40].

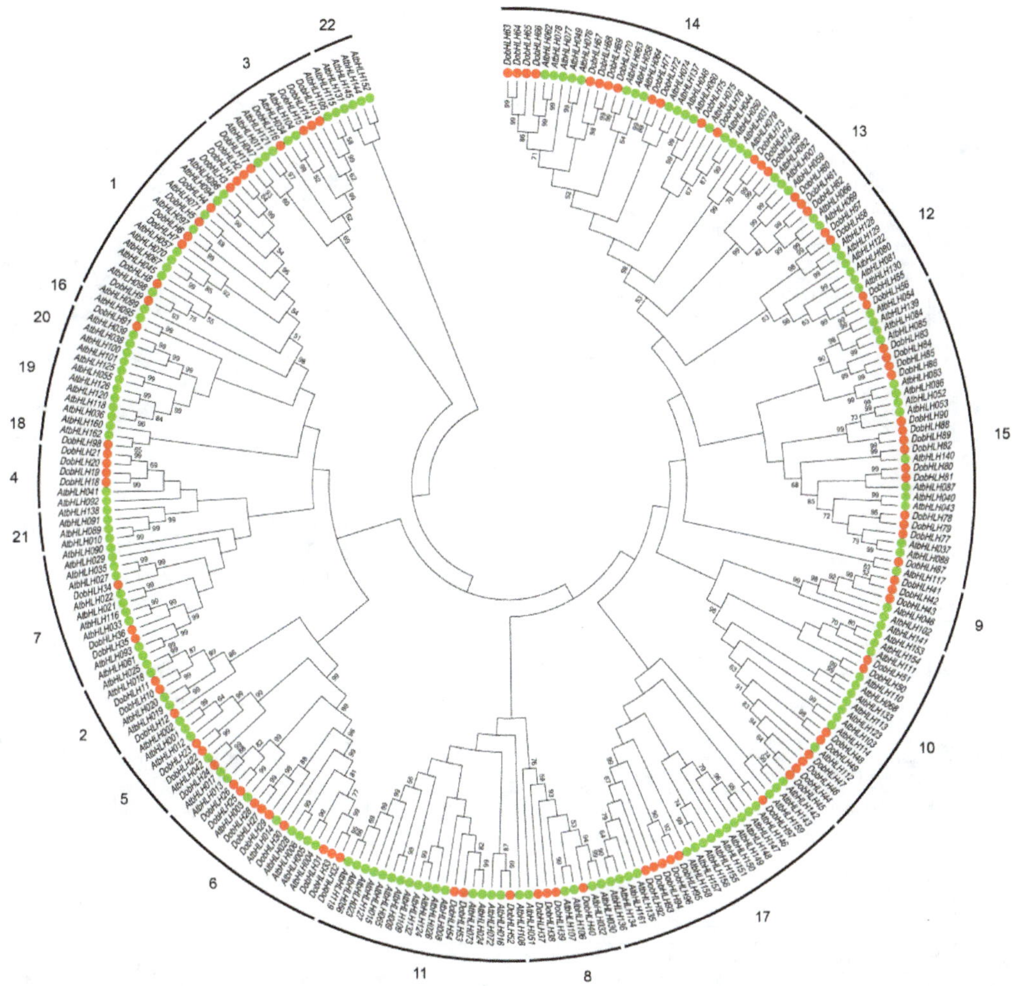

Figure 1. Phylogenetic analysis of bHLH proteins from *D. officinale* and Arabidopsis. A total of 98 DobHLH proteins and 162 AtbHLH from Arabidopsis (Arabidopsis thaliana) were used to construct the unrooted neighbor-joining (NJ) tree with a bootstrap of 1000 replicates. The bHLH proteins are grouped into 22 clades, and marked in different colors: *AtbHLH* genes are in green while *DobHLH* genes are in red.

2.3. Gene Structures and Conserved Motifs of DobHLH Genes

To reveal the evolution of *DobHLH* genes, the exon-intron structures were investigated by comparing CDS sequences with genomic sequences. The results showed that the intron numbers varied greatly among the DobHLH family members, from some completely lacking intron to some having up to 11 introns (Figure 2). Most of the members within the same clade shared the same number of introns. For example, clades 1, 2, 8, 10, 12, 16, 17 had two, three, one, six, five, two, and one intron, respectively, while all members in clade 6 and eight members in clade 15 had no introns. Appropriately 25% of the members (25) had one or two introns, mainly including members from clades 1, 5, 8, 16, 17 and 18. Most members from clades 3, 5, 7, 10, 11, 12, 13, and 14 including 43 DobHLH genes contained three to seven introns. Four genes (*DobHLH24, 42, 43,* and *64*) had more than seven introns. The lengths of introns imbedded in the *DobHLH* genes ranged from 44 bp to 21,971 bp, with the longest intron in the *DobHLH53* gene.

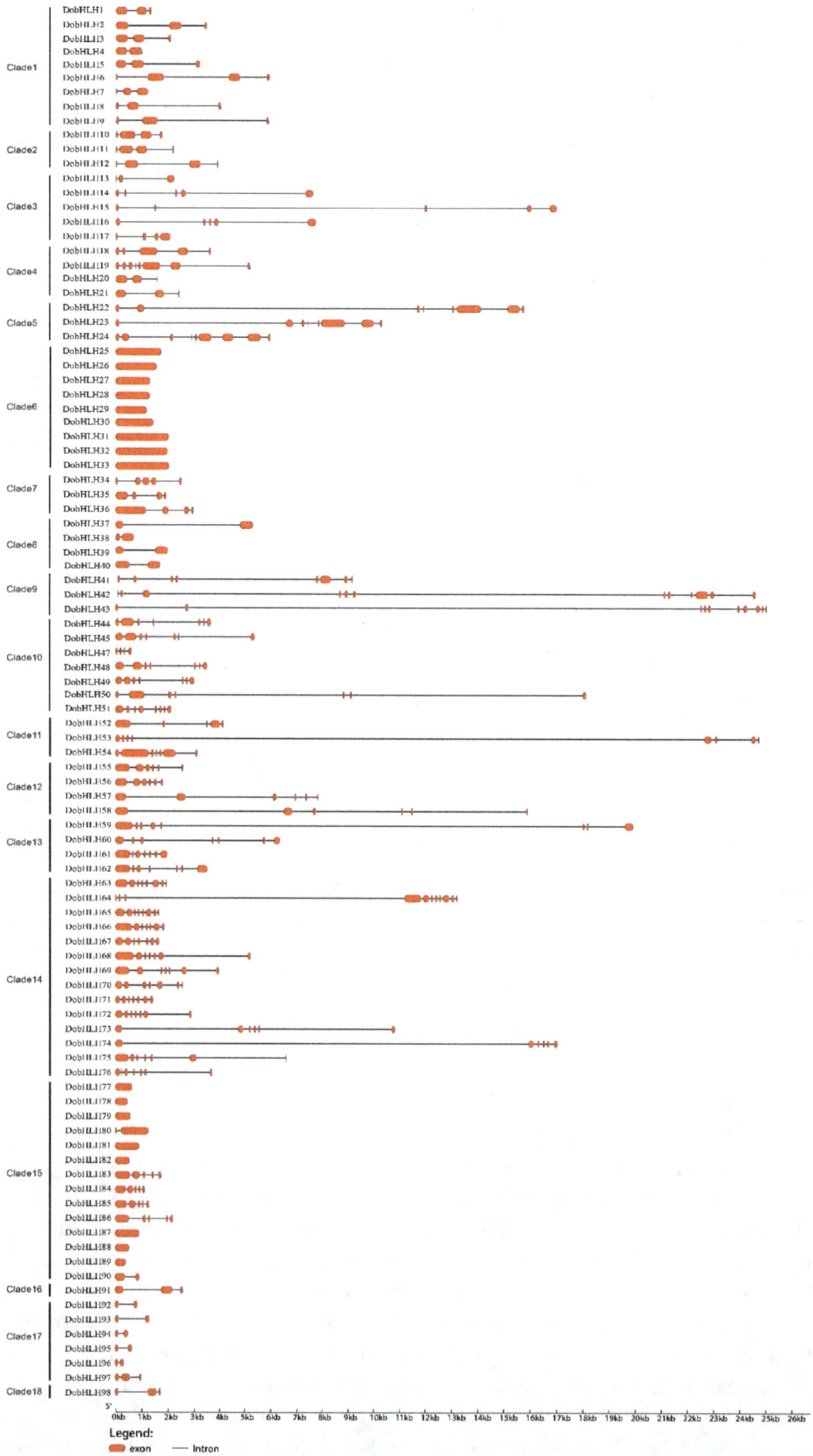

Figure 2. The gene structures of *DobHLH* genes in *D. officinale*. The exon-intron structures of 98 *DobHLH* genes were obtained with GSDS 2.0. The red rectangles and black lines indicate exons and introns, respectively, according to their lengths.

To further understand the structure of DobHLH proteins, we searched 10 conserved motifs in DobHLHs with MEME software. As shown in Figure 3, motif 1 and motif 2 were the two most conserved motifs, which were widely present in DobHLH members. The latter ten residues in the basic region and the first helix were motif 1, while the second helix was the main part of motif 2. Motif 1 was present in 88 DobHLH proteins while motif 2 was present in all the 98 DobHLH proteins. Ten DobHLH proteins (including DobHLH13, 14, 16, 17, 43, 92, 94, 95, 96, 98) had motif 2 only and lacked the basic region of bHLH domain. The other 88 DobHLH proteins had both motif 1 and motif 2, and 30 DobHLH members contained only motif 1 and motif 2, mainly from clades 4, 7, 8, 9, 11, 15, 16, and 17. Clades 5 and 6 contained the most motifs (five to seven). Motif 3 was present in all the members from clades 12, 13, 14 and some members from clade 15. Motif 3 was distributed in clade 1, 2, 5, and 6, with the exception of DobHLH25 and DobHLH26. Motif 5 and motif 6 were both present in the genes of clade 5 and clade 6. Motif 7 was limited in two members from clade 5 (DobHLH22 and DobHLH23) and seven members from clade 6. Motif 8 was present in all genes from clade 10 and clade 13. Motif 9 was limitedly present in the members of clade 6 without DobHLH29. Motif 10 was present in nine members from clades 1 and 6 (DobHLH1-6, 8, 9, 25, 26). There were similar motifs in most members from each clade. Eight of nine DobHLH members from clade 1 contained four motifs (motif 1, 2, 4, 10), and three members from clade 2 had motif 1, 2, and 4. Four of five members in clade 3 included motif 2. Ten DobHLH members from clade 10 had motifs 1, 2, and 8 while 14 members from clade 14 included motifs 1, 2, and 3.

2.4. Conserved Amino Acids in the DobHLH Domains and DNA-Binding Ability

As the bHLH domain composed of motif 1 and motif 2 is especially important for bHLH TF to function, we analyzed the amino acids in bHLH domains from 98 DobHLH proteins (Figure 4). It was found that the *D. officinale* bHLH domain included 55 AAs, which was similar in Arabidopsis (56, [22]) but shorter than in tomato (61, [27]). The basic region in the *D. officinale* bHLH domain included 17 AAs and five AAs, showing over 50% identity in sequences. The first helix and the second helix both contained 15 AAs and six and eight AAs had more than 50% identity, respectively. The loop in the *D. officinale* bHLH domain had AAs and only Asp-40 was conserved with an identity of 60%.

As seen in Table S2, there were 20 AAs in the *D. officinale* bHLH domain that showed over 50% conservation and three residues (Arg-16, Leu-27, and Leu-55) were highly conserved in *D. officinale*, up to 89%, 96% and 91%, respectively. It has been found that these three residues were conserved both in plants and animals [22], suggesting their core sites in the bHLH domain. Moreover, seven residues including Ile-20, Leu-24, Gln-28, Met-44, Ile-49, Val-52, and Leu-55 were more conserved in plants such as *D. officinale* and Arabidopsis than in animals, which was also observed in eggplant [22]. This could imply their important role in plants. Compared with residues in tomato and Arabidopsis, Glu-13, Arg-14, Gln-28, Lys-36, Ala-48, and Ile-52 were less conserved.

As TFs, the most important function for bHLH proteins is to bind the promoter region of their target genes. Based on previous studies on the ability to bind DNA, we distinguished the types of DobHLH proteins (Table 2). It was found that 98 DobHLH proteins could be divided into two types: 22 non-DNA binding DobHLHs and 76 DNA binding DobHLHs. The 22 non-DNA binding DobHLHs mainly included members from clades 2, 10, 12, 15, and 17, and there were less than six residues in the basic region of 17 AAs in the bHLH domain. According to differences in the DNA binding sequences, 76 DNA binding DobHLHs could be further classified into 65 E-box binders and 11 non-E-box binders. When the 13 and 16 residues in the basic region are Glu and Arg, respectively, the bHLH protein has the capacity to bind E-box. Otherwise, the DobHLH protein belongs to non-E-box binders. Eleven non-E-box binders were from clades 10, 15, 17, and 18. The E-box binders could be further discriminated into G-box binders and non-G-box binders. When the 9, 13, and 17 residues in the basic region are His/Lys, Glu and Arg, respectively, the bHLH protein is considered to bind the classic G-box. In *D. officinale*, there were four DobHLH proteins that belong to non-G-box binders including DobHLH34, 35, and 36 from clade 7 and DobHLH97 from clade 17. The remaining 61 DobHLH proteins were G-box binders, accounting for 62.24% of all the DobHLH members.

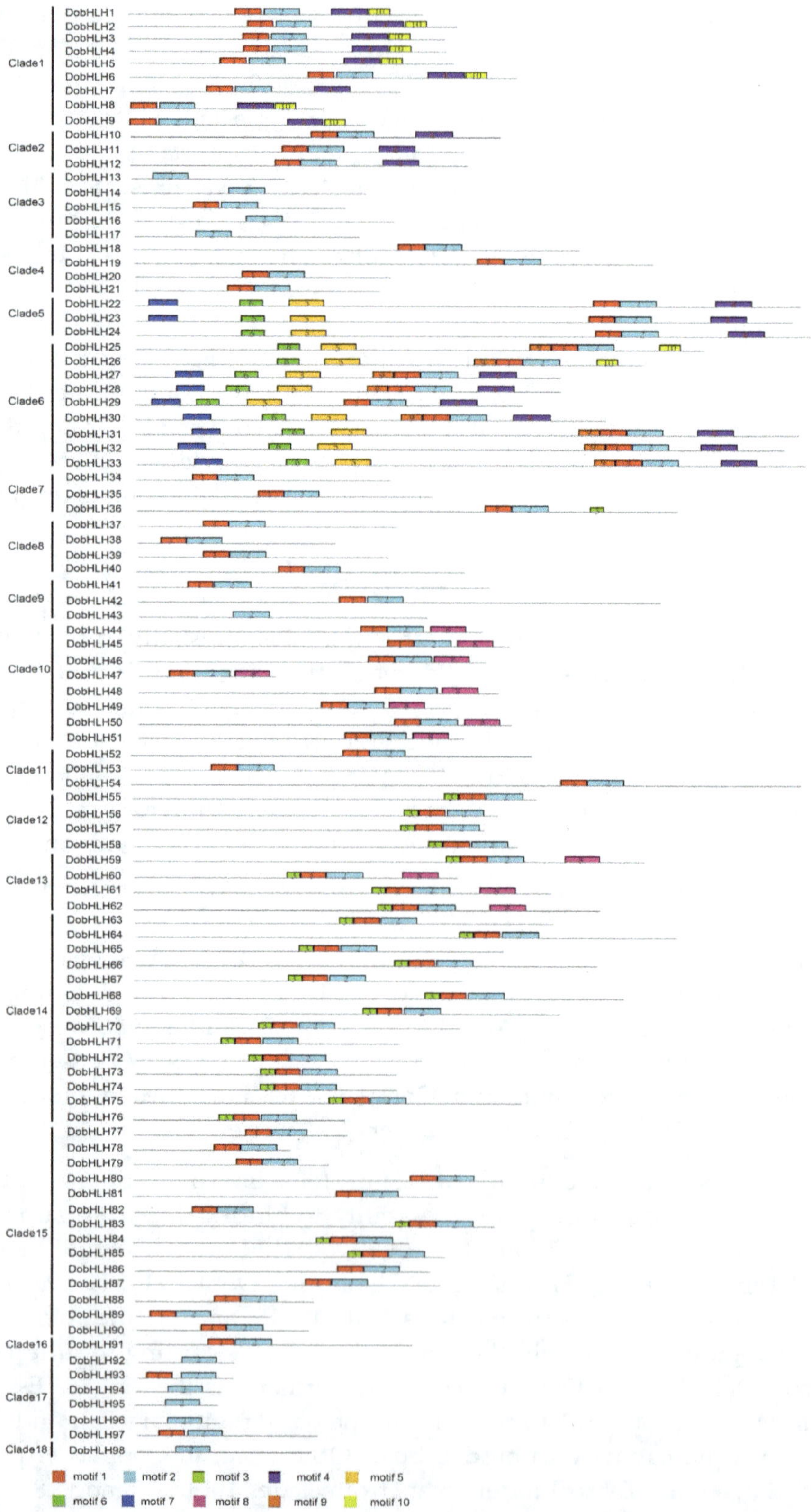

Figure 3. The conserved motifs of DobHLH proteins predicted by MEME. The colored boxes with numbers represent 10 conserved motif and the grey lines indicate non-conserved lines.

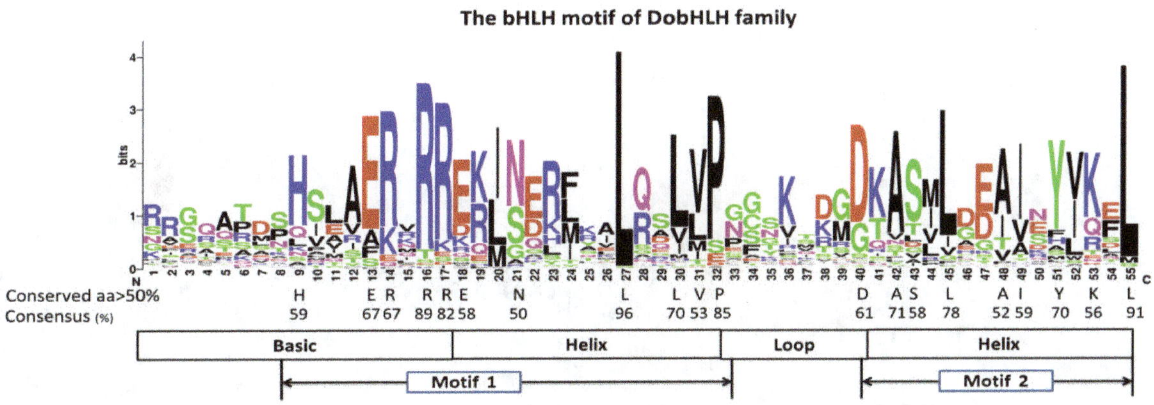

Figure 4. Analysis of the conserved amino acids in all DobHLH domains. The overall height of each amino acid represents the sequence conservation at that position. The conserved AAs and the consensus of more than 50% among 98 DobHLH domains are indicated in black letters.

Table 2. The predicted types of DNA-binding based on DobHLH domain sequences.

Predicted DNA-Binding Types Based on the bHLH Domain	Number of DobHLHs	The Ratio
DNA binding		
E-box	65	62.25%
G-box	61	62.24%
Non-G-box	4	4.08%
Non-E-box	11	11.22%
Total	76	77.55%
Non-DNA binding	22	22.45%

2.5. Cis-Elements in the Promoter Regions of DobHLH Genes

Due to functional divergence, the members in the same gene family could show diverse expression patterns. To investigate regulatory patterns, we detected the cis-elements in the promoter regions of *DobHLH* genes (Table S3). Many cis-elements related to responses to stresses were predicted by PlantCARE. The most frequent cis-element was G-box, which was present in 72 *DobHLH* genes, suggesting that a light signal could be vital in transcriptionary regulation. Other cis-elements such as the CGTCA-motif (or TGACG-motif), ABRE, and ERE were distributed in 65, 69, and 78 *DobHLH* genes, respectively, which are responsible for responses to methyl jasmonate (MeJA), abscisic acid (ABA), and ethylene, respectively. There were 166 CGTCA-motifs (or TGACG-motifs), 201 ABRE cis-elements, and 188 ERE cis-elements. The cis-elements related to stress responses were also found in *DobHLH* gene promoter regions such as TC-rich repeats in 47 *DobHLH* genes, WUN-motif in 50 DobHLH genes, LTR in 39 *DobHLH* genes, and MBS in 42 *DobHLH* genes. These cis-elements can respond to defense and stress, wounding, low temperature, and drought. Other cis-elements involved in plant growth and development were observed in some DobHLH genes including ARE, CAT-box, and O2-Site.

2.6. Expression Patterns of DobHLH Genes in Eight Tissues

The expression patterns of *DobHLH* genes were analyzed based on the transcriptomic data of eight tissues including column, flower buds, lip, sepal, leaf, stem, white root, and green root tip. There were 23 *DobHLH* genes that were hardly detected in any of the eight tissues The expression of 23 *DobHLH* genes was not detected in any of the eight tissues (when FPKM values were < 2, the gene was considered to be not expressed). Those 23 *DobHLH* genes were members from clades 1, 4, 10, 14, and 15. The other 75 *DobHLH* genes were expressed in at least one tissue tested with FPKM > 2. A heatmap of the expression patterns including 75 *DobHLH* genes was constructed by R, as shown in Figure 5. The genes could be classified into five groups.

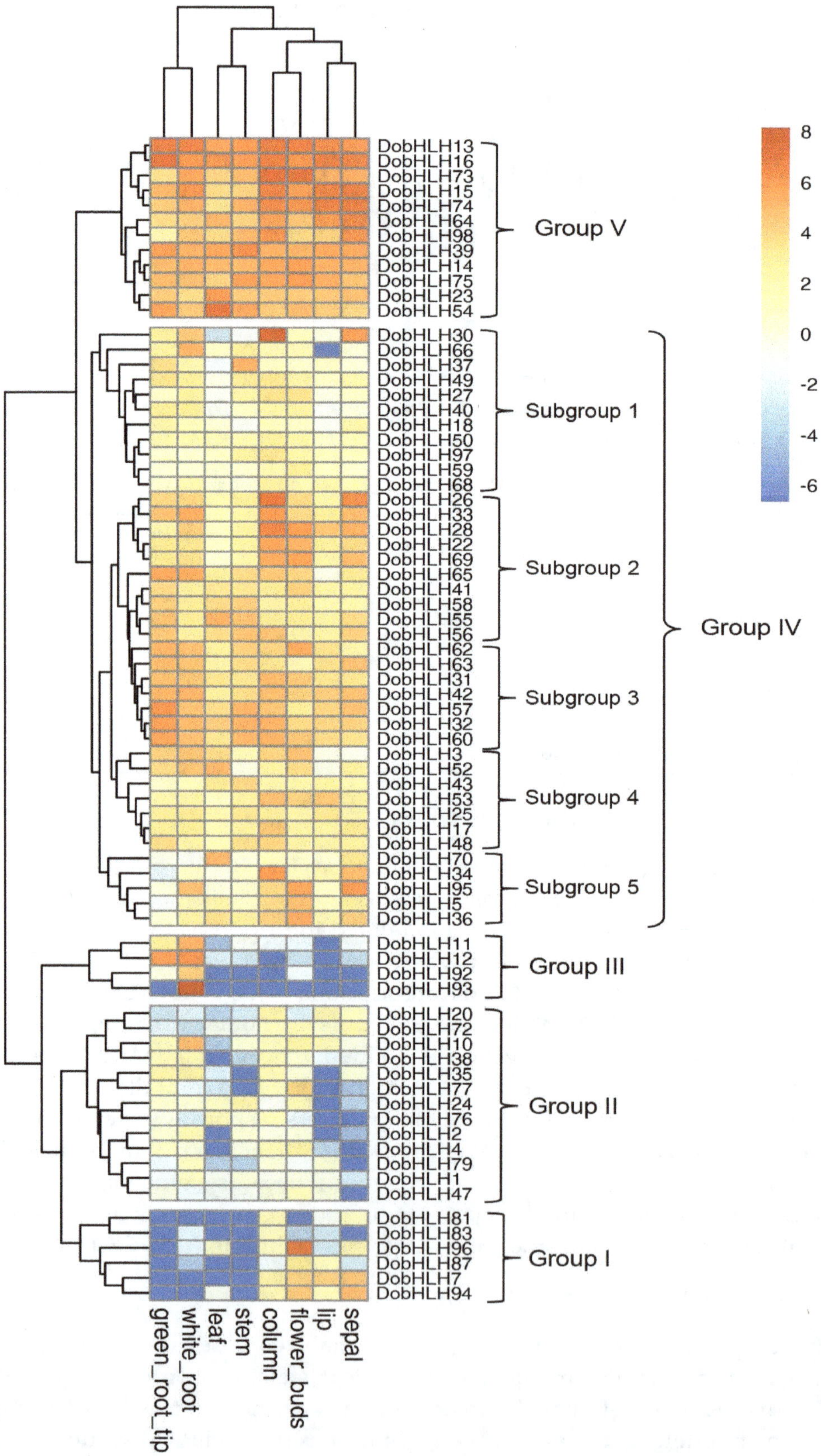

Figure 5. Expression patterns of 75 *DobHLH* genes in eight tissues. The expression levels of 75 *DobHLH* genes were from the RNA-Seq data. The eight samples included the column, flower buds, lip, sepal, leaf, stem, white root and green root tip. The color scale represents the values of Log_2(FPKM + 0.01).

Group I included six genes (*DobHLH7, DobHLH81, DobHLH83, DobHLH87, DobHLH94,* and *DobHLH96*). Their transcripts could not be detected across four vegetative tissues (leaf, stem, white root, and green root tip) and were expressed with various levels in four flower tissues. *DobHLH81* and *DobHLH83* were lowly expressed only in the column. *DobHLH87* was moderately expressed in flower buds while the transcripts of *DobHLH96* were abundant in flower buds with FPKM of 124.68. *DobHLH7* and *DobHLH87* were expressed in four flower tissues with different levels.

Group II included 13 genes (*DobHLH1, 2, 4, 10, 20, 24, 35, 38, 47, 72, 76, 77,* and *79*). Only two genes (*DobHLH10* and *DobHLH77*) were moderately expressed in certain tissues (white root and flower buds, respectively). The other genes were not expressed in most tissues and expressed only in one or two tissues with low FPKM values (<7.50).

Group III included four genes (*DobHLH11, DobHLH12, DobHLH92, DobHLH93*) that showed a strong tissue-specific expression pattern. These four genes were low expressed in root tissues. *DobHLH11* and *DobHLH12* were expressed in both white root and green root tip. *DobHLH12* was highly expressed in both white root and green root tip, while *DobHLH11* was highly expressed in the white root, but low in the green root tip. The transcripts of *DobHLH92* and *DobHLH93* were limitedly detected in the white root, but not the green root tip. *DobHLH93* showed the highest expression level in the white root with FPKM of up to 286.63.

Group IV contained 40 *DobHLH* genes. According to their expression differences, these genes could be further divided into five subgroups. Subgroup 1 included 11 genes (*DobHLH18, 27, 28, 30, 37, 40, 49, 50, 59, 66, 68,* and *97*). *DobHLH30, 37,* and *66* were not expressed or lowly expressed in most tissues and highly expressed in some tissues. For example, *DobHLH30* showed the highest expression with FPKM of 231.28 in the column, followed by FPKM of 61.66 in the sepal. *DobHLH37* was the most abundant with FPKM of 42.56 in the stem while *DobHLH66* was expressed with the highest FPKM of 40.61 in the white root. The remaining eight *DobHLH* genes were not expressed or expressed with low FPKM values. Subgroup 2 contained five members (*DobHLH22, DobHLH26, DobHLH28, DobHLH33,* and *DobHLH69*). With the exception of *DobHLH28*, which was moderately expressed in the lip, all these genes were not expressed or lowly expressed in the leaf, lowly expressed in the stem and lip (5 < FPKM < 10), moderately expressed in the white root and green root tip, and moderately or abundantly expressed in the column, flower buds, and sepal. Subgroup 3 contained 12 genes (*DobHLH31, 32, 41, 42, 55, 56, 57, 58, 60, 62, 63,*and *65*) that were widely expressed in all tested tissues with various levels. All these genes were moderate or abundant with FPKM values of more than 14.68 in the green root tip. For six genes (*DobHLH32, 42, 57, 60, 63,* and *65*), there were more transcripts in the white root and green root tip than in other tissues tested. Except for *DobHLH42*, other genes had higher expression levels in the green root tip than in the white root. *DobHLH31, 41, 55, 56, 58, 62,* and *63* genes showed the highest expression levels in non-root tissues. Subgroup 4 contained seven genes (*DobHLH3, 17, 25, 43, 48, 52,* and *53*). Most of them were ubiquitously expressed in eight tissues tested with low or moderate levels. *DobHLH52* showed the highest expression level with FPKM of 35.73 in the leaf. Subgroup 5 contained five genes (*DobHLH5, 34, 36, 70,* and *95*). These genes showed no or low expression levels in the lip, leaf, stem, white root, and green root tip, exhibiting moderate levels in the column, flower buds, and sepal tissues. *DobHLH95* showed the highest level with FPKM of 64.94 in the sepal and *DobHLH34* contained a maximum FPKM of 63.41 in the column.

Group V contained 12 genes (*DobHLH13, 14, 15, 16, 23, 39, 54, 64, 73, 74, 75,* and *98*). Only *DobHLH98* was expressed in the green root tip with a low level. All other genes were moderately or highly expressed in all tested tissues with FPKM of more than 11.89, suggesting their various roles in the eight tissues. *DobHLH13* and *DobHLH16* had the most abundant transcripts in the green root tip with FPKM values of 122.75 and 140.77, respectively. *DobHLH73* was highly expressed in the column and flower buds. DobHLH54 had the most transcripts in the leaf with FPKM of 151.78, while *DobHLH39* showed the highest expression level in the stem with FPKM of 80.19.

To further confirm the expression profiles of *DobHLH* obtained from RNA-Seq data, eight genes (including *DobHLH11-16, 23,* and *52*) were randomly selected for the performance of qRT-PCR assays.

The results showed that most of the genes exhibited similar expression patterns to RNA-Seq data (Figure S2). For example, *DobHLH11* and *DobHLH12* genes were expressed highest in roots and *DobHLH13, 14, 15,* and *16* were expressed in the root, stem, and leaf tissues, consistent with RNA-Seq data. However, the expression pattern of *DobHLH23* showed a slight difference from the RNA-Seq data.

2.7. *Expressional Changes of DobHLH Genes Concerned with ABA and JA Signals under MeJA and ABA Treatments*

To investigate whether *DobHLH* genes that contain stress-related regulatory cis-elements could respond to different stresses, we simulated the stress treatment experiments using exogenous MeJA and ABA treatments and inspected the expressional changes of 16 *DobHLH* genes concerned with ABA and JA signals by the qRT-PCR technique. As shown in Figure 6, eight (including *DobHLH* 3, 10, 15, 16, 23, 31, 33, and 37) of 12 *DobHLH* genes that contained JA-responsive elements (CGTCA-motif cis-element) exhibited expressional changes. In contrast to the controls, four *DobHLH* genes (including *DobHLH12, 13, 14* and *32*) exhibited no expressional changes in response to MeJA treatments. Similarly, with ABA treatment, 11 (including *DobHLH* 3, 11, 12, 15, 16, 23, 30, 32, 33, 37, and 39) of 13 *DobHLH* genes that contained ABA-responsive elements (ABRE-motif cis-element) exhibited expressional changes compared to the controls, whereas two *DobHLH* genes (*DobHLH13, 14*) exhibited no expressional changes in response to ABA treatments. In particular, we noted that six *DobHLH* genes (*DobHLH* 3, 15, 16, 23, 33, and 37) of the ten *DobHLH* genes (including *DobHLH* 3, 12, 13, 14, 15, 16, 23, 32, 33, and 37) contained both JA and ABA-responsive elements, which showed expressional responses to both ABA and MeJA treatments. However, *DobHLH13* and *14* did not respond to both ABA and MeJA treatments at the transcription level. These results suggest that most of the *DobHLH* genes might function in respond to stress signals at the transcription level, likely through specific cis-elements in promoter regions, though the processes of regulatory transcription might be complex.

2.8. *Protein–Protein Interaction Networks*

It has been reported than bHLH proteins can form a homodimer or a heterodimer to bind DNA sequences in the promoter region to regulate the transcription of the target gene. Therefore, interactions between bHLH proteins play an important role in their function. The predicted protein–protein interaction of DobHLH proteins based on homologs with Arabidopsis bHLH proteins is shown in Figure 7. It was observed than most DobHLH proteins could interact with one or more DobHLH proteins. For example, it was predicted that DobHLH17 can interact with multiple DobHLH proteins including DobHLH6, 8, 9, 13, 14, 15, 16, and 60. The homolog of DobHLH17 in Arabidopsis is POPEYE (PYE), which is involved in the regulation of responses to iron deficiency in Arabidopsis roots. It has been demonstrated that PYE can interact with multiple proteins such as UNE12, AtbHLH104, ILR3, FMA, MUTE, and SPCH. ILR3 plays a role in the development of root hair. The interaction between DobHLH17 and DobHLH13 or DobHLH14 could have a similar function in *D. officinale*. Moreover, FMA, SPCH, and MUTE together regulated the stomata formation in Arabidopsis and ICE1 could interact with FMA, SPCH, and MUTE [25], implying that their homologs in *D. officinale* could potentially participate in the regulation of stomatal differentiation. HEC2 and SPT both play a role in floral development [25], and interactions between their homologs in *D. officinale* (DobHLH77, 78, 79, and DobHLH53) could influence pistil development through the control of hormones. These predicted interactions suggest that DobHLH proteins could function together to regulate plant growth, development, and other biological processes that merit further investigation.

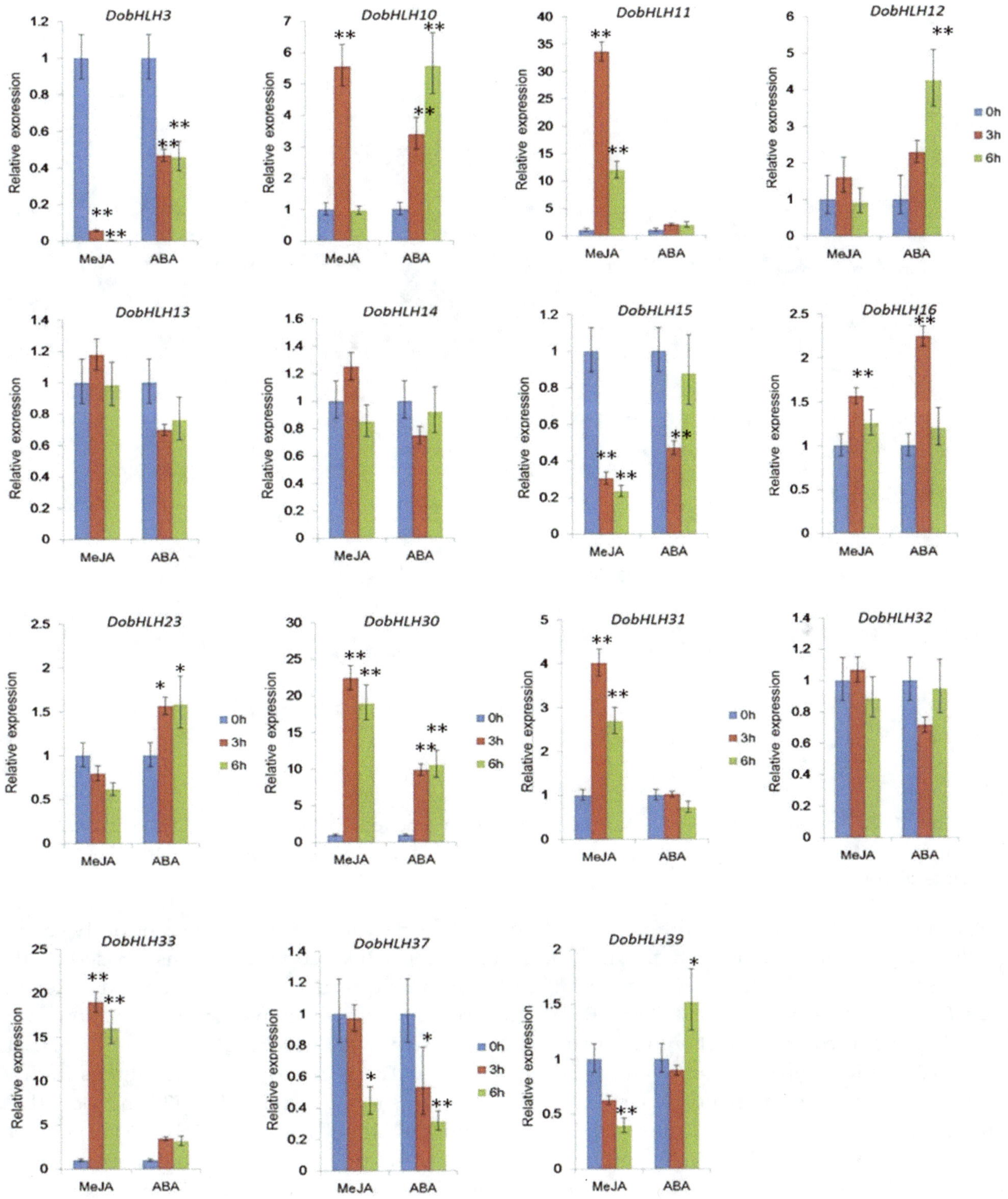

Figure 6. Expression patterns of 15 stress-related *DobHLH* genes under exogenous MeJA and ABA treatments by the qRT-PCR technique. After exogenous MeJA and ABA treatments in leaf tissues, the relative expression levels at 0, 3, and 6 h were tested with three biological duplicates. The relative expression levels at the beginning (0 h) were applied as the control. The bars denote the standard deviation. The star indicates the significance difference (* at $p < 0.05$, ** at $p < 0.01$) with the Student's -test.

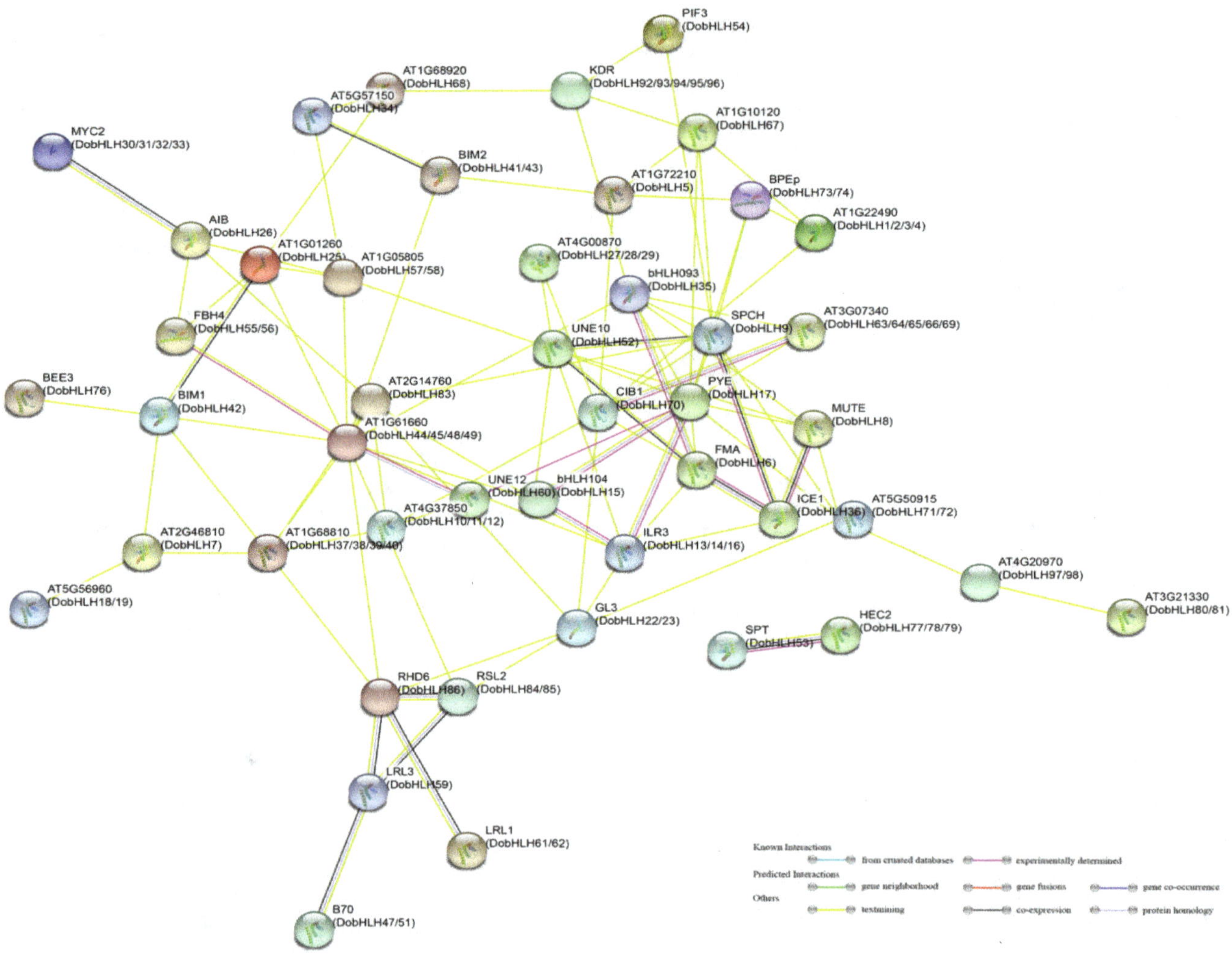

Figure 7. The predicted network of protein–protein interactions between DobHLHs by STRING. The different colors represent different types of interactions. Arabidopsis bHLH names are marked while their homologs in *D. officinale* are in parentheses.

To further verify protein–protein interactions in the predicted network, we randomly selected 12 *DobHLH* genes to investigate their interactions using the Y2H technique in yeast. Based on their co-expressions in a given tissue, we made different combinations in Y2H experiments. The selected *DobHLH* genes were cloned into AD and BD vectors, respectively. As shown in Figure 8, none of the DobHLH proteins could interact with the empty BD vector, while three DobHLH proteins (DobHLH36, 48, and 70) could interact with the empty AD on SD/-Leu-Trp-His-Ade medium, suggesting their self-activation. In the combination of DobHLH proteins, it was found that DobHLH25 and DobHLH26 showed an interaction on the SD/-Leu-Trp-His medium and DobHLH60 and DobHLH17 showed a strong interaction on the SD/-Leu-Trp-His-Ade medium. Other protein–protein combinations had no interaction. These results can partially confirm the predicted interaction networks and indicate that they could play similar functions in *D. officinale*.

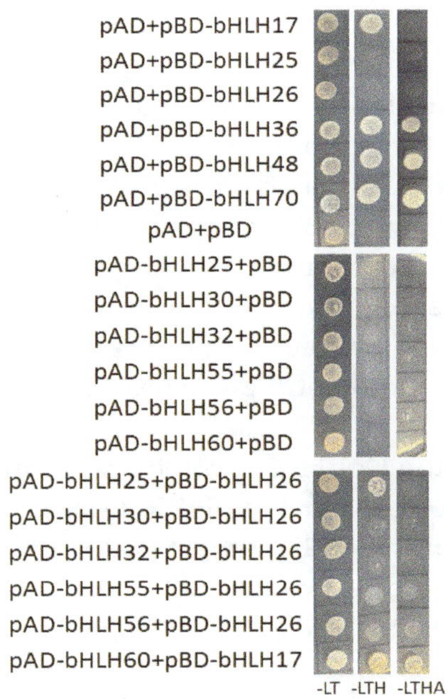

Figure 8. The protein–protein interactions between DobHLHs detected by Y2H. The sequences of DobHLH25, 30, 32, 55, 56, and 60 genes were ligated to the GAL4 activation domain (AD) while the sequences of DobHLH17, 26, 36, 48, and 70 genes were fused to the GAL4 DNA-binding domain (BD). SD-LT, SD-LTH, and SD-LTHA represent the SD-LeuTrp, SD-Leu-Trp-His, and SD-Leu-Trp-His-Ade medium, respectively.

3. Discussion

TFs can bind cis-element sequences in downstream genes to activate or repress their expression, which confers vital roles in plant growth, development, and responses to stresses [16]. The genome sequences of *D. officinale* have been reported [2,39], which makes it possible to identify the TF family at the genome level. The bHLH family is the second largest family of TFs and plays an essential role in plants. In the present study, based on the published genome data, bHLH family members were first identified in *D. officinale*, which provides valuable information to dissect their potential function in molecular and physiological processes.

A total of 98 DobHLH members were identified according to the genome data [39]. In previous transcriptome data, some bHLH members were predicted [13–15,39,45] and the number of bHLH members varied amongst these publications. The number of bHLH members identified by the genome in the present study was greater than the numbers by the transcriptome data. This was because some bHLH genes were limitedly expressed in the specific tissues and could not be detected in the specific tissues used for RNA-Seq. Only in three transcriptomes from the root, stem, and leaf were there as many as 140 *bHLH* transcripts, which could result from alternative splicing events [13]. Compared with other TF families identified in *D. officinale*, the number of bHLH family members ranked second, after the MYB family [46], and was greater than other families such as GRAS [10] and MADS [47]. The number of bHLH members in *D. officinale* was similar to that in jujube [25] and peach [44] and was more than in *Carthamus tinctorius* [48]. It has been found that many crops contain more than 150 bHLH members including maize [28], wheat [42,49], and *Panax ginseng* [50].

The bHLH members in plants could be divided into different clades (or subfamilies) and the numbers of clades in different species varied from 15 to 25 [22]. In the present study, the 98 DobHLH genes in *D. officinale* could be classified into 18 clades. We found that four clades in Arabidopsis were not present in *D. officinale*. The similar results were found in other plants such as *C. tinctorius* [48], apple [51], and wheat [42], which suggests those genes were lost during their evolution. Clades 14 and 15 of the bHLH family in *D. officinale* contained 14 and 15 members, respectively, which accounted for as many as 28.57% members of the family. The expansion of two clades in *D. officinale* could potentially imply their important functions. However, the numbers of members from clade 7 and 11 in *D. officinale* were lower than those in Arabidopsis and rice, which suggests that these clades shrunk in *D. officinale*.

It was found that the bHLH family in *D. officinale* showed conservation in both gene structure and protein motif. The gene structure of bHLH family in *D. officinale* showed highly conserved patterns. Most members in the same clade had the same number of introns and all numbers from clade 6 had no intron. Similar results were observed in *Nelumbo nucifera* [52], apple [51], cotton [30] and rice [53], which further suggested conserved gene organization during the evolution of different species. However, there were some exceptions where some members in the same clade contained variable intron numbers such as clade 14, which had five, six, seven, or 10 introns. The events of exon loss and gain occurred during the evolution of these genes, which could result in functional diversities between closely-related genes.

The most conserved motifs in the DobHLH proteins were motif 1 and motif 2, which consisted of the bHLH domain. As a core domain for the bHLH family, the bHLH domain was also highly conserved in other species such as tomato [27], wheat [42], and maize [28]. In the bHLH domain of *D. officinale*, the most conservative residues were Leu-27 and Leu-54, which were in consistence with the residues in Arabidopsis, tomato, and rice [22,27,53]. These results suggest that the two residues may be essential for their functions, especially for the formation of dimers [27]. Moreover, most residues in the domain had similar conservation as in Arabidopsis and tomato, while some residues in *D. officinale* showed less conservation than those in Arabidopsis and tomato including Glu-13, Arg-14, Gln-28, Lys-36, Ala-48, and Ile-52.

As TFs, their most important function is to bind the specific DNA sequences to regulate the gene expression. According to previous reports and the bHLH domain, the potential types of DNA-binding for DobHLHs were predicted in *D. officinale*. Compared with predicted DNA-binding types in Arabidopsis and tomato, the number of DNA bindings (77.55%) in *D. officinale* was lower than that in Arabidopsis (81.63%), but higher than in tomato (69.18%). Among them, the number of G-box DNA binding bHLHs (61/98, 62.24%) in *D. officinale* was similar to that in Arabidopsis (89/147, 60.54%), which was much higher than that in tomato (72/159, 45.28%). The number of the non-E-box binders in *D. officinale* was slightly higher than that in Arabidopsis as well as in tomato. The comparative results suggest that a large number of bHLH proteins in *D. officinale* could bind DNA similar to other species. Notably, most members from the same clade showed the same type of DNA-binding, which was also observed in other species such as Arabidopsis and tomato [27]. Furthermore, the bHLH proteins from *D. officinale* and Arabidopsis in the same clade exhibited the same type of DNA-binding, which confirmed their conservation during evolution.

The cis-regulatory elements are located at the 5′ upstream of the gene and are responsible for gene transcription, which plays an essential role in the regulatory network and control of plant growth and development. Many cis-elements related to the development, responses to stress as well as hormone were predicted in the promoter regions of *DobHLH* genes, which suggest their potential roles in the responding conditions. The most common cis-element observed in *DobHLH* gene promoters was the G-box, which can respond to light. Similar results were found in other species such as *C. tinctorius* [48] and maize [28]. The second most common types of cis-elements in DobHLH genes were the hormone-responsive motifs that responded to MeJA, ABA, and ethylene. Among 98 *DobHLH* genes, there were as many as 201 ABRE elements in 69 *DobHLH* genes, suggesting their putative roles in response to ABA. Notably, a total of 188 ethylene-responsive ERE elements were predicted in 78 *DobHLH* genes, while few ERE elements were observed in *bHLH* genes from other plants. The rest of the cis-elements including TC-rich repeats, LTR element, and MBS were also predicted in *DobHLH* gene promoters, which were related to abiotic stresses such as low temperature and drought. The results showed that *bHLH* genes can respond to various environmental changes through their cis-elements.

There has almost been no report about bHLH functions in *D. officinale* despite many reports on other plants such as Arabidopsis and rice. The potential gene function could be predicted based on its expression pattern. In the present study, the comprehensive expression patterns of *DobHLH* genes were detected in eight tissues according to the previous RNA-Seq data, which strongly showed tissue-specific expression. A total of 23 *DobHLH* genes were barely expressed in these tissues. Similarly, some *bHLH* genes from other species were also not detected in any tissue [49]. For example, there were 30 *bHLH* genes in potato that were not expressed in any of the 12 tested tissues [54]. However, some *bHLH* genes could widely be expressed in various tissues such as the 12 *bHLH* genes in *D. officinale* including *DobHLH13* and *DobHLH16*. These ubiquitous and abundant transcripts of *bHLH* genes could be found in jujube [25] and potato [54]. The widespread expression of *bHLH* genes in multiple tissues suggests their multiple roles in these tissues.

The strongly tissue-specific expression patterns of 75 *DobHLH* genes were observed. A group of *bHLH* genes in *C. tinctorius* was found to exhibit high transcripts in petals [48]. During the flower development of jujube, *ZjbHLH62* and *ZjbHLH53* genes were stably expressed while the expression levels of *ZjbHLH4, 12, 23, 78,* and *87* genes decreased [25]. There were seven *bHLH* genes with high expression levels in the petals of wheat [49]. *DobHLH96* was specifically and abundantly expressed in the flower bud, which suggests its potential role in the development of the flower bud. Moreover, the expression levels of *DobHLH74* in flower organs were much higher than that in the vegetative tissues. Its homolog in Arabidopsis is AtbHLH31 (At1G59640, BIGPETAL, BPE), which participated in the control of petal size and under the regulation of APETALA3, PISTILLATA, APETALA1, PISTILLATA3, and AGAMOUS [55].

Some *bHLH* genes could be highly expressed in roots such as *DobHLH12* and *DobHLH93*. In other species, some *bHLH* genes were also found to show abundant expression levels in roots such as five *bHLH* genes in wheat [49] and as many as 78 *bHLH* genes in maize [28]. These results suggest bHLH could play a conserved role in root development. Moreover, we found that some *bHLH* genes were abundant in the green root tip, implying that it could function to maintain root apical meristem. Some *DobHLH* genes had abundant transcripts in leaf. For example, *DobHLH54* had three cis-elements in its promoter region and may play a role in the process of photosynthesis.

MYC2 is an important regulator in the JA signaling pathway. In the present study, four putative MYC2 homologs were identified. A homolog of MYC2 (DobHLH33) in *D. officinale* was reported [56], which could be induced by MeJA and was involved in the biosynthesis of terpenoid indole alkaloids through the key enzyme genes in the MVA pathway. Moreover, four putative MYC2 homologs were expressed in roots as well as other tissues, suggesting their various functions in multiple tissues. Notably, DobHLH30 had the highest expression in the column (231.28). *MYC2* (*DobHLH33*) also had abundant expression in flower organs, which could have implications for functioning in flower development.

Usually, bHLH proteins function through the formation of homodimers or heterodimers to co-bind with their target genes to regulate their expressions. Many studies have reported the interactions of bHLH proteins to regulate the expressions of targeted genes [25,31,42,48,57]. Our current study found that many DobHLH proteins did have potential interactions via model prediction. In addition, we verified protein interactions between several DobHLH proteins such as DobHLH 25 and 26, and DobHLH17 and 60. We verified protein interactions between limited DobHLH proteins in this study. Many DobHLH proteins may function together to regulate plant growth and development, responses to environmental changes, and secondary metabolite processes in *D. officinale*. Additionally, we noted that DobHLH17 and DobHLH60 are homologous to UNE12 and PYE in Arabidopsis, and to ZmbHLH114 and ZmbHLH163 in maize, whose interactions have been verified in previous studies [28,57]. It is likely that protein interactions between given bHLHs proteins might be, at least partially, conserved in plants.

4. Materials and Methods

4.1. Plant Materials

The seedling plants of *D. officinale* in vitro (provided by Prof. Shibao Zhang's group) were well grown in the tissue culture room of the Kunming Institute of Botany, Chinese Academy of Sciences

(Kunming, Yunnan, China). The medium was Murashige and Skoog (MS) medium with 30 g/L sucrose, and the growth was 25 °C with a constant photoperiod (12 h light/12 h dark) with an illumination of 40 μmol quanta/(m^2 s). The fresh and healthy leaves, stem, and roots were harvested, immediately frozen in liquid nitrogen and stored at −80 °C for RNA extraction.

Three individual *D. offcinale* plants of six months old in vitro were used for each treatment. MeJA (Coolaber, China) and ABA (Sangon, China) solutions were prepared with ethanol and water. The treated plants were sprayed with 100 μM MeJA or 100 μM ABA for three or six hours while the control plants were sprayed with ethanol and water. The leaves were harvested and quickly frozen at −80 °C for further RNA extraction.

4.2. Identification of the bHLH TF Family in *Dendrobium officinale*

Although two research groups reported the genome of *D. officinale* (*D. catenatum*), respectively [2,39], at present only one genome (PRJNA262478) is available. The protein data of *D. officinale* were downloaded from NCBI. HMMER was used to identify the potential bHLH proteins with the bHLH domain (PF00010) [58]. The presence of the bHLH domain was manually confirmed by CDD and SMART. After repeated or uncomplete proteins were discarded, the remaining proteins with the bHLH domain were considered as the bHLH candidate members in *D. officinale*. The isoelectric point and molecular weight of candidate DobHLH proteins were analyzed with the ProtParam tool [59]. The subcellular localization of DobHLH proteins was predicted by WOLF PSORT (https://www.genscript.com/psort.html).

4.3. Phylogenetic Analysis of DobHLH Proteins

The AtbHLH proteins of Arabidopsis were downloaded from TAIR. The proteins of rice bHLHs were also downloaded [40]. The sequences of bHLH proteins from these species were aligned with ClustalW software [60], and then an un-rooted tree was obtained with MEGA 7.0 software using a neighbor-joining method with a bootstrap of 1000 replicates [61].

4.4. Gene Structure and Conserved Motif Analysis of DobHLHs

The gene structures of *DobHLH* genes were analyzed by comparing cDNA sequences with the correspondent genomic DNA sequences with GSDS [62]. MEME was used to search 10 conserved motifs in DobHLH proteins [63] and the InterPro database was used to further analyze those motifs [64].

4.5. Expression Patterns of DobHLH Genes in Different Tissues

The previously reported RNA-Seq data of eight *D. officinale* tissues were used to analyze the expression patterns of *DobHLH* genes. First, the raw sequencing reads (PRJNA348403) were downloaded from the NCBI [65]. After low-quality reads and adapter sequences were filtered, all the clean reads were mapped to the reference genome of *D. officinale*, resulting in unique reads using HISAT2 software [66]. Then, transcript abundances were estimated with the FPKM method using Stringtie software [66]. The heatmap of DobHLH genes was obtained by the R package pheatmap, v1.0.10 (https://rdrr.io/cran/pheatmap/).

4.6. Predicted Protein–Protein Interaction of DobHLH Members

The potential protein–protein interactions were predicted with the online STRING server (https://string-db.org). The protein sequences of 98 DobHLHs were uploaded into the server and *Arabidopsis thaliana* was chosen as the comparative organism. After the BLAST analysis was finished, the interaction internet of DobHLHs was generated with genes showing the highest bitscore.

4.7. *Cis-Element Analysis in Promoter Regions of DobHLH Genes*

The 2000 bp sequences at the upstream of the transcription start site for each *DobHLH* gene were downloaded from the NCBI. Cis-regulatory elements were analyzed with PlantCARE.

4.8. *RNA Extraction and cDNA Preparation*

The total RNA was extracted with the RNAprep Pure Plant Plus Kit (Cat DP441, Tiangen, China) according to the manufacturer's protocol. The concentration and quality of RNA samples were detected on a Nanodrop 2000 (Thermo Scientific, Waltham, MA, USA). cDNA was prepared with the one-step cDNA Synthesis Kit (Cat AT311, Transgen, China), which was used for DobHLH gene amplification in yeast two hybrid assays. A sample of 200 ng of total RNA in a reaction system of 20 μL was converted to cDNA with TransScript All-in-One First-Strand cDNA Synthesis SuperMix for qPCR (Cat AT341, Transgen, China), which was used for qRT-PCR assays.

4.9. *Yeast Two Hybrid Assays*

The CDS sequences of 11 *DobHLH* genes were amplified with PCR using cDNA prepared as templates. The primers used are shown in Table S4. All the sequences were confirmed to be correct by sequencing. Then, the sequences of *DobHLH25*, *30*, *32*, *55*, *56*, and *60* genes were cloned into the pDest22 construct (Invitrogen) with the GAL4 activation domain (AD) while the sequences of *DobHLH17*, *25*, *26*, *36*, *48*, and *70* genes were cloned into pDest32 (Invitrogen) with the GAL4 DNA-binding domain (BD) through recombination reaction. Yeast two hybrid (Y2H) was performed as previously described [67].

4.10. *qRT-PCR Assays*

One μL of prepared cDNA mixture was used for qRT-PCR in a 20 μL volume with SuperReal PreMix Plus SYBR Green Kit (TIANGEN Biotech, China), according to the manufacturer's instructions. The reaction was carried out on an Applied Biosystems QuantStudio 6 Flex Real-Time PCR System (ThemoFisher Scientific, USA) and the cycling conditions was as follows: first at 95 °C for 15 min, then 40 cycles of at 95 °C for 10 s, at 60 °C for 20 s, and finally at 72 °C for 32 s. The melt curves were performed at 95 °C for 15 s, at 60 °C for 1 min, and at 95 °C for 15 s. The $2^{-\Delta\Delta CT}$ method was used to calculate Log_2fold change with *DoActin* gene as an internal reference [13,68]. CT values stand for the average of cycle times from three technical replicates. The primer sequences used for qRT-PCR are listed in Table S4.

5. Conclusions

This research presents a genome-wide identification and characterization of the bHLH TF family in *D. officinale*. The phylogenetic analysis found that 98 DobHLH members were classified into 18 clades, which showed conserved gene structures and motifs. The characteristics of the DobHLH domains for their function were predicted. Furthermore, the transcript data of *DobHLH* genes revealed their universal or specific expression profiles, which may contribute to understanding their function in *D. officinale*. Therefore, the present study provides additional insights into the bHLH TF family and will be helpful for the further investigation of DobHLH functions in *D. officinale*.

Supplementary Materials:
Figure S1: The phylogenetic tree of bHLH proteins from *D. officinale* and rice. Figure S2: Expression patterns of eight *DobHLH* genes in leaf, stem, and root tissues using the qRT-PCR technique. The relative expression levels were obtained from the mean of three biological duplicates. The bars denote the standard deviation. Table S1: The additional information on 98 *DobHLH* genes in *D. officinale*. Table S2: The consensus amino acids of bHLH domains in different species. Table S3: The predicted cis-elements in *DobHLH* genes. Table S4: The primer sequences used in the study.

Author Contributions: Conceptualization, A.L.; Data curation, Y.W.; Investigation, Y.W.; Methodology, Y.W.; Supervision, A.L.; Writing–original draft, Y.W.; Writing–review & editing, A.L. All authors have read and agreed to the published version of the manuscript.

Acknowledgments: We are grateful to senior engineer Hua Wang of the Kunming Institute of Botany, Chinese Academy of Sciences for his kind donation of in vitro *D. officinale* materials.

References

1. Ng, T.B.; Liu, J.; Wong, J.H.; Ye, X.; Sze, S.C.W.; Tong, Y.; Zhang, K.Y. Review of research on Dendrobium, a prized folk medicine. *Appl. Microbiol. Biotechnol.* **2012**, *93*, 1795–1803. [CrossRef] [PubMed]
2. Yan, L.; Wang, X.; Liu, H.; Tian, Y.; Lian, J.; Yang, R.; Hao, S.; Wang, X.; Yang, S.; Li, Q.; et al. The Genome of *Dendrobium officinale* Illuminates the Biology of the Important Traditional Chinese Orchid Herb. *Mol. Plant* **2015**, *8*, 922–934. [CrossRef] [PubMed]
3. Tang, H.; Zhao, T.; Sheng, Y.; Zheng, T.; Fu, L.; Zhang, Y. *Dendrobium officinale* Kimura et Migo: A Review on Its Ethnopharmacology, Phytochemistry, Pharmacology, and Industrialization. *Evid. Based Complement. Alternat. Med.* **2017**, *2017*, 7436259. [CrossRef] [PubMed]
4. Shen, C.; Guo, H.; Chen, H.; Shi, Y.; Meng, Y.; Lu, J.; Feng, S.; Wang, H. Identification and analysis of genes associated with the synthesis of bioactive constituents in *Dendrobium officinale* using RNA-Seq. *Sci. Rep.* **2017**, *7*, 187. [CrossRef] [PubMed]
5. Xing, S.; Zhang, X.; Ke, H.; Lin, J.; Huang, Y.; Wei, G. Physicochemical properties of polysaccharides from *Dendrobium officinale* by fractional precipitation and their preliminary antioxidant and anti-HepG2 cells activities in vitro. *Chem. Cent. J.* **2018**, *12*, 100. [CrossRef]
6. Liang, J.; Wu, Y.; Yuan, H.; Yang, Y.; Xiong, Q.; Liang, C.; Li, Z.; Li, C.; Zhang, G.; Lai, X.; et al. *Dendrobium officinale* polysaccharides attenuate learning and memory disabilities via anti-oxidant and anti-inflammatory actions. *Int. J. Biol. Macromol.* **2019**, *126*, 414–426. [CrossRef]
7. Tao, S.C.; Lei, Z.X.; Huang, K.W.; Li, Y.R.; Ren, Z.Y.; Zhang, X.F.; Wei, G.; Chen, H.M. Structural characterization and immunomodulatory activity of two novel polysaccharides derived from the stem of *Dendrobium officinale* Kimura et Migo. *J. Funct. Foods.* **2019**, *57*, 121–134. [CrossRef]
8. Kuang, M.T.; Li, J.Y.; Yang, X.B.; Yang, L.; Xu, J.Y.; Yan, S.; Lv, Y.F.; Ren, F.C.; Hu, J.M.; Zhou, J. Structural characterization and hypoglycemic effect via stimulating glucagon-like peptide-1 secretion of two polysaccharides from *Dendrobium officinale*. *Carbohydr. Polym.* **2020**, *241*, 116326. [CrossRef]
9. Pan, L.H.; Li, X.F.; Wang, M.N.; Zha, X.Q.; Yang, X.F.; Liu, Z.J.; Luo, Y.B.; Luo, J.P. Comparison of hypoglycemic and antioxidative effects of polysaccharides from four different *Dendrobium* species. *Int. J. Biol. Macromol.* **2014**, *64*, 420–427. [CrossRef]
10. Zeng, X.; Ling, H.; Chen, X.; Guo, S. Genome-wide identification, phylogeny and function analysis of GRAS gene family in *Dendrobium catenatum* (Orchidaceae). *Gene* **2019**, *705*, 5–15. [CrossRef]
11. Huang, K.; Li, Y.; Tao, S.; Wei, G.; Huang, Y.; Chen, D.; Wu, C. Purification, Characterization and Biological Activity of Polysaccharides from *Dendrobium officinale*. *Molecules* **2016**, *21*, 701. [CrossRef] [PubMed]
12. Yu, Z.; Liao, Y.; da Silva, J.A.T.; Yang, Z.; Duan, J. Differential Accumulation of Anthocyanins in *Dendrobium officinale* Stems with Red and Green Peels. *Int. J. Mol. Sci.* **2018**, *19*, 2857. [CrossRef] [PubMed]
13. Yuan, Y.; Zhang, J.; Liu, X.; Meng, M.; Wang, J.; Lin, J. Tissue-specific transcriptome for *Dendrobium officinale* reveals genes involved in flavonoid biosynthesis. *Genomics* **2020**, *112*, 1781–1794. [CrossRef] [PubMed]
14. Lei, Z.; Zhou, C.; Ji, X.; Wei, G.; Huang, Y.; Yu, W.; Luo, Y.; Qiu, Y. Transcriptome Analysis Reveals genes involved in flavonoid biosynthesis and accumulation in *Dendrobium catenatum* From Different Locations. *Sci. Rep.* **2018**, *8*, 6373. [CrossRef] [PubMed]
15. Guo, X.; Li, Y.; Li, C.; Luo, H.; Wang, L.; Qian, J.; Luo, X.; Xiang, L.; Song, J.; Sun, C.; et al. Analysis of the *Dendrobium officinale* transcriptome reveals putative alkaloid biosynthetic genes and genetic markers. *Gene* **2013**, *527*, 131–138. [CrossRef] [PubMed]
16. Sun, X.; Wang, Y.; Sui, N. Transcriptional regulation of bHLH during plant response to stress. *Biochem. Biophys. Res. Commun.* **2018**, *503*, 397–401. [CrossRef]
17. Jones, S. An overview of the basic helix-loop-helix proteins. *Genome Biol.* **2004**, *5*, 226. [CrossRef]
18. Pires, N.; Dolan, L. Origin and diversification of basic-helix-loop-helix proteins in plants. *Mol. Biol. Evol.* **2010**, *27*, 862–874. [CrossRef]
19. Moore, A.W.; Barbel, S.; Jan, L.Y.; Jan, Y.N. A genomewide survey of basic helix-loop-helix factors in Drosophila. *Proc. Natl. Acad. Sci. USA* **2000** *97*, 10436–10441. [CrossRef]

20. Riechmann, J.L.; Heard, J.; Martin, G.; Reuber, L.; Jiang, C.; Keddie, J.; Adam, L.; Pineda, O.; Ratcliffe, O.J.; Samaha, R.R.; et al. Arabidopsis transcription factors: Genome-wide comparative analysis among eukaryotes. *Science* **2000**, *290*, 2105–2110. [CrossRef]
21. Murre, C.; McCaw, P.S.; Baltimore, D. A new DNA binding and dimerization motif in immunoglobulin enhancer binding, daughterless, MyoD, and myc proteins. *Cell* **1989**, *56*, 777–783. [CrossRef]
22. Toledo-Ortiz, G.; Huq, E.; Quail, P.H. The Arabidopsis basic/helix-loop-helix transcription factor family. *Plant Cell* **2003**, *15*, 1749–1770. [CrossRef] [PubMed]
23. Atchley, W.R.; Terhalle, W.; Dress, A. Positional dependence, cliques, and predictive motifs in the bHLH protein domain. *J. Mol. Evol.* **1999**, *48*, 501–516. [CrossRef] [PubMed]
24. Massari, M.E.; Murre, C. Helix-loop-helix proteins: Regulators of transcription in eucaryotic organisms. *Mol. Cell Biol.* **2000**, *20*, 429–440. [CrossRef] [PubMed]
25. Li, H.; Gao, W.; Xue, C.; Zhang, Y.; Liu, Z.; Zhang, Y.; Meng, X.; Liu, M.; Zhao, J. Genome-wide analysis of the bHLH gene family in Chinese jujube (*Ziziphus jujuba* Mill.) and wild jujube. *BMC Genom.* **2019**, *20*, 568. [CrossRef]
26. Bailey, P.C.; Martin, C.; Toledo-Ortiz, G.; Quail, P.H.; Huq, E.; Heim, M.A.; Jakoby, M.; Werber, M.; Weisshaar, B. Update on the basic helix-loop-helix transcription factor gene family in Arabidopsis thaliana. *Plant Cell* **2003**, *15*, 2497–2502. [CrossRef] [PubMed]
27. Sun, H.; Fan, H.J.; Ling, H.Q. Genome-wide identification and characterization of the bHLH gene family in tomato. *BMC Genom.* **2015**, *16*, 9. [CrossRef]
28. Zhang, T.; Lv, W.; Zhang, H.; Ma, L.; Li, P.; Ge, L.; Li, G. Genome-wide analysis of the basic Helix-Loop-Helix (bHLH) transcription factor family in maize. *BMC Plant Biol.* **2018**, *18*, 235. [CrossRef]
29. Wang, P.; Su, L.; Gao, H.; Jiang, X.; Wu, X.; Li, Y.; Zhang, Q.; Wang, Y.; Ren, F. Genome-Wide Characterization of bHLH Genes in Grape and Analysis of their Potential Relevance to Abiotic Stress Tolerance and Secondary Metabolite Biosynthesis. *Front. Plant Sci.* **2018**, *9*, 64. [CrossRef]
30. Lu, R.; Zhang, J.; Liu, D.; Wei, Y.L.; Wang, Y.; Li, X.B. Characterization of bHLH/HLH genes that are involved in brassinosteroid (BR) signaling in fiber development of cotton (*Gossypium hirsutum*). *BMC Plant Biol.* **2018**, *18*, 304. [CrossRef]
31. Niu, X.; Guan, Y.; Chen, S.; Li, H. Genome-wide analysis of basic helix-loop-helix (bHLH) transcription factors in *Brachypodium distachyon*. *BMC Genom.* **2017**, *18*, 619. [CrossRef] [PubMed]
32. Ludwig, S.R.; Wessler, S.R. Maize R Gene Family: Tissue-Specific Helix-Loop-Helix Proteins. *Cell* **1990**, *62*, 849–851. [CrossRef]
33. Xu, W.; Dubos, C.; Lepiniec, L. Transcriptional control of flavonoid biosynthesis by MYB-bHLH-WDR complexes. *Trends Plant Sci.* **2015**, *20*, 176–185. [CrossRef] [PubMed]
34. Van Moerkercke, A.; Steensma, P.; Schweizer, F.; Pollier, J.; Gariboldi, I.; Payne, R.; van den Bossche, R.; Miettinen, K.; Espoz, J.; Purnama, P.C.; et al. The bHLH transcription factor BIS1 controls the iridoid branch of the monoterpenoid indole alkaloid pathway in *Catharanthus roseus*. *Proc. Natl. Acad. Sci. USA* **2015**, *112*, 8130–8135. [CrossRef]
35. Zhang, X.; Luo, H.; Xu, Z.; Zhu, Y.; Ji, A.; Song, J.; Chen, S. Genome-wide characterisation and analysis of bHLH transcription factors related to tanshinone biosynthesis in *Salvia miltiorrhiza*. *Sci. Rep.* **2015**, *5*, 11244. [CrossRef] [PubMed]
36. Li, C.; Qiu, J.; Ding, L.; Huang, M.; Huang, S.; Yang, G.; Yin, J. Anthocyanin biosynthesis regulation of DhMYB2 and DhbHLH1 in Dendrobium hybrids petals. *Plant Physiol. Biochem.* **2017**, *112*, 335–345. [CrossRef] [PubMed]
37. Oh, E.; Yamaguchi, S.; Kamiya, Y.; Bae, G.; Chung, W.I.; Choi, G. Light activates the degradation of PIL5 protein to promote seed germination through gibberellin in Arabidopsis. *Plant J.* **2006**, *47*, 124–139. [CrossRef]
38. Chinnusamy, V.; Ohta, M.; Kanrar, S.; Lee, B.H.; Hong, X.; Agarwal, M.; Zhu, J.K. ICE1: A regulator of cold-induced transcriptome and freezing tolerance in Arabidopsis. *Genes Dev.* **2003**, *17*, 1043–1054. [CrossRef]
39. Zhang, G.Q.; Xu, Q.; Bian, C.; Tsai, W.C.; Yeh, C.M.; Liu, K.W.; Yoshida, K.; Zhang, L.S.; Chang, S.B.; Chen, F.; et al. The *Dendrobium catenatum* Lindl. genome sequence provides insights into polysaccharide synthase, floral development and adaptive evolution. *Sci. Rep.* **2016**, *6*, 19029. [CrossRef]
40. Li, X.; Duan, X.; Jiang, H.; Sun, Y.; Tang, Y.; Yuan, Z.; Guo, J.; Liang, W.; Chen, L.; Yin, J.; et al. Genome-wide analysis of basic/helix-loop-helix transcription factor family in rice and Arabidopsis. *Plant Physiol.* **2006**, *141*, 1167–1184. [CrossRef]

41. Zhao, K.; Li, S.; Yao, W.; Zhou, B.; Li, R.; Jiang, T. Characterization of the basic helix-loop-helix gene family and its tissue-differential expression in response to salt stress in poplar. *PeerJ* **2018**, *6*, e4502. [CrossRef] [PubMed]
42. Guo, X.J.; Wang, J.R. Global identification, structural analysis and expression characterization of bHLH transcription factors in wheat. *BMC Plant Biol.* **2017**, *17*, 90. [CrossRef] [PubMed]
43. Shen, W.; Cui, X.; Li, H.; Teng, R.-M.; Wang, Y.-X.; Liu, H.; Zhuang, J. Genome-wide identification and analyses of bHLH family genes in Brassica napus. *Can. J. Plant Sci.* **2019**, *99*, 589–598. [CrossRef]
44. Zhang, C.; Feng, R.; Ma, R.; Shen, Z.; Cai, Z.; Song, Z.; Peng, B.; Yu, M. Genome-wide analysis of basic helix-loop-helix superfamily members in peach. *PLoS ONE* **2018**, *13*, e0195974. [CrossRef] [PubMed]
45. Chen, Y.; Wang, Y.; Lyu, P.; Chen, L.; Shen, C.; Sun, C. Comparative transcriptomic analysis reveal the regulation mechanism underlying MeJA-induced accumulation of alkaloids in *Dendrobium officinale*. *J. Plant Res.* **2019**, *132*, 419–429. [CrossRef] [PubMed]
46. He, C.; da Silva, J.A.T.; Wang, H.; Si, C.; Zhang, M.; Zhang, X.; Li, M.; Tan, J.; Duan, J. Mining MYB transcription factors from the genomes of orchids (Phalaenopsis and Dendrobium) and characterization of an orchid R2R3-MYB gene involved in water-soluble polysaccharide biosynthesis. *Sci. Rep.* **2019**, *9*, 1381. [CrossRef]
47. Chen, Y.; Shen, Q.; Lyu, P.; Lin, R.; Sun, C. Identification and expression profiling of selected MADS-box family genes in *Dendrobium officinale*. *Genetica* **2019**, *147*, 303–313. [CrossRef]
48. Hong, Y.; Naveed, A.; Tian, Y.; Liu, J.; Wang, L.; Wang, G.; Liu, X.; Dong, Y.; Wang, F.; Liu, W.; et al. Genome-Wide Identification, Expression Analysis, and Subcellular Localization of *Carthamus tinctorius* bHLH Transcription Factors. *Int. J. Mol. Sci.* **2019**, *20*, 3044. [CrossRef]
49. Wang, L.; Xiang, L.; Hong, J.; Xie, Z.; Li, B. Genome-wide Analysis of bHLH Transcription Factor Family Reveals Their Involvement in Biotic and Abiotic Stress Responses in Wheat (*Triticum aestivum* L.). *3 Biotech* **2019**, *9*, 236. [CrossRef]
50. Chu, Y.; Xiao, S.; Su, H.; Liao, B.; Zhang, J.; Xu, J.; Chen, S. Genome-wide characterization and analysis of bHLH transcription factors in *Panax ginseng*. *Acta Pharm. Sin. B* **2018**, *8*, 666–677. [CrossRef]
51. Yang, J.; Gao, M.; Huang, L.; Wang, Y.; van Nocker, S.; Wan, R.; Guo, C.; Wang, X.; Gao, H. Identification and expression analysis of the apple (*Malus* × *domestica*) basic helix-loop-helix transcription factor family. *Sci. Rep.* **2017**, *7*, 28. [CrossRef] [PubMed]
52. Mao, T.Y.; Liu, Y.Y.; Zhu, H.H.; Zhang, J.; Yang, J.X.; Fu, Q.; Wang, N.; Wang, Z. Genome-wide analyses of the bHLH gene family reveals structural and functional characteristics in the aquatic plant *Nelumbo nucifera*. *PeerJ* **2019**, *7*, e7153. [CrossRef] [PubMed]
53. Carretero-Paulet, L.; Galstyan, A.; Roig-Villanova, I.; Martínez-García, J.F.; Bilbao-Castro, J.R.; Robertson, D.L. Genome-wide classification and evolutionary analysis of the bHLH family of transcription factors in Arabidopsis, poplar, rice, moss, and algae. *Plant Physiol.* **2010**, *153*, 1398–1412. [CrossRef] [PubMed]
54. Wang, R.; Zhao, P.; Kong, N.; Lu, R.; Pei, Y.; Huang, C.; Ma, H.; Chen, Q. Genome-Wide Identification and Characterization of the Potato bHLH Transcription Factor Family. *Genes* **2018**, *9*, 54. [CrossRef] [PubMed]
55. Varaud, E.; Brioudes, F.; Szécsi, J.; Leroux, J.; Brown, S.; Perrot-Rechenmann, C.; Bendahmane, M. AUXIN RESPONSE FACTOR8 regulates Arabidopsis petal growth by interacting with the bHLH transcription factor BIGPETALp. *Plant Cell* **2011**, *23*, 973–983. [CrossRef]
56. Zhu, Y.; Meng, C.; Zhu, L.; Li, D.; Jin, Q.; Song, C.; Cai, Y.; Fan, H.; Lin, Y. Cloning and characterization of DoMYC2 from *Dendrobium officinale*. *Plant Cell Tiss. Organ. Cult.* **2017**, *129*, 533–541. [CrossRef]
57. Braun, P.; Carvunis, A.-R.; Charloteaux, B.; Dreze, M.; Ecker, R.J.; Hill, E.D.; Roth, P.F.; Vidal, M.; Galli, M.; Balumuri, P. Evidence for network evolution in an Arabidopsis interactome map. *Science* **2011**, *333*, 601–607.
58. Finn, R.D.; Clements, J.; Arndt, W.; Miller, B.L.; Wheeler, T.J.; Schreiber, F.; Bateman, A.; Eddy, S.R. HMMER web server: 2015 update. *Nucleic Acids Res.* **2015**, *43*, W30–W38. [CrossRef]
59. Wilkins, M.R.; Gasteiger, E.; Bairoch, A.; Sanchez, J.C.; Williams, K.L.; Appel, R.D.; Hochstrasser, D.F. Protein identification and analysis tools in the ExPASy server. *Methods Mol. Biol.* **1999**, *112*, 531–552. [CrossRef]
60. McWilliam, H.; Li, W.; Uludag, M.; Squizzato, S.; Park, Y.M.; Buso, N.; Cowley, A.P.; Lopez, R. Analysis Tool Web Services from the EMBL-EBI. *Nucleic Acids Res.* **2013**, *41*, W597–W600. [CrossRef]
61. Kumar, S.; Stecher, G.; Tamura, K. MEGA7: Molecular Evolutionary Genetics Analysis Version 7.0 for Bigger Datasets. *Mol. Biol. Evol.* **2016**, *33*, 1870–1874. [CrossRef] [PubMed]

62. Hu, B.; Jin, J.; Guo, A.Y.; Zhang, H.; Luo, J.; Gao, G. GSDS 2.0: An upgraded gene feature visualization server. *Bioinformatics* **2015**, *31*, 1296–1297. [CrossRef] [PubMed]
63. Bailey, T.L.; Boden, M.; Buske, F.A.; Frith, M.; Grant, C.E.; Clementi, L.; Ren, J.; Li, W.W.; Noble, W.S. MEME SUITE: Tools for motif discovery and searching. *Nucleic Acids Res.* **2009**, *37*, W202–W208. [CrossRef] [PubMed]
64. Apweiler, R.; Attwood, T.K.; Bairoch, A.; Bateman, A.; Birney, E.; Biswas, M.; Bucher, P.; Cerutti, L.; Corpet, F.; Croning, M.D.; et al. The InterPro database, an integrated documentation resource for protein families, domains and functional sites. *Nucleic Acids Res.* **2001**, *29*, 37–40. [CrossRef] [PubMed]
65. Zhang, G.Q.; Liu, K.W.; Li, Z.; Lohaus, R.; Hsiao, Y.Y.; Niu, S.C.; Wang, J.Y.; Lin, Y.C.; Xu, Q.; Chen, L.J.; et al. The Apostasia genome and the evolution of orchids. *Nature* **2017**, *549*, 379–383. [CrossRef]
66. Pertea, M.; Kim, D.; Pertea, G.M.; Leek, J.T.; Salzberg, S.L. Transcript-level expression analysis of RNA-seq experiments with HISAT, StringTie and Ballgown. *Nat. Protoc.* **2016**, *11*, 1650–1667. [CrossRef]
67. Wang, Y.; Xu, W.; Chen, Z.; Han, B.; Haque, M.E.; Liu, A. Gene structure, expression pattern and interaction of Nuclear Factor-Y family in castor bean (*Ricinus communis*). *Planta* **2018**, *247*, 559–572. [CrossRef]
68. Livak, K.J.; Schmittgen, T.D. Analysis of relative gene expression data using real-time quantitative PCR and the 2(-Delta Delta C(T)) Method. *Methods* **2001**, *25*, 402–408. [CrossRef]

3

Comparative Transcriptome Analysis of Different *Dendrobium* Species Reveals Active Ingredients-Related Genes and Pathways

Yingdan Yuan [1,2], Bo Zhang [1,2,3], Xinggang Tang [1,2], Jinchi Zhang [1,2,*] and Jie Lin [1,2,*]

1 Co-Innovation Center for Sustainable Forestry in Southern China, Nanjing Forestry University, Nanjing 210037, China; yuanyingdan@126.com (Y.Y.); bozhangophelia@gmail.com (B.Z.); xinggangtang@126.com (X.T.)
2 Jiangsu Province Key Laboratory of Soil and Water Conservation and Ecological Restoration, Nanjing Forestry University, Nanjing 210037, China
3 Department of Environmental Science and Policy, University of California, Davis, CA 95616, USA
* Correspondence: zhang8811@njfu.edu.cn (J.Z.); jielin@njfu.edu.cn (J.L.)

Abstract: *Dendrobium* is widely used in traditional Chinese medicine, which contains many kinds of active ingredients. In recent years, many *Dendrobium* transcriptomes have been sequenced. Hence, weighted gene co-expression network analysis (WGCNA) was used with the gene expression profiles of active ingredients to identify the modules and genes that may associate with particular species and tissues. Three kinds of Dendrobium species and three tissues were sampled for RNA-seq to generate a high-quality, full-length transcriptome database. Based on significant changes in gene expression, we constructed co-expression networks and revealed 19 gene modules. Among them, four modules with properties correlating to active ingredients regulation and biosynthesis, and several hub genes were selected for further functional investigation. This is the first time the WGCNA method has been used to analyze *Dendrobium* transcriptome data. Further excavation of the gene module information will help us to further study the role and significance of key genes, key signaling pathways, and regulatory mechanisms between genes on the occurrence and development of medicinal components of *Dendrobium*.

Keywords: *Dendrobium*; comparative transcriptome; active ingredients; different tissues

---

1. Introduction

Dendrobium is a perennial epiphytic herb of the genus *Orchidaceae*, with more than 1500 species in the world which are mostly growing in tropical and subtropical Asia and eastern Australia [1,2]. In China, there are more than 80 species of *Dendrobium* that had been reported in studies, mainly distributed in south of the Qinling Mountains [3,4]. As a kind of ornamental and medicinal plants, more and more researchers have paid attention to it in recent years [5]. Also, *Dendrobium* is a traditional Chinese herbal medicine, that has been used for benefiting stomach and clearing heat, nourishing yin, and promoting fluid [6]. The medicinal ingredients of *Dendrobium* are very complex, not only including polysaccharides and alkaloids, but also flavonoids, various amino acids, and trace mineral elements. However, the most two important components are polysaccharides and alkaloids [7–10].

With the development of high-throughput technology, a variety of data sources have been derived, including gene expression microarray, RNA-seq, metabolomic and CHIP-seq, these have become powerful tools for studying plant growth, development, and physiology at the transcriptional and metabolic levels [11–13]. In order to explore the key genes and regulatory pathways for the synthesis of its medicinal ingredients, more researchers have done different kinds of *Dendrobium* transcriptome research [14–16]. However, it has been difficult to systematically explain the relationship between gene expression or metabolite changes and trait differences [17]. The analysis of correlation networks can bridge the gap between single gene interpretation and systematic biology research by mining the link between genes and gene products [18–20], such as integrating a single gene into a co-expression network based on pairwise gene expression correlation [21]. Gene co-expression analysis has been used to discover new candidate genes [22,23], identify key modulators of immune responses [24], and reconstruct regulatory pathways [25]. Genes belonging to the same co-expression sub-network (or module) are likely to be functionally related [26–28], participate in similar biological processes, or be part of the same pathway [25].

Weighted gene co-expression network analysis (WGCNA) is one of the most useful methods based on gene co-expression networks. It focuses on the set of genes other than on a single gene in the observed gene expression data, and it alleviates the multiple detection problems inherent in chip data analysis and can be used in unweighted correlation networks [29]. Compared with many other analysis methods, WGCNA has the advantages of summarizing and standardizing the methods and functions of integrated R packages, including methods of weighted and unweighted correlation networks [30]. WGCNA is used in combination with gene chip data, transcriptome data, and metabolome data for metabolic regulation network simulation, mining inter-genetic interactions, screening functional genes, etc. It has been extensively studied in plant growth, tissue and organ development, pigmentation and fragrance synthesis [31–34].

To reveal the underlying molecular mechanism of the active ingredients of *Dendrobium*, we downloaded three datasets from the NCBI Sequence Reading Archive (SRA) to identify highly connected hub genes and important modules. This study used three different *Dendrobium* species and different tissues as materials to perform transcriptome sequencing data, combined with polysaccharide and alkaloid content data, and used WGCNA analysis to construct a co-expression gene network. Correlation analysis was performed between the gene module and the polysaccharide and alkaloid data, and hub genes related to the main medicinal ingredients were discovered, in order to provide new clues for further research on the molecular mechanism of medicinal ingredients of *Dendrobium*. For the first time, a co-expression network analysis of transcriptome genes in *Dendrobium* was constructed, and modules with high correlation in secondary metabolism were analyzed, laying a foundation for the discovery of functional genes of medicinal ingredients.

2. Results

2.1. Determination of Total Alkaloid and Polysaccharide Contents in Different Species and Different Tissues

Polysaccharides and alkaloids are the main medicinal components of *Dendrobium*. Therefore, we determined the polysaccharide and total alkaloid in three different tissues of three *Dendrobium* species (Figure 1). The biennial *Dendrobium officinale* has the highest content in the comparison of different tissues and species. For polysaccharide content, the stem is the main enrichment tissue, while for total alkaloid, the leaf is the main enrichment tissue. In the stem, the polysaccharide content of *Dendrobium* varied from 23.34 to 37.41%. In the leaf, the total alkaloid content of *Dendrobium* varied from 0.0291 to 0.0421%.

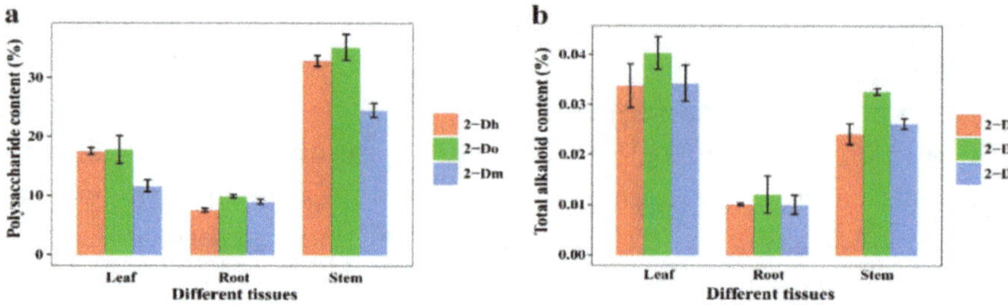

Figure 1. Polysaccharide and total alkaloid contents variation in different tissues of different species. (**a**) Polysaccharide content. (**b**) Total alkaloid content. There are three replicates of each sample. 2-Dh: two-year-old *Dendrobium huoshanese*, 2-Do: two-year-old *Dendrobium officinale*, 2-Dm: two-year-old *Dendrobium moniliforme*.

2.2. Differential Expression Analysis

In this study, we calculated the FPKM of each sample to regulate the expression and investigate the gene expression differences among different tissues of different *Dendrobium* species. We compared the three data sets from different comparison groups using a Venn diagram (Figure 2a–c). In addition, we compared the number of DEGs between different species in the same tissue, and many genes were differentially expressed only in one or two comparisons. Then the edgeR was used to test the differential expression of the repeated count data. We used "FDR < 0.05 & | $\log_2$FC | ≥ 1" as the criterion for significant differences in gene expression. When $\log_2$FC > 1, DEG is considered to be up-regulated. In contrast, for $\log_2$FC < -1, it is considered a downward adjustment. We did up and down-regulated analysis of DEGs for three different tissues and three different *Dendrobium* species, and the results are shown in Figure 3a–i.

According to the transcriptome data, all DEGs were used for hierarchical cluster analysis of transcription abundance in three different tissues. The heatmap of DEGs between different tissues and different *Dendrobium* species show similar transcriptome profiles for Dh_R, Dh_L, Dh_S, Do_R, Do_L, Do_S, Dm_R, Dm_L, and Dm_S (Figure 4a). A total of 35,159 DEGs were identified and analyzed using criteria of $\log_{10}$ (FPKM+1) and $p < 0.05$. The trend of the specific expression level is shown in the number under the color bar at the top left. On the left is the gene cluster tree. The closer the degree of separation between the two genes, the closer their expression is. In order to reflect the main trends and tissue-specific expression of different *Dendrobium* species, all DEGs were clustered into ten expression profiles (Figure 4b) using the K-means method and hierarchical clustering with similar regulation model and $\log_2$ (foldchange). DEGs belonging to cluster 3 were more highly expressed in leaves than in other tissues. Except for subcluster 1 and subcluster 3, all the other subclusters have just one gene trend.

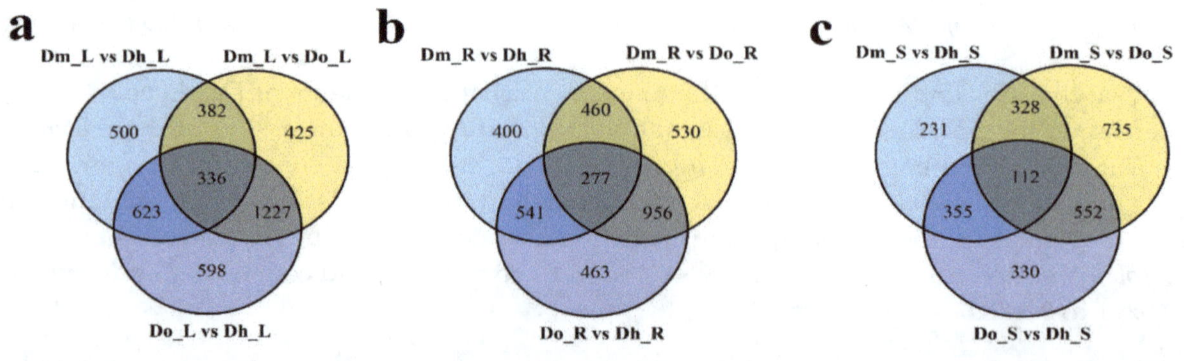

Figure 2. Venn diagram of differentially expressed genes (DEGs) in different comparisons. All DEGs are grouped into three comparison groups represented by three circles. The overlapping portions of the different circles represent the number of DEGs common to these comparison groups. (**a**) Venn diagram of three kinds of *Dendrobium* leaves. (**b**) Venn diagram of three kinds of *Dendrobium* roots. (**c**) Venn diagram of three kinds of *Dendrobium* stems.

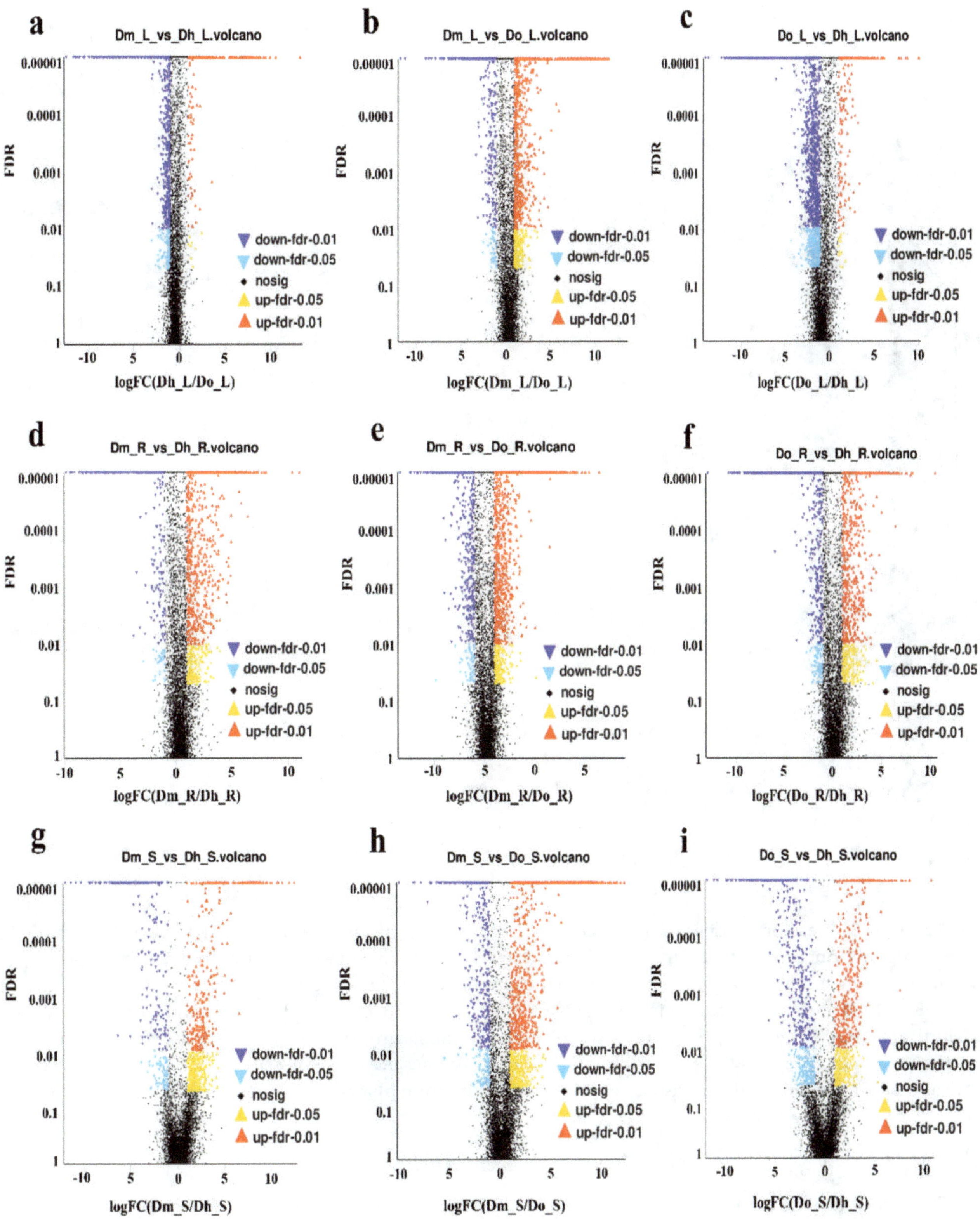

Figure 3. Volcano plots of the DEGs in different comparisons. Red dots indicate significant up-regulation of genes, and blue dots indicate significant down-regulation of genes. Black dots represent non-DEGs. (**a**) Dm_L vs. Dh_L volcano; (**b**) Dm_L vs. Do_L volcano; (**c**) Do_L vs. Dh_L volcano; (**d**) Dm_R vs. Dh_R volcano; (**e**) Dm_R vs. Do_R volcano; (**f**) Do_R vs. Dh_R volcano; (**g**) Dm_S vs. Dh_S volcano; (**h**) Dm_S vs. Do_S volcano; (**i**) Do_S vs. Dh_S volcano.

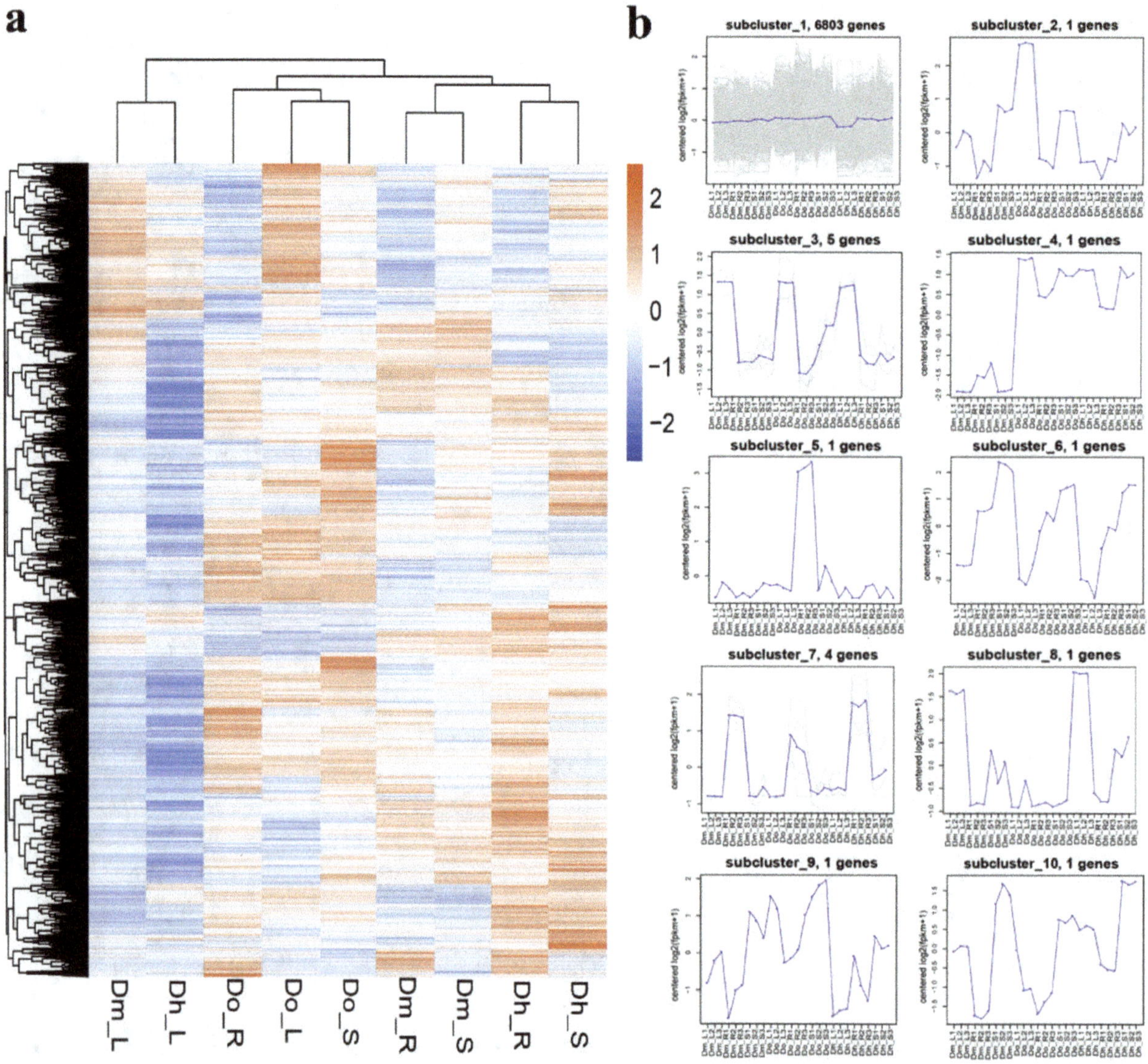

Figure 4. Expression profiles of all DEGs. (**a**) The heatmap of DEGs and the FPKM distribution of all unigenes obtained by hierarchical cluster analysis. Each column in the figure represents a sample, and each row represents a gene. The colors in the graph indicate the magnitude of gene expression ($\log_{10}$ (FPKM + 1)) in the sample. Red indicates that the gene is highly expressed in the sample, and the blue indicates that the gene expression is low. (**b**) K-means clustering analysis of gene expression profiles. The blue line represents the expression model. The gray lines are the expression profiles of each DEGs. The x-axis represents different tissues of different Dendrobium plants. The y-axis represents $\log_2$ (ratio).

2.3. GO and KEGG of DEGs Annotation and Enrichment Analysis

GO is a database established by the Gene Ontology Consortium. Its purpose is to standardize biological terms about genes and gene products in different databases, to define and describe gene and protein functions. Using the GO database, genes can be classified according to biological process, cellular component, and molecular function. The GO annotation statistics were performed on the differentially expressed genes in pairs, and one of the samples was used as a control. The obtained results can be used to plot the GO annotation bar graph of up and down DEGs (Figure 5). Figure 5 only shows the comparison of the stems of different *Dendrobium* species, and the other two tissues are shown in the Supplementary Figure S1.

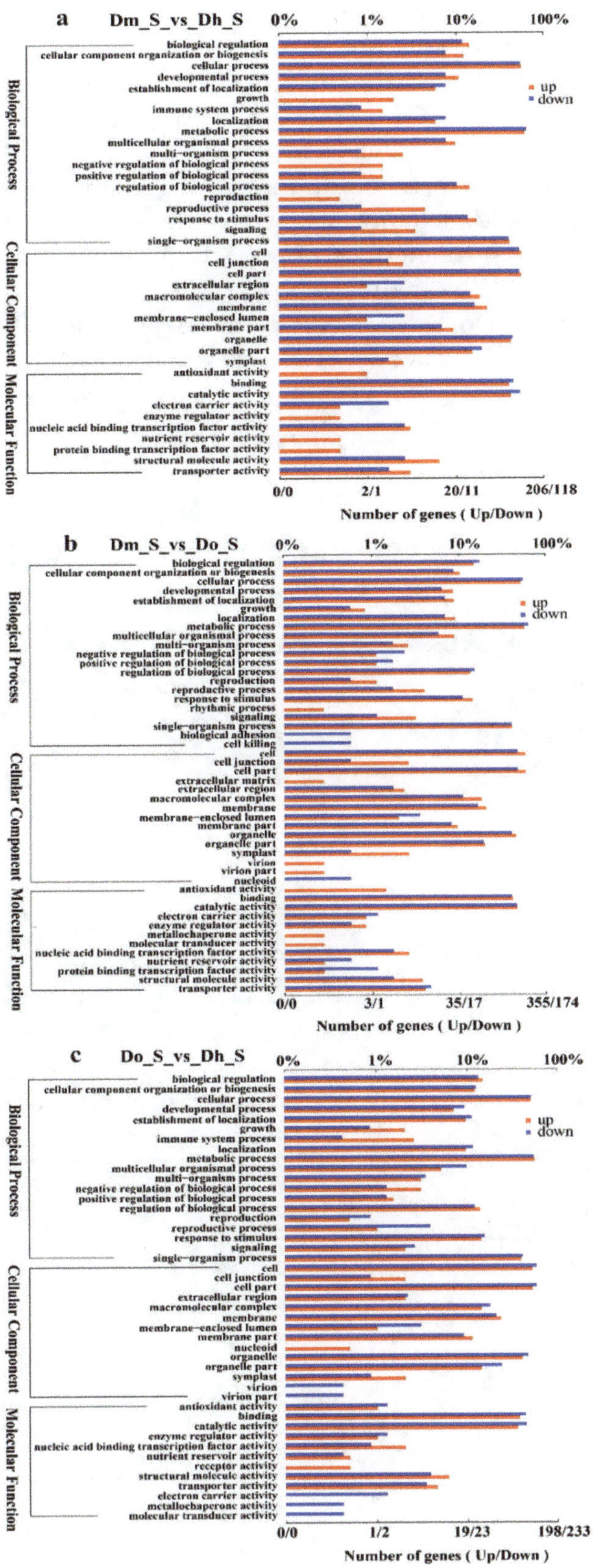

Figure 5. Gene Ontology (GO) annotations of up and down regulated DEGs. The bottom x-axis indicates the number of DEGs annotated to a GO term, the upper x-axis indicates the proportion of DEGs annotated to a GO terms to the total number of all GO annotated DEGs; and the y-axis represents each detailed classification of GO. (**a**) Dm_S vs Dh_S GO annotation; (**b**) Dm_S vs. Do_S GO annotation; (**c**) Do_S vs. Dh_S GO annotation.

KEGG focuses on biochemical pathways, especially genes involved in protein, carbohydrate, and energy metabolism. In the KEGG enrichment analysis, we mapped all these genes to a reference pathway in the KEGG database to determine the biological pathways in which these genes may be involved (Figure 6). In terms of the KEGG pathways, "Dm_S vs. Dh_S" comparisons are involved in 235 pathways (Figure 6a). In "Dm_S vs. Do_S" comparisons, 223 pathways were involved (Figure 6b). Finally, in "Do_S vs. Dh_S" comparisons, 237 pathways were involved (Figure 6c). Among these pathways, "flavonoid biosynthesis" was enriched in the Dm_S vs. Dh_S comparison. "Lipopolysaccharide biosynthesis" related to polysaccharide was the most enriched pathways in the Do_S vs. Dh_S comparison. Notably, we observed terpenoid biosynthesis in the secondary metabolites also enriched in the Do_S vs. Dh_S comparison. The KEGG pathways of different *Dendrobium* species of the other two tissues are shown in the Supplementary Figure.

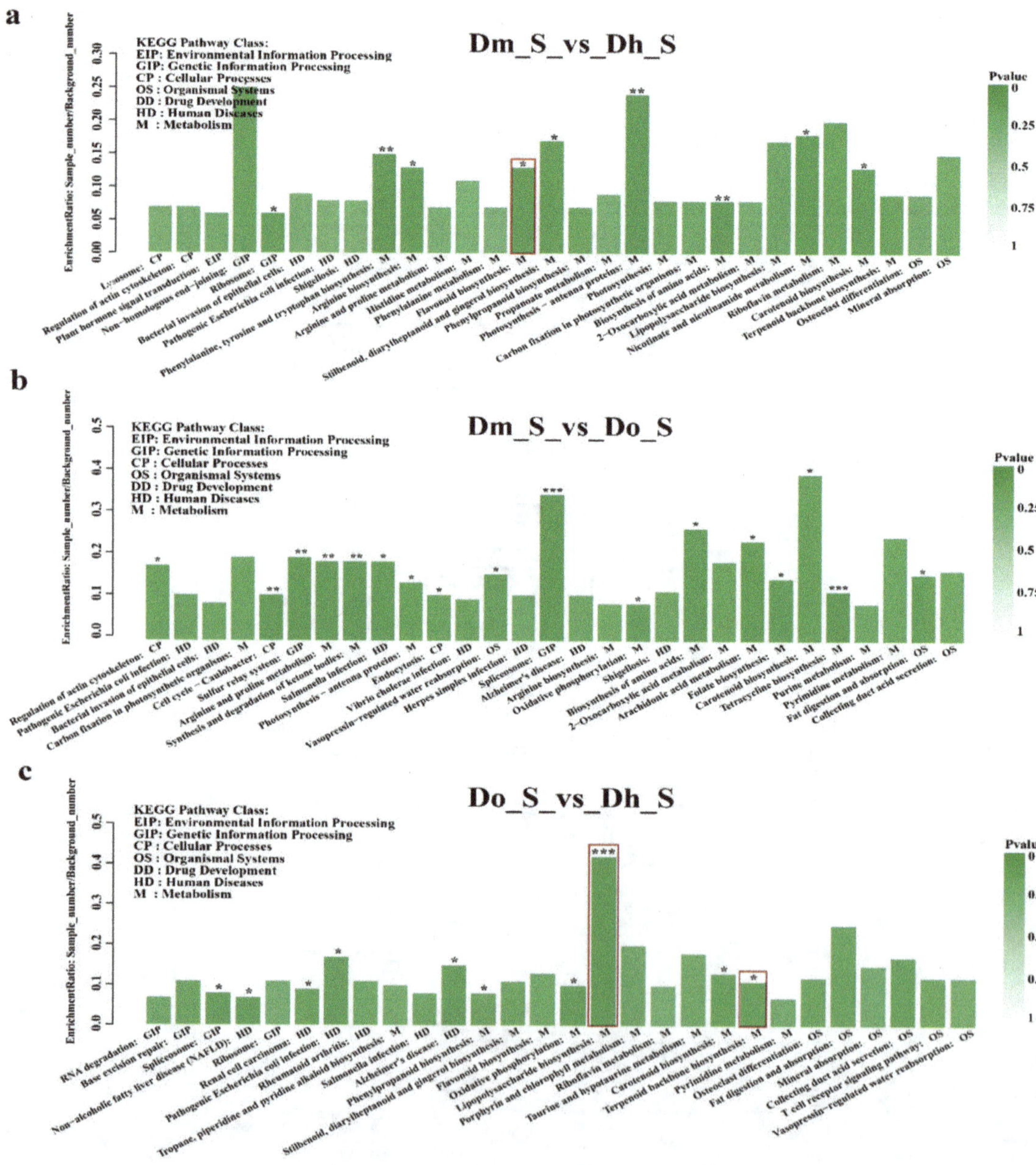

Figure 6. Kyoto Encyclopedia of Genes and Genomes (KEGG) pathway enrichment of DEGs. The x-axis represents the pathway name, and the y-axis represents the enrichment ratio (sample number/background number). (**a**) Dm_S vs. Dh_S; (**b**) Dm_S vs. Do_S; (**c**) Do_S vs. Dh_S. All pathways in the figure with asterisks indicate significant KEGG enrichment, with three asterisks indicating p-value < 0.001, two asterisks indicating p-value < 0.01, one asterisks indicating p-value < 0.05.

2.4. *WGCNA Analysis*

2.4.1. Construction of Gene Co-Expression Network of Dendrobium

Consequently, we used the WGCNA package tool to construct a co-expression module that expressed the expression of 8056 genes in 27 *Dendrobium* samples. The heat map of the cluster dendrogram and tissues and species traits are shown in Figure 7a. One of the most critical parameters is the power value, which mainly affects the independence and average connectivity of the co-expression modules. First, we selected the appropriate power value. When the power value was 18, the independence reached 0.6, and the average connectivity was higher. Therefore, the power values and results of constructing co-expression modules indicate that 19 different co-expression modules were identified in *Dendrobium*. These co-expression modules were constructed and are shown in different colors (Figure 7). The size of these modules depends on the number of genes they contain. The number of genes and module names are shown in Table 1.

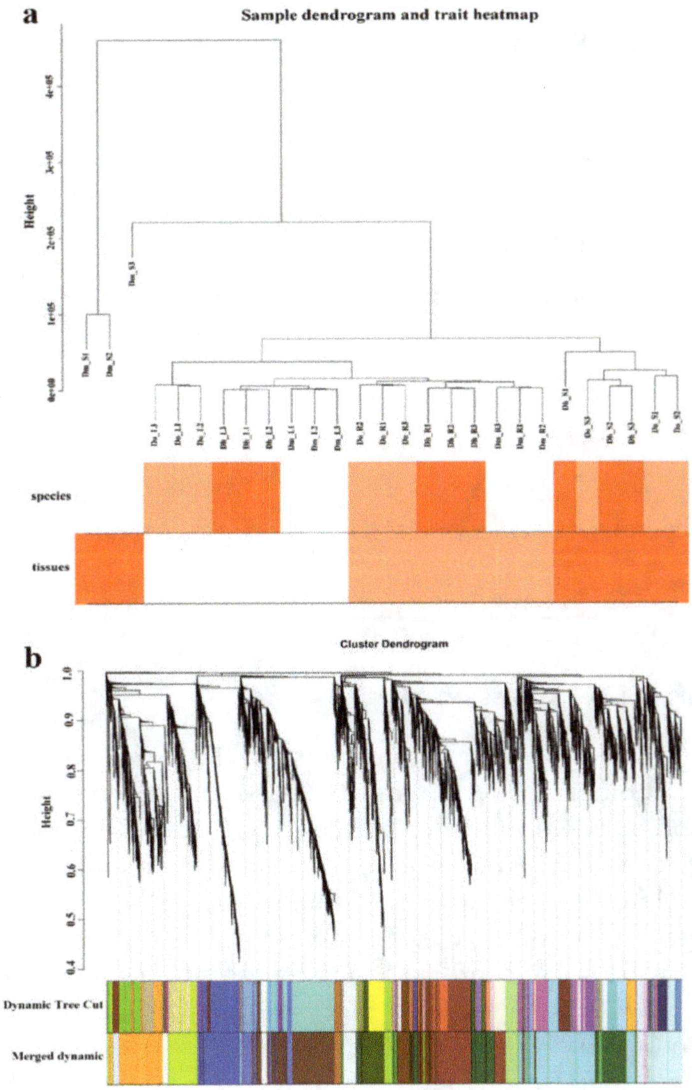

Figure 7. Clustering dendrogram. (**a**) Clustering dendrogram of 27 samples and heatmaps of species and tissues traits. The expression is from low to high, and the color transitions from white to red. (**b**) Clustering dendrogram of DEGs, with dissimilarity based on the topological overlap, together with assigned module colors. The clustered branches represent different modules, and each line represents one gene.

Table 1. The number of genes in 19 constructed modules.

Module Colors	Node Number
Blue	562
Brown	892
Cyan	1132
darkgreen	1016
darkolivegreen	362
darkorange	665
floralwhite	184
greenyellow	440
lightcyan	180
lightgreen	178
lightpink4	40
lightsteelblue1	193
paleturquoise	470
plum2	57
royalblue	275
saddlebrown	179
salmon4	1015
sienna3	109
yellowgreen	107

2.4.2. Interaction Analysis of Co-Expression Module

Subsequently, we analyzed the interactions between the 19 co-expression modules (Figure 8). The heatmap shows the topological overlap matrix (TOM) of all genes in the analysis. Light color indicates low overlap, and dark red indicates high overlap. Except for some high-brightness areas, the overall difference between the modules is not significant, which indicates that the gene expression between the modules is relatively independent and has a high scale independence. The correlations between module eigengene and traits: tissues and species, were analyzed and data are shown in Figure 8b. Two modules are significantly associated with species: Lightgreen (p-value = 1×10^{-5}, cor = 0.73) and Lightsteelblue1 (p-value = 2×10^{-6}, cor = −0.76). The modules that highly correlated to tissues were Yellowgreen (p-value = 1×10^{-6}, cor = 0.79), Salmon4 (p-value = 9×10^{-7}, cor = 0.79), Blue (p-value = 3×10^{-6}, cor = 0.77), and Sienna3 (p-value =2×10^{-10}, cor = 0.9).

We analyzed the connectivity of eigengenes to find the interactions between these constructed co-expression modules. First of all, cluster analysis of these eigengenes was performed (Figure 9a). These 19 clusters were divided into two clusters, including 5 modules (modules 3, 8, 9, 11, and 14) and the remaining 14 modules. According to that, the connectivity effect between different modules is obviously different. (Figure 9b).

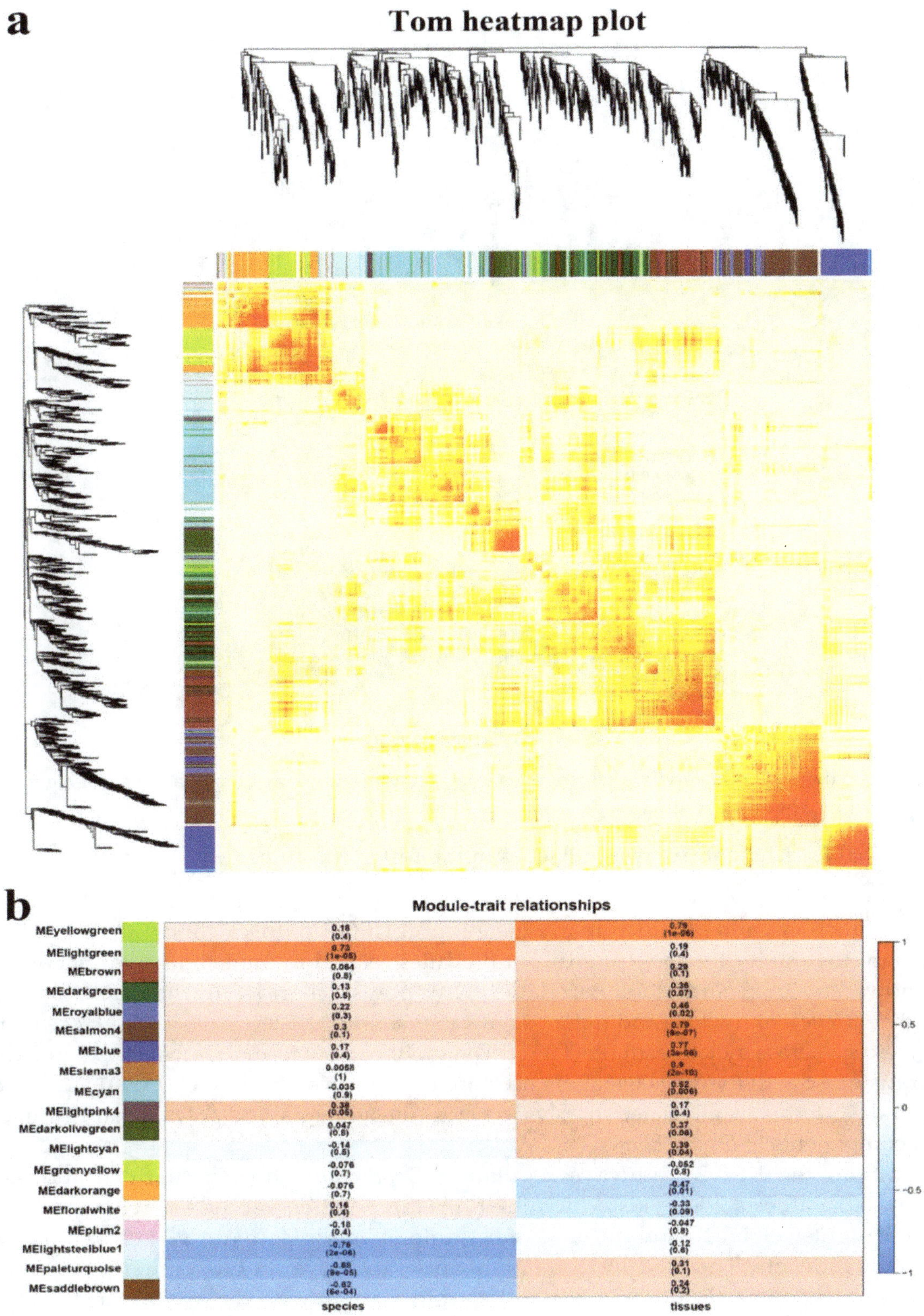

Figure 8. Co-expression network analysis across different tissues and different species. (**a**) Visualizing the gene network using a heatmap plot. The heatmap depicts the topological overlap matrix (TOM) among all genes in the analysis. (**b**) Module-trait associations. Each row corresponds to a module characteristic gene (eigengene), and each column corresponds to a trait. Each cell contains a corresponding correlation and p-value. According to the color legend, the table is color-coded by correlation.

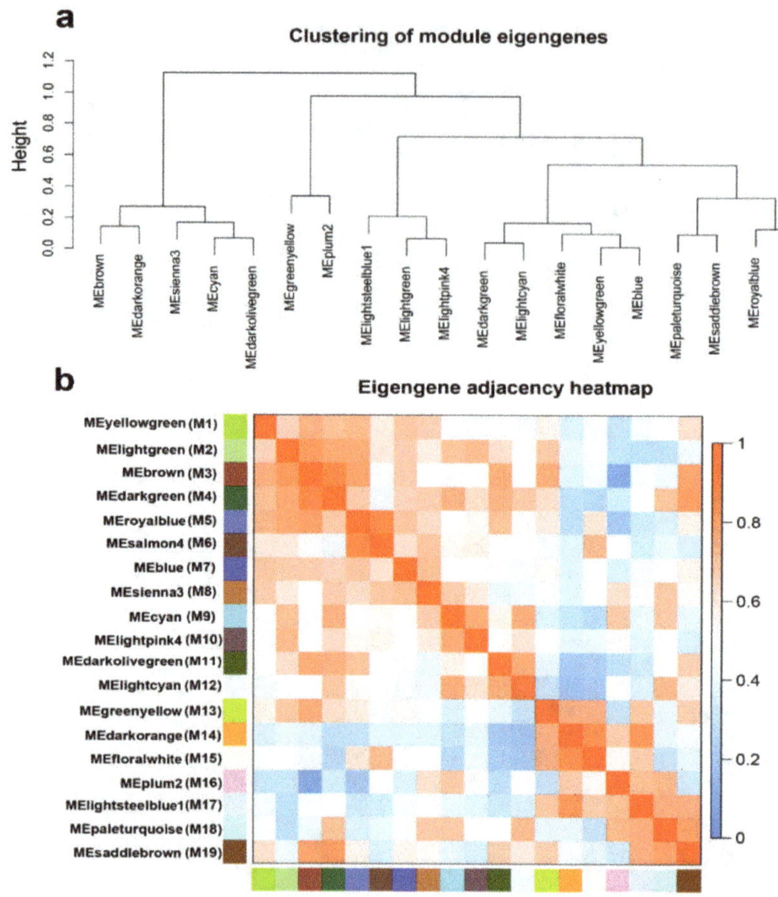

Figure 9. Analysis of connectivity of eigengenes in different module. (**a**) Cluster analysis of eigengenes. (**b**) The heatmap of connectivity of eigengenes.

2.4.3. Functional Enrichment Analysis of Genes in Interested Modules

We performed pathway enrichment analysis on the genes of each module separately, explored their related functions, and obtained many meaningful KEGG pathway and GO terms (Figure 10). It is important to identify the most significant modules related to medicinal components. Sienna3 module showed a significantly high correlation with polysaccharide. In the KEGG pathway, it is significantly enriched to fructose and mannose metabolism (ko00051), starch and sucrose metabolism (ko00500), and galactose metabolism (ko00052). The top three GO terms, shown in expanded size, are glucosyltransferase activity (GO:0046527) and starch synthase activity (GO:0009011). These KEGG pathways and GO terms are all related to polysaccharide biosynthesis, and polysaccharides are the main medicinal components of *Dendrobium*.

In addition, the KEGG pathways regulating *Dendrobium* polysaccharides and secondary metabolism biosynthesis were also identified in three other modules, which are the blue module, the lightsteelblue1 module, and the salmon4 module (Figure 11). In blue module, genes were significantly enriched in these pathways: anthocyanin biosynthesis (ko00942), drug metabolism—cytochrome P450 (ko00982) and steroid biosynthesis (ko00100). In lightsteelblue1, the significantly enriched pathways are as follows: terpenoid backbone biosynthesis (ko00900), diterpenoid biosynthesis (ko00904), galactose metabolism (ko00052), and glycolysis/gluconeogenesis (ko00010). In salmon4, genes were significantly enriched in these pathways: the pentose phosphate pathway (ko00030), terpenoid backbone biosynthesis (ko00900), and diterpenoid biosynthesis (ko00904). The enriched pathways in these three modules are significantly related to the synthesis of polysaccharides and secondary metabolites of *Dendrobium*. Therefore, sienna3, blue, lightsteelblue1 and salmon4 modules were defined as important module and extracted for further analysis.

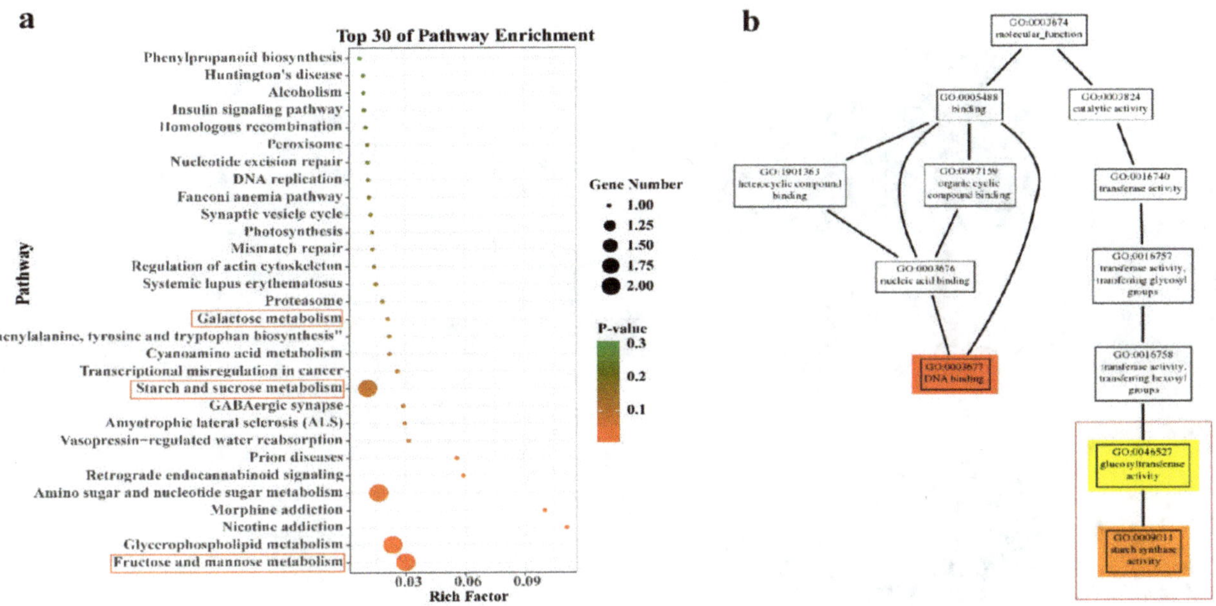

Figure 10. GO and Kyoto Encyclopedia of Genes and Genomes (KEGG) pathway enrichment of module sienna3. (**a**) Top 30 of KEGG pathway enrichment. (**b**) Thumbnails view of directed acyclic graph (DAG) on molecular function of module sienna3.

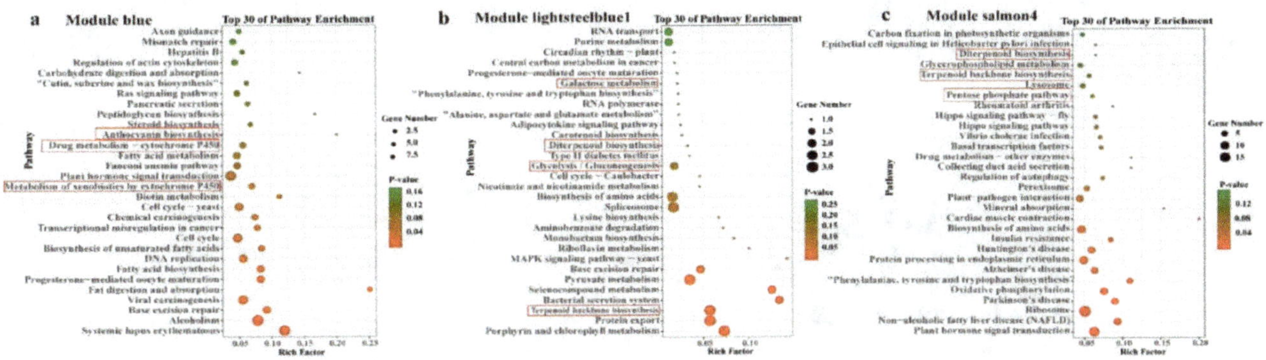

Figure 11. Kyoto Encyclopedia of Genes and Genomes (KEGG) pathway enrichment of interested module. (**a**) Module blue; (**b**) module lightsteelblue1; (**c**) module salmon4.

2.4.4. Module Visualize and Hub Genes

We performed genes trend expression analysis of four modules of interest (Figure 12). Interestingly, by analyzing the trend expression of genes in sienna3 module, we found that they are specifically expressed only in the stem, which suggests that they play an important role in the stem. Through further analysis of this module, several hub genes were discovered, such as beta-glucosidase and granule-bound starch synthase (*GBSSI*). This indicates that the specific expression of these polysaccharide-related genes in stems may be related to the content of polysaccharides in stems significantly higher than in other tissues.

In the blue module, Do_S shows specific expression in this module, several hub genes were detected in each important module, with each gene interacting with many other genes. These genes include bHLH, beta-glucosidase, alpha-L-fucosidase, and other genes related to *Dendrobium* polysaccharide and secondary metabolism. In salmon4 module, Do_S shows specific expression in this module, hub genes include *CCR4* (cinnamoyl -CoA reductase), bZIP, and MYB transcription factors were identified. In lightsteelblue1, a large number of genes were annotated to *ketohexokinase* (*KHK*). *KHK*, also known as fructokinase, which is a key enzyme in fructose metabolism. Fructose is a kind of plant polysaccharide, which suggests that this module is related to polysaccharide biosynthesis.

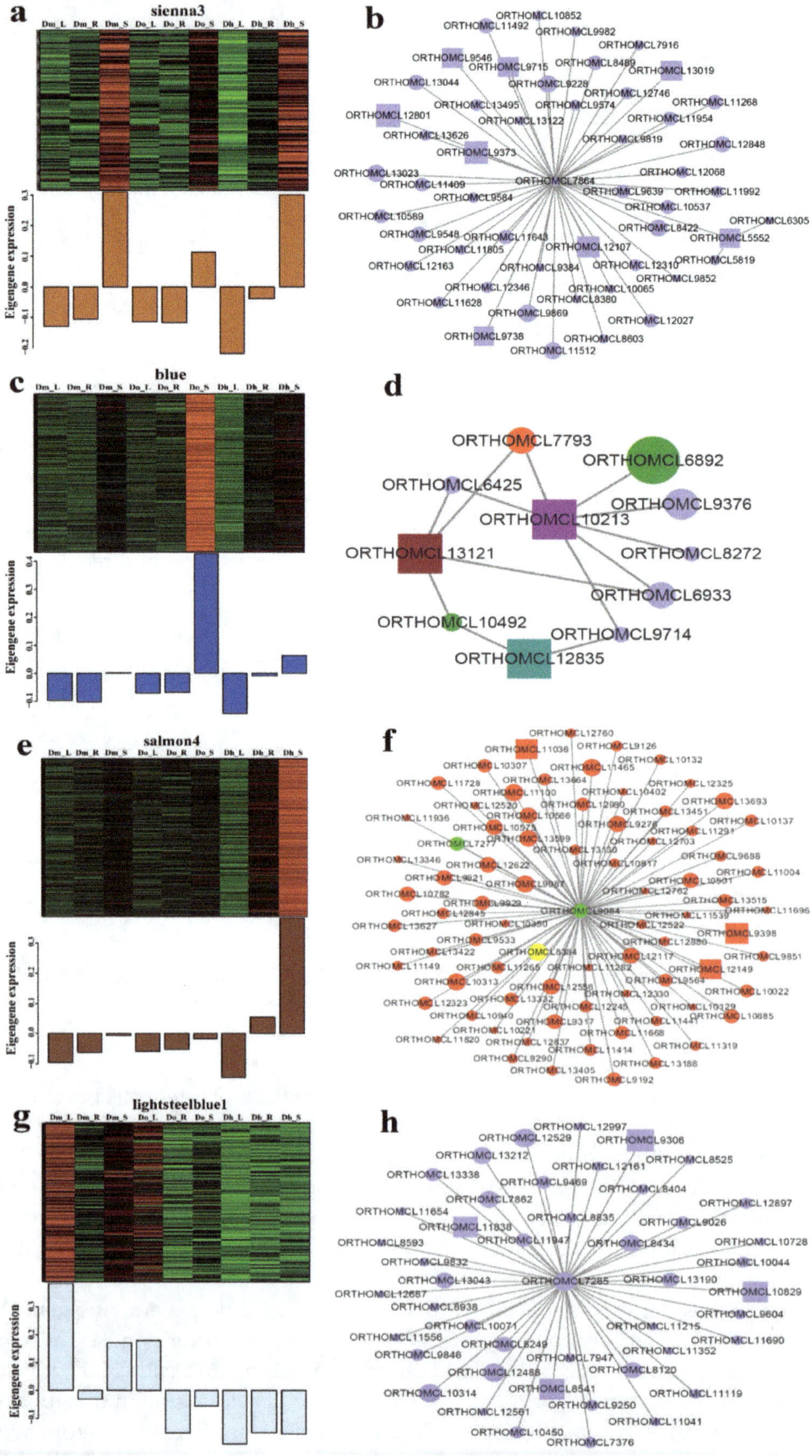

Figure 12. Expression profile and transcriptional regulatory network associated with the tissue-specific modules. (**a**), (**c**), (**e**), and (**g**) heatmaps showing genes in module that were significantly over-represented in Dm_L, Dm_R, Dm_S, Do_L, Do_R, Do_S, Dh_L, Dh_R, Dh_S and predicted transcriptional regulatory

network associated with the gene sets showing expression patterns at Dm_L, Dm_R, Dm_S, Do_L, Do_R, Do_S, Dh_L, Dh_R, and Dh_S. Heatmaps show the expression profile of all the co-expressed genes (number given on the top) in the modules (labelled on top). Candidate hub genes are shown in rectangular shapes. Purple in figure (**b**, **d** and **h**) represents genes related to polysaccharides. Red in figure (**d** and **f**) represents genes related to secondary metabolites. The shapes, from big to small, indicate the weights from big to small. Green in figure d represents bHLH transcription factors. Green in figure (**f**) represents bZIP transcription factors and yellow represents MYB transcription factors.

3. Discussion

Traditional biological research focuses on elucidating the effects of individual functional elements (such as DNA, mRNA, and protein) on life activities at the molecular level. Although those methods are of great significance for revealing the genetic mechanisms of specific traits, it can only partially explain the cause of a certain life activities. With the rapid development of sequencing technology, traditional biological research cannot fully and effectively explore the biological significance contained in massive data. As a research method of systems biology, the network is widely used in the exploration of life sciences with the help of data of genome, transcriptome, and metabolome. Compared with other regulatory networks, WGCNA can screen for genes related to specific traits and perform modular classification from large samples to obtain highly biologically significant co-expression modules, which has proven to be an efficient data mining method [35].

WGCNA has been widely used in plants in recent years. In order to obtain key expression modules and key hub genes related to drought resistance in *Brassica napus* L., WGCNA was used to analyze *Brassica napus* transcriptome data in multiple samples (48 transcriptome data), the well-watered and droughted networks contained 17 and 20 modules, respectively, suggesting that there are additional expression patterns in the droughted network because of rearrangement of the transcriptome in response to the drought treatment. [36]. In the study of *Fragaria* L. flowers [37], researchers generated different tissue- and stage-transcriptomic profiling of woodland strawberry (*Fragaria vesca*) flower development, they discovered a developing receptacle-specific module exhibiting similar molecular features to those of young floral meristems and hub genes of the strawberry homologs of a number of meristem regulators, including LOST MERISTEM and WUSCHEL in the developing receptacle network. [37]. Analysis of the pollen transcriptome of three male sterile lines using weighted gene co-expression network analysis revealed that two modules were significantly associated with male sterility and many hub genes that were differentially expressed in the sterile lines [38]. Farcuh et al. used WGCNA to investigate sugar metabolism during leaf and fruit development of two Japanese plum varieties, and identified 11 key sugar metabolism-related genes, the results showed that sugar metabolism was reprogrammed in a non-climacteric bud mutant of a climacteric plum fruit and showed an increase in sorbitol synthesis [39]. In *Ginkgo biloba*, a total of 12 gene modules were revealed to be involved in flavonoid metabolism structural genes and transcription factors by constructing co-expression networks, they reveal that some hub genes operate during the biosynthesis by identifying transcription factors (TFs) and structure genes and seven key hub genes were also identified by analyzing the correlation between gene expression level and flavonoids content [40]. Through these studies, it was found that many of the hub genes obtained by WGCNA analysis were indeed very important genes. In order to obtain the hub gene related to the synthesis and regulation of Dendrobium polysaccharides and alkaloids, we also performed WGCNA analysis.

Therefore, we constructed a *Dendrobium* gene co-expression network using a WGCNA approach and identified co-expression modules using transcriptome data from three kinds of *Dendrobium* species and three different tissues. Correlation analysis between co-expression modules and two traits (species and tissues) was carried out, and four highly significant active ingredients-related modules (p-value < 0.05) were identified. These modules consist of highly connected functional genes, and different modules appear to be involved in individual functions [41]. Meanwhile, KEGG pathway enrichment analysis of modules associated with polysaccharide and secondary metabolism indicated that these pathways in different *Dendrobium* species and different tissues are related to each other at the transcriptomic level. In sienna3 module, we found that several hub genes related to polysaccharide biosynthesis, such as *KHK* (*ketohexokinase*) which is a key enzyme in fructose metabolism. In the expression trend of this module, the stem has a clear advantage, indicating that the polysaccharide content is indeed concentrated on the *Dendrobium* stem, which is consistent with our previous determination of the polysaccharide content. Meanwhile, in blue module, we found that not only the hub genes related to polysaccharide, but also related to secondary metabolism. In the expression trend, we found that the stem of *Dendrobium officinale* to be the highest. This result is consistent with the trend in the determination of polysaccharides and alkaloids. The results also indicated that one component can be regulated by multiple modules, and one module can simultaneously be associated with multiple components.

Module hub genes are generally considered representative of a given module in a biological network. Previous studies reported that MYB-bHLH-WDR (MBW) ternary complexes comprise the essential regulatory machinery for catechin and anthocyanin biosynthesis [42]. In the present study, transcription factors MYB and bHLH were identified as hub genes in modules related to secondary metabolism. In addition, several genes involved in polysaccharide and secondary metabolism biosynthesis (*CCR4: cinnamoyl -CoA reductase* and *KHK: ketohexokinase*) were identified in modules. MYB transcription factors play an important role in the regulation of phenylpropane biosynthesis. Phenylpropane synthesis is upstream of the regulation of flavonoid biosynthesis, indicating that MYB is also an important transcription factor for the synthesis of flavonoids. In gentian, both GtMYBP3 and GtMYBP4 can activate the gene expression of flavonol synthesis, and then significantly increase the flavonol content in seedlings [43]. Ginkgo GbMYBF2 inhibits the expression of *CHS* (*chalcone synthase*), *F3H* (*flavanone 3-hydroxylase*), *FLS* (*flavonol synthase*, Flavonol synthase and ANS genes) on the phenylpropane synthesis pathway, thereby reducing the content of flavonoids and anthocyanins [44]. *Salvia miltiorrhiza* SmMYB39 affects the synthesis of rosmarinic acid by regulating the expression of key enzyme genes of the phenylpropane metabolic pathway [45]. Current studies indicate that plant bHLH transcription factors are involved in regulating various signal transduction and anabolic pathways, such as light signal transduction, hormone synthesis, glandular and root hair development, and stress [46–49].

In this study, the WGCNA method was used for the first time in *Dendrobium*, and the modules related to specific tissues and genes related to specific traits were identified. Hub genes were further analyzed to find related genes and predict gene functions [30]. Combining the WGCNA method and RNA-Seq data can be used to better mine the genes and transcription factors related to traits. In strawberry, modules related to tissue specificity such as torus were found in strawberry, and 7 hub genes were identified in torus tissue [37]; in tomato, genes related to vitamin C biosynthesis were found [50]; co-expression modules related to acidity and genes related to anthocyanin synthesis were found in apple [33,51]. In addition, specific modules in other tissues such as roots, leaves, flowers, and fruits at other periods can be excavated to find relevant metabolic processes and important genes and potential transcription factors.

4. Materials and Methods

4.1. Plant Material

Dendrobium plants were cultivated artificially in a greenhouse located in Anhui Tongjisheng Biotechnology Company, Lu'an, China. Seed germination, the growth of protocorm-like bodies protocols and the condition of pots used for planting were described by our previous study [15]. Two-year-old *D. huoshanense*, *D. officinale*, and *D. moniliforme* were selected to provide three replicates of each sample. Stems, leaves, and roots were collected from three different kinds of *Dendrobium* plants for RNA extraction. Then, after drying, they were used to determine polysaccharide and total alkaloid contents in March, 2017. The same samples were used for transcriptome sequencing and polysaccharide and total alkaloid contents determination.

4.2. *Polysaccharide and Total Alkaloid Contents Determination*

The method for determining the polysaccharide content is consistent with the Chinese Pharmacopoeia (version 2010), and the details are described in our previous study [15]. The method of total alkaloids has been modified based on the Bush et al. method and has been described in detail in our previous studies [52,53]. A part of data of this part has been published [14,15].

4.3. *Identification of Orthologous Genes*

Datasets for three kinds of *Dendrobium* species were obtained from the NCBI Sequence Read Archive (SRA) (https://www.ncbi.nlm.nih.gov/sra) with accessing number SRP122499, SRP139000, and SRP150489. There was a total of 27 samples of *Dendrobium* that were taken from three different species (*Dendrobium huoshanense, Dendrobium officinale,* and *Dendrobium moniliforme*) and three different tissues (stems, leaves, and roots). The data for these samples were submitted by our own laboratory. The OrthoMCL pipeline [54] was used with standard settings to identify potential orthologous genes in *Dendrobium*, and subsequently aligned using the MUSCLE algorithm [55].

4.4. *Differential Gene Expression Analysis*

The gene expression levels of all samples were estimated by RSEM version 1.2.15 [56], and the bowtie2 parameter setting mismatch was 0. The R Bioconductor package edgeR [57] was used to identify differentially expressed genes (DEGs) in two samples. DEGs were used for KEGG and GO enrichment analyses, which were performed using the KOBAS version 2.0.12 with default settings and Goatools version 0.5.9 (https://github.com/tanghaibao/Goatools) [58,59].

4.5. *WGCNA Analysis*

4.5.1. Construction of Weighted Gene Co-Expression Networks and Identification of Modules

We use thresholds for different expression data to calculate the number of genes. The WGCNA algorithm was applied to the evaluation of gene expression. The flashClust toolkit (R language) was used to perform cluster analysis on samples and set appropriate thresholds. After background correction and quantile normalization, the top 50% of genes (8056 genes) with the most variant in the analysis of variance were selected for WGCNA analysis. The WGCNA package in R was used to construct the co-expression network of the 8056 genes after verification [30].

The gradient independence method was used to test the scale independence and average connectivity of different power modules (the power value ranging from 1 to 20). According to the scale-independent conditions, the appropriate power value was 0.6. After the power value determined, the module was constructed by the WGCNA algorithm, and the genetic information corresponding to each module was extracted. For high reliability of the results, the minimum number of genes per module was set to 50.

4.5.2. Interactions Analysis of Co-Expression Modules

The interaction relationship between different co-expression modules was calculated by WGCNA, and the topological overlap matrix (TOM) was constructed using the correlation expression values. The resulting topological overlap is a biologically meaningful measure of gene similarity based on the co-expression relationship between two genes. Using each TOM as input, the flashClust function was used for hierarchical cluster analysis. Then the dynamicTreeCut algorithm (minModuleSize = 30) was used to detect the module as a branch of the dendrogram [30]. Random colors were assigned to the modules, and the first principal component was used to calculate the module characteristic genes of each module. The module characteristic genes can be regarded as the representative of the gene expression pattern in this module, and the module combining highly relevant characteristic genes (merge-CutHeight = 0.07). By calculating the Pearson correlation between the module characteristic genes and the interest characteristics, and using the module characteristic genes to estimate the module-trait relationship, the samples are classified according to the corresponding traits, types, and tissues, and the module with a correlation coefficient $\geq | 0.75 |$ and p-value ≤ 0.01 was selected for further analysis. Through the heat map of the gene network topological overlap, the structure of co-expression module was visualized. Then through the hierarchical clustering tree diagram of the characteristic genes and the heat map of the corresponding characteristic gene network, the relationship between the modules was summarized.

4.5.3. Functional Enrichment Analysis of the Key Module

We performed GO and KEGG enrichment analysis for key modules with most genes by using the Database for Annotation, Visualization and Integrated Discovery (DAVID, https://david.ncifcrf.gov/summary.jsp) [60] and KEGG Orthology Based Annotation System (KOBAS, https://bio.tools/kobas). Analysis results were extracted out under the condition of $p < 0.05$ after correction [61,62].

4.5.4. Identification of Hub Genes

The hub gene is usually used as an abbreviation for a highly connected gene, which has a high degree of connectivity in co-expression modules. Module membership (MM) is defined as the correlation between the expression of each gene and its module eigengenes (MEs). Gene significance (GS) is defined as the correlation of each gene to the trait of interest. In the module of interest, the genes with the highest MM and GS were set as candidate genes for further analysis [63]. We chose the intramodular hub genes by external traits based GS > 0.2 and MM > 0.8 with a threshold of p-value < 0.05 [64]. In our study, based on the size of the module, we classified the top 10% of the most connected genes in the module as hub genes. We construct and visualize gene–gene interaction networks using Cytoscape tool [65].

5. Conclusions

Weighted gene co-expression network associated with traits was constructed, and 19 tissue and species-specific modules were obtained. The biological significance of the tissue and species-specific module was revealed, the regulatory genes related to the medicinal components of *Dendrobium* were identified, and a local biologically significant network was constructed. The results of this study can provide important experimental data and theoretical basis for the further analysis of the genetic mechanism of the medicinal ingredients of *Dendrobium*.

Author Contributions: Conception and design of the research: Y.Y. and X.T.; acquisition of data: B.Z.; statistical analysis: X.T.; drafting the manuscript: Y.Y. and J.Z.; revision of manuscript for important intellectual content: J.L. All authors read and approved the final manuscript.

References

1. Wood, H.P. *The Dendrobiums*; ARG Gantner Verlag Timber Press: Portland, OR, USA, 2006.
2. Zotz, G. The systematic distribution of vascular epiphytes—A critical update. *Bot. J. Linn. Soc.* **2013**, *171*, 453–481. [CrossRef]
3. Lian, W.; Xu, J. The chromosome number in *Dendrobium* I. Ten species. *J. Wuhan Bot. Res.* **1989**, *7*, 112–114.

4. Zhitao, N.; Shuying, Z.; Jiajia, P.; Ludan, L.; Jing, S.; Xiaoyu, D. Comparative analysis of *Dendrobium* plastomes and utility of plastomic mutational hotspots. *Sci. Rep.* **2017**, *7*, 2073. [CrossRef] [PubMed]
5. Tang, H.; Zhao, T.; Sheng, Y.; Zheng, T.; Fu, L.; Zhang, Y. *Dendrobium officinale* kimura et migo: A review on its ethnopharmacology, phytochemistry, pharmacology, and industrialization. *Evid. Based Complement. Altern. Med.* **2017**, *2017*, 19. [CrossRef] [PubMed]
6. Chang, C.-C.; Ku, A.F.; Tseng, Y.-Y.; Yang, W.-B.; Fang, J.-M.; Wong, C.-H. 6, 8-di-c-glycosyl flavonoids from *Dendrobium huoshanense*. *J. Nat. Prod.* **2010**, *73*, 229–232. [CrossRef] [PubMed]
7. Ng, T.B.; Liu, J.; Wong, J.H.; Ye, X.; Sze, S.C.W.; Tong, Y.; Zhang, K.Y. Review of research on *Dendrobium*, a prized folk medicine. *Appl. Microbiol. Biotechnol.* **2012**, *93*, 1795–1803. [CrossRef]
8. Jin, Q.; Jiao, C.; Sun, S.; Song, C.; Cai, Y.; Lin, Y.; Fan, H.; Zhu, Y. Metabolic analysis of medicinal *Dendrobium officinale* and *Dendrobium huoshanense* during different growth years. *PLoS ONE* **2016**, *11*, e0146607. [CrossRef]
9. Hsieh, Y.S.; Chien, C.; Liao, S.K.; Liao, S.F.; Hung, W.T.; Yang, W.B.; Lin, C.C.; Cheng, T.J.; Chang, C.C.; Fang, J.M.; et al. Structure and bioactivity of the polysaccharides in medicinal plant *Dendrobium huoshanense*. *Bioorg. Med. Chem.* **2008**, *16*, 6054–6068. [CrossRef]
10. Luo, J.P.; Deng, Y.Y.; Zha, X.Q. Mechanism of polysaccharides from *Dendrobium huoshanense*. On streptozotocin-induced diabetic cataract. *Pharm. Biol.* **2008**, *46*, 243–249. [CrossRef]
11. Kaufmann, K.; Wellmer, F.; Muino, J.M.; Ferrier, T.; Wuest, S.E.; Kumar, V.; Serrano-Mislata, A.; Madueno, F.; Krajewski, P.; Meyerowitz, E.M.; et al. Orchestration of floral initiation by apetala1. *Science* **2010**, *328*, 85–89. [CrossRef]
12. Lou, Q.; Liu, Y.; Qi, Y.; Jiao, S.; Tian, F.; Jiang, L.; Wang, Y. Transcriptome sequencing and metabolite analysis reveals the role of delphinidin metabolism in flower colour in grape hyacinth. *J. Exp. Bot.* **2014**, *65*, 3157–3164. [CrossRef] [PubMed]
13. Verdonk, J.C.; Haring, M.A.; Tunen, A.J.V.; Schuurink, R.C. Odorant1 regulates fragrance biosynthesis in petunia flowers. *Plant. Cell* **2005**, *17*, 1612–1624. [CrossRef] [PubMed]
14. Yuan, Y.; Yu, M.; Jia, Z.; Song, X.; Liang, Y.; Zhang, J. Analysis of *Dendrobium huoshanense* transcriptome unveils putative genes associated with active ingredients synthesis. *BMC Genom.* **2018**, *19*, 978. [CrossRef] [PubMed]
15. Yuan, Y.; Zhang, J.; Kallman, J.; Liu, X.; Meng, M.; Lin, J. Polysaccharide biosynthetic pathway profiling and putative gene mining of *Dendrobium moniliforme* using RNA-seq in different tissues. *BMC Plant. Biol* **2019**, *19*, 521. [CrossRef] [PubMed]
16. Guo, X.; Li, Y.; Li, C.; Luo, H.; Wang, L.; Qian, J.; Luo, X.; Xiang, L.; Song, J.; Sun, C.; et al. Analysis of the *Dendrobium officinale* transcriptome reveals putative alkaloid biosynthetic genes and genetic markers. *Gene* **2013**, *527*, 131–138. [CrossRef] [PubMed]
17. Khaitovich, P.; Weiss, G.; Lachmann, M.; Hellmann, I.; Enard, W.; Muetzel, B.; Wirkner, U.; Ansorge, W.; Pääbo, S. A neutral model of transcriptome evolution. *PloS Biol.* **2004**, *2*, e132. [CrossRef]
18. Barabási, A.L.; Oltvai, Z.N. Network biology: Understanding the cell's functional organization. *Nat. Rev. Genet.* **2004**, *5*, 101–113. [CrossRef]
19. Carter, S.L.; Brechbuhler, C.M.; Griffin, M.; Bond, A.T. Gene co-expression network topology provides a framework for molecular characterization of cellular state. *Bioinformatics* **2004**, *20*, 2242–2250. [CrossRef]
20. Yook, S.H.; Oltvai, Z.N.; Barabási, A.L. Functional and topological characterization of protein interaction networks. *Proteomics* **2004**, *4*, 928–942. [CrossRef]
21. Langfelder, P.; Horvath, S. Eigengene networks for studying the relationships between co-expression modules. *BMC Syst. Biol.* **2007**, *1*, 54. [CrossRef]
22. Cai, B.; Li, C.H.; Huang, J. Systematic identification of cell-wall related genes in populus based on analysis of functional modules in co-expression network. *PLoS ONE* **2014**, *9*, e95176. [CrossRef] [PubMed]
23. Van Dam, S.; Cordeiro, R.; Craig, T.; van Dam, J.; Wood, S.H.; de Magalhães, J.P. Genefriends: An online co-expression analysis tool to identify novel gene targets for aging and complex diseases. *BMC Genom.* **2012**, *13*, 535. [CrossRef] [PubMed]
24. Tully, J.P.; Hill, A.E.; Ahmed, H.M.; Whitley, R.; Skjellum, A.; Mukhtar, M.S. Expression-based network biology identifies immune-related functional modules involved in plant defense. *BMC Genom.* **2014**, *15*, 421. [CrossRef] [PubMed]
25. Ma, S.; Gong, Q.; Bohnert, H.J. An Arabidopsis gene network based on the graphical gaussian model. *Genome Res.* **2007**, *17*, 1614–1625. [CrossRef]

26. Oldham, M.C.; Horvath, S.S.; Geschwind, D.H. Conservation and evolution of gene co-expression networks in human and chimpanzee brains. *Proc. Natl. Acad. Sci. USA* **2006**, *103*, 17973–17978. [CrossRef]
27. Takeshi, O.; Shinpei, H.; Masayuki, S.; Motoshi, S.; Hiroyuki, O.; Kengo, K. Coxpresdb: A database of coexpressed gene networks in mammals. *Nucleic Acids Res.* **2007**, *36*, D77–D82.
28. Lee, J.M.; Sonnhammer, E.L. Genomic gene clustering analysis of pathways in eukaryotes. *Genome Res.* **2003**, *13*, 875–882. [CrossRef]
29. Presson, A.P.; Sobel, E.M.; Papp, J.C.; Suarez, C.J.; Whistler, T.; Rajeevan, M.S.; Vernon, S.D.; Horvath, S. Integrated weighted gene co-expression network analysis with an application to chronic fatigue syndrome. *BMC Syst. Biol.* **2008**, *2*, 95. [CrossRef]
30. Langfelder, P.; Horvath, S. Wgcna: An R package for weighted correlation network analysis. *BMC Bioinform.* **2008**, *9*, 559. [CrossRef]
31. Amrine, K.C.; Blanco-Ulate, B.; Cantu, D. Discovery of core biotic stress responsive genes in Arabidopsis by weighted gene co-expression network analysis. *PLoS ONE* **2015**, *10*, e0118731. [CrossRef]
32. Borges, R.M.; Bessiere, J.M.; Ranganathan, Y. Diel variation in fig volatiles across syconium development: Making sense of scents. *J. Chem. Ecol.* **2013**, *39*, 630–642. [CrossRef]
33. El-Sharkawy, I.; Liang, D.; Xu, K. Transcriptome analysis of an apple (Malus × Domestica) yellow fruit somatic mutation identifies a gene network module highly associated with anthocyanin and epigenetic regulation. *J. Exp. Bot.* **2015**, *66*, 7359–7376. [CrossRef]
34. Smita, S.; Katiyar, A.; Pandey, D.M.; Chinnusamy, V.; Bansal, K.C. Identification of conserved drought stress responsive gene-network across tissues and developmental stages in rice. *Bioinformation* **2013**, *9*, 72–78. [CrossRef] [PubMed]
35. Zhao, W.; Langfelder, P.; Fuller, T.; Dong, J.; Li, A.; Hovarth, S. Weighted gene coexpression network analysis: State of the art. *J. Biopharm. Stat.* **2010**, *20*, 281–300. [CrossRef] [PubMed]
36. Greenham, K.; Guadagno, C.R.; Gehan, M.A.; Mockler, T.C.; Weinig, C.; Ewers, B.E.; McClung, C.R. Temporal network analysis identifies early physiological and transcriptomic indicators of mild drought in *Brassica rapa*. *Elife* **2017**, *6*, e29655. [CrossRef] [PubMed]
37. Hollender, C.A.; Kang, C.; Darwish, O.; Geretz, A.; Matthews, B.F.; Slovin, J.; Alkharouf, N.; Liu, Z. Floral transcriptomes in woodland strawberry uncover developing receptacle and anther gene networks. *Plant. Physiol.* **2014**, *165*, 1062–1075. [CrossRef]
38. Hu, J.; Chen, G.; Zhang, H.; Qian, Q.; Ding, Y. Comparative transcript profiling of alloplasmic male-sterile lines revealed altered gene expression related to pollen development in rice (*Oryza sativa* l.). *BMC Plant. Biol.* **2016**, *16*, 175. [CrossRef]
39. Farcuh, M.; Li, B.; Rivero, R.M.; Shlizerman, L.; Sadka, A.; Blumwald, E. Sugar metabolism reprogramming in a non-climacteric bud mutant of a climacteric plum fruit during development on the tree. *J. Exp. Bot.* **2017**, *68*, 5813–5828. [CrossRef]
40. Ye, J.; Cheng, S.; Zhou, X.; Chen, Z.; Kim, S.U.; Tan, J.; Zheng, J.; Xu, F.; Zhang, W.; Liao, Y.; et al. A global survey of full-length transcriptome of *Ginkgo biloba* reveals transcript variants involved in flavonoid biosynthesis. *Ind. Crop. Prod.* **2019**, *139*, 11547. [CrossRef]
41. Oldham, M.C.; Konopka, G.; Iwamoto, K.; Langfelder, P.; Kato, T.; Horvath, S.; Geschwind, D.H. Functional organization of the transcriptome in human brain. *Nat. Neurosci.* **2008**, *11*, 1271. [CrossRef]
42. Xu, W.; Dubos, C.; Lepiniec, L. Transcriptional control of flavonoid biosynthesis by MYB–bHLH–WDR complexes. *Trends Plant. Sci.* **2015**, *20*, 176–185. [CrossRef] [PubMed]
43. Nakatsuka, T.; Saito, M.; Yamada, E.; Fujita, K.; Kakizaki, Y.; Nishihara, M. Isolation and characterization of gtmybp3 and gtmybp4, orthologues of R2R3-MYB transcription factors that regulate early flavonoid biosynthesis, in gentian flowers. *J. Exp. Bot.* **2012**, *63*, 6505–6517. [CrossRef] [PubMed]
44. Xu, F.; Ning, Y.; Zhang, W.; Liao, Y.; Li, L.; Cheng, H.; Cheng, S. An R2R3-MYB transcription factor as a negative regulator of the flavonoid biosynthesis pathway in ginkgo biloba. *Funct. Integr. Genom.* **2014**, *14*, 177–189. [CrossRef] [PubMed]
45. Zhang, S.; Ma, P.; Yang, D.; Li, W.; Liang, Z.; Liu, Y.; Liu, F. Cloning and characterization of a putative R2R3 MYB transcriptional repressor of the rosmarinic acid biosynthetic pathway from salvia miltiorrhiza. *PLoS ONE* **2013**, *8*, e73259. [CrossRef]

46. Qi, T.; Huang, H.; Wu, D.; Yan, J.; Qi, Y.; Song, S.; Xie, D. Arabidopsis della and jaz proteins bind the WD-repeat/bHLH/MYB complex to modulate gibberellin and jasmonate signaling synergy. *Plant. Cell* **2014**, *26*, 1118–1133. [CrossRef]
47. Nakata, M.; Mitsuda, N.; Herde, M.; Koo, A.J.; Moreno, J.E.; Suzuki, K.; Howe, G.A.; Ohme-Takagi, M. A bhlh-type transcription factor, ABA-inducible bhlh-type transcription factor/ja-associated myc2-like1, acts as a repressor to negatively regulate jasmonate signaling in Arabidopsis. *Plant. Cell* **2013**, *25*, 1641–1656. [CrossRef]
48. Fan, M.; Bai, M.-Y.; Kim, J.-G.; Wang, T.; Oh, E.; Chen, L.; Park, C.H.; Son, S.-H.; Kim, S.-K.; Mudgett, M.B. The bHLH transcription factor hbi1 mediates the trade-off between growth and pathogen-associated molecular pattern–triggered immunity in arabidopsis. *Plant. Cell* **2014**, *26*, 828–841. [CrossRef]
49. Liu, W.; Tai, H.; Li, S.; Gao, W.; Zhao, M.; Xie, C.; Li, W.X. B hlh 122 is important for drought and osmotic stress resistance in Arabidopsis and in the repression of aba catabolism. *N. Phytol.* **2014**, *201*, 1192–1204. [CrossRef]
50. Gao, C.; Ju, Z.; Li, S.; Zuo, J.; Fu, D.; Tian, H.; Luo, Y.; Zhu, B. Deciphering ascorbic acid regulatory pathways in ripening tomato fruit using a weighted gene correlation network analysis approach. *J. Integr. Plant. Biol.* **2013**, *55*, 1080–1091. [CrossRef]
51. Bai, Y.; Dougherty, L.; Cheng, L.; Zhong, G.-Y.; Xu, K. Uncovering co-expression gene network modules regulating fruit acidity in diverse apples. *BMC Genom.* **2015**, *16*, 612. [CrossRef]
52. Bush, L.P.; Wilkinson, H.H.; Schardl, C.L. Bioprotective alkaloids of grass-fungal endophyte symbioses. *Plant. Physiol.* **1997**, *114*, 1–7. [CrossRef] [PubMed]
53. Yuan, Y.; Yu, M.; Zhang, B.; Liu, X.; Zhang, J. Comparative nutritional characteristics of the three major chinese dendrobium species with different growth years. *PLoS ONE* **2019**, *14*, e0222666. [CrossRef] [PubMed]
54. Li, L.; Stoeckert, C.J.; Roos, D.S. Orthomcl: Identification of ortholog groups for eukaryotic genomes. *Genome Res.* **2003**, *13*, 2178–2189. [CrossRef] [PubMed]
55. Edgar, R.C. Muscle: Multiple sequence alignment with high accuracy and high throughput. *Nucleic Acids Res.* **2004**, *32*, 1792–1797. [CrossRef]
56. Dewey, C.N.; Li, B. Rsem: Accurate transcript quantification from RNA-seq data with or without a reference genome. *BMC Bioinform.* **2011**, *12*, 323.
57. Robinson, M.D.; Mccarthy, D.J.; Smyth, G.K. Edger: A bioconductor package for differential expression analysis of digital gene expression data. *Bioinformatics* **2010**, *26*, 139. [CrossRef]
58. Tang, H.; Klopfenstein, D.; Pedersen, B.; Flick, P.; Sato, K.; Ramirez, F.; Yunes, J.; Mungall, C. Goatools: Tools for gene ontology. *Zenodo* **2015**. [CrossRef]
59. Mao, X.; Cai, T.; Olyarchuk, J.G.; Wei, L. Automated genome annotation and pathway identification using the KEGG orthology (ko) as a controlled vocabulary. *Bioinformatics* **2005**, *21*, 3787. [CrossRef]
60. Dennis, G.; Sherman, B.T.; Hosack, D.A.; Yang, J.; Gao, W.; Lane, H.C.; Lempicki, R.A. David: Database for annotation, visualization, and integrated discovery. *Genome Biol.* **2003**, *4*, R60. [CrossRef]
61. Ashburner, M.; Ball, C.A.; Blake, J.A.; Botstein, D.; Butler, H.; Cherry, J.M.; Davis, A.P.; Dolinski, K.; Dwight, S.S.; Eppig, J.T. Gene ontology: Tool for the unification of biology. *Nat. Genet.* **2000**, *25*, 25. [CrossRef]
62. Xie, C.; Mao, X.; Huang, J.; Ding, Y.; Wu, J.; Dong, S.; Kong, L.; Gao, G.; Li, C.-Y.; Wei, L. Kobas 2.0: A web server for annotation and identification of enriched pathways and diseases. *Nucleic Acids Res.* **2011**, *39*, W316–W322. [CrossRef] [PubMed]
63. Fuller, T.F.; Ghazalpour, A.; Aten, J.E.; Drake, T.A.; Lusis, A.J.; Horvath, S. Weighted gene coexpression network analysis strategies applied to mouse weight. *Mamm. Genome* **2007**, *18*, 463–472. [CrossRef] [PubMed]
64. Horvath, S.; Dong, J. Geometric interpretation of gene coexpression network analysis. *PLoS Comput. Biol.* **2008**, *4*, e1000117. [CrossRef]
65. Shannon, P.; Markiel, A.; Ozier, O.; Baliga, N.S.; Wang, J.T.; Ramage, D.; Amin, N.; Schwikowski, B.; Ideker, T. Cytoscape: A software environment for integrated models of biomolecular interaction networks. *Genome Res.* **2003**, *13*, 2498–2504. [CrossRef] [PubMed]

4

Molecular Identification of Endophytic Fungi and their Pathogenicity Evaluation Against *Dendrobium nobile* and *Dendrobium officinale*

Surendra Sarsaiya [1,2,†], Archana Jain [1,†], Qi Jia [2], Xiaokuan Fan [3], Fuxing Shu [2], Zhongwen Chen [2], Qinian Zhou [2], Jingshan Shi [1,*] and Jishuang Chen [1,2,3,*]

1 Key Laboratory of Basic Pharmacology and Joint International Research Laboratory of Ethnomedicine of Ministry of Education, Zunyi Medical University, Zunyi 563003, China; sarsaiya.s@gmail.com (S.S.); assjain80@gmail.com (A.J.)

2 Bioresource Institute for Healthy Utilization, Zunyi Medical University, Zunyi 563003, China; jiaqi_1003@163.com (Q.J.); ncusfx@163.com (F.S.); 15195884364@163.com (Z.C.); zhouqnzmu@163.com (Q.Z.)

3 College of Biotechnology and Pharmaceutical Engineering, Nanjing Tech University, Nanjing 211800, China; biofanxk@163.com

* Correspondence: shijs@zmu.edu.cn (J.S.); bihu_zmu@zmu.edu.cn (J.C.)

† These authors contributed equally to this work.

Abstract: *Dendrobium* are tropical orchid plants that host diverse endophytic fungi. The role of these fungi is not currently well understood in *Dendrobium* plants. We morphologically and molecularly identified these fungal endophytes, and created an efficient system for evaluating the pathogenicity and symptoms of endophytic fungi on *Dendrobium nobile* and *Dendrobium officinale* though in vitro co-culturing. ReThe colony morphological traits of *Dendrobium* myco-endophytes (DMEs) were recorded for their identification. Molecular identification revealed the presence of *Colletotrichum tropicicola, Fusarium keratoplasticum, Fusarium oxysporum, Fusarium solani,* and *Trichoderma longibrachiatum*. The pathogenicity results revealed that *T. longibrachiatum* produced the least pathogenic effects against *D. nobile* protocorms. In seedlings, *T. longibrachiatum* showed the least pathogenic effects against *D. officinale* seedlings after seven days. *C. tropicicola* produced highly pathogenic effects against both *Dendrobium* seedlings. The results of histological examination of infected tissues revealed that *F. keratoplasticum* and *T. longibrachiatum* fulfill Koch's postulates for the existence of endophytes inside the living tissues. The DMEs are cross-transmitted inside the host plant cells, playing an important role in plant host development, resistance, and alkaloids stimulation.

Keywords: Dendrobium; molecular identification; endophytic fungi; pathogenicity; protocorm; seedling

---

1. Introduction

The Orchidaceae is the chief family of plants, with over 25,000 plant species worldwide. They are also some of the most vulnerable flowering plants, as numerous genera are endangered and nearly all genera are at risk of habitat harm and over-assortment [1]. China is a rich source of orchid plants, with over 1447 species, mostly located in the subtropical and tropical provinces in the southwest and south [1,2]. Many *Dendrobium* orchids are horticultural plants and have been used for profitable trade due to their flowering profusion, extensive variety of flower colors, shapes, sizes, and year-round producibility, along with long lifespan [3]. Most species are in danger of extinction. Orchids extend their lives as herbaceous plants using two evolutionary methods: sympodial growth and monopodial development, which are influenced by the many endophytic fungal species as main pathogens of orchids [4,5]. *Dendrobium officinale* has been extensively used in traditional medicines for over 2000 years to decrease fever, inhibit tumors, increase antioxidant activity, treat hypoglycemia, recover from loss of eyesight, and control the immune system, according to the best of China Pharmacopeia [6,7].

Dendrobium plants parts, for example, roots, stems, buds, and leaves of tropical orchid, harbor diverse fungal taxa, including mutualistic mycorrhiza, and endophytic fungi and considerably diverse nonmycorrhizal fungal associates. The role of the root-allied fungi is not well understood. They typically originate in the velamen, without causing any disease symptoms [8]. They may encourage the growth of *Dendrobium* by activating soil chemicals in the rhizosphere. The impact of quantities or variations of secondary metabolite have been investigated. At large, they act as a supply for bioactive molecules that defend the host from rhizospheric pathogens [8]. Endophytic fungal groups commonly establish a sole host specificity at the species level, which can be additionally encouraged by microclimatic conditions and microhabitat [9]. The relationship is chiefly stimulated by the endophyte fungi, yielding an overabundance of natural compounds as soon as endophytic fungi are cultured in the external environment of their ordinary hosts or environmental niches under in vitro test conditions [10].

Protocorm and seedling expansion is an important step in commercial orchid production, and its conservation is crucial tool for maintaining the genetic diversity of the orchid plant [11]. Orchid (*Dendrobium nobile* and *D. officinale*) protocorms initiating from the orchid seeds are typically very small, similar to dust, and deficient an endosperm. Subsequently, orchid seed incubation and seedling expansion require well-matched endophytic fungi to deliver the carbon, nutrients, and water to the seeds under usual plant conditions [3]. Consequently, an in vitro proliferation method could be a useful for the mass measure proliferation of these orchids for their marketization. The protocorms and seedlings rarely survive after relocation into nature from well mature sterile culture. The orchid protocorm may be reliant on appropriate endophytic fungi for seedling existence [12].

Most examinations of *Dendrobium* myco-endophytes (DMEs) focused on symbiotic in vitro practices using fungal endophyte strains obtained from fully-grown dendrobium roots, buds, stems, and leaves [3]. Understanding whether these endophytic fungi are pathogenic, conditional pathogenic, or non-pathogenic for the host plant is important. DMEs can be transmitted horizontal or vertically. Vertical transmission occurs when the seeds are contacted through the fungal endophytes and are transmitted to the host plant. Horizontal transmission involves the formation of exterior spores and their airborne dispersal infects many other hosts [13].

A steady state between the fungal pathogen and its host plant is achieved when the pathogen resides in equilibrium with the surrounding host tissues and causes little damage, which is also called least pathogenic. The virulence expression of the pathogen is dependent on the particular host environment. When the pathogen is isolated from an asymptomatic host and introduced into a new host, strong pathogenic reactions may be observed [14]. Commonly, fungal endophytes have functions ranging from latent pathogens to mutualistic symbionts. Reliant on the host genome type, some endophytic fungi may be pathogenic in stressed hosts, whereas they can be helpful in other conditions due to conditional pathogenic properties [15]. In the environment, orchids are chiefly dependent on these fungal endophytes for their nourishment and propagation along with the succeeding seedling (protocorm) phases. However, only imperfect quantitative approaches for assessing the *Dendrobium*–fungus connections at the protocorm and seedling phase are available at present, which places major constraints on understanding the host–endophyte relationships [16].

In this study, we focused on species-specific endophytic fungi pre-isolated from *D. nobile* plant parts, which were inoculated onto well-developed protocorms and seedlings of *D. nobile* and *D. officinale* for the evaluation of pathogenicity using in vitro inoculation and histopathological examination of infected tissues. The test endophytic fungus was re-isolated from the leaves tissues for examining Koch's postulates.

2. Results

2.1. Isolation and Molecular Identification of Endophytic Fungi

Five DME isolates were recovered (ZF01, ZF02, ZF03, ZF04, and ZF05) from leaf segments of the wild *D. nobile*. The colony morphological traits of the DME isolates, such as hyphal structures and spore arrangements, were used for their identification. The isolated endophytic fungal colony characteristics and microscopic characteristics findings were found as follows: (1) ZF01: fast grower (35 ± 0.2 mm), flat and buff growth surface, whitish color, reverse whitish to grey/orange, margin irregular mycelium with conidiogenous cells hyaline to pale brown, conidia hyaline with round ends, smooth-walled; (2) ZF02: moderate grower (16 ± 0.2 mm), colony flat velvety, white color with no diffusing pigment coloration, margin irregular, white rims, mycelium with microconidia oval and elongated and clavate, sickle-shaped conidia containing numerous septations, septate hyaline hyphae; (3) ZF03: fast grower (25 ± 0.2 mm), colony cottony white to light violet color, aerial white margins septate and hyaline, macroconidia slightly curved and thick with an attenuated apical cell, three to five septa, short aerial conidiophore, unbranched, microconidia non-septate, aerial conidia, ellipsoidal to cylindrical, slightly curved; (4) ZF04: fast grower (28 ± 0.2 mm), colony cottony white to light white with creamy to white-greyish and reverse light yellow color, irregular white margins with cream yellow edges with long and narrow macroconidia, cylindrical, dorsal and ventral surface parallel, three to five septa, phyalide long and thin; and (5) ZF05: fast grower (34 ± 0.2 mm), colony cottony off white to greenish, later green tufts sporulation, irregular white margins with very hyaline hyphae, long conidiophores, flask-shaped single phialides, small-walled ellipsoidal conidia, formed chlamydospores (smooth walled) in terminal or intercalary (Table S1). Molecular characterization was used to confirm isolate identification in which the rDNA (recombinant deoxyribose nucleic acid)-ITS (internal transcribed spacer) region was amplified. The sequences data were aligned using BLAST (basic local alignment search tool). The endophytic fungi were found to be the closest homologs of *Colletotrichum tropicicola* (ZF01), *Fusarium keratoplasticum* (ZF02), *Fusarium oxysporum* (ZF03), *Fusarium solani* (ZF04), and *Trichoderma longibrachiatum* (ZF05) (Figure 1). Results from BLAST were categorized for the isolated endophytes under ascomycota, which coincided with the morphological identification. These were represented by class Sordariomycetes, including isolates belonging to *Colletotrichum* and *Trichoderma*, and Hypocreales included isolates belonging to *Fusarium*. These were the most frequent and dominant endophytic fungi isolated from *D. nobile*.

2.2. Evaluation of Protocorm and Seedling Pathogenicity

The DME pathogenicity results with *D. nobile* and *D. officinale* protocorm revealed that *T. longibrachiatum* (ZF05) showed the least pathogenicity (11%) against the *D. nobile* protocorm after 7, 14, and 21 days, whereas it showed moderate pathogenic (33%) effects at 21 days of incubation against *D. officinale*. We found that most of the plant pathogens are latent pathogens or conditional pathogens, causing symptoms some weeks later when the plant defense mechanism has weakened. Many environmental factors function together to trigger the virulence factors or genes of the endophytic fungi. *T. longibrachiatum* (ZF05) started to change the color of the *D. officinale* protocorm from green to light yellowish at 14 days, which changed to brown to blackish at 21 days; whereas no color change symptoms were observed in the *D. nobile* protocorm. The second least pathogenic strain was *F. keratoplasticum* (ZF02), which was found to be non-pathogenic against *D. nobile* after 7 and 14 days of incubation, and least pathogenic (11%) against *D. officinale* after seven days of incubation, but showed moderate pathogenic effect after 21 days incubation on both *D. nobile* and *D. officinale*. The symptoms on the protocorms of *D. nobile* and *D. officinale* by *F. keratoplasticum* (ZF02) included the color of *D. officinale* protocorm starting to change from green to light yellowish at 14 days, then turning yellow at 21 days, whereas no color change symptoms were observed in the *D. nobile* protocorm due to *F. keratoplasticum* (ZF02). *F. oxysporum* (ZF03) and *F. solani* (ZF04) were moderate pathogens at seven days of incubation and strong pathogens (100%) at 21 days of incubation. *C. tropicicola* (ZF01) showed high pathogenicity

(77–100%) effects against both *D. nobile* and *D. officinale* seedlings. *F. oxysporum* (ZF03) produced no symptoms on the protocorms of *D. nobile* and *D. officinale* at 7 and 14 days, indicating these are latent pathogens because they demonstrated high pathogenicity at 21 days and their symptoms included color change of the green protocorm to brownish for both *D. nobile* and *D. officinale*. *F. solani* (ZF04) produced moderate symptoms on the protocorms of both *D. nobile* and *D. officinale* by changing their green color to light reddish to brown in the initial days and then turned dark reddish-brown at 21 days. *C. tropicicola* (ZF01) significantly affected the protocorms of *D. nobile* and *D. officinale* by changing their green color to dark brown at 7, 14, and 21 days (Figures 2 and 3).

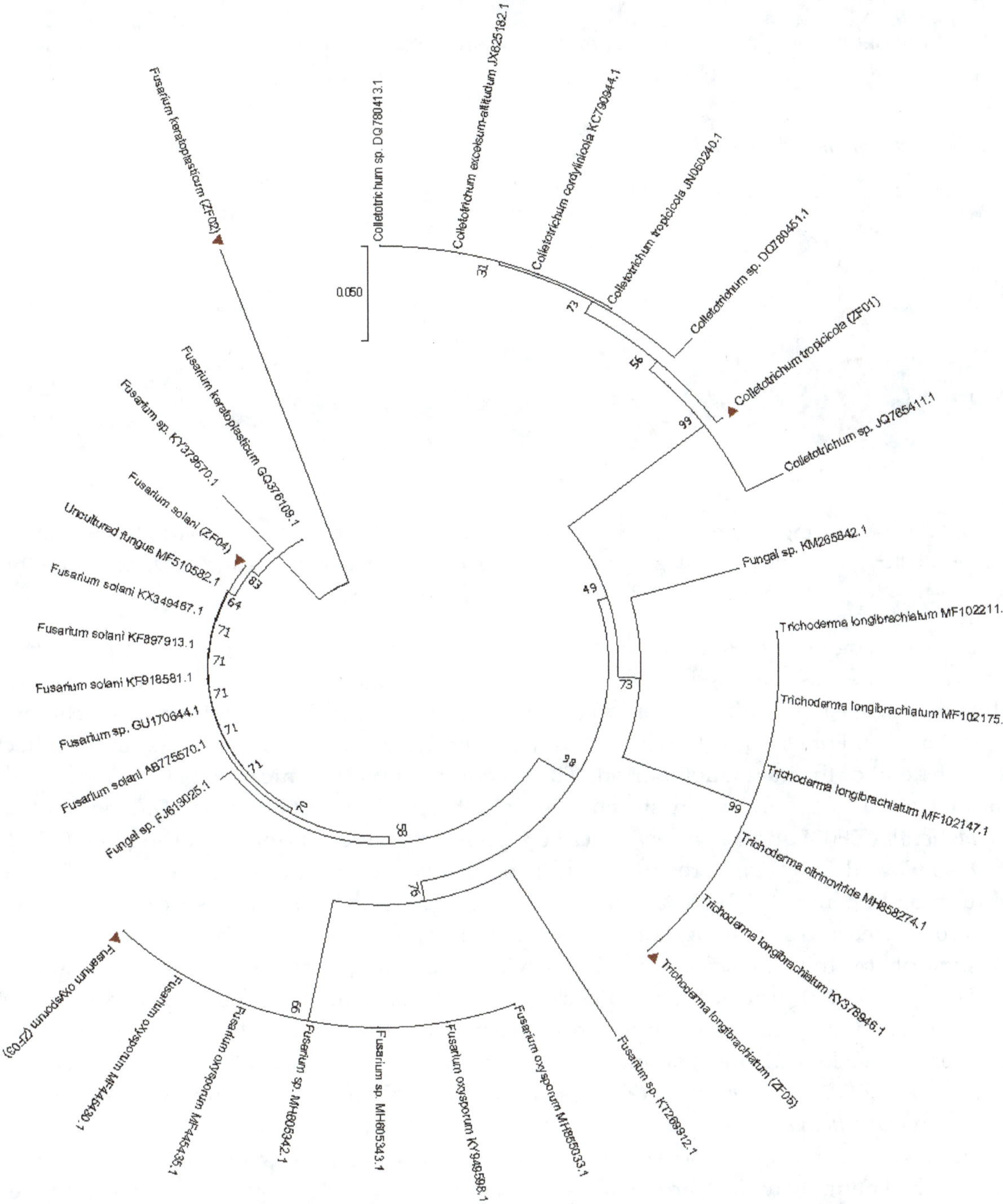

Figure 1. Phylogenetic identification of the endophytic fungi based on ITS4 and ITS5 regions of DNA sequences. The evolutionary detachments were calculated using the Kimura two-parameter method.

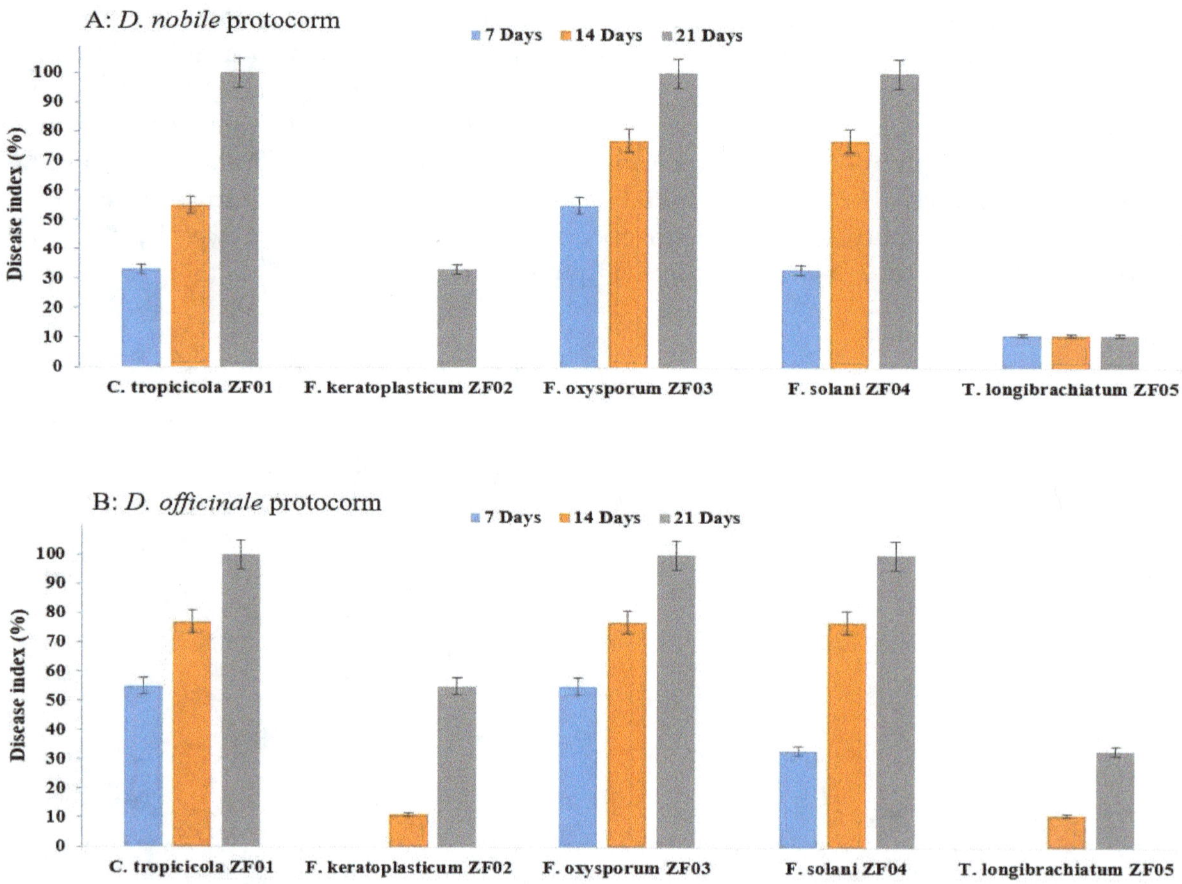

Figure 2. Co-culturing pathogenic effects on protocorm of (**A**) *D. nobile* and (**B**) *D. officinale* with endophytic fungi. Means values are significantly different at $p = 0.05$ according to Duncan's multiple range test.

The DME pathogenicity results for *D. nobile* and *D. officinale* seedling revealed that *T. longibrachiatum* (ZF05) produced the least pathogenic effects on *D. officinale* protocorm after seven days and increased gradually with increasing incubation period (at 14 days; 33% disease index), which was stable at 21 days (33% disease index). For *D. nobile*, the fungus showed highly pathogenic (77% disease index) effects from the start to the end of the incubation period. The second least pathogenic strains were *F. keratoplasticum* (ZF02) and *F. solani* (ZF04), which showed least pathogenic effects against *D. officinale*, whereas *F. keratoplasticum* (ZF02) and *F. oxysporum* (ZF03) showed stable moderate pathogenic (33%) effects against *D. nobile* and *D. officinale*, respectively, throughout the incubation period. *C. tropicicola* (ZF01) showed high pathogenicity (77–100%) against both *D. nobile* and *D. officinale* seedlings. The seedlings symptoms included a color change of *D. nobile* seedlings from green to black spots and brownish to blackish spots on the *D. officinale* seedlings with *T. longibrachiatum* (ZF05) at 7, 14, and 21 days. *F. keratoplasticum* (ZF02) produced symptoms on the *D. nobile* and *D. officinale* seedlings including color change of the *D. officinale* seedlings from green to yellowish at 7 days, then to light brown at 14 days, then to black at 21 days; similar symptoms were observed in the *D. nobile* seedlings with *F. keratoplasticum* (ZF02). *F. oxysporum* (ZF03) produced light brown color symptoms on the seedlings of *D. nobile* and *D. officinale* at 7 and 14 days, which later changed to brownish to yellowish. *F. solani* (ZF04) produced moderate symptoms on the seedlings by changing its green color to light yellowish to brown at 7 days, changing to dark brown at 14 days, then to blackish brown at 21 days on the seedlings of *D. nobile* and *D. officinale*. *C. tropicicola* (ZF01) produced significant symptoms on the seedlings of *D. nobile* and *D. officinale* by changing the seedling color from green to brown at 7, 14, and 21 days (Figures 4 and 5).

Figure 3. Least pathogenicity by *Trichoderma longibrachiatum* (ZF05) against the *D. nobile* protocorm after 21 days. (**A**) Control (*D. nobile* protocorm with water agar media); (**B**) pure culture of *Trichoderma longibrachiatum* (ZF05); (**C**) *D. nobile* protocorm with *Trichoderma longibrachiatum* (ZF05) after 21 days; (**D**) Single *D. nobile* protocorm with *Trichoderma longibrachiatum* (ZF05) growth indicated by the arrow.

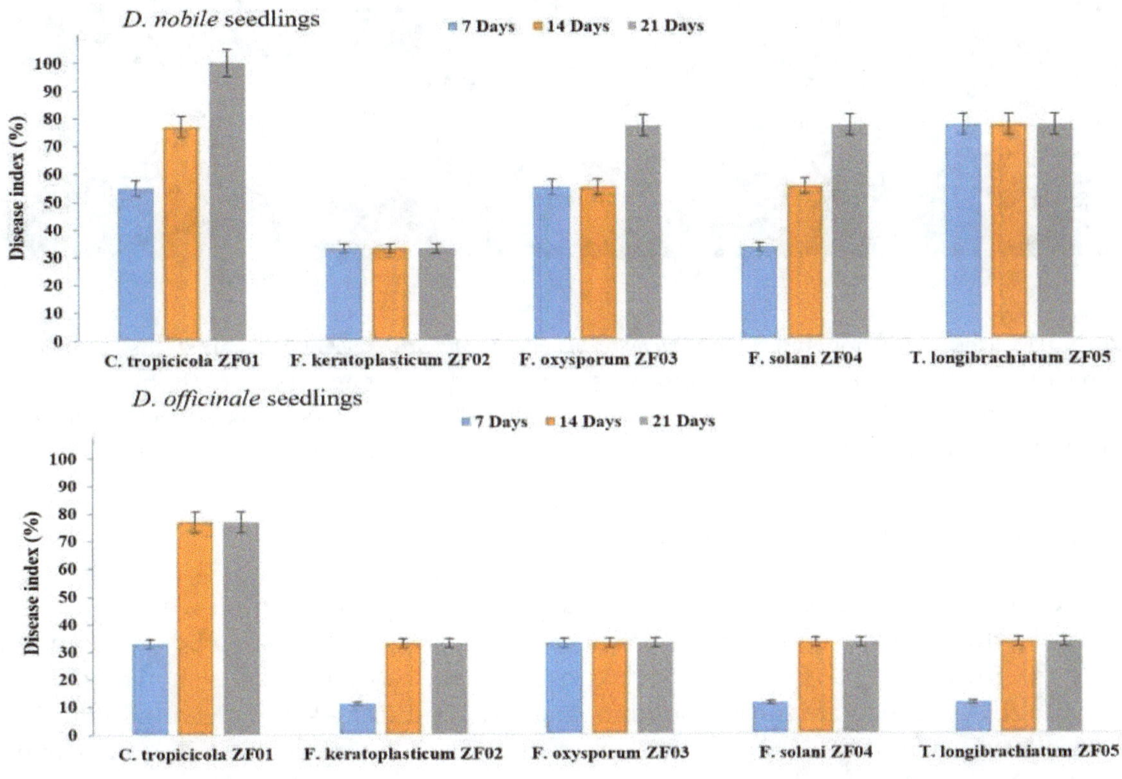

Figure 4. Co-culturing pathogenic effects of endophytic fungi on seedlings of *D. nobile* and *D. officinale*. Means values are significantly different at $p = 0.05$ according to Duncan's multiple range test.

Figure 5. Moderate pathogenicity by *Trichoderma longibrachiatum* (ZF05) against the *D. officinale* seedling after 21 days. (**A**) Control (*D. officinale* seedling with water agar media); (**B**) pure culture of *T. longibrachiatum* (ZF05); (**C**) *D. officinale* seedling with *T. longibrachiatum* (ZF05) after 21 days; (**D**) Single *D. officinale* seedling with *T. longibrachiatum* (ZF05) growth indicated by the arrow.

2.3. *Re-Isolation of Endophytic Fungi from Co-Culturing Seedling Tissues*

After the coculturing incubation periods (21 days), the endophytic fungi were re-isolated from the plant tissues to confirm the presence of fungi inside the plant tissues by re-inoculating the samples into the growth media (PDA: potato dextrose agar media). The results showed that the same endophytic fungi were re-isolated from the uninoculated plant tissues, which were identified based on characteristics: *C. tropicicola* (ZF01), *F. keratoplasticum* (ZF02), *F. oxysporum* (ZF03), *F. solani* (ZF04), and *T. longibrachiatum* (ZF05). All isolated endophytic fungi were confirmed as the same fungi via microscopic examination as described in Table S1 and Figure 6.

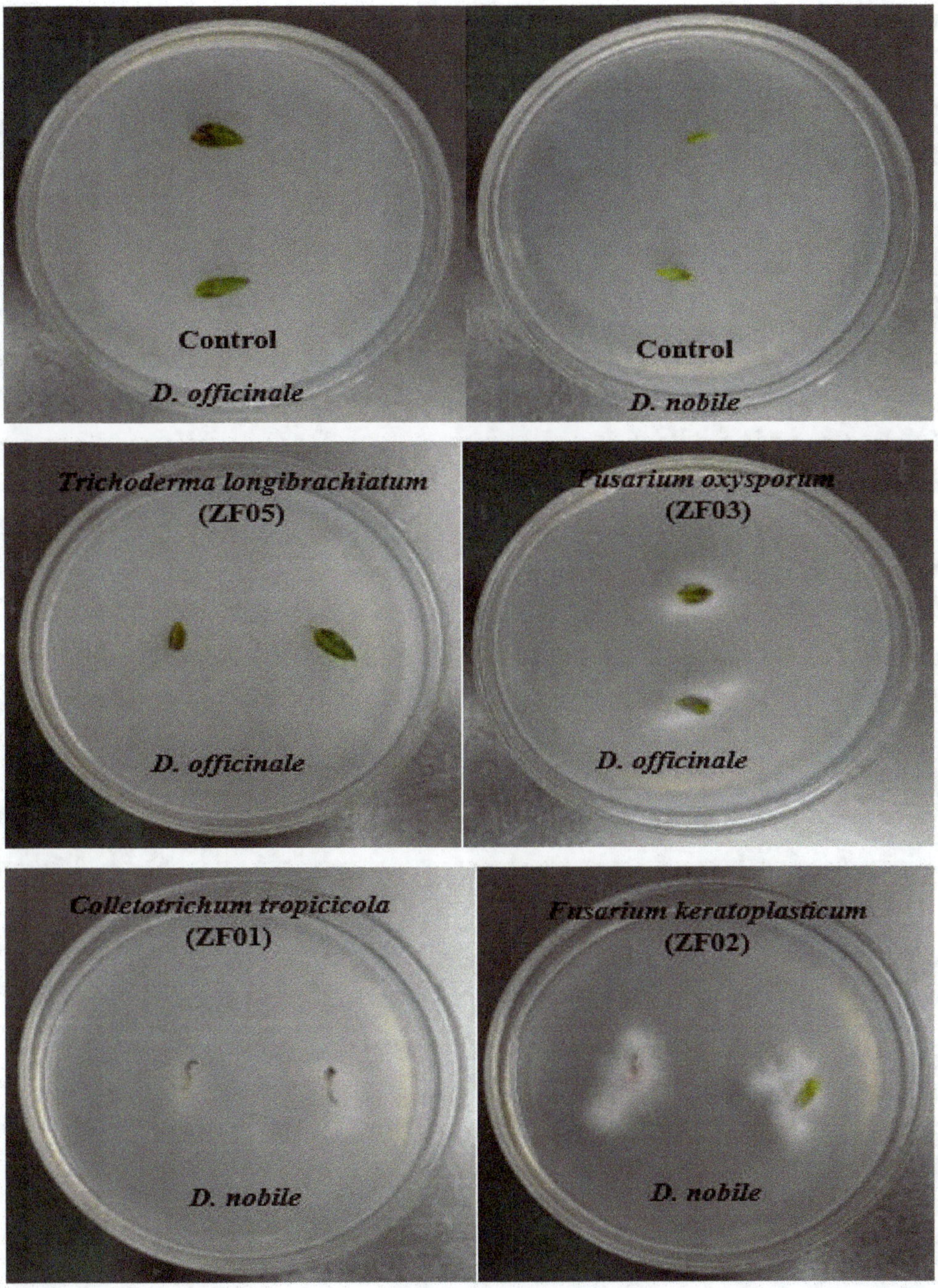

Figure 6. Re-isolation of test endophytic fungi from the uninoculated tissues of *D. nobile* and *D. officinale*.

2.4. Histological Examination of the Least Pathogenic Fungal Endophyte Seedling Samples

The results showed that the thin sections stained with lactophenol cotton blue showed the presence of endophytic fungi in the intercellular spaces of *D. nobile* and *D. officinale* seedling tissues (Figure 7). The transverse section of *D. nobile* seedling stems at lower magnification showed dense blue *Trichoderma longibrachiatum* mycelia in the epidermis and mesophyll region (Figure 7A) and, at high magnification, we observed sprinkled colonies proximal to vascular bundles (Figure 7B). The transverse section of the *D. officinale* seedling stem at lower magnification showed *T. longibrachiatum* mycelia in the intercellular mycelia and a lower amount in epidermal region (Figure 7C); at higher magnification, we observed a large amount of *T. longibrachiatum* mycelium in the phloem region with slightly less in the epidermis (Figure 7D).

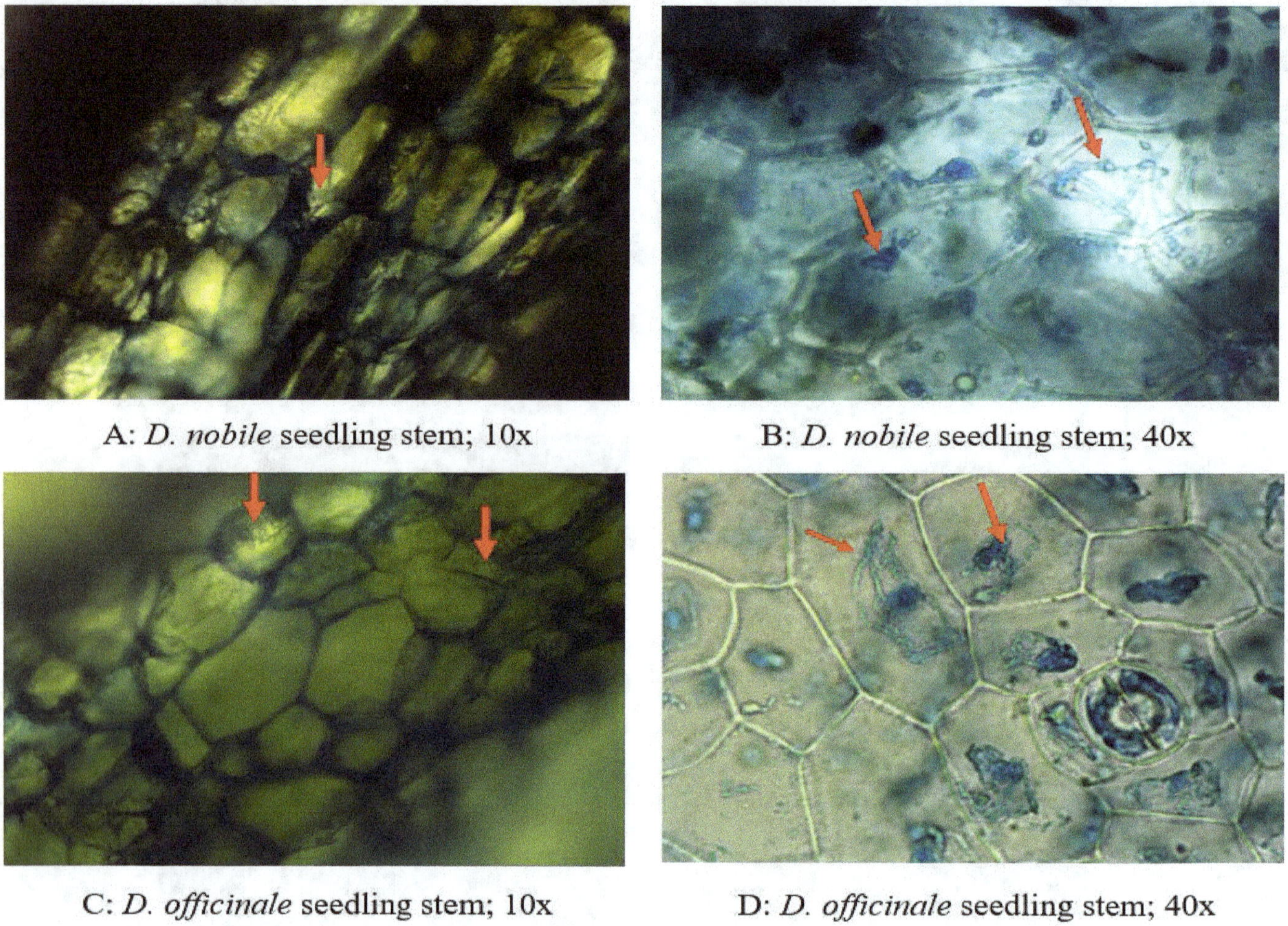

Figure 7. Histopathological examination of *D. nobile* and *D. officinale* stem infected by the least pathogenic test endophytic fungi, *T. longibrachiatum* (ZF05). (**A**,**B**) Transverse section of *D. nobile* stem showing the presence of *T. longibrachiatum* (ZF05). (**A**) *T. longibrachiatum* in epidermal region (10×). (**B**) Dense colonies of *T. longibrachiatum* (ZF05) in phloem with spores (40×). (**C**,**D**) Transverse section of *D. officinale* stem showing the presence of *T. longibrachiatum* (ZF05). (**C**) *T. longibrachiatum* (ZF05) in intercellular mycelia with less in epidermal region (10×). (**D**) Dense colonies of *T. longibrachiatum* (ZF05) in phloem with spores (40×).

3. Discussion

Dendrobium is largest genera of Orchidaceae, with more than 1000 species globally [3]. *D. nobile* and *D. officinale* are a wild epiphytic orchid found in the tropical rain forests in China, especially in Guizhou province, China. The solitary and attractive inflorescence of these orchids is slightly exclusive and valued amongst orchid cultivators. Due to their high commercial demand, they are being removed from their usual habitat, placing them at high risk of becoming extinct. To preserve and reinstate threatened and rare orchid plants, these plants must be reintroduced with fungal co-culturing. Hence, some pure endophytic fungal strains that stimulate or decrease *Dendrobium* protocorm and seedling growth must be determined in vitro [17]. For this purpose, co-culturing will be essential for *D. nobile* and *D. officinale* success. Biosynthesis of many active compounds will be necessary, predominantly where specific fungal strains are vital for actively providing molecules to the protocorm with or without producing symptoms.

This study provides valuable information about the orchid bionetwork with endophytic fungi associated with *D. nobile* and *D. officinale* under laboratory environments. The in vitro pathogenicity evaluation process was used to effectively assess well-suited and species-specific endophytic fungi in terms of their symptoms and pathogenicity index for *D. nobile* and *D. officinale*. For the first time, we found that *T. longibrachiatum* (ZF05) produced the least pathogenic effects on *D. nobile* and *D. officinale*, which provided the plants with nutrition and helped them build an active defense mechanism to survive without the presence of any nutrients in the media. This phenomenon was observed as asymptomatic colonization due to the balanced antagonism between the host and the endophyte [13]. Endophytes also possess various virulence factors that are contradicted by host plant defense mechanisms. If endophyte virulence and host *Dendrobium* defense are well balanced, the relationship is avirulent and asymptomatic. This stage is only a transitory time where environmental influences play a key role in destabilizing the delicate equilibrium of antagonisms. [18]. Endophytic fungi can deliver appropriate carbon sources, amino acids, vitamins, and hormones that are important for seedling and protocorm development [19]. Khamchatra et al. [3] also stated that *Beauvaria* and *Fusarium* species are endophytic fungi, which we also recovered, that may play a role in the growth and survival of the plants like *D. friedericksianum*. Though grown in orchid stems and roots, the fungi may be non-casual and non-mycorrhizal endophytes. Many of these endophytes were reported in many white rot fungi and are incapable of phytopathogenicity [3]. In another study, Meng et al. [20] found that several fungal species recovered from some species of *Dendrobium* protocorms and seedlings have the ability to cause disease symptoms. Zi et al. [21] described *Epulohiza*, an anamorphic phase of *Tulasnella*, in the *Dendrobium* protocorms. Athipunyakom et al. [22] reported that *Trichosporiella multisporum* is present in *Paphiopedilum* roots. In an ecosystem, orchids depend on orchid fungal endophytes to provide needed nutrients for growth, a process termed an asymptomatic relationship. The symbiotic plant relationship, specifically in vitro approaches, is adopted because it enables higher growth rates and/or symbiotic seedlings and protocorms progress faster than without this relationship [11]. Although extensive information about the co-culturing expansion is not available, the process was endorsed as an effective process for improving the growth of many orchids [19].

From this study, we found that *C. tropicicola* (ZF01) is a highly pathogenic strain for the protocorms and seedlings of both *D. nobile* and *D. officinale*. *Dendrobium* endophyte contact might, in addition to balancing between defense and virulence, might more precisely control this complex contact [10]. Plant–pathogen association may be responsible for the growth conditions for plant disease. Because several fungal endophytes may be latent plant pathogens, certain inherent or environmental influences may prompt pathogenic effects [18]. Many fungal endophytes are silent/conditional pathogens, only resulting in disease as the plant ages or under stress conditions. Fungal endophytes accomplish asymptomatic colonization through a stable neutral antagonism between the fungal virulence and its response to the plant defense mechanism [23]. We hypothesized from our research that *T. longibrachiatum* (ZF05) mycelium may migrate from inoculated to uninoculated tissues of host plants. The histological image of *D. nobile* and *D. officinale* seedlings clarified that the fungus infects tissues. This paper is the first report on the cross-transmission of *T. longibrachiatum* (ZF05) from inoculated tissues to uninoculated tissues confirmed by re-isolated from host plant segments that fulfilled the Koch postulates. *Colletotrichum*, *Fusarium*, and *Trichoderma* include morphologically similar taxa that are commonly found as endophytes, saprobes, and plant pathogens [13–15].

As endophytes exist in within plant tissues and endlessly network with their host tissues, fungal endophytes may be linked intracellularly, which is responsible for the cross-transmission of the fungus into the new cells. Along this line, we strongly suggest that endophytic fungus is re-isolated from the uninoculated infected segments of the *D. nobile* and *D. officinale* seedlings. This means that these endophytic fungi are cross-transmitted intracellularly from one cell to another healthy cell, creating comprehensive endophytic molecular connections, cross-species appearance, and on/off switching of the compulsory gene cascades that constantly yield a chosen molecule [10].

4. Materials and Methods

4.1. Isolation of Endophytic Fungal Strains

Endophytic fungi were isolated from leaves parts. Five healthy wild *D. nobile* plants were collected (from agricultural farm single location) from Jinshishi, Chishui, Guizhou, China and processed in the Bioresource Institute for Health Utilization, Zunyi Medical University, Zunyi, China, and then inoculated aseptically into PDA media. The samples were processed according to a method modified from Novotná et al. [8]. All the leaves segments of wild *D. nobile* were used for surface sterilization processing. All the leaves samples were washed with tap water to remove the dirt spots and connective tissue of stems. Thereafter, the leaves were washed with tap water to further clean the surface. After cleaning, all the plant samples were placed into laminar air flow and exposed to UV rays for 20 min. Under sterilized conditions in the laminar air flow, the samples (1.5 cm pieces) were inoculated onto PDA media plates. The sample plates were placed in a fungal incubator at 25 °C for 5 days. After 5 day of incubation, the fungal hyphal tips emerging from the samples were recovered and purified by sub-culturing from the grown hyphal tips.

4.2. Morphological and Molecular Identification of Plant Endophytic Fungi

The endophytic fungal genus was identified using lactophenol cotton blue staining according to the protocol followed by Barnett and Hunter [24] and Ellis [25]. The DNA was extracted from approximately 100 mg of fungal mycelia using the DNAsecure Plant Kit according to the manufacturer's instructions (Tianjin Biotech Co. Ltd., Beijing, China). PCR was performed using a standard protocol [26]. The reactions were prepared in 25 µL volumes constituting 2.0 µL primer, 12.5 µL 2× TsingKe (Blue) Master Mix (TSINGKE Biological Technology Co., Ltd, Beijing, China), 9.5 µL RNase-free distilled water, and template DNA (2 µL). A non-template tube was used as a negative control during PCR amplification to monitor purity. Amplifications were completed using a PCR thermal cycler equipment (Bio-Rad C1000 Thermal Cycler, Minnesota, USA). PCR cycling was performed using the conditions and primer sequences [27] presented in Table S2.

4.3. Electrophoresis of PCR Products

The PCR samples were analyzed using electrophoresis on a 1% (w/v) agarose gel (Agarose G-10, Gene Company Ltd., Hong Kong, China) [28]. A DNA marker (100 to 1000 bp molecular weights) (Beijing TransGen Biotech Co. Ltd., Beijing, China) was added to the first well of the gel. Electrophoresis was conducted in a horizontal electrophoresis system (Bio-Rad PowerPac™ Basic, Henderson, Singapore) for 30 min at 120 V and 250 mA using 1× TAE (Tris-acetate-EDTA) buffer. The gel was marked with ethidium bromide (0.1 µg/mL). A Molecular Imager Gel Doc™ XR+ with Image Lab Software system (Bio-Rad Gel Doc™ XR+ Imaging System, Hercules, CA, United States) was used to capture images using Image Lab software (version 4.1).

4.4. Sequence Analysis of PCR Products

The amplified ITS regions of PCR samples were sequenced by Tsingke, Beijing, China. Sequence data were subjected to BLAST analysis on the NCBI (national center for biotechnology information) web tool to validate the characteristics of the amplified sequences along with isolates. All the sequences have been deposited in the GenBank under accession numbers i.e., *Colletotrichum tropicicola* ZF01: MN826680; *Fusarium keratoplasticum* ZF02: MN832906; *Fusarium oxysporum* ZF03: MN826681; *Fusarium solani* ZF04: MN826682; and *Trichoderma longibrachiatum* ZF05: MN826683. MEGA 7.0 software was used for phylogeny tree study.

4.5. Collected Samples (Protocorm and Seedling) and Sample Processing

Water agar (0.65% w/v) was used as media and relocated into the glass bottles for the coculturing experiments. No additional nutriments were added into the media; only twofold distilled water and agar were used. *D. nobile* and *D. officinale* (both protocorms aged 4 months) and seedlings (aged 8 months) were obtained from the Bioresource Institute for Healthy Utilization (BIHU), Zunyi Medical University (ZMU; Zunyi, Guizhou, China) for experiments and authenticated per previous work [29–32]. After the sterilization of the media bottles, the protocorms of the *D. nobile* and *D. officinale* were inoculated into separate bottles under aseptic conditions. For seedlings, we cut marks into the top leaf of the seedlings where it joined then stem using a sterilized scalpel blade over sterilized paper, as shown in Figure 8. Three seedlings were transferred into each bottle under aseptic conditions for pathogenicity and symptoms evaluation.

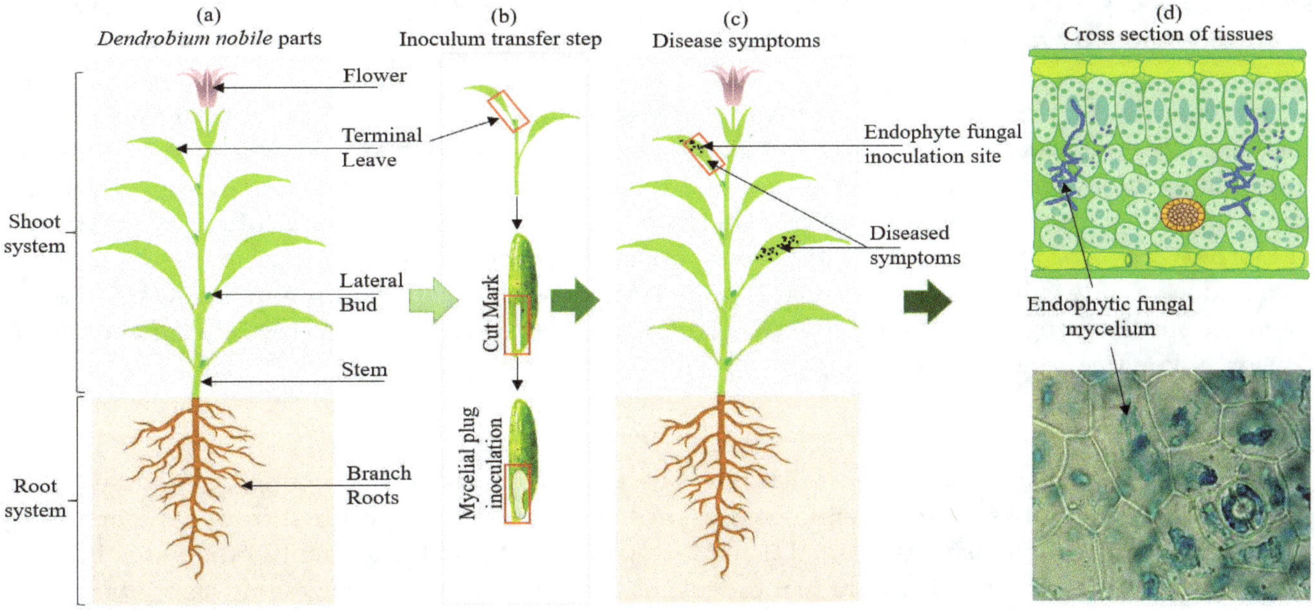

Figure 8. Different phases of coculturing experiment for pathogenicity evaluation. (**a**) Healthy seedlings; (**b**) cut marks at the top leaf and inoculation of mycelial plug of endophytic fungi; (**c**) symptoms developed for pathogenicity evaluation; and (**d**) histopathological examination of infected tissues.

4.6. In Vitro Inoculation for Seedling Pathogenicity Assay

For pathogenicity experiments, the identified endophytic fungal mycelium was used for the pathogenicity evaluation. For protocorms, 0.5-cm fungal discs were transferred into the protocorm bottles of *D. nobile* and *D. officinale*. The control was set in triplicates under the same conditions without added any endophytic fungal cultures. After the inoculation, the test co-culturing bottles were transferred into the plant growth chamber instrument and maintained at 25 °C along with a 14/10 h light/dark cycle with 2000 lux light intensity. All the test bottles were examined in 7-day intervals up to 21 days. For *D. nobile* and *D. officinale* seedlings, the contents of endophytic fungal flask cultures were filtered to separate the mycelium suspension. The mycelium was concentrated and injected onto cut marks at terminal leaves at the link to the stem. The sterilized cotton plug served as a control for seedlings co-culturing pathogenicity and symptoms evaluation experiments. The disease index was measured using the disease scale and disease index formula. The scale of the disease index is described as follows: 0 indicates no disease (non-pathogens), 1 indicates small spots < 1% (pin brown spots: fewest pathogens), 3 indicates spots (1–11%: least pathogens), 5 indicates spots with 12–25%

coverage (moderate pathogens), 7 indicates circular spots (26–55% coverage: moderate pathogens), and 9 indicates circular to irregular spots (>75% coverage: highly pathogenic) [33].

$$Disease\ index\ percentage = \frac{Sum\ of\ disease\ rating}{Total\ number\ of\ rating \times Maximum\ disease\ grade} \times 100\%$$

4.7. Re-Isolation and Identification of Endophytic Fungi from Seedlings

All the samples (seedlings) were washed with sterilized distilled water to eliminate surface fungal growth. Thereafter, the seedlings were placed into laminar air flow and exposed UV rays for 20 min to eliminate surface fungi. After UV exposure, 75% alcohol was added into the sample bottles for 30 s and mixed using gentle shaking, which was then washed twice with sterilized distilled water under laminar airflow. Thereafter, $HgCl_2$ was immersed into the glass bottle for 3 min with continuous shaking. The $HgCl_2$ water was removed after 3 min and then the samples were washed 3 times using sterilized distilled water. Brown paper sterilized in a hot air oven at 160 °C for 1.5 h was used for cutting the samples into smaller pieces. The samples (seedling) were moved onto the potato dextrose agar media plates. Two pieces of samples were transferred onto each Petri plate with the surface touching the media. The plates were incubated for five days at 25 °C in the incubator. After the fifth day of incubation, fungal hyphal tips emerging from the samples were recovered and purified using sub-culturing from growing hyphal tips. The growth, colony properties, and microscopic features of DMEs were recorded from the strains grown on the PDA media [34,35]. Microscopic details were obtained via a slide culture practice. Fungal hyphae stained with lactophenol cotton blue were assessed microscopically using light microscopy (Nikon Eclipse E200MV R with DS-1600 Panasonic CMOS Sensor, Nikon Corporation, Tokyo, Japan).

4.8. Histological Examination of Least Pathogenic Fungal Endophyte Samples

Infected *D. nobile* and *D. officinale* samples were used for visual histopathological assessment of endophytic fungi by staining the infected tissues using a method modified from Ding et al. [36]. Infected seedling segments were stained using lactophenol cotton blue to identify the tissue infected with endophytic fungi. The distribution of fungal endophytes and their localization was studied using a microscope (Nikon Eclipse E200, Model Eclipse E200MV R, Nikon Corporation, Tokyo, Japan) with DS-1600 Panasonic CMOS sensor. The endophytic fungal colonies were observed as blue after staining with lactophenol cotton blue in vascular bundles and cortex region of *D. nobile* and *D. officinale* leaves and stems. Photographs were captured under different magnifications (10× and 40×).

4.9. Statistical Analysis

Statistical calculations were completed using SPSS11 software (USA) and the Fisher's least significant difference procedure. Correlations of disease index caused by *C. tropicicola* (ZF01), *F. keratoplasticum* (ZF02), *F. oxysporum* (ZF03), *F. solani* (ZF04), and *T. longibrachiatum* (ZF05) were calculated using SAS CORR procedure.

5. Conclusions

In this study, we evaluated fungal pathogenicity and colonization inside plant tissues under in vitro conditions. Firstly, *T. longibrachiatum* (ZF05) was found to be the least pathogenic or a conditional pathogen that supports the development of the *D. nobile* protocorms and *D. officinale* seedlings without the presence of any nutrients in the media. *C. tropicicola* (ZF01) is a highly pathogenic strain, responsible for the host death. We concluded that endophytic fungi were cross-transmitted from host plant inoculated to uninoculated cells, which was confirmed by histopathological examination and re-isolation of the same endophytic fungi from uninoculated plant tissues. Future investigations should determine what role, if any, the plant host specificity plays in the interior plant passage and differential tissues establishment by test fungal endophytes. How fungi are able to precisely move within tissue (host plant) should also be examined. The symbiotic seedling and protocorm growth are advantageous and expedient method to improve orchid growth under the experimental conditions and could help with the reintroduction of *Dendrobium* orchids to the natural environment.

Author Contributions: S.S., A.J., X.F. performed experiments. J.C. and J.S. designed and supervised the project. Q.J., F.S., Z.C. and Q.Z. arranged the materials for experiments. S.S. and J.C. wrote and edited the manuscript. All authors have read and agreed to the published version of the manuscript.

Acknowledgments: The authors are grateful for the financial support under Distinguished High-Level Talents Research Grant from a Guizhou Science and Technology Corporation Platform Talents Fund (Grant No.: [2017]5733-001 & CK-1130-002), National Natural Science Foundation of China (31560079), the Jiangsu Synergetic Innovation Center for Advanced Bio-Manufacture (XTD1825), Cultivation of Academic New Seedlings and Exploration of Innovative Specialties(CK-1130-012), and Scientific and Technological Projects in Honghuagang District of Zunyi City (2018):10. We are also thanks to our all laboratory colleagues especially Xiwu Pan for their experimental help.

Abbreviations

DME	Dendrobium myco-endophytes
PDA	Potato dextrose agar
DNA	Deoxyribose nucleic acid
PCR	Polymerase chain reaction
ITS	Internal transcribed spacer
NCBI	National centre for biological information
BLAST	Basic local alignment search tool
UV	Ultra Violet
HgCl2	Mercury chloride

References

1. Huang, H.; Zi, X.M.; Lin, H.; Gao, J.Y. Host-specificity of symbiotic mycorrhizal fungi for enhancing seed germination, protocorm formation and seedling development of over-collected medicinal orchid, *Dendrobium devonianum*. *J. Microbiol.* **2018**, *56*, 42. [CrossRef]
2. Zhang, B.; Sarsaiya, S.; Pan, X.; Jin, L.; Xu, D.; Zhang, B.; Duns, G.J.; Shi, J.; Chen, J. Optimization of nutritional conditions using a temporary immersion bioreactor system for the growth of Bletilla striata pseudobulbs and accumulation of polysaccharides. *Sci. Hortic.* **2018**, *240*, 155–161. [CrossRef]
3. Khamchatra, N.M.; Dixon, K.; Chayamarit, K.; Apisitwanich, S.; Tantiwiwat, S. Using in situ seed baiting technique to isolate and identify endophytic and mycorrhizal fungi from seeds of a threatened epiphytic orchid, *Dendrobium friedericksianum* Rchb.f. (Orchidaceae). *Agric. Nat. Resour.* **2016**, *50*, 8–13. [CrossRef]
4. Srivastava, S.; Kadooka, C.; Uchida, J.Y. *Fusarium* species as pathogen on orchids. *Microbiol. Res.* **2018**, *207*, 188–195. [CrossRef]
5. Hew, C.S.; Yong, J.W.H. *Physiology of Tropical Orchids in Relation to the Industry*, 2nd ed.; World Scientific Publishing Co. Pte. Ltd.: Toh Tuck Link, Singapore, 1997.
6. Zhang, Y.; Zhang, L.; Liu, J.; Liang, J.; Si, J.; Wu, S. *Dendrobium officinale* leaves as a new antioxidant source. *J. Funct. Foods* **2017**, *37*, 400–415. [CrossRef]
7. Sarsaiya, S.; Shi, J.; Chen, J. Bioengineering tools for the production of pharmaceuticals: Current perspective and future outlook. *Bioengineered* **2019**, *10*, 469–492. [CrossRef]
8. Novotná, A.; Benítez, Á.; Herrera, P.; Cruz, D.; Filipczyková, E.; Suárez, J.P. High diversity of root-associated fungi isolated from three epiphytic orchids in southern Ecuador. *Mycoscience* **2018**, *59*, 24–32. [CrossRef]
9. Deng, Z.; Cao, L. Fungal endophytes and their interactions with plants in phytoremediation: A review. *Chemosphere* **2017**, *168*, 1100–1106. [CrossRef] [PubMed]
10. Kusari, S.; Singh, S.; Jayabaskaran, C. Biotechnological potential of plant-associated endophytic fungi: Hope versus hype. *Trends Biotechnol.* **2014**, *32*, 297–303. [CrossRef] [PubMed]

11. Mala, B.; Kuegkong, K.; Sa-ngiaemsri, N.; Nontachaiyapoom, S. Effect of germination media on in vitro symbiotic seed germination of three Dendrobium orchids. *S. Afr. J. Bot.* **2017**, *112*, 521–526. [CrossRef]
12. Zhao, P.; Wu, F.; Feng, F.S.; Wang, W.J. Protocorm-like body (PLB) formation and plant regeneration from the callus culture of *Dendrobium candidum* Wall *ex* Lindl. *In Vitr. Cell. Dev. Biol. Plant.* **2008**, *44*, 178. [CrossRef]
13. Sarsaiya, S.; Shi, J.; Chen, J. A comprehensive review on fungal endophytes and its dynamics on Orchidaceae plants: Current research, challenges, and future possibilities. *Bioengineered* **2019**, *10*, 316–334. [CrossRef] [PubMed]
14. Sarsaiya, S.; Jia, Q.; Fan, X.; Jain, A.; Shu, F.; Lu, Y.; Shi, J.; Chen, J. First report of leaf black circular spots on *Dendrobium nobile* caused by *Trichoderma longibrachiatum* in Guizhou Province, China. *Plant Dis.* **2019**, *103*, 3275. [CrossRef]
15. Jain, A.; Sarsaiya, S.; Wu, Q.; Lu, Y.; Shi, J. A review of plant leaf fungal diseases and its environment speciation. *Bioengineered* **2019**, *10*, 409–424. [CrossRef]
16. Yamamoto, T.; Miura, C.; Fuji, M.; Nagata, S.; Otani, Y.; Yagame, T.; Yamato, M.; Kaminaka, H. Quantitative evaluation of protocorm growth and fungal colonization in Bletilla striata (Orchidaceae) reveals less-productive symbiosis with a non-native symbiotic fungus. *BMC Plant Biol.* **2017**, *17*, 50. [CrossRef]
17. Tan, X.M.; Wang, C.L.; Chen, X.M.; Zhou, Y.Q.; Wang, Y.Q.; Luo, A.X.; Liu, Z.H.; Guo, S.X. In vitro seed germination and seedling growth of an endangered epiphytic orchid, *Dendrobium officinale*, endemic to China using mycorrhizal fungi (Tulasnella sp.). *Sci. Hortic.* **2014**, *165*, 62–68. [CrossRef]
18. Kusari, P.; Spiteller, M.; Kayser, O.; Kusari, S. Recent Advances in Research on *Cannabis sativa* L. Endophytes and Their Prospect for the Pharmaceutical Industry. In *Microbial Diversity and Biotechnology in Food Security*; Kharwar, R., Upadhyay, R., Dubey, N., Raghuwanshi, R., Eds.; Springer: New Delhi, India, 2014; pp. 3–16.
19. Decruse, S.W.; Neethu, R.S.; Pradeep, N.S. Seed germination and seedling growth promoted by a Ceratobasidiaceae clone in Vanda thwaitesii Hook. f., an endangered orchid species endemic to South Western Ghats, India and Sri Lanka. *S. Afr. J. Bot.* **2018**, *116*, 222–229. [CrossRef]
20. Meng, Y.Y.; Shao, S.C.; Liu, S.J.; Gao, J.Y. Do the fungi associated with roots of adult plants support seed germination? A case study on *Dendrobium exile* (Orchidaceae). *Glob. Ecol. Conserv.* **2019**, *17*, e00582. [CrossRef]
21. Zi, X.M.; Sheng, C.L.; Goodale, U.M.; Shao, S.C.; Gao, J.Y. In situ seed baiting to isolate germination-enhancing fungi for an epiphytic orchid, *Dendrobium aphyllum* (Orchidaceae). *Mycorrhiza* **2014**, *24*, 487–499. [CrossRef]
22. Athipunyakom, P.; Manoch, L.; Piluek, C. Isolation and identification of mycorrhizal fungi from eleven terrestrial orchids. *Kasetsart J. Nat. Sci.* **2004**, *38*, 216–228.
23. Alurappa, R.; Chowdappa, S.; Narayanaswamy, R.; Sinniah, U.R.; Mohanty, S.K.; Swamy, M.K. Endophytic Fungi and Bioactive Metabolites Production: An Update. In *Microbial Biotechnology*; Patra, J., Das, G., Shin, H.S., Eds.; Springer: Singapore, 2018; pp. 455–482. [CrossRef]
24. Barnett, J.L.; Hunter, B.B. *Illustrated Genera of Imperfect Fungi*; Burgess Publishing Company: Minneapolis, MN, USA, 1972; p. 90.
25. Ellis, M.B. *Demataceous Hyphomycetes*; Commonwealth Mycological Institute: Kew, UK, 1976.
26. Manganyi, M.C.; Regnier, T.; Kumar, A.; Bezuidenhout, C.C.; Ateba, C.N. Biodiversity and antibacterial screening of endophytic fungi isolated from *Pelargonium sidoides*. *S. Afr. J. Bot.* **2018**, *116*, 192–199. [CrossRef]
27. Abliz, P.; Fukushima, K.; Takizawa, K.; Nieda, N.; Miyaji, M.; Nishimura, K. Rapid identification of the genus fonsecaea by PCR with specific oligonucleotide primers. *J. Clin. Microbiol.* **2013**, *41*, 873–876. [CrossRef] [PubMed]
28. Sambrook, J.; Fritsch, E.F.; Maniatis, T. *Molecular Cloning: A Laboratory Manual*; Cold Spring Harbor Laboratory Press: New York, NY, USA, 1989.
29. Cao, J.; Fan, X.; Sarsaiya, S.; Pan, X.; Yang, N.; Jin, L.; Zhang, B.; Shi, J.; Chen, J. Accumulative and component analysis of polysaccharide in protocorm of *Dendrobium Candidum*. *J. Biobased Mater. Bioenergy* **2018**, *12*, 348–355. [CrossRef]
30. Hu, Y.; Zhang, B.; Jia, M.; Chen, J. Tissue Culture of *Dendrobium candidum* by Batch Immersion Bioreactor. *China J. Agric. Sci. Technol.* **2016**, *18*, 190–194.
31. Yu, J.; Jiang, H.; Zhang, B.; Jin, L.; Xu, D.; Zhang, B.; Chen, J. Effects of two endophytic fungi from *Dendrobium candidum* on the growth of plantlets and protocorms. *Acta Plant Pathol.* **2017**, *47*, 541–550.

32. Zhang, B.; Hu, Y.; Jin, L.; Chen, J. Comparison of tissue culture medicinal and nutrient content of *Dendrobium candidum*. *Jiangsu Agric. Sci.* **2015**, *43*, 324–327.
33. Chiang, K.S.; Liu, H.I.; Bock, C.H. A discussion on disease severity index values. Part I: Warning on inherent errors and suggestions to maximise accuracy. *Ann. Appl. Biol.* **2017**, *171*, 139–154. [CrossRef]
34. Currah, R.S.; Zelmer, C.D.; Hambleton, S.; Richardson, K.A. Fungi from Orchid Mycorrhizas. In *Orchid Biology: Reviews and Perspectives*; Arditti, J., Pridgeon, A., Eds.; Kluwer Academic Publisher: Dordrecht, The Netherlands, 1997; pp. 117–170.
35. Sneh, B.; Burpee, L.; Ogoshi, A. *Identification of Rhizoctonia Species*; APS Press: St. Paul, MN, USA, 1991.
36. Ding, C.H.; Wang, Q.B.; Guo, S.; Wang, Z.Y. The improvement of bioactive secondary metabolites accumulation in Rumex gmelini Turcz through co-culture with endophytic fungi. *Braz. J. Microbiol.* **2018**, *49*, 362–369. [CrossRef]

5

Coelonin, an Anti-Inflammation Active Component of *Bletilla striata* and its Potential Mechanism

Fusheng Jiang [1,†], Meiya Li [2,†], Hongye Wang [3], Bin Ding [1], Chunchun Zhang [4], Zhishan Ding [5], Xiaobo Yu [3,*] and Guiyuan Lv [4,*]

1. College of Life Science, Zhejiang Chinese Medical University, Hangzhou 310053, China
2. Academy of Chinese Medical Sciences, Zhejiang Chinese Medical University, Hangzhou 310053, China
3. State Key Laboratory of Proteomics, Beijing Proteome Research Center, National Center for Protein Sciences (PHOENIX Center, Beijing), Beijing Institute of Lifeomics, Beijing 102206, China
4. College of Pharmaceutical Science, Zhejiang Chinese Medical University, Hangzhou 310053, China
5. College of Medical Technology, Zhejiang Chinese Medical University, Hangzhou 310053, China

* Correspondence: xiaobo.yu@hotmail.com (X.Y.); zjtcmlgy@163.com (G.L.)

† These authors contributed equally to this work.

Abstract: Ethanol extract of *Bletilla striata* has remarkable anti-inflammatory and anti-pulmonary fibrosis activities in the rat silicosis model. However, its active substances and molecular mechanism are still unclear. To uncover the active ingredients and potential molecular mechanism of the *Bletilla striata* extract, the lipopolysaccharide (LPS)-induced macrophage inflammation model and phospho antibody array were used. Coelonin, a dihydrophenanthrene compound was isolated and identified. It significantly inhibited LPS-induced interleukin-1β (IL-1β), interleukin-6 (IL-6) and tumor necrosis factor-α (TNF-α) expression at 2.5 μg/mL. The microarray data indicate that the phosphorylation levels of 32 proteins in the coelonin pre-treated group were significantly down-regulated. In particular, the phosphorylation levels of the key inflammatory regulators factor nuclear factor-kappa B (NF-κB) were significantly reduced, and the negative regulator phosphatase and tensin homologue on chromosome ten (PTEN) was reduced. Moreover, the phosphorylation level of cyclin dependent kinase inhibitor 1B ($p27^{Kip1}$), another downstream molecule regulated by PTEN was also reduced significantly. Western blot and confocal microscopy results confirmed that coelonin inhibited LPS-induced PTEN phosphorylation in a dose-dependent manner, then inhibited NF-κB activation and $p27^{Kip1}$ degradation by regulating the phosphatidylinositol-3-kinases/ v-akt murine thymoma viral oncogene homolog (PI3K/AKT) pathway negatively. However, PTEN inhibitor co-treatment analysis indicated that the inhibition of IL-1β, IL-6 and TNF-α expression by coelonin was independent of PTEN, whereas the inhibition of $p27^{Kip1}$ degradation resulted in cell-cycle arrest in the G1 phase, which was dependent on PTEN. The anti-inflammatory activity of coelonin in vivo, which is one of the main active ingredients of *Bletilla striata*, deserves further study.

Keywords: coelonin; *Bletilla striata*; anti-inflammation; signal pathway; cell-cycle arrest; PTEN

---

1. Introduction

Bletilla striata (Thunb.) Reichb.f is a famous traditional Chinese herb that is widely used in the treatment of lung and stomach diseases such as pneumogastric hemorrhage, silicosis, tuberculosis, and gastric ulcer; it can also be used for the treatment of skin cracks, burns and freckles when combined with other traditional Chinese medicines. Numerous compounds have been identified from *Bletilla striata*, such as benzyls, phenanthrenes, dihydrophenanthrenes, anthracene, phenolic acid and polysaccharides [1,2]. Among these, polysaccharides are the most extensively and deeply studied, and their anti-ulcer [3], wound healing [4], homeostasis [5] and immune regulation [6] effects have represented most of the efficacy of *Bletilla striata*. However, in recent years, the pharmacological activities of the small molecular components in *Bletilla striata* have also attracted much attention. Liu [7] reported that the 80% ethanol elunt fraction of D101 macroporous resin significantly reduced bleeding time and increased the maximum platelet aggregation rate. Our previous research showed that the ethanol extract of *Bletilla striata* dose dependently inhibited alcohol induced gastric ulcer and silica induced silicosis in rats [8,9]. Furthermore, the ethanol extract of *Bletilla striata* significantly down regulated the serum level of IL-1β, TNF-α, transforming growth factor-β (TGF-β) and other inflammatory factors in rats with silicosis [9], thereby reducing the degree of pulmonary fibrosis, and this effect is far more effective than the polysaccharide of *Bletilla striata* [10]. However, its active components and underlying molecular mechanisms are unclear.

Silicosis is a type of systemic disease, characterized by chronic persistent inflammation and progressive fibrosis in lung tissue. The innate immune response mediated by alveolar macrophage plays a very important role in inflammatory reaction during the process of silicosis. The activated macrophages release proinflammatory mediators such as IL-6, IL-1β, TNF-α, TGF-β and platelet-derived growth factor (PDGF), etc. [11]. These inflammatory factors are recognized as key factors in pulmonary fibrosis, and the interruption of these factor pathways can alleviate or prevent fibrosis [12–14]. The classic LPS-induced RAW264.7 macrophage model can mimic the process of macrophage activation in vitro. One active compound 2,7-dihydroxy-4-methoxy-9,10-dihydrophenanthrene (coelonin) from *Bletilla striata* was separated and identified under the guidance of this cell model and combined with column chromatography.

Although few studies have described the anti-inflammatory effect of coelonin, but we found that this compound significantly down regulated IL-1β and IL-6 expression at 2.5 μg/mL on LPS-induced RAW264.7 cell. Hence, coelonin may be one of the main active components contributing to the anti-silicosis effect of *Bletilla striata*. In this study, we used a Phospho Explorer Antibody Array PEX100 to discover the potential target of the anti-inflammation effect of coelonin. The microarray results imply that coelonin may play an anti-inflammatory and cell-cycle regulation role through the PTEN/AKT pathway. Examining the down-stream signaling profile and cytokines secretion in RAW264.7 cells induced by LPS with or without coelonin or the PTEN inhibitor SF1670 suggests that coelonin blocked RAW264.7 cells in the G1 phase cell cycle in a PTEN- dependent manner, and PTEN may partially participated in coelonin inhibition on the secretion of inflammatory factors. Therefore, the potential molecular mechanism of the anti-inflammatory effect of coelonin remains to be addressed. Furthermore, as one of the main active ingredients of *Bletilla striata*, the anti-inflammatory activity of coelonin in vivo deserves further study.

2. Results

2.1. Separation, Purification and Identification of Active Components from Bletilla striata

The ethanol extraction of *Bletilla striata* tuber was separated into five fractions using the polyamide adsorption method, then they were characterized by the high performance liquid chromatography (HPLC) method (see Figure 1A). The results indicated that there were few common peaks in each fraction, which shows the effective enrichment effect of the polyamide column. The anti-inflammation activity of the five fractions was screened on the LPS-induced RAW264.7 cell model, and the real-time polymerase chain reaction (RT-PCR) results indicate that except F0 and F80, the fractions dose-dependently inhibited IL-1β expression, whereas F80 showed inhibition activity at low dosage, but the messenger RNA (mRNA) expression level of IL-1β was dose-dependently increased to even higher than the LPS-treated group at 30 μg/mL (see Supplementary Figure S1). F40 showed remarkable inhibition activity and 83.07% of IL-1β mRNA expression was inhibited at a concentration of 10 μg/mL (see Figure 1B).

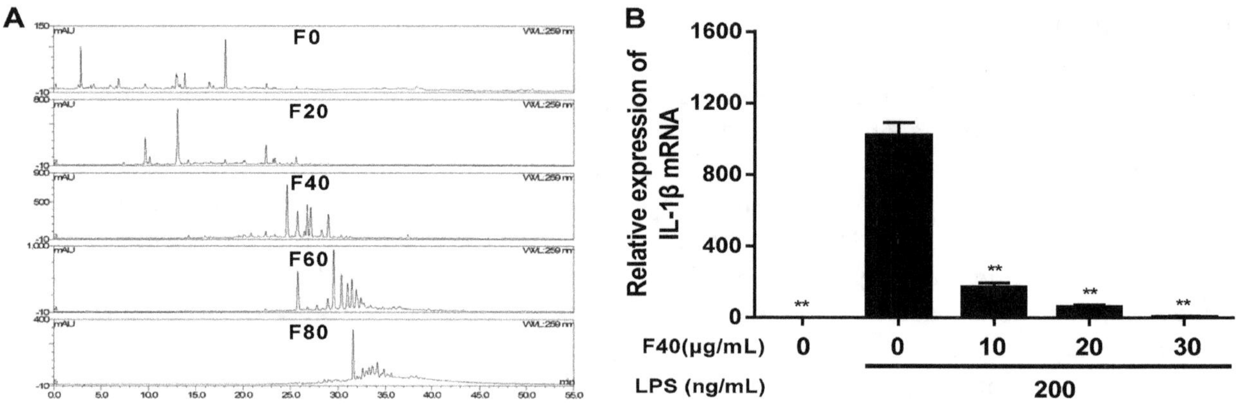

Figure 1. (**A**) HPLC characterization of the five fractions. A total of 10 μL each sample (1 mg/mL) was injected and analyzed using a Dionex UltiMate™ 3000 HPLC system with photodiode array detection (PAD) at 259 nm. A Symmetrix ODS-RC18 (25 × 4.6 mm, 5 mm) HPLC column protected with a Phenomenex security guard column (C18, 4 × 3.0 mm) operated at 30 °C was used, and the flow rate was maintained at 1 mL/min. The elution solvents were acetonitrile (a) and 0.1% acetic acid (b). Samples were eluted according to the following gradient: 0–35 min 30% a isocratic, 35–45 min 30% to 40% a, 45–55 min 40% a isocratic, and finally washing and recondition of the column. (**B**) Relative expression of IL-1β mRNA after treatment with F40. RAW264.7 cells were pretreated with different concentration of F40 for 1 h and then treated with 200 ng/mL LPS for 6 h. Total RNA was extracted and genes expression level were analyzed by RT-PCR in triplicate. The expression level of each gene was normalized to glyceraldehyde-3-phosphate dehydrogenase (GAPDH) mRNA. Data are expressed as mean ± SD ($n = 6$). ** $p < 0.01$ vs. LPS treatment group.

The F40 fraction was further separated by silica gel chromatography. Dry silica gel was packed into a glass column (diameter via height ratio 1:10), then dry sample (sample silica gel ratio 1:3) was loaded and eluted with chloroform-methanol (50:1, V/V) at 2 mL/min. Fractions of 10 mL were collected and monitored by thin-layer chromatography (TLC), visualized by iodine vapor and those possessing similar Rf values were combined and six sub-fractions of F40 were obtained (see Figure 2A). Anti-inflammation assay manifested that all sub-fractions inhibited the expression of IL-1β in a dose-dependent manner (see Supplementary Figure S2), and the sub-fractions of F40-3 and F40-4 showed significant inhibition ratio at 20 μg/mL (see Figure 2B). HPLC analysis revealed that F40-3 and F40-4 were mainly composed of two identical compounds (see Figure 3A), and the two compounds were purified by Dionex UltiMate™ 3000 semi-prepared HPLC system. A Welch Ultimate® XB-C18 (10 × 250 mm, 10 μm) HPLC column operated at 30 °C was used, and the flow rate was maintained at 5 mL/min. Samples were isocratic eluted with acetonitrile (30%) and 0.1% acetic acid (70%).

The results of mass spectrometry (MS) and nuclear magnetic resonance (NMR) spectra are as follows: Compound I (HPLC > 98%) electrospray ionization-mass spectrometry (ESI-MS) *m/z*: 245.1108 $(M + H)^+$. ^{1}H-NMR (CD_3OD) δ: 2.79 (4H, m, CH_2), 3.71 (3H, s, OCH_3), 6.19 (1H, dd, J = 1.8, 2.4 Hz, H-4), 6.25 (2H, dd, J = 1.8, 1.8 Hz, H-2, 6), 6.63 (3H, m, H-2′, 4′, 6′), 7.08 (1H, dd, J = 7.8, 8.4 Hz, H-5′); ^{13}C-NMR (CD_3OD) δ: 160.83 (C-5), 157.98 (C-3), 156.93 (C-3′), 144.06 (C-1), 143.29 (C-1′), 128.84 (C-5′), 119.43 (C-6′), 114.93 (C-2′), 112.36 (C-4′), 107.63 (C-2), 105.12 (C-6), 98.54 (C-4), 54.10 (OCH_3), 37.80 (CH2-a′), 37.47 (CH_2-a). Compared with the data shown in literature [15], the compound was identified as batatasin III. Compound II (HPLC > 98%) ESI-MS *m/z*: 243.1006 $(M + H)^+$. ^{1}H-NMR (CD_3OD) δ: 2.64 (4H, s, H-9, 10), 3.83 (3H, s, 4-OCH_3), 6.32 (1H, d, J = 2.4 Hz, H-1), 6.41 (1H, d, J = 2.4 Hz, H-3), 6.64 (1H, d, J = 1.8 Hz, H-6), 6.63 (1H, dd, J = 6, 2.4 Hz, H-8), 8.02 (H, d, J = 9 Hz, H-5). ^{13}C-NMR (CD_3OD) δ: 157.68 (C-2), 154.63 (C-7), 156.04 (C-4), 140.42 (C-10a), 139.08 (C-8a), 128.63 (C-5), 113.61 (C-4a), 124.77 (C-5a), 112.17 (C-6), 115.37 (C-8), 106.90 (C-1), 97.83 (C-3), 54.45 (-OCH_3), 29.80 (C-9), 30.38 (C-10). Comparison with the data shown in literature [16], led to the compound being identified as coelonin.

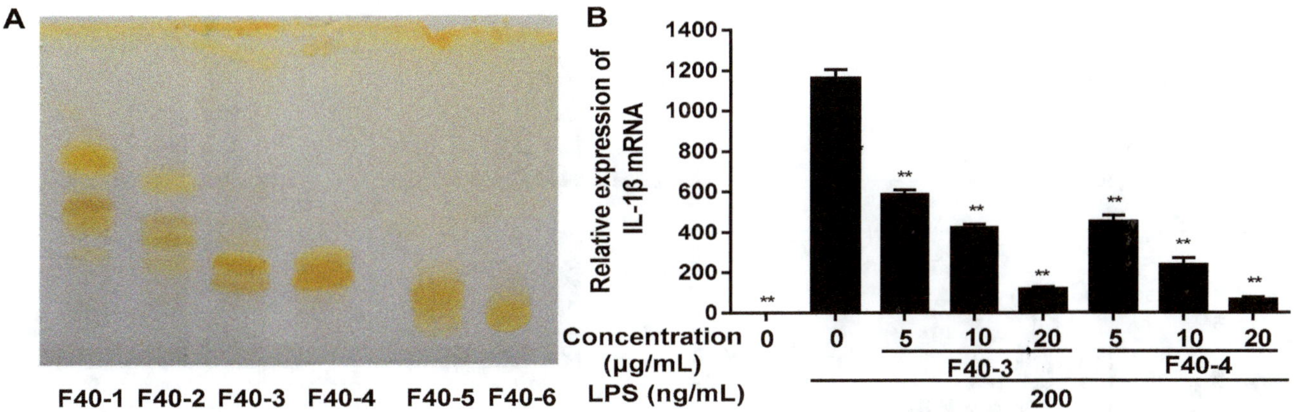

Figure 2. (**A**) Sub-fractions separated from F40 by silica gel chromatography. TLC was performed on precoated silica gel 60 F254 plates (Qingdao Haiyang Chemical Co., Ltd., Qingdao, China), developed with chloroform-methanol (4:0.1, V/V) and then exposed to the iodine vapor in a dark enclosed chamber for 10 min. (**B**) The relative expression of IL-1β mRNA after treatment with sub-fractions of F40-3 and F40-4. RAW264.7 cells were pretreated with different concentration of different sub-fractions of F40 for 1 h, then treated with 200 ng/mL LPS for 6 h. Total RNA was extracted and genes expression levels were analyzed by RT-PCR in triplicate. The expression level of each gene was normalized to GAPDH mRNA. Data are expressed as mean ±SD ($n = 6$). ** $p < 0.01$ vs. LPS treatment group.

The anti-inflammation activity was verified (see Figure 3B) and both coelonin and batatasin III showed dose-dependent inhibition activity. A total of 93.1% of IL-1β mRNA expression was inhibited by coelonin at 2.5 μg/mL, which was significantly better than that of batatasin III at 10 μg/mL (62.3%). This result implies that coelonin is probably the main anti-inflammatory component of *Bletilla striata*.

2.2. Inhibitory Effect of Coelonin on LPS-Induced IL-1β, IL-6, TNF-α Gene Expression and Protein Secretion in RAW264.7 Macrophages

LPS can induce the expression and secretion of a variety of inflammatory factors, such as IL-1β, IL-6, TNF-α, monocyte chemo-attractant protein-1 (MCP-1), inducible nitric oxide synthase (iNOS) and cyclooxygenase 2 (COX2) et al. Using IL-1β as an index, a highly active ingredient, coelonin, was isolated from *Bletilla striata* by the RAW264.7 cell inflammation model. Here, two other important inflammatory factors, IL-6 and TNF-α were examined as well. We confirmed that coelonin was without a significant cytotoxic at a concentration of up to 5 μg/mL [17] (see Supplementary Figures S3 and S4). LPS markedly elevated IL-1β, IL-6 and TNF-α mRNA expression and protein secretion, but coelonin dose-dependently lowered these levels in macrophages (see Figure 3C,D). These results confirmed the anti-inflammation effect of coelonin. However, there was almost no anti-inflammation report of this compound; therefore, the underlying mechanism is worth further research.

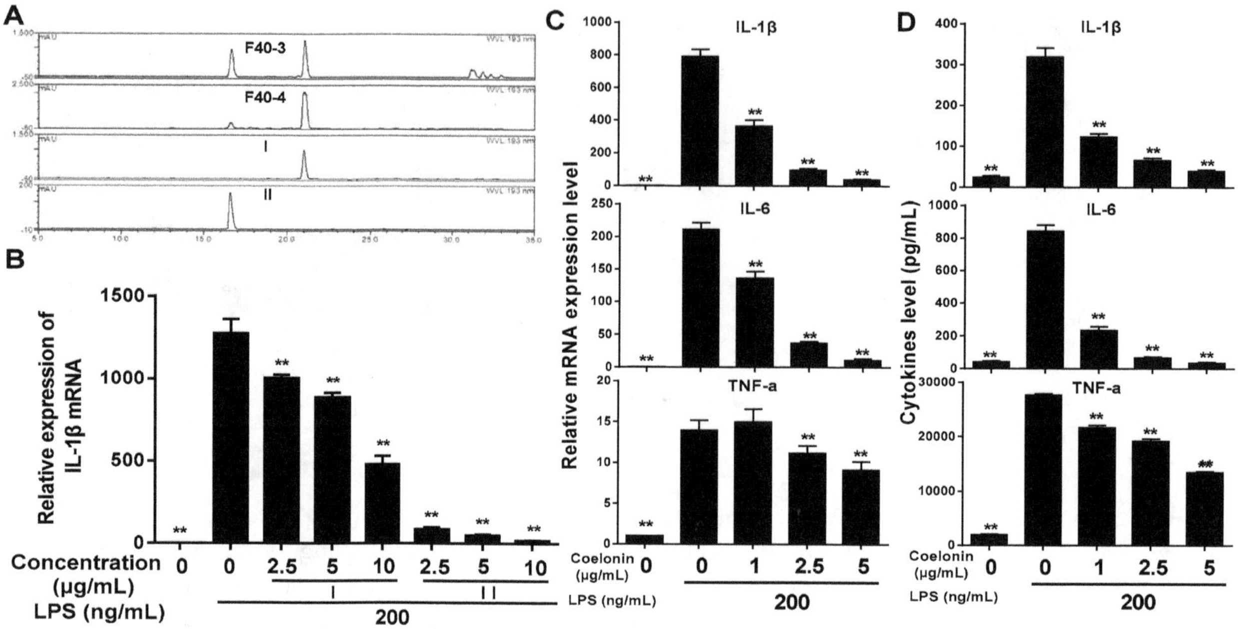

Figure 3. (**A**) HPLC characterization of the sub-fractions F40-3, F40-4 and two purified compounds. 10 μL each sample (0.1 mg/mL) was injected and analyzed using a Dionex UltiMate^{TM} 3000 HPLC system with PAD at 193 nm. A Symmetrix ODS-RC18 (25 × 4.6 mm, 5 mm) HPLC column protected with a Phenomenex security guard column (C18, 4 × 3.0 mm) operated at 30 °C was used, and the flow rate was maintained at 1 mL/min. The elution solvents were acetonitrile (a) and 0.1% acetic acid (b). Samples were eluted according to the following gradient: 0–5 min 10% to 35% a, 5–12 min 35% a isocratic, 12–16 min 35% to 45% a, 16–22 min 45% a isocratic, 22–35 min 45% to 80% a, and finally washing and recondition of the column. (**B**) The relative expression of IL-1β mRNA treated by active compounds. (**C**) Inhibitory effect of coelonin on LPS-induced gene expression in RAW264.7 macrophages. RAW264.7 cells were pretreated with different concentration of compound I or II for 1 h, then stimulated with 200 ng/mL of LPS for 6 h. Total RNA was extracted and genes expression level were analyzed by RT-PCR in triplicate. (**D**) Inhibitory effect of coelonin on LPS-induced cytokine secretion in RAW264.7 macrophages. RAW264.7 cells were pretreated with coelonin for 1 h and then treated with 200 ng/mL LPS. 12 h after LPS stimulation, culture supernatants was collected for IL-6 and TNF-α detection by cytometric bead array (CBA) method; for IL-1β detection, the remaining cells were following treated by 1 mM adenosine triphosphate (ATP) for an additional 15 min at 37 °C [18], then supernatants were collected and analyzed by the CBA method in triplicate. Data are expressed as mean ± SD (n = 6). ** $p < 0.01$ vs. LPS treatment group.

2.3. Identifying Differentially Expressed Protein Phosphorylation Sites Induced by Coelonin Treatment

To identify differentially expressed signaling-associated phosphorylated proteins between coelonin-treated and untreated RAW264.7 cells induced by LPS, the expression levels of phospho-antibody specific proteins were compared. Of the 1318 antibodies analyzed in microarray experiments, a total of 32 different phosphorylation proteins showed downregulated expression using a fold ratio ≥2 as the cutoff criterion (Figure 4, Table 1). In addition, we performed a protein network analysis using REACTOME (http://reactome.ncpsb.org/) to identify major interactions, three major cellular processes were significantly enriched (p value < 0.05) as follows: (1) Immune system, (2) signal transduction, and (3) cell cycle. LPS is the component of the outer membrane of Gram-negative bacteria, and is one of the most well characterized pathogen-associated molecular patterns (PAMPs), which can be recognized by Toll-like receptor 4 (TLR4), and trigger innate responses [19], such as enhancing the secretion of cytokines and chemokines, and promoting macrophages migration and proliferation [20]. Therefore, it is evident that the protein network analysis results imply coelonin may block LPS induced RAW264.7 cell signal transduction, immune response, and proliferation.

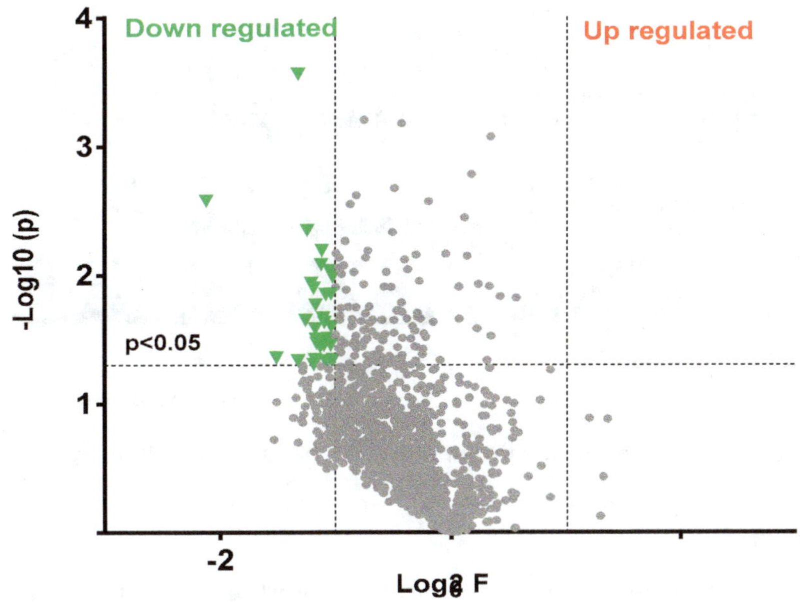

Figure 4. Protein and phosphorylation altered after coelonin treatment. From a total of 1318 differentially phosphorylated proteins, 32 were selected as significantly downregulated by coelonin using a volcano plot analysis.

Table 1. The phosphorylation of 32 proteins were significantly downregulated by coelonin treatment ($p < 0.05$).

Name	Phosphorylation Site or Antibody	Gene ID	Name	Phosphorylation Site or Antibody	Gene ID
HSP90B	Ser254	3326	Trk A	Tyr791	4914
NFκB-p65	Ser536	5970	PKC α/β II	Ab-638	5578
Ezrin	Thr566	7430	MAP3K8/COT	Thr290	1326
FLT3	Ab-599	2322	NFκB-p105/p50	Ab-932	4790
p53	Ser33	7157	$p27^{Kip1}$	Thr187	1027
CK2-b	Ab-209	1460	Shc	Tyr349	6464
ATF1	Ab-63	466	Smad1	Ab-187	4086
AXL	Tyr691	558	ERK3	Ab-189	5597
Rac1/cdc42	Ser71	5879	Caveolin-1	Tyr14	857
PTEN	Ser380	5728	MARCKS	Ser163	4082
CDK2	Ab-160	1017	GRK2	Ser685	156
DNA-PK	Ab-2056	5591	Ephrin B1	Ab-317	1947
EGFR	Ab-1069	1956	Estrogen Receptor-α	Ser104	2099
Keratin 8	Ser73	3875	BRCA1	Ser1524	672
ATPase	Ab-16	476	PTEN	Ser380/Thr382/Thr383	5728
CDC25A	Ab-75	993	eIF4B	Ser422	1975

Moreover, we entered the ENTREZ Gene IDs of the 32 genes into the DAVID Bioinformatics Resources 6.8 database. As depicted in Figure 5, the numbers of changed genes in the PI3K/AKT signaling pathway, the mitogen-activated protein kinase (MAPK) signaling pathway, the neurotrophin signaling pathway, the Sphingolipid signaling pathway and Ras signaling pathway were ranked in the top 15. In addition, we found that within these genes, the maximum number of genes belonged to the PI3K/AKT signal pathway. It is noteworthy that the phosphorylation of both subunits p65 and p105/50 of transcription factor NF-κB in this signaling pathway was reduced, which was closely related to the regulation of inflammatory gene expression [21].

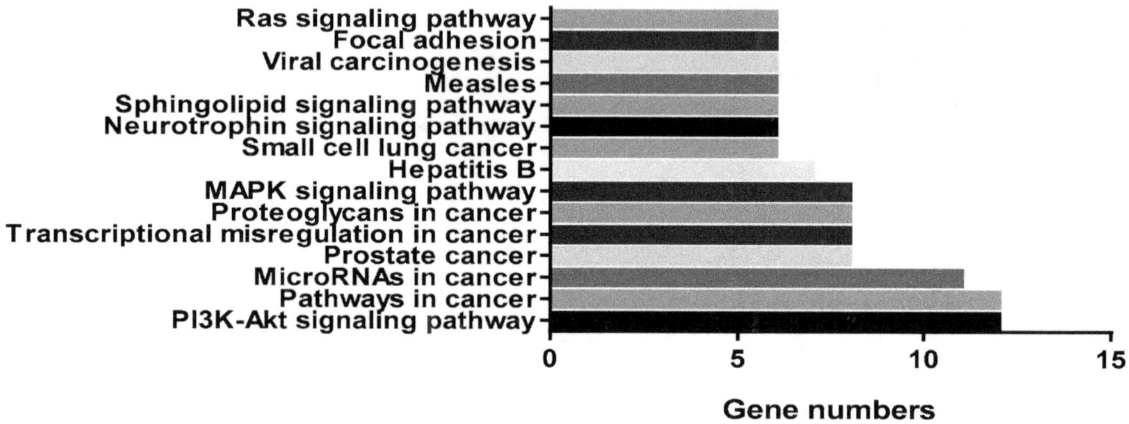

Figure 5. Number of genes in signaling pathways changed by coelonin pretreatment.

2.4. Validation of Proteomic Findings–Coelonin Treatment Inhibits Inflammatory Cytokines Secretion by Blocking NF-κB Activation

NF-κB is the most important signaling molecule induced by LPS through TLR4. The phosphorylated NF-κB can translocate to the nucleus, interacts with the κB elements and cause numerous cytokines secretion such as IL-1β, MCP-1 and TNF-α et al. [21]. As observed in Table 1, both NF-κB p65 and p105/50 phosphorylation levels were significantly reduced by 2.5 μg/mL coelonin treatment. Therefore, we performed western blot and immunofluorescence assays by nucleus translocation of p65 for validation. As shown in Figure 6A, the western blot results indicated that LPS stimulation significantly increased p65 accumulation in the nucleus, but the coelonin dose dependently reduced the effect of LPS. Moreover, confocal microscopic analysis reconfirmed that LPS stimulation significantly induced p65 translocation from the cytoplasm to the nucleus, which was remarkably inhibited by 2 μM of ammonium pyrrolidine dithiocarbamate (APDC), an inhibitor of NF-κB, and 5.0 μg/mL of coelonin pre-treatment (Figure 6B). In addition to cytokines, numerous studies have shown that NF-κB also regulates the expression of iNOS and COX2 genes [22,23]. Furthermore, over expression of iNOS and COX2 can lead to inflammation, tissue damage and even tumorigenesis [24,25]. As observed in Figure 6A, 200 ng/mL LPS treatment for 24 h caused a remarkable increased expression of iNOS and COX2. Pre-treatment with coelonin dose dependently reversed LPS-induced iNOS and COX2 expression. These results were consistent with the microarray results and confirmed that coelonin exerts its anti-inflammatory effect by inhibiting NF-κB activity.

2.5. Colonin May Partially Inhibit the Activation of NF-κB through PTEN/AKT Pathway

It is well know that LPS can activate NF-κB through the Toll-like receptor 4/myeloid differentiation factor 88/IL-1 receptor associated kinase/TNF receptor associated factor 6/TGF beta-Activated Kinase 1/inhibitor of nuclear factor-κB kinase (TLR4/MyD88/IRAK/TRAF6/TAK1/IKKs) pathway [26] and now, several lines of evidence suggest that LPS can also activate NF-κB through the TLR4/MyD88/PI3K/AKT/IKKs pathway [27,28]. However, the PI3K/AKT pathway is negatively regulated by the PTEN [29,30]. Previous research has shown that the down regulation of PTEN can activate NF-κB activity by increasing p65 nucleus translocation in mouse mesangial cells and bovine alveolar macrophages, contrary, activate PTEN would reverse the effect [29,31]. The microarray results indicated that coelonin treatment significantly down-regulated the phosphorylation of PTEN at Ser380/Thr382/Thr383 (see Table 1), which implies that coelonin may inactivate NF-κB by restoring the activity of PTEN, as phosphorylation of PTEN will make it inactivated [30]. Thus, western blotting was carried out to verify this presumption. As show in Figure 6A, LPS significantly increased the phosphorylation of PTEN, AKT and inhibitor of NF-κB (IκBa), which was dose-dependently reduced by coelonin pre-treatment (Figure 6A). In order to further confirm the possible inhibition activity of coelonin against LPS-induced NF-κB activation mediated by PTEN/AKT pathway, RAW264.7 cells were pre-treated with PTEN inhibitor SF1670. As show in Figure 7A, pre-treatment with SF1670 significantly increased LPS-induced AKT phosphorylation, which could not be downregulated by additional coelonin. Meanwhile, SF1670 also dramatically promoted LPS-induced secretion of IL-1β, IL-6 and TNF-α, and significantly reduced but could not completely block the inhibitory activity of coelonin. This result further indicates that PTEN did participate in PI3K/AKT/NF-κB activation pathway. However, it is note worthy that most levels of IL-1β and IL-6 were still significantly inhibited by coelonin co-treated with SF1670, suggesting that more critical pathways need to be identified besides the PTEN/AKT pathway.

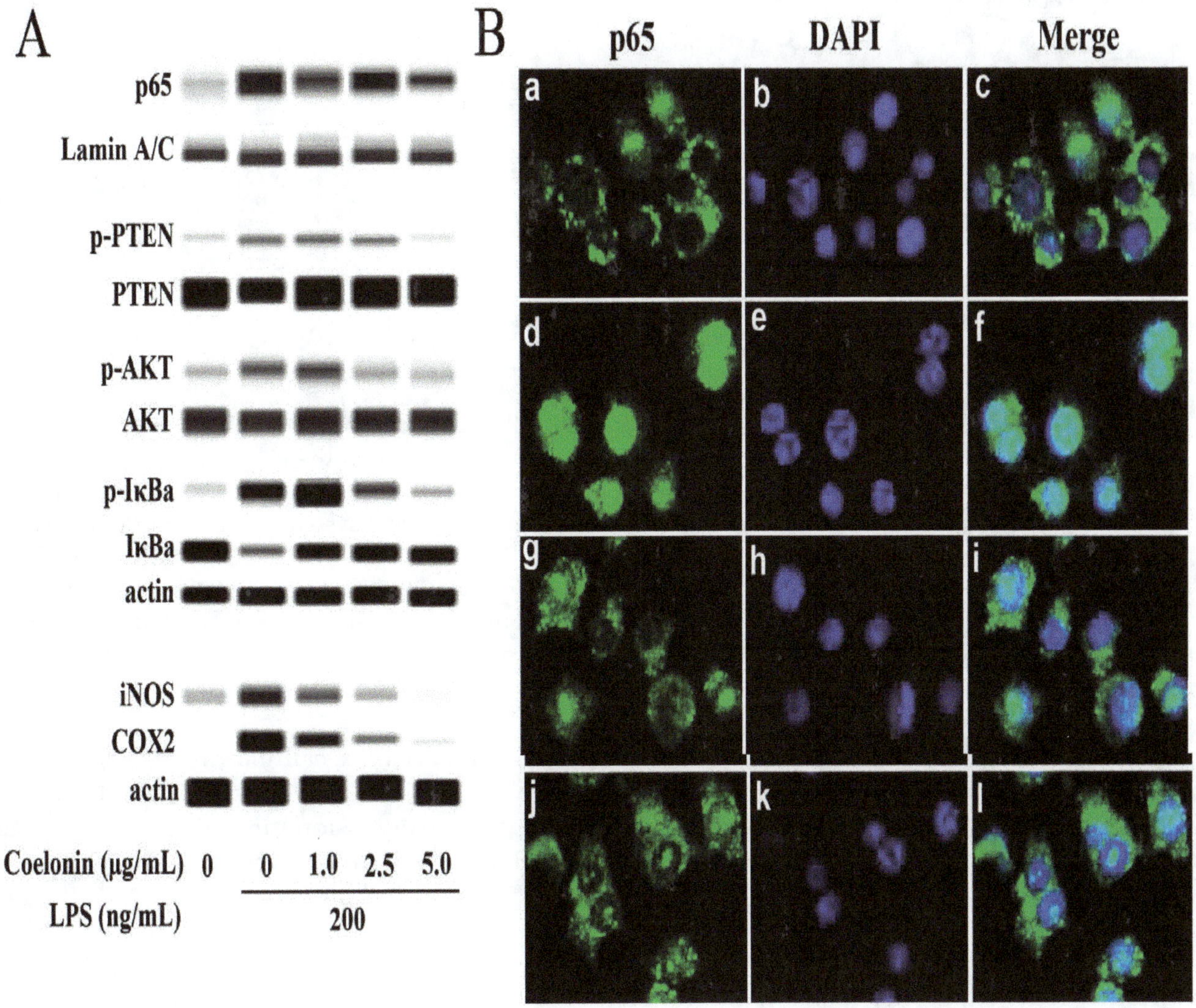

Figure 6. (**A**) Coelonin inhibits LPS induced NF-κB activation in macrophage. RAW264.7 cells were pretreated with coelonin for 1 h and then treated with 200 ng/mL LPS for 30 min. Then, the nuclear protein was extracted, the content of p65 in the nucleus was determined, the total cellular protein was extracted, and the levels of phosphorylated PTEN, AKT and IκBα were determined. For iNOS and COX2 detection, RAW264.7 cells were pretreated with coelonin for 1 h, then treated with 200 ng/mL LPS for 24 h, then total cellular protein was extracted and analyzed. (**B**) Confocal microscopy analysis of p65 nucleus translocation. RAW264.7 cells were incubated with solvent (a–c) or 200 ng/mL LPS for 1 h in the absence (d–f) or presence 2 μM of the NF-κB inhibitor APDC (g–i) or 5 μg/mL of coelonin (j–l).

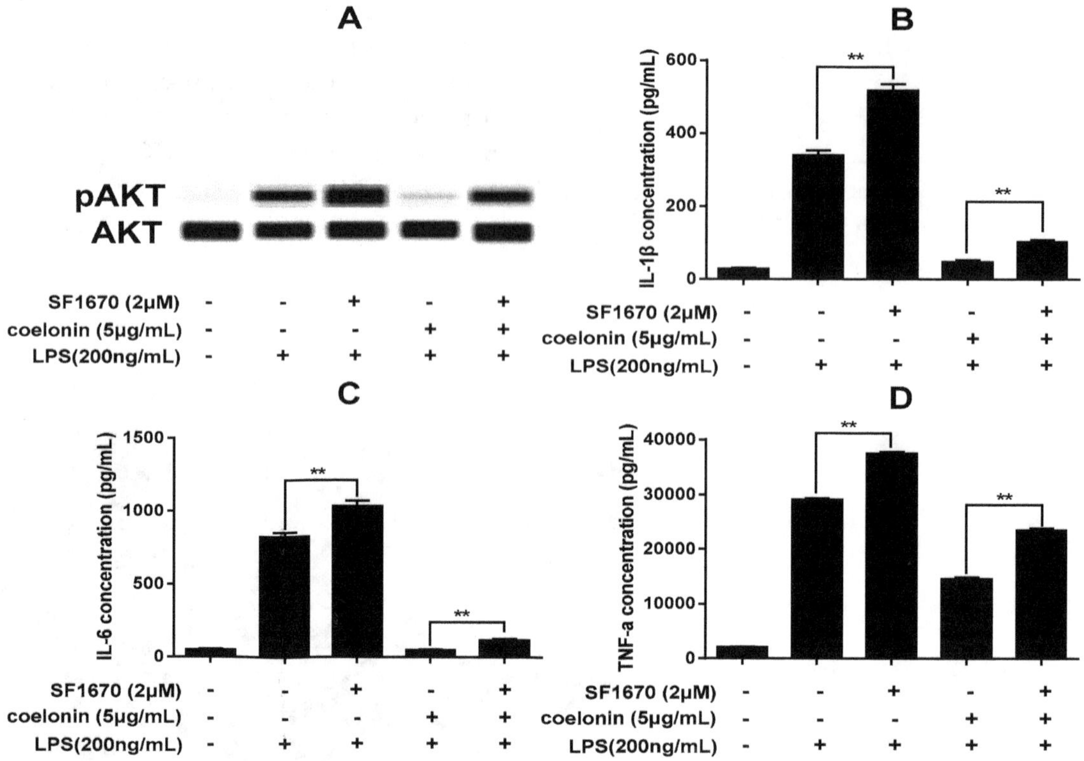

Figure 7. (**A**) Western blotting result of pAKT. RAW264.7 cells were incubated with solvent or 200 ng/mL LPS for 30 min in the absence or presence 5 μg/mL of coelonin or 2 μM of the PTEN inhibitor SF1670 or 2 μM of SF1670 combined with 5 μg/mL of coelonin. Then total cellular protein was extracted and analyzed by a simple western immunoblotting technique on a Peggy Sue system. (**B–D**) Effect of PTEN inhibitor SF1670 on anti-inflammatory activity of coelonin. RAW264.7 cells were incubated with solvent or 200 ng/mL LPS for 12 h in the absence or presence 2 μM of the PTEN inhibitor SF1670 or 5 μg/mL of coelonin or 2 μM of SF1670 combined with 5 μg/mL of coelonin. Culture supernatants was collected for IL-1β, IL-6 and TNF-α detection by CBA method. Data are expressed as mean ± SD ($n = 6$). ** $p < 0.01$.

2.6. Coelonin Treatment Leads to G1 Cell Cycle Arrest through PTEN

The PI3K-Akt pathway has been shown be involved in a variety of cellular processes, including inflammation response [27], cell survival and proliferation [32]. Many reports indicate that the PI3K/AKT pathway plays a pivotal role in regulating cell cycle progression through phosphorylation and degradation of cell cycle regulator $p27^{Kip1}$ [33,34]. $P27^{Kip1}$ can interact with cyclin-dependent kinase 2 (CDK2) and cyclinE to prevent cell entry into the S phase, and over-expression $p27^{Kip1}$ would promote cell G1 phase cell cycle arrest [35]. In many cancer cells, such as cervical cancer cells, oral squamous cell carcinoma cells, and prostatic carcinoma cells, the expression protein of $p27^{Kip1}$ was significantly reduced, the invasion and migration ability was enhanced, and the expression of $p27^{Kip1}$ was negatively correlated with survival [33,36]. We noticed that the microarray results showed that coelonin significantly down regulated the phosphorylation of $p27^{kip1}$, and interestingly, a previous study indicated that the up-regulation of PTEN prevents $p27^{Kip1}$ phosphorylation and proteolysis [37]. Hence, we speculate that coelonin may inhibit $p27^{Kip1}$ phosphorylation and degradation mediated by PTEN. Western blot confirmed that 5 μg/mL coelonin treatment significantly recovered the $p27^{Kip1}$ level, which was even higher than the un-treatment group, while LPS stimulation significantly reduced the $p27^{Kip1}$ level, and PTEN inhibitor SF1670 almost completely abrogated the effect of coelonin (see Figure 8A). Theoretically, high levels of $p27^{Kip1}$ can block cells in the G1 phase cell-cycle, so we used flow cytometry to verify whether coelonin could induce G1 phase cell cycle arrest in RAW264.7 cells. As observed in Figure 8B, coelonin pre-treatment remarkably induced G1 phase cell cycle arrest and SF1670 completely inhibited the effect of coelonin. These data suggest that coelonin inhibits $p27^{Kip1}$

degradation in a PTEN-dependent manner, thereby exerting its G1 phase cell cycle arrest effect in RAW264.7 cells.

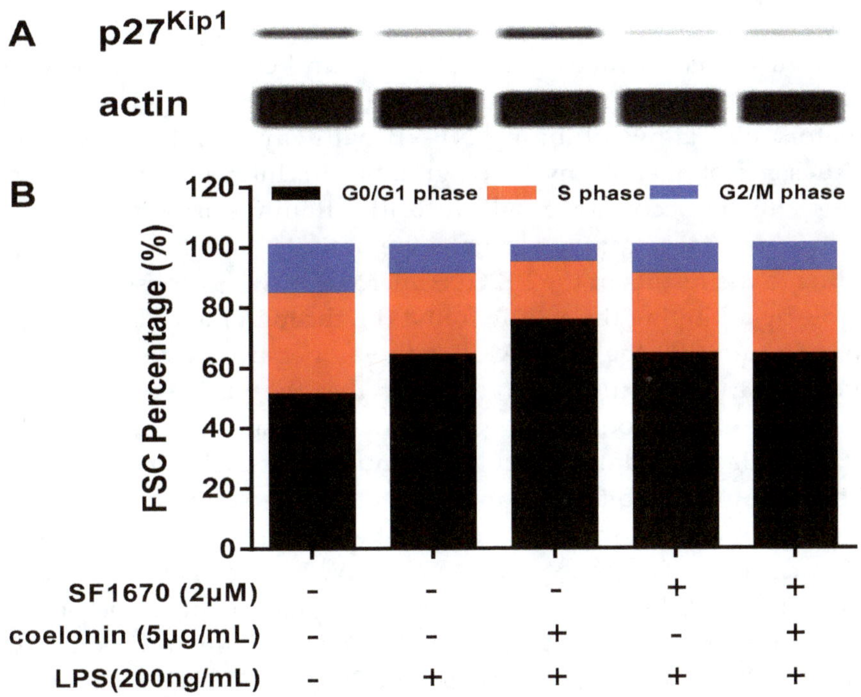

Figure 8. (**A**) Western blotting result of p27^{kip1}. RAW264.7 cells were incubated with solvent or 200 ng/mL LPS for 12 h in the absence or presence 5 μg/mL of coelonin or 2 μM of the PTEN inhibitor SF1670 or 2 μM of SF1670 combined with 5 μg/mL of coelonin. Then total cellular protein was extracted and analysed by a simple western immunoblotting technique on a Peggy Sue system. (**B**) Coelonin induce G1 cell cycle arrest through PTEN analysed by flow cytometry. RAW264.7 cells were incubated with solvent or 200 ng/mL LPS for 12 h in the absence or presence 5 μg/mL of coelonin or 2 μM of the PTEN inhibitor SF1670 or 2 μM of SF1670 combined with 5 μg/mL of coelonin. Cells were harvested, fixed and stained with PI, and cell cycle distributions were analyzed on a BD Accuri™ C6 flow cytometer in triplicate.

3. Discussion

To date, over 200 phenanthrene compounds have been isolated and identified, most of which come from the Orchidaceae family [38]. In *Bletilla striata*, a member of Orchidaceae, more than 30 phenanthrenes have been isolated from its pseudobulbs and fibrous roots, most of which showed antitumor and antimicrobial activities [39,40]. In addition, numerous studies have shown that phenanthrene and dihydrophenanthrene derivatives with remarkable anti-inflammation activity [38,41,42]. In this study, coelonin, a compound with a strong anti-inflammatory activity, was isolated and identified from the pseudobulbs of *Bletilla striat* under the guidance of biological activity screening. Few studies have reported its anti-inflammatory activity, and our studies show that coelonin can dose-dependently inhibit LPS-induced expression and secretion of IL-1β, IL-6 and TNF-α in RAW264.7 cells. Furthermore, we found that *Bletilla striata* has a high content of coelonin (0.020%–0.301% in different samples) [43]. Therefore, coelonin is probably one of the main anti-inflammatory active components of *Bletilla striata*.

In order to better elucidate the potential anti-inflammatory molecular mechanism of coelonin, PEX100 protein microarrays containing 1318 antibodies were used. The results indicated that 32 different phosphorylated proteins were significantly downregulated by pre-treatment with coelonin, which were closely related to the response of LPS-induced signal transduction, immune response and cell proliferation. Additionally, most of these genes were focused on the PI3K/AKT signal pathway, and we performed verification on the phosphorylation levels of PTEN, NF-κB and p27^{Kip1}. NF-κB is considered as a central regulator of LPS-induced pro-inflammatory response in macrophage activation [44], which is usually formed as p65:p50 heterodimer. In addition, PEX100 microarray results indicated that the phosphorylation levels of both p65 and p50 were significantly downregulated by coelonin pre-treatment (Table 1), which was further confirmed by western blot and confocal microscopic analysis on p65 nucleus translocation (Figure 6). Studies have shown that numerous dihydrophenanthrene derivatives exert anti-inflammatory effects by inhibiting NF-κB pathway [38,41,42], which was significantly correlated with the presence of phenolic hydroxyl groups [45]. Interestingly, coelonin has two phenolic hydroxyl groups, which is in accordance with its anti-inflammatory effect. However, few reports have studied the upstream targets that regulate the activity of NF-κB by dihydrophenanthrene derivatives. According to the results of the PEX100 microarray, we propose that coelonin may play an anti-inflammatory role by inhibiting the activity of NF-κB through the PTEN/AKT pathway. However, some studies have indicated that the PI3K/AKT pathway negatively regulates LPS induced NF-κB activation [46]. In contrast, other studies have shown that the PI3K/AKT pathway does positively regulate LPS-induced gene expression [27]. We found that LPS stimulation could dramatically induce AKT phosphorylation, IκBa degradation and NF-κB nucleus translocation (Figure 6), and the PI3K inhibitor LY294002 significantly inhibited LPS inducted IL-1β, IL-6 and TNF-α expression in RAW264.7 cells (see Supplementary Figure S5), which is consistent with the findings that the PI3K/AKT pathway is required for LPS induction of gene expression in RAW264.7 cells. Meanwhile, PTEN inhibitor SF1670 significantly increased LPS-induced AKT phosphorylation and cytokines secretion (Figure 7), which is consistent with the report of Zhao et al. [47]. These results indirectly supported that both PI3K/AKT and PTEN/AKT pathways are involved in LPS inducted NF-κB activation. However, the phosphorylation level of Akt did not decrease after co-treatment with PTEN inhibitor SF1670 and coelonin, but coelonin still could significantly down-regulate the expression of inflammatory cytokines induced by LPS. This contradictory result not only indicated that coelonin has other pathways to exert its anti-inflammation activity, but also indicated that AKT pathway has complex mechanisms in the regulation of inflammation. In fact, the exact mechanism of activation of NF-κB by AKT pathway remains controversial. Some reports indicated that activation AKT lead to IKK-dependent IκBα degradation and nucleus translocation of NF-κB, while others shown that AKT-dependent activation of NF-κB by stimulating the transactivation potential of the p65 subunit, rather than inducing IκBα degradation [48]. Obviously, although coelonin could inhibit IκBα phosphorylation and degradation in a dose-dependent manner, it is not clear whether coelonin plays a role through the AKT pathway or TLR4/MyD88/TRAF6/TAK1 pathway, the canonical pathway of NF-κB activation (Figure 9). Therefore, it is difficult to draw a definite conclusion only through the intervention of inhibitors, and more scientific and reasonable experiments need to be designed to elucidate the alternative mechanism of coelonin inactivating NF-κB.

In addition, our study also detected the degradation of p27^{Kip1} after the LPS challenge, which is a downstream target of PTEN/AKT. This effect was dramatically inhibited by coelonin pre-treatment. However, the PTEN inhibitor SF1670 completely abrogated the effect of coelonin (Figure 8). We noticed that LPS treatment also induced RAW264.7 cells G1 phase arrest (Figure 8B), which contradicts a previous report [20], but is consistent with report [49]. The exact molecular mechanism underlying these contradictory results need to be further studied, but it is clear that LPS-induced G1 phase arrest in RAW264.7 cells did not occur through the up-regulation of p27^{Kip1}. However, LPS treatment did promote p27^{Kip1} degradation (Figure 8A), and which may be dependent on PTEN inactivation (Figure 9). While coelonin could block the LPS-induced degradation of p27^{Kip1} by restoring PTEN activity (Figure 9). Studies have shown that alveolar macrophage proliferation may play an important role in the formation of multinucleated giant cells and granuloma induced by silica or asbestos [50]. These results suggest that, in addition to inhibiting the secretion of inflammatory factors by macrophages, inhibiting the proliferation of macrophages may also play a role in alleviating silica or asbestos-induced lung pathological changes to an extent.

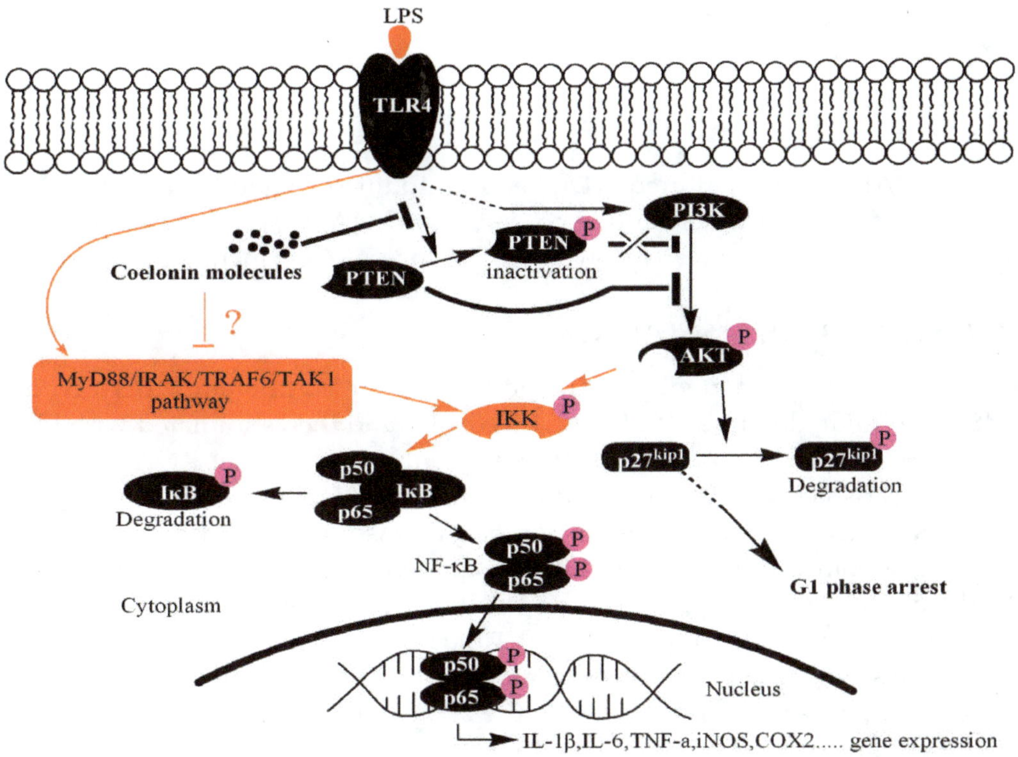

Figure 9. Proposed mechanism of coelonin inhibiting LPS-stimulated activation of NF-κB and inducing G1 phase cell-cycle arrest in RAW264.7 cells. Coelonin inhibits the expression of inflammatory cytokines IL-1β, IL-6 and TNF-α in RAW264.7 cells treated by LPS may partially through the PTEN/AKT pathway. However, it induces G1 cell cycle arrest of RAW264.7 cells by inhibiting the degradation of $p27^{Kip1}$ in a PTEN-dependent manner. The black arrow section has been verified, while the red arrow section needs further validation. The question mark indicated that coelonin may inactivate NF-κB by inhibiting TLR4/MyD88/TRAF6/TAK1 signal pathway, the canonical pathway of NF-κB activation.

4. Materials and Methods

4.1. Active Component Separation, Purification and Identification

A modified polyamide adsorption separation method was used [51]. An amount of 100.0 g tuber powder of *Bletilla striata* (collected from Meichuan, Wuxue, Hubei province, China) was reflux extracted by 1 L 80% ethanol, and the process was repeated three times. The filtrate was concentrated under vacuum to the volume of 600 mL, and an equal volume of distilled water and 30.0 g polyamide (100–200 mesh, Taizhou Luqiao Sijia Biochemical Plastics Factory, Taizhou, China) was added, then the suspension was concentrated under vacuum to the volume of 600 mL. Finally, the mix suspension was packed, and successively eluted with water, 20%, 40%, 60% and 80% ethanol, and each elution was concentrated and dried under vacuum to obtain the fractions F0, F20, F40, F60 and F80, respectively. The IL-1β mRNA expression level, detected by a RT-PCR as the index of anti-inflammation activity on the LPS-induced RAW264.7 cell model was carried out to evaluate the anti-inflammatory effect of the fractions. Then, the active fraction was further separated using silica gel (200–300 mesh, Qingdao Haiyang Chemical Co., Ltd., Qingdao, China) column chromatography, the subfractions were collected after monitoring by TLC (Qingdao Haiyang Chemical Co., Ltd., Qingdao, China), and the anti-inflammation effect was analyzed. The final active compound was purified by Dionex UltiMate™ 3000 semi-prepared HPLC system (Dionex, Sunnyvale, CA, USA) following the guidance of anti-inflammation assay. Furthermore, the molecular weight of the active compounds were determined on an ACQUITY UPLC system coupled to a SYNAPT-G2-Si high-definition mass spectrometer (Waters,

Milford, MA USA), and their NMR spectra were measured on a Bruker DRX-600 spectrometer (Bruker, Rheinstetten, Germany) with CD_3OD as the solvent.

4.2. Cell Culture

RAW264.7 cells (ATCC) were cultured in Dulbecco's Modified Eagle's Medium (DMEM) containing 10% heat inactivated fetal bovine serum (Gibco, Waltham, MA, USA), 100 units/mL penicillin and 100 μg/mL streptomycin. The cells were grown at 37 °C in a 5% CO_2 incubator.

4.3. RNA Isolation, cDNA Synthesis and RT-PCR

RAW264.7 cells were pre-treated with fractions or active components (dissolved in Dimethyl Sulfoxide, DMSO and diluted with DMEM medium) for 1 h and then stimulated with LPS (200 ng/mL) for 6 h, 0.05% DMSO was applied as the parallel solvent control. Total RNA was extracted using TRIzol (Invitrogen, Carlsbad, CA, USA) according to the manufacturer's protocol, and subsequently used to obtain cDNA with a reverse transcription polymerase chain reaction following the protocol of PrimeScript reverse transcription reagent kit with genomic DNA (gDNA) Eraser (TaKaRa, Dalian, China) in a 20 μL volume. Levels of IL-1β, IL-6 and TNF-α were determined using specific primers (see Supplementary Table S1) by RT-PCR on a 7500 Real-Time PCR System (Applied Biosystems, Foster City, CA, USA). The expression of each gene was normalized relative to the GAPDH expression level, and relative expression levels were determined using the $2^{-\Delta\Delta Ct}$ method.

4.4. Cytokine Assays

RAW264.7 cells were pre-treated with coelonin for 1 h and then stimulated with LPS (200 ng/mL) for 12 h, 0.05% DMSO was applied as the parallel solvent control. The culture supernatant was collected for IL-6 and TNF-α detection. Then, the remaining cells were treated with 1 mM ATP for an additional 15 min at 37 °C [18], and supernatants were collected for IL-1β detection. IL-6, TNF-α and IL-1β levels were quantified using the cytometric beads array (CBA) method, according to the manufacturer's protocols (BD, Franklin Lakes, NJ, USA), performed on a BD Accuri™ C6 flow cytometer (BD, Ann Arbor, MI, USA).

4.5. Protein Extraction

The cells were washed three times with phosphate-buffered saline (PBS) chilled to 4 °C. Whole-cell proteins were extracted with M-PER Mammalian Protein Extraction Reagent (78503, Thermo Fisher Scientific, Waltham, MA, USA), containing protease and phosphatase inhibitor (Roche, Mannheim, Germany), at 4 °C for 30 min. Then, the samples were centrifuged at 14,000× *g* for 10 min, and the supernatant was transferred to a new tube for analysis. Nuclear proteins were extracted in accordance with the instructions of the Nuclear and Cytoplasmic Protein Extraction Kit (P0027, Beyotime Biotechnology, Shanghai, China) for detection of the activity of NF-κB.

4.6. Signal Pathway Phosphorylation Antibody Array Screening

Cell lysates obtained from RAW264.7 cells treated with LPS (200 ng/mL) and with or without coelonin (2.5 μg/mL) were applied to a Phospho Explorer antibody Array PEX100, which was designed and manufactured by Full Moon Biosystems, Inc. (Sunnyvale, CA, USA). The microarray contains 1318 antibodies [52], each of which has two replicates that are printed on a coated glass microscope slide, along with multiple positive and negative controls. The antibody microarray experiment was performed according to the manufacturer's protocol. The protein phosphorylation level was measured as a ratio of the phospho and unphospho values.

4.7. Automated Western Immunoblotting

Before blotting, the protein was quantified using the bicinchoninic acid (BCA) method. Simple western immunoblotting was performed on a simple wes system (ProteinSimple, San Jose, CA, USA) using a Size Separation Master Kit with Split Buffer (12–230 kDa) according to the manufacturer's standard instruction and using specific antibody. Anti-PTEN (phospho S380) antibody [EP2138Y] (ab76431), Anti-PTEN antibody [Y184] (ab32199), Anti-AKT1 (phospho S473) antibody [EP2109Y] (ab81283), Anti-AKT1/2/3 antibody [EPR16798] (ab179463), Anti-IκBα (phospho S32+S36) antibody (ab12135), Anti-IκBα antibody (ab32518), Anti-NF-κB-p65 antibody (ab32536), Anti-iNOS antibody (ab178945), Anti-COX2 antibody (ab179800) were purchased from Abcam(Cambridge, USA), Anti-$p27^{kip1}$ antibody (2552), Anti-Lamin A/C antibody (2032S,) and anti-β-actin (4970S) antibody were obtained from CST (Danvers, MA, USA). Compass software (version 4.0.0, ProteinSimple, San Jose, CA, USA) was used to program the Peggy Sue and for presentation (and quantification) of the western immunoblots. Output data were displayed from the software calculated average of seven exposures (5–480 s).

4.8. Confocal Microscopy Analysis

RAW264.7 cells were grown on glass coverslips. After the treatment with LPS (200 ng/mL) with or without pre-treatment with coelonin or NF-κB-inhibitor APDC, cells were fixed with 4% formalin in PBS for 20 min at room temperature, permeabilized with 0.1% Triton X-100 for 15 min, and nonspecific protein binding sites were blocked with 10% FBS at room temperature for 1 h. Then samples were incubated overnight at 4 °C with a p65 (ab32536, Abcam, Cambridge, MA, USA) specific Antibody (1:100 dilution in PBS), washed three times with PBS, and then incubated with a Alexa Fluor® 488-labeled secondary antibody (ab60314, Abcam; 1:600 dilution in PBS) for 2 h at room temperature. The coverslips were rinsed three times with PBS, then stained with 4′,6-diamidino-2-phenylindole (DAPI) for 10 min. Again, the coverslips were rinsed with PBS three times and then mounted on glass slides using Antifluorescence Quenching Sealing Solution. The coverslips were analyzed by confocal microscopy (LSM880; ZEISS, Upper Cohen, Germany).

4.9. Cell Cycle Analysis

After treatment, the RAW264.7 cells were washed three times with chilled 1 × PBS (pH 7.4) and fixed overnight with 70% ethanol at 4 °C followed by centrifugation at 2,000 rpm for 8 min. The cells were re-suspended in 1 × PBS (pH 7.4) with PI/RNase Staining Buffer (BD, 550825, San Diego, CA, USA) for 30 min. The cell cycle distributions were analyzed on a BD Accuri™ C6 flow cytometer (BD, Ann Arbor, MI, USA).

4.10. Statistical Analysis

Data are presented as the mean ± SD derived from at least three independent experiments. ANOVA analysis was used to examine the statistical significance of the differences between the groups, and the criterion for significance for all the experiments was $p < 0.05$.

5. Conclusions

Our study indicated that coelonin is one of the active components of *Bletilla striata*. Furthermore, we showed that using a PEX100 antibody microarray, a total of 32 different phosphorylation proteins were downregulated by coelonin pre-treatment on LPS-induced RAW264.7 cells. The maximum number of proteins belonged to the PI3K/AKT signal pathway, and three of them, PTEN, p65 and p27 Kip1 were confirmed by western blot, and more proteins and signaling pathways need to be verified.

Western blot and confocal microscopy analysis revealed that coelonin inhibits the expression of IL-1β, IL-6 and TNF-α in a dose-dependent manner by eliminating lipopolysaccharide-induced NF-κB activity. However, besides inhibiting IκBα degradation, which pathways coelonin mainly inhibits the activation of NF-κB still need to be further studied. While, we did confirm that coelonin inhibit LPS-induced p27 ^{Kip1} degradation and block RAW264.7 cells in the G1 phase in a PTEN dependent manner (Figure 9). Overall, our results suggest that traditional Chinese medicine *Bletilla striata* has anti-inflammatory activity, and coelonin is one of the main active components. It may play a potential role in treating silicosis by inhibiting the proliferation of macrophages and the secretion of inflammatory factors. Additionally, PTEN may play an important role during this process, our following work will use modified RAW264.7 cells to address this issue.

Author Contributions: Conceptualization: G.Y.L., methodology: X.B.Y. and Z.S.D.; software: F.S.J. and H.Y.W.; formal analysis: M.Y.L. and B.D.; writing—original draft preparation: F.S.J. and M.Y.L.; writing—review and editing: F.S.J., M.Y.L. and C.C.Z.; supervision: G.Y.L. and X.B.Y.; funding acquisition: F.S.J., M.Y.L. and Z.S.D. All authors have read and approved the final version of the manuscript.

Acknowledgments: We thank the public platform of the Medical Research Center, Academy of Chinese Medical Sciences, Zhejiang Chinese Medical University for its instrumentation and equipment, and Jiangjiang Qin for his guidance in revising the manuscript.

Abbreviations

TLR4	Toll-like receptor 4
MyD88	myeloid differentiation factor 88
IRAK	IL-1 receptor associated kinase
TRAF6	TNF receptor associated factor 6
TAK1	TGF beta-Activated Kinase 1
IKKs	inhibitor of nuclear factor-κB kinase
PTEN	phosphatase and tensin homologue on chromosome ten
PI3K	phosphatidylinositol-3-kinases
AKT	v-akt murine thymoma viral oncogene homolog
p27Kip1	cyclin dependent kinase inhibitor 1B
NF-κB	nuclear factor-kappa B
IκBa	inhibitor of NF-κB
IL-1β	interleukin-1β
iNOS	inducible nitric oxide synthase
COX2	cyclooxygenase 2

References

1. Pérez, G.; Martha, R. Orchids: A review of uses in traditional medicine, its phytochemistry and pharmacology. *J. Med. Plants Res.* **2010**, *4*, 592–638. [CrossRef]
2. Sun, D.F.; Shi, J.S.; Zhang, W.M.; Gu, G.P.; Zhu, C.L. Study on the extraction of polysaccharides from *Blettila striata* by the continuous counter-current equipment (in Chinese). *Chin. Wild Plant. Resour.* **2006**, *13*, 1797–1801. [CrossRef]
3. Lv, X.B.; Huang, C.Q.; Wu, Z.C.; Yang, D.J.; Den, L. The therapeutic effects of polysaccharides from *Bletilla Striata* on gastric ulcer rats. *J. Yunnan Univ. Tradit. Chin. Med.* **2012**, *35*, 30–32.
4. Luo, Y.; Diao, H.J.; Xia, S.H.; Dong, L.; Chen, J.N.; Zhang, J.F. Physiologically active polysaccharide hydrogel promotes wound healing. *J. Biomed. Master Res. A* **2010**, *94A*, 193–204. [CrossRef] [PubMed]
5. Dong, Y.X.; Liu, X.X.; Dong, L.; Wang, X.; Huang, Y.; Wang, Y.L. Study on hemostatic effect and mechanism of polysaccharides from *Bletilla striata* in blood heat and hemorrhage model rats. *China Pharm.* **2016**, *27*, 4347–4350.

6. Zhang, Y.; Zhou, Q.; Lai, S. The effects of *Bletilla striata* polysaccharide on proliferation of hematopoietic cells and immunological function in mice treated by cyclophosphamide. *Pharmacol. Clin. Chin. Mater. Med.* **2009**, *4*.
7. Liu, X.X. Hemostatic effects, mechanism and biopotency of *Bletillae rhizoma*. Master's Thesis, Guiyang Medical College, Guiyang, China, 2015.
8. Shi, Z.Z.; Xu, Z.H.; Fu, Y.H.; Yu, H.S.; Jiang, F.S.; Ding, Z.S. Study of *Rhizoma bletillae* fibrous root alcohol extract on anti gastric ulcer. *J. Shanxi Coll. Tradit. Chin. Med.* **2015**, *28*, 63–89.
9. Deng, Y.Z.; Jin, L.X.; Gao, C.X.; Qian, C.D.; Jiang, F.S.; Ding, Z.S.; Li, M.Y. Research on the Anti-Pulmonary Fibrosis effect of the small molecule components of *Bletilla striata* in rat silicosis model. *J. Chin. Med. Mater.* **2016**, *39*, 2615–2619.
10. Li, H.Y.; Shi, Z.Z.; Shu, L.F.; Wang, J.; Li, M.Y.; Ding, Z.S.; Jiang, F.S. Research on the Anti-Pulmonary Fibrosis effect of the *Bletilla striata* polysaccharide in rat silicosis model. *J. Chin. Med. Mater.* **2016**, *39*, 1638–1642.
11. Young, K.S.; Woo, P.S.; Ran, L.M.; Young, K.E.; Taek, U.S.; Hoon, K.Y.; Sik, P.C.; Bal, L.H. Silica induced expression of IL-1β, IL-6, TNF-β, TGF-α, in the experimental murine lung fibrosis. *Tuberc. Respir. Dis.* **1998**, *45*, 835–845. [CrossRef]
12. Piguet, P.F.; Vesin, C.; Grau, G.E.; Thompson, R.C. Interleukin 1 receptor antagonist (IL-1ra) prevents or cures pulmonary fibrosis elicited in mice by bleomycin or silica. *Cytokine* **1993**, *5*, 57–61. [CrossRef]
13. Piguet, P.F.; Collart, M.A.; Grau, G.E.; Sappino, A.P.; Vassalli, P. Requirement of tumour necrosis factor for development of silica-induced pulmonary fibrosis. *Nature* **1990**, *344*, 245–247. [CrossRef] [PubMed]
14. Smoktunowicz, N.; Alexander, R.E.; Franklin, L.; Williams, A.; Holman, B.; Mercer, P. The anti-fibrotic effect of inhibition of TGFβ-ALK5 signalling in experimental pulmonary fibrosis in mice is attenuated in the presence of concurrent γ-herpesvirus infection. *Dis. Model. Mech.* **2015**, *8*, 1129–1139. [CrossRef] [PubMed]
15. Yang, J.Z.; Jiang, H.; Wang, W.J.; Zhang, Y.M.; Liu, Y.; Chen, Y.G. Isolation and Characterization of Batatasin III and 3,4′-Dihydroxy-5-methoxybibenzyl: A Pair of Positional Isomers from *Sunipia scariosa*. *Trop. J. Pharm. Res.* **2014**, *13*, 533–535. [CrossRef]
16. Zhang, H.L.; Tian, L.; Fu, H.W.; Pei, Y.H.; Hua, H.M. Studies on constituents from the fermentation of Alternalia sp. *China J. Chin. Mater. Med.* **2005**, *30*, 351.
17. Deng, Y.Z.; Jin, L.X.; Gao, C.X.; Qian, C.D.; Jiang, F.S.; Ding, Z.S.; Li, M.Y. Study on the Active Components and Molecular Mechanism of *Bletilla striata* on Suppressing Pulmonary Fibrosis. *J. Chin. Med. Mater.* **2016**, *39*, 2618.
18. Stoffels, M.; Zaal, R.; Kok, N.; Van der Meer, J.W.; Dinarello, C.A.; Simon, A. ATP-Induced IL-1β Specific Secretion: True Under Stringent Conditions. *Front. Immunol.* **2015**, *6*, 54. [CrossRef] [PubMed]
19. Beutler, B.; Jiang, Z.; Georgel, P.; Crozat, K.; Croker, B.; Rutschmann, S.; Du, X.; Hoebe, K. Genetic analysis of host resistance: Toll-like receptor signaling and immunity at large. *Annu. Rev. Immunol.* **2006**, *24*, 353–389. [CrossRef] [PubMed]
20. Jiao, H.W.; Jia, X.X.; Zhao, T.J.; Rong, H.; Zhang, J.N.; Cheng, Y.; Zhu, H.P.; Xu, K.L.; Guo, S.Y.; Shi, Q.Y.; et al. Up-regulation of TDAG51 is a dependent factor of LPS-induced RAW264.7 macrophages proliferation and cell cycle progression. *Immunopharm. Immunot.* **2016**, *38*, 124–130. [CrossRef]
21. Zhu, Y.; Tong, Q.; Ye, J.; Ning, Y.; Xiong, Y.; Yang, M.; Xiao, H.; Lu, J.; Xu, W.; Li, J.; et al. Nogo-B Facilitates LPS-Mediated Immune Responses by Up-Regulation of TLR4-Signaling in Macrophage RAW264.7. *Cell Physiol. Biochem.* **2017**, *41*, 274–285. [CrossRef]
22. Surh, Y.J.; Chun, K.S.; Cha, H.H.; Han, S.S.; Keum, Y.S.; Park, K.K.; Lee, S.S. Molecular mechanisms underlying chemopreventive activities of anti-inflammatory phytochemicals: Down-regulation of COX-2 and iNOS through suppression of NF-kappa B activation. *Mutat. Res.* **2001**, *480-481*, 243–268. [CrossRef]
23. Wang, Q.S.; Xiang, Y.; Cui, Y.L.; Lin, K.M.; Zhang, X.F. Dietary Blue Pigments Derived from Genipin, Attenuate Inflammation by Inhibiting LPS-Induced iNOS and COX-2 Expression via the NF-κB Inactivation. *PLoS ONE* **2012**, *7*, e34122. [CrossRef] [PubMed]
24. Cohen, G.; Tretiakova, M.; Carroll, R.; Bissonnette, M. COX-2 and iNOS are overexpressed in human colonic aberrant crypt foci. *Gastroenterol* **2003**, *124*, A605. [CrossRef]
25. Pan, M.H.; Lai, C.S.; Wang, Y.J.; Ho, C.T. Acacetin suppressed LPS-induced up-expression of iNOS and COX-2 in murine macrophages and TPA-induced tumor promotion in mice. *Biochem. Pharmacol.* **2006**, *72*, 1293–1303. [CrossRef] [PubMed]

26. Lu, Y.C.; Yeh, W.C.; Ohashi, P.S. LPS/TLR4 signal transduction pathway. *Cytokine* **2008**, *42*, 145–151. [CrossRef] [PubMed]
27. Ojaniemi, M.; Glumoff, V.; Harju, K.; Liljeroos, M.; Vuori, K.; Hallman, M. Phosphatidylinositol 3-kinase is involved in Toll-like receptor 4-mediated cytokine expression in mouse macrophages. *Eur. J. Immunol.* **2003**, *33*, 597–605. [CrossRef] [PubMed]
28. He, Z.Y.; Gao, Y.; Deng, Y.X.; Li, W.; Chen, Y.M.; Xing, S.P.; Zhao, X.Y.; Ding, J.; Wang, X.R. Lipopolysaccharide Induces Lung Fibroblast Proliferation through Toll-Like Receptor 4 Signaling and the Phosphoinositide3-Kinase-Akt Pathway. *PLoS ONE* **2012**, *7*, e35926. [CrossRef] [PubMed]
29. Zhang, L.; Huang, C.Q.; Guo, Y.J.; Gou, X.X.; Hinsdale, M.; Lloyd, P.; Liu, L. MicroRNA-26b Modulates the NF-kB Pathway in Alveolar Macrophages by Regulating PTEN. *J. Immunol.* **2015**, *195*, 5404–5414. [CrossRef] [PubMed]
30. Yang, Z.; Cao, X.M.; Xu, W.T.; Xie, C.; Chen, J.; Zhu, Y.; Lu, N.H. Phosphorylation of phosphatase and tensin homolog induced by Helicobacter pylori promotes cell invasion by activation of focal adhesion kinase. *Oncol. Lett.* **2018**, *15*, 1051–1057. [CrossRef] [PubMed]
31. Feng, X.J.; Liu, S.X.; Wu, C.; Kang, P.P.; Liu, Q.J.; Hao, J.; Li, F.; Zhang, Y.J.; Fu, X.H.; Zhang, S.B.; et al. The PTEN/PI3K/Akt signalling pathway mediates HMGB1-induced cell proliferation by regulating the NF-κB/cyclin D1 pathway in mouse mesangial cells. *Am. J. Physiol. Cell Physiol.* **2014**, *306*, C1119–C1128. [CrossRef] [PubMed]
32. Cantley, L.C. The phosphoinositide 3-kinase pathway. *Science* **2002**, *296*, 1655–1657. [CrossRef]
33. Prasad, S.B.; Yadav, S.S.; Das, M.; Modi, A.; Kumari, S.; Pandey, L.K.; Singh, S.; Pradhan, S.; Narayan, G. PI3K/AKT pathway-mediated regulation of p27Kip1is associated with cell cycle arrest and apoptosis in cervical cancer. *Cell Oncol.* **2015**, *38*, 215–225. [CrossRef] [PubMed]
34. Liang, J.; Slingerland, J.M. Multiple roles of the PI3K/PKB (Akt) pathway in cell cycle progression. *Cell Cycle* **2003**, *2*, 339–345. [CrossRef] [PubMed]
35. Li, J.; Yang, X.K.; Yu, X.X.; Ge, M.L.; Wang, W.L.; Zhang, J.; Hou, Y.D. Overexpression of p27(KIP1) induced cell cycle arrest in G1 phase and subsequent apoptosis in HCC-9204 cell line. *World J. Gastroenterol.* **2000**, *6*, 513–521. [CrossRef] [PubMed]
36. Kuo, M.Y.P.; Hsu, H.Y.; Kok, S.H.; Kuo, R.C.; Yang, H.; Hahn, L.J.; Chiang, C.P. Prognostic role of p27Kip1expression in oral squamous cell carcinoma in Taiwan. *Oral Oncol.* **2002**, *38*, 172–178. [CrossRef]
37. Mamillapalli, R.; Gavrilova, N.; Mihaylova, V.T.; Tsvetkov, L.M.; Wu, H.; Zhang, H.; Sun, H. PTEN regulates the ubiquitin-dependent degradation of the CDK inhibitor $p27^{KIP1}$ through the ubiquitin E3 ligase SCF^{SKP2}. *Curr. Biol.* **2001**, *11*, 263–267. [CrossRef]
38. Kovács, A.; Vasas, A.; Hohmann, J. Natural phenanthrenes and their biological activity. *Phytochemistry* **2008**, *69*, 1084–1110. [CrossRef] [PubMed]
39. He, X.; Fang, J.C.; Wang, X.X.; Zhao, Z.F.; Chang, Y.; Guo, H.; Zheng, X. *Bletilla striata*: Medicinal uses, phytochemistry and pharmacological activities. *J. Ethnopharmacol.* **2017**, *195*, 20–38. [CrossRef]
40. Qian, C.D.; Jiang, F.S.; Yu, H.S.; Shen, Y.; Fu, Y.H.; Cheng, D.Q.; Gan, L.S.; Ding, Z.S. Antibacterial Biphenanthrenes from the Fibrous Roots of *Bletilla striata*. *J. Nat. Prod.* **2015**, *78*, 939–943. [CrossRef] [PubMed]
41. Lin, Y.; Wang, F.; Yang, L.; Chun, Z.; Bao, J.; Zhang, G. Anti-inflammatory phenanthrene derivatives from stems of Dendrobium denneanum. *Phytochemistry* **2013**, *95*, 242–251. [CrossRef] [PubMed]
42. Ma, W.; Zhang, Y.; Ding, Y.Y.; Liu, F.; Li, N. Cytotoxic and anti-inflammatory activities of phenanthrenes from the medullae of *Juncus effusus* L. *Arch. Pharm. Res.* **2015**, *39*, 154–160. [CrossRef]
43. Jiang, F.S.; Shen, X.T.; Ding, B.; Li, M.Y.; Ding, Z.S.; Lv, G.Y. Comparison of the contents of three active ingredients in *Bletilla striata* from different sources. *China J. Chin. Mater. Med.* **2019**, *44*, 115–120.
44. Lawrence, T.; Natoli, G. Transcriptional regulation of macrophage polarization: Enabling diversity with identity. *Nat. Rev. Immunol.* **2011**, *11*, 750–761. [CrossRef] [PubMed]
45. Kanekar, Y.; Basha, K.; Duche, S.; Gupte, R.; Kapat, A. Regioselective synthesis of phenanthrenes and evaluation of their anti-oxidant based anti-inflammatory potential. *Eur. J. Med. Chem.* **2013**, *67*, 454–463. [CrossRef] [PubMed]
46. Luyendyk, J.P.; Schabbauer, G.A.; Tencati, M.; Holscher, T.; Pawlinski, R.; Mackman, N. Genetic Analysis of the Role of the PI3K-Akt Pathway in Lipopolysaccharide-Induced Cytokine and Tissue Factor Gene Expression in Monocytes/Macrophages. *J. Immunol.* **2008**, *180*, 4218–4226. [CrossRef] [PubMed]

47. Zhao, M.; Zhou, A.; Xu, L.; Zhang, X. The role of TLR4-mediated PTEN/PI3K/AKT/NF-κB signaling pathway in neuroinflammation in hippocampal neurons. *Neuroscience* **2014**, *269*, 93–101. [CrossRef]
48. Mayo, M.W.; Madrid, L.V.; Westerheide, S.D.; Jones, D.R.; Yuan, X.J.; Baldwin, A.S., Jr.; Whang, Y.E. PTEN Blocks Tumor Necrosis Factor-induced NF-κB-dependent Transcription by Inhibiting the Transactivation Potential of the p65 Subunit. *J. Biol. Chem.* **2002**, *277*, 11116–11125. [CrossRef] [PubMed]
49. Vadiveloo, P.; Keramidaris, E.; Morrison, W.; Stewart, A. Lipopolysaccharide-induced cell cycle arrest in macrophages occurs independently of nitric oxide synthase II induction. *BBA - Mol. Cell Res.* **2001**, *1539*, 140–146. [CrossRef]
50. Prieditis, H.; Adamson, I.Y.R. Alveolar macrophage kinetics and multinucleated giant cell formation after lung injury. *J. Leukocyte Biol.* **1996**, *59*, 534–538. [CrossRef]
51. Xu, M.; Shen, Y.; Zhang, K.; Liu, N.N.; Jiang, F.S.; Ding, Z.S. Antioxidant activity of total flavonoid aglycones and the main compound pinostrobin chalcone separated from leaves of Carya cathayensis. *Chin. J. ETMF* **2013**, *19*, 204–208.
52. Mitchell, S.J.; Martin-Montalvo, A.; Mercken, E.M.; Palacios, H.H.; Ward, T.M.; Abulwerdi, G.; Minor, R.K.; Vlasuk, G.P.; Ellis, J.L.; Sinclair, D.A.; et al. The SIRT1 activator SRT1720 extends lifespan and improves health of mice fed a standard diet. *Cell Rep.* **2014**, *6*, 836–843. [CrossRef]

6

Comparative Metabolomic Analysis Reveals Distinct Flavonoid Biosynthesis Regulation for Leaf Color Development of *Cymbidium sinense* 'Red Sun'

Jie Gao †, Rui Ren †, Yonglu Wei, Jianpeng Jin, Sagheer Ahmad, Chuqiao Lu, Jieqiu Wu, Chuanyuan Zheng, Fengxi Yang * and Genfa Zhu *

Guangdong Key Laboratory of Ornamental Plant Germplasm Innovation and Utilization, Environmental Horticulture Research Institute, Guangdong Academy of Agricultural Sciences, Guangzhou 510640, China; gaojie@gdaas.cn (J.G.); Renruinjau@163.com (R.R.); hjyylab@126.com (Y.W.); 13424050551@163.com (J.J.); sagheerhortii@gmail.com (S.A.); chuqiaolu18@163.com (C.L.); jiebaoqiqiqiu@163.com (J.W.); 15014323983@163.com (C.Z.)

* Correspondence: yangfengxi@gdaas.cn (F.Y.); zhugenfa@gdaas.cn (G.Z.)

† These authors contributed equally to this work.

Abstract: The colorful leaf is an important ornamental character of *Cymbidium sinense* (*C. sinense*), especially the red leaf, which has always been attracted by breeders and consumers. However, little is documented on the formation mechanism of the red leaf of *C. sinense*. In this study, the changing patterns of flavonoid-related metabolites, corresponding enzyme activities and genes expression in the leaves of *C. sinense* 'Red Sun' from red to yellow and finally to green was investigated. A total of 196 flavonoid-related metabolites including 11 anthocyanins metabolites were identified using UPLC-MS/MS-based approach. In the process of leaf color change, 42 metabolites were identified as having significantly different contents and the content of 28 differential metabolites turned to zero. In anthocyanin biosynthetic pathway, content of all 15 identified metabolites showed downregulation trend in the process of leaf color change. Among the 15 metabolites, the contents of Naringenin chalcone, Pelargonidin O-acetylhexoside and Anthocyanin 3-O-beta-D-glucoside decreased to zero in the green leaf stage. The changing pattern of enzyme activity of 10 enzymes involved in the anthocyanin biosynthetic pathway showed different trends from red leaves that have turned yellow and finally green, while the expression of genes encoding these enzymes was all down-regulated in the process of leaf color change. The results of this study revealed the types of flavonoid-related metabolites and the comprehensive analysis of metabolites content, enzyme activities and genes expression providing a new reference for breeders to improve the leaf color of *C. sinense* 'Red Sun'.

Keywords: metabolomic analysis; differential metabolites; enzyme activity; gene expression; leaf color; *Cymbidium sinense*

---

1. Introduction

As the largest family of monocotyledons, Orchidaceae has a long history of cultivation and contains abundant varieties. Among these varieties, *C. sinense* is a unique variety produced in China, which has been loved by consumers all around the world. The most important phenotypic feature of *C. sinense* is that it has a very rich variation in leaf color, which improves its ornamental and economic values. *C. sinense* contains a number of varieties based on the location and color of leaf variegation, including chimera art, claw art, crown art, crane art, middle penetration art, spot art and treasure art and so on [1]. At present, the leaf variegation varieties of *C. sinense* are basically yellow-green leaves, but there are few reports about red leaves. *C. sinense* 'Qihei' is the most common variety of green leaves.

In the process of culture, the color of 'Qihei' leaf buds changes to red, and then the leaf color changes from red to yellow, finally to green, leaving only a little yellow at the tip of the leaf during the growth and development of leaf buds. The leaf variegation variety is named *C. sinense* 'Red Sun'. So far, the metabolic changes associated with the formation of red leaves in *C. sinense* 'Red Sun' are not known.

Compared with flower color, the influencing factors of leaf color are more complex. Generally, the greening of leaves is mainly due to the absolute proportion of chlorophyll, and the formation of yellow leaves is mostly due to the degradation of chlorophyll, leaving the color of carotenoids to dominate in leaves [1,2]. Anthocyanins are responsible for the red color of leaves, which is helpful for plants against various biotic and abiotic stresses [3,4]. In *Cymbidium* orchids, the activation of anthocyanin synthesis can be restored by introducing MYB and bHLH anthocyanin regulators simultaneously [5]. The synthesis pathway of anthocyanins begins with the precursor phenylalanine, resulting in the formation of dihydrokaempferol (DHK) by the catalysis of six enzymes. After the catalysis of four enzymes, DHK produces three kinds of steady-state anthocyanins, including pelargonidin, cyanidin and delphinidin, respectively [6–8]. At present, more than 600 anthocyanins found in nature are derived from these three substances and the accumulation of anthocyanins in plants has two main functions: one is to produce rich and colorful visual signals to promote pollination or seed transmission, and the other is to resist a series of biotic or abiotic stresses [9,10].

Plant metabolomics carries out a qualitative and quantitative analysis of small molecular metabolites in plants, so as to help researchers understand the synthesis and accumulation patterns of metabolites [11]. At present, the study of plant metabolites mainly involves the identification of metabolites, variety differentiation and auxiliary breeding and so on [12–15]. Flavonoids-targeted metabolomics refers to the analysis of small molecular metabolites of flavonoids (anthocyanins), which is often used to analyze the formation mechanism of plant color. Analysis of Flavonoids-targeted metabolomics in the white and purple flowers of *Phalaenopsis* identified 142 different flavonoid-related metabolites, of which the most important anthocyanin was the derivative of cyaniding [16]. A study on green and purple asparagus suggested that the difference in color was mainly caused by the contents of peonidin and cyanidin and their glycoside derivatives [17]. Metabolomic analysis the color difference of fig peel showed that the content of anthocyanin derivatives in purple peel was significantly higher than that in green fruit [18]. These results indicate that metabolomics is an important and effective method to analyze the formation mechanism of plant color.

In order to understand the key metabolites involved in the process of color change from red to green in the leaves of *C. sinense* 'Red Sun', flavonoids-targeted metabolomic analysis was used to analyze the change pattern of flavonoid-metabolites in the process of leaf color change. Physiology, enzyme kinetics and molecular biology experiments were carried out to explore the mechanism of leaf color difference. To the best of our knowledge, this study is the first effort to analyze the mechanism of the formation of *C. sinense* leaf variegation from the perspective of metabolites, and the results provide a new train of thought and basis for the study of *C. sinense* leaf variegation.

2. Results

2.1. Pigments Content Analysis

The leaves of *C. sinense* 'Red Sun' show red at the leaf bud stage. With the development of the leaves, the leaves gradually change from red to yellow, and finally the mature leaves develop into green with a little yellow color at the tip, as seen in Figure 1. The color of leaves was affected by a variety of pigments, including the contents of chlorophyll, carotenoids, total flavonoids and total anthocyanins in *C. sinense* 'Red Sun'. The results showed that the maximum contents of the four pigments were found at the red leaf stage, while the minimum contents were shown by leaves at yellow stage. The contents of all four pigments in green leaves were higher than those in yellow leaves, but lower than those in red leaves (Figure 2).

Figure 1. Phenotypes of *Cymbidium sinense* 'Red Sun' red leaves (RL) (**a**), yellow leaves (YL) (**b**) and green leaves (GL) (**c**).

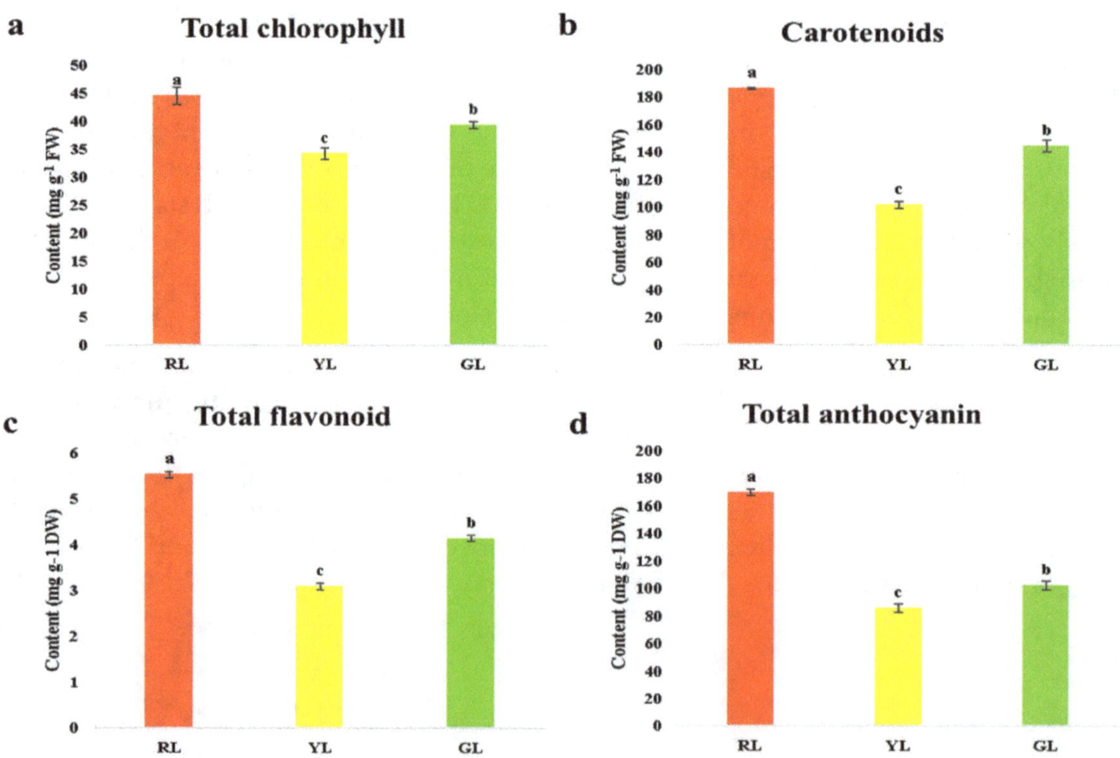

Figure 2. Chlorophyll (**a**), carotenoids content (**b**), total flavonoid (**c**) and total anthocyanin (**d**) content of different color leaves of *Cymbidium sinense* 'Red Sun'. Bars represent the mean of three biological replicates ±SE. Lowercase letters indicate significant differences at $p < 0.05$.

2.2. *Qualitative and Quantitative Analyses of Metabolites and Quality Control (QC) Analysis of Sample*

Anthocyanin is the cause of red leaf color. From the pigment contents, it can be seen that the contents of total flavonoids and total anthocyanins change significantly in the process of leaf color change. To compare the differences of flavonoid-metabolites, the UPLC (Shim-pack UFLC SHIMADZU CBM30A)–MS/MS (Applied Biosystems 4500 QTRAP) technique was used to detect the flavonoid-related metabolites in RL, YL and GL. The total ion flow map of the mixed sample quality control (QC) sample (Total ions current, TIC, is the map obtained by adding the intensities of all ions in each time point mass spectrum) is shown in Figure S1a. The multi-peak map of MRM metabolite detection of multi-substance extraction is illustrated in Figure S1b. Based on the local metabolic database, the metabolites of the samples were qualitatively and quantitatively analyzed by mass spectrometry. The multi-peak map of MRM metabolite detection in the multi-reaction monitoring mode shows the substances that can be detected in the sample, and each mass peak of different color represents a detected metabolite. A total of 196 flavonoid-related metabolites were detected (Table S1), including 11 anthocyanins, 3 Chalcone, 9 Dihydroflavonoid, 5 Dihydroflavonol, 12 flavanols, 11 Flavone, 59 Flavonoid, 20 Flavonoidcarbonoside, 3 Flavonoids, 59 Flavonols and 4 Isoflavones. In the process of instrumental analysis, one QC sample is inserted into every 10 test and analysis samples to monitor the repeatability of the analysis process. Through the overlap display analysis of the TIC map of different quality control QC samples (Figure S1c), the results showed that the curve overlap of metabolite detection of total ion current was high, that is, the retention time and peak intensity were the same, indicating that the signal stability was good when the same sample was detected by mass spectrometry at different time points. All the detected metabolite content data were normalized by range method, and the accumulation patterns of metabolites among different samples were analyzed by cluster analysis (Hierarchical cluster analysis, HCA) by R software (www.r-project.org/) (Figure 3).

2.3. Formatting of Mathematical Components Principal Component Analysis (PCA) and Orthogonal Projections to Latent Structures-Discrimination Analysis (OPLS-DA)

Principal Component Analysis (PCA) is a multidimensional data statistical analysis method of unsupervised pattern recognition. Through principal component analysis of samples (including QC samples), we can preliminarily understand the overall metabolic differences among samples and the degree of variability between samples within groups. From the analysis results, it can be observed that there are significant differences among RL, YL and GL groups, but there is no significant difference within groups (Figure S2).

Although PCA can effectively extract the main information, it is not sensitive to variables with small correlation, and Partial Least Squares-Discriminant Analysis (PLS-DA) can solve this problem. Compared with PCA, PLS-DA can maximize the distinction between groups and facilitate the search for differential metabolites. Through the analysis of PLS-DA, the orthogonal variables, which are not related to the classification variables of metabolites are first eliminated, and then the differences of correlation between groups and within groups are analyzed. According to the OPLS-DA model, we analyzed the metabolic group data, draw the score chart of each group, and further showed the differences between each group. The prediction parameters of the evaluation model are R2X, R2Y and Q2, in which R2X and R2Y represent the interpretation rate of the model to X and Y matrix respectively, and Q2 indicates the prediction ability of the model. The closer these three indexes are to 1, the more stable and reliable the model is. $Q2 > 0.5$ can be regarded as an effective model, and $Q2 > 0.9$ is an excellent model. From the results, there are significant differences among the three groups of data, but there is no significant difference between groups (Figure S3a). The alignment verification of OPLS-DA was carried out ($n = 200$, that is, 200 permutation experiments were carried out). In the model verification, the horizontal lines correspond to R2 and Q2 of the original model, and the red dots and blue dots represent R2′ and Q2′ of the model after Y replacement, respectively. The results showed that R2′ and Q2′ of each group were smaller than R2 and Q2 of the original model, which indicated that the model was meaningful and the differential metabolites could be screened according to VIP value analysis (Figure S3b).

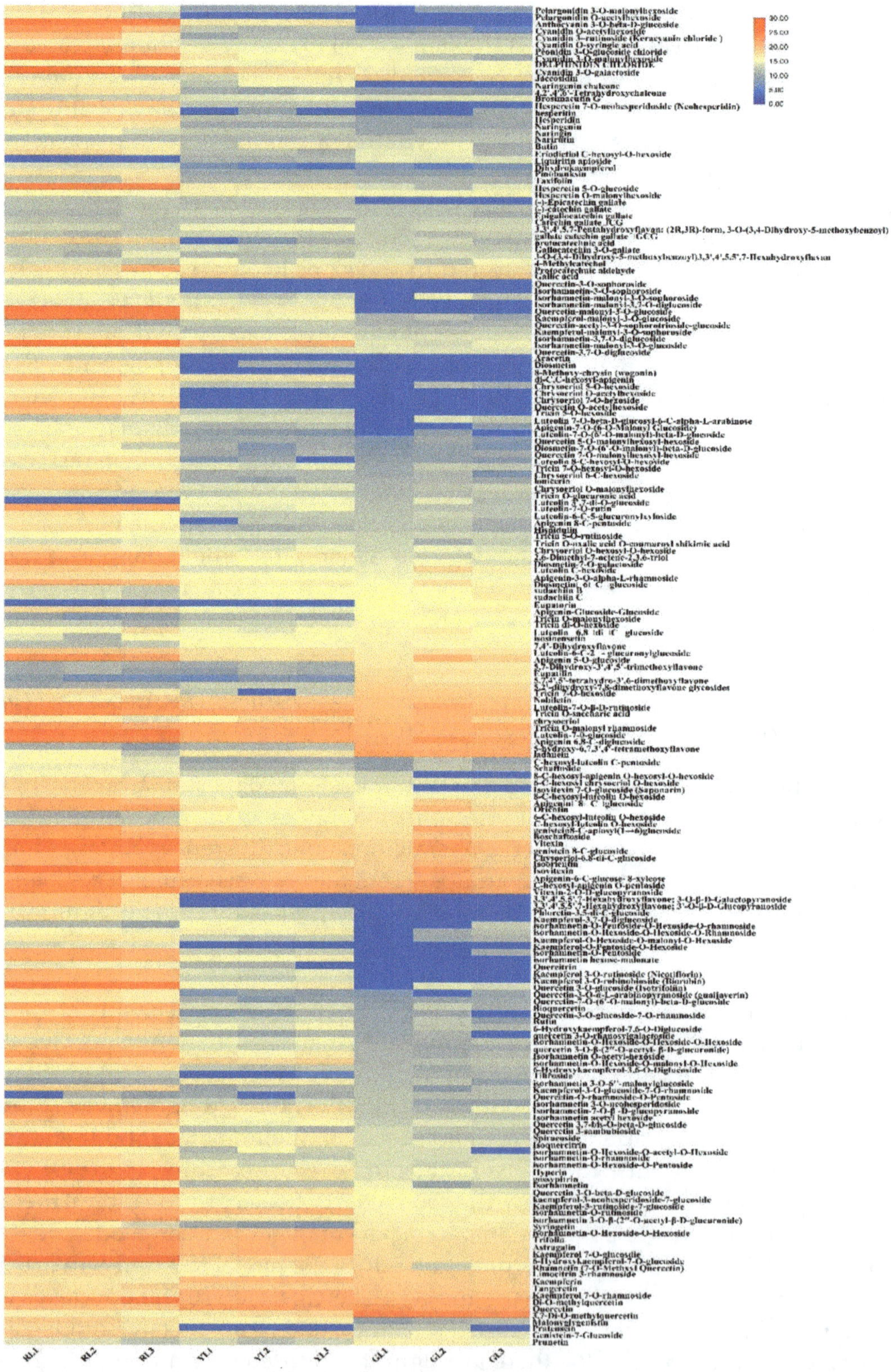

Figure 3. Hierarchical clustering analysis of all metabolites detected in this study. The abscissa indicates three biological replicates of red leaves (RL1, RL2, and RL3), yellow leaves (YL1, YL2, and YL3) and green leaves (GL1, GL2, and GL3), and the ordinate indicates the metabolites detected in this study. The red segments indicate a relatively high content of metabolites, while the blue segments indicate a relatively low content of metabolites. The relative metabolite contents represented by color segments at the corresponding locations are listed in Table S1.

2.4. *Screening Differential Metabolites in the Process of Leaf Color Change*

The differential metabolites were screened by combining the VIP values of fold change and OPLS-DA model. Following was the screening criteria: (1) If the difference of metabolites content between the control group and the experimental group is more than 2 times or less than 0.5, the difference is considered to be significant; (2) On the basis of the above, the metabolites with VIP ≥ 1 are selected. Volcano Plot was used to show the difference of expression level between the two groups of samples, and the difference was statistically significant (Figure 4). Compared with red leaves, 85 metabolites in green leaves were significantly different (10 up-regulated and 75 down-regulated), and 61 metabolites in green leaves were significantly different from yellow leaves (16 up-regulated and 45 down-regulated) (Tables S2 and S3). Compared with red leaves, 86 metabolites showed significant change in yellow leaves (3 up-regulated and 83 down-regulated) (Table S4). Through the analysis of Vennn diagram, it was found that there were 42 metabolites with different contents in the three periods of leaf color change. Except for the content of Liquiritinapioside, the other 41 substances decreased significantly with the change of leaf color (Figures S4 and S5). After a comprehensive analysis of the contents of all the differential metabolites, we found that the contents of 28 metabolites turned to zero in the process of leaf color change, as shown in Figure 5. Among these metabolites with drastic changes, the contents of Naringenin chalcone, an important intermediate in anthocyanin synthesis, and two important anthocyanin metabolites, Pelargonidin O-acetylhexoside and Anthocyanin 3-O-beta-D-glucoside, all decreased to zero in the green leaf stage.

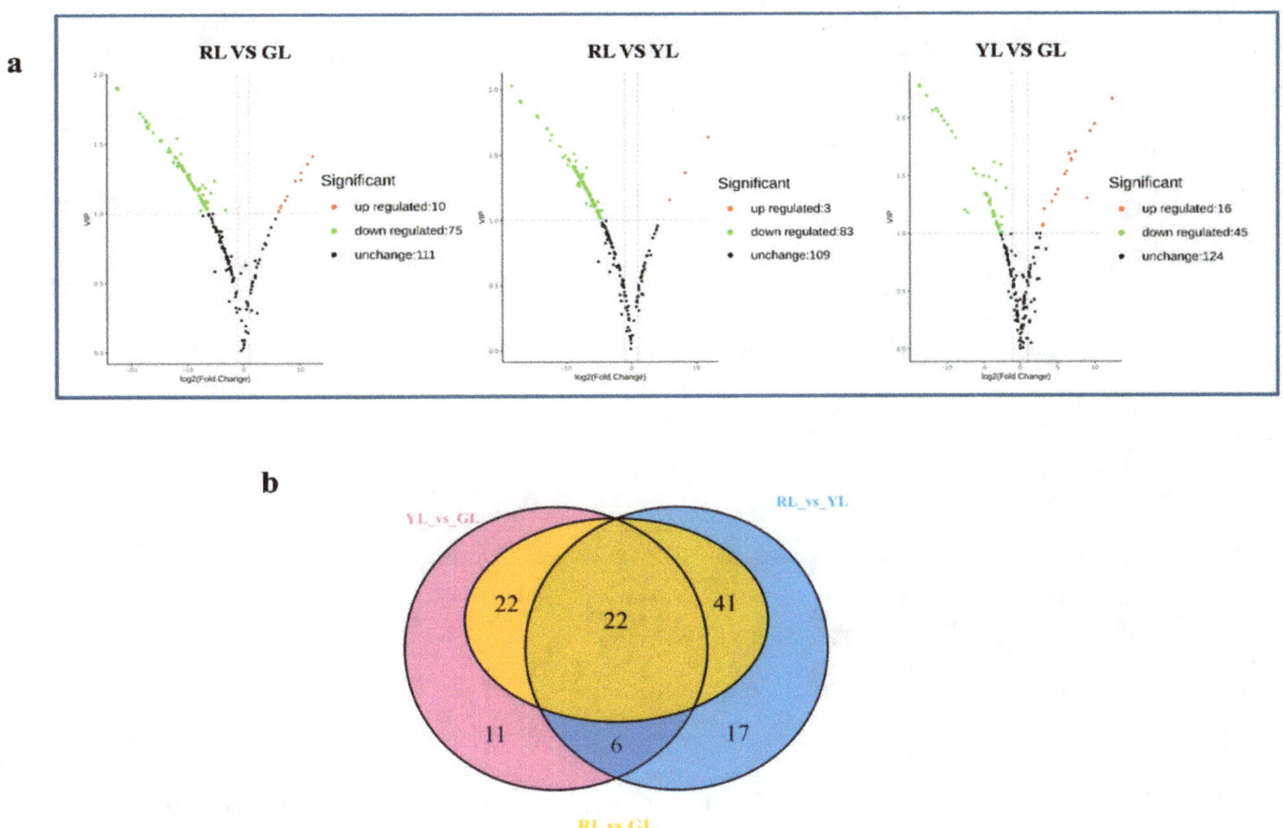

Figure 4. (**a**) Volcano plot of differential metabolites for RL vs. GL, RL vs. YL and YL vs. GL. The colors of the scatter points in Figure 4a indicate the final screening results: red indicates metabolites that were significantly up-regulated; green indicates metabolites that were significantly down-regulated; grey indicates metabolites with no significant difference. (**b**) Venn diagram analysis of differential metabolites. RL, red leaves; YL, yellow leaves; GL, green leaves.

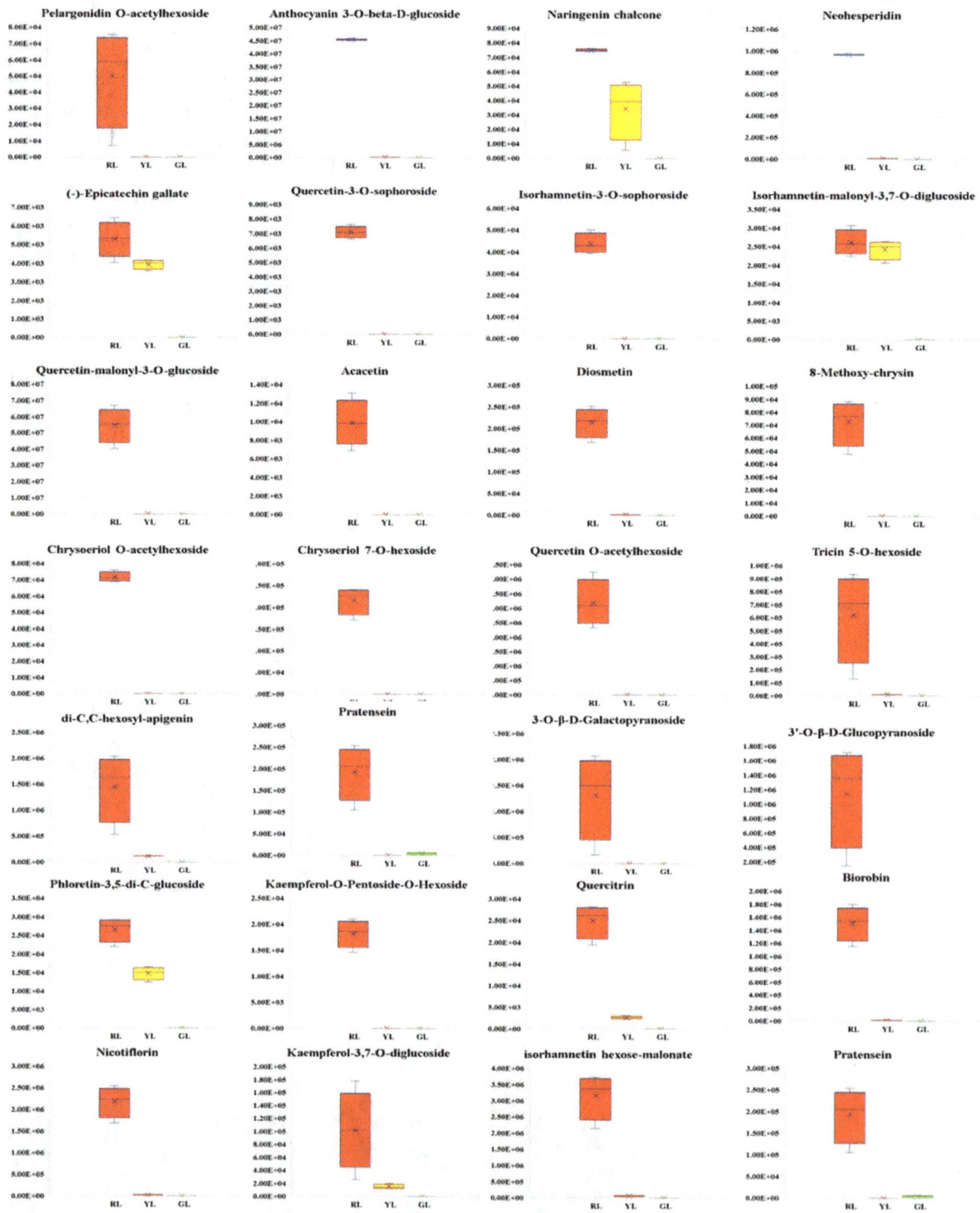

Figure 5. Differences in the content of 28 metabolites in the process of leaf color change. Y-scale represent the integral value of chromatographic peak area.

2.5. Intermediates Content, Enzymes Activities and Genes Expression Associated with the Anthocyanin Biosynthetic Pathway

In order to further analyze the regulation mechanism of anthocyanin in the process of *C. sinense* 'Red Sun' leaf color change, the detected intermediates of anthocyanin synthesis pathway and all the enzyme activities involved in anthocyanin synthesis were analyzed [19] (Figure 6). It can be observed that all 15 detected intermediates show a downward trend in the process of leaf color change. In total, 11 kinds of anthocyanins metabolites were detected, and the contents of eight of them (Cyanidin 3-rutinoside, Cyanidin 3-O-galactoside, Peonidin 3-O-glucoside chloride, Cyanidin 3-O-malonylhexoside, Cyanidin O-acetylhexoside, Anthocyanin 3-O-beta-D-glucoside, Pelargonidin 3-O-malonylhexoside and Pelargonidin O-acetylhexoside) changed significantly during the change of leaf color, especially the content of Pelargonidin O-acetylhexoside (molecular weight: 474.096 Da), a derivative of Pelargonidin, and Anthocyanin 3-O-beta-D-glucoside (molecular weight: 449.089 Da), a derivative of Cyanidin finally decreased to 0 in green leaves. After measuring the activity of all the enzymes involved in anthocyanin synthesis, it was found that the activity of henylalanine ammonia-lyase (PAL), trans-cinnamate 4-monooxygenase (C4H), 4-coumarate-CoA ligase (4CL) and flavonoid 3′,5′-hydroxylase (F3′5′H) was up-regulated, the activity of chalcone synthase (CHS), chalcone isomerase (CHI), naringenin 3-dioxygenase (F3H) and flavonoid 3′-monooxygenase (F3′H) was down-regulated. The activity of DFR was down-regulated at first and then up-regulated, while the activity of ANS was up-regulated and then decreased. QRT-PCR analysis of the genes encoding these enzymes showed that all genes were significantly down-regulated (Figure 7). Comprehensive analysis of enzyme activity and the corresponding gene expression pattern showed that the changing trend of enzyme activity of CHS, CHI, F3H, F3′H was consistent with that of gene expression pattern, while that of the other six enzymes was different from that of gene expression pattern.

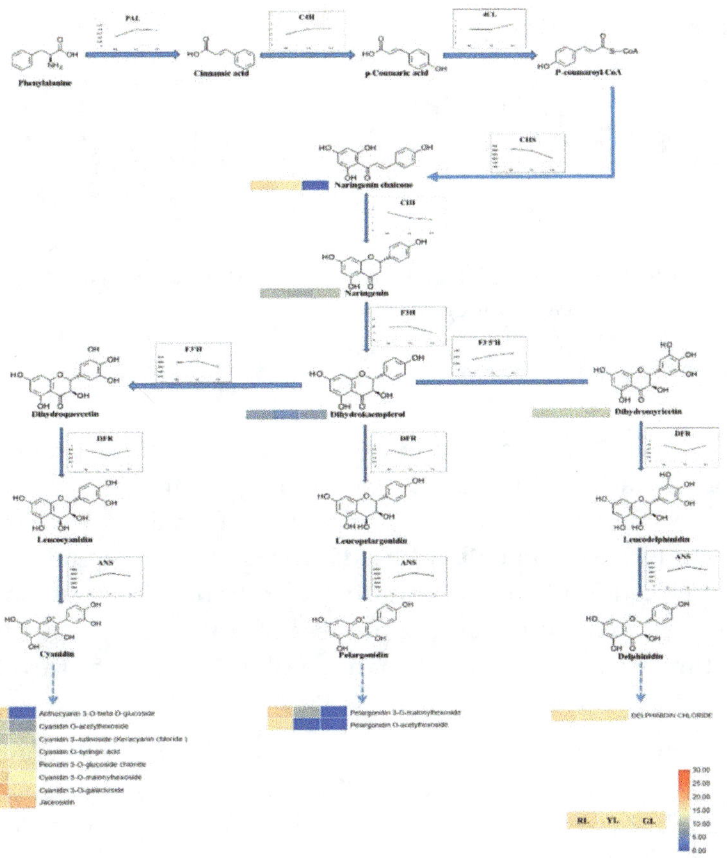

Figure 6. The changing patterns of enzymes activity and intermediate products contents related to anthocyanin synthesis in the process of *Cymbidium sinense* 'Red Sun' leaf color change. Red and blue shading in the lower right corner indicates the relatively high-or low content, respectively. PAL, Phenylalanine ammonia lyase, C4H, cinnamate 4-hydroxylase, 4CL, 4-coumarate CoA ligase, CHS, chalcone synthase, CHI, chalcone isomerase, F3H, flavone 3-hydroxylase, F3′H, flavonoid 3′-hydroxylase, F3′5′H, flavonoid 3′,5′-hydroxylase, DFR, dihydroflavonol reductase, ANS, anthocyanidin synthase. Units on y-scale of enzymes activity is U/g.

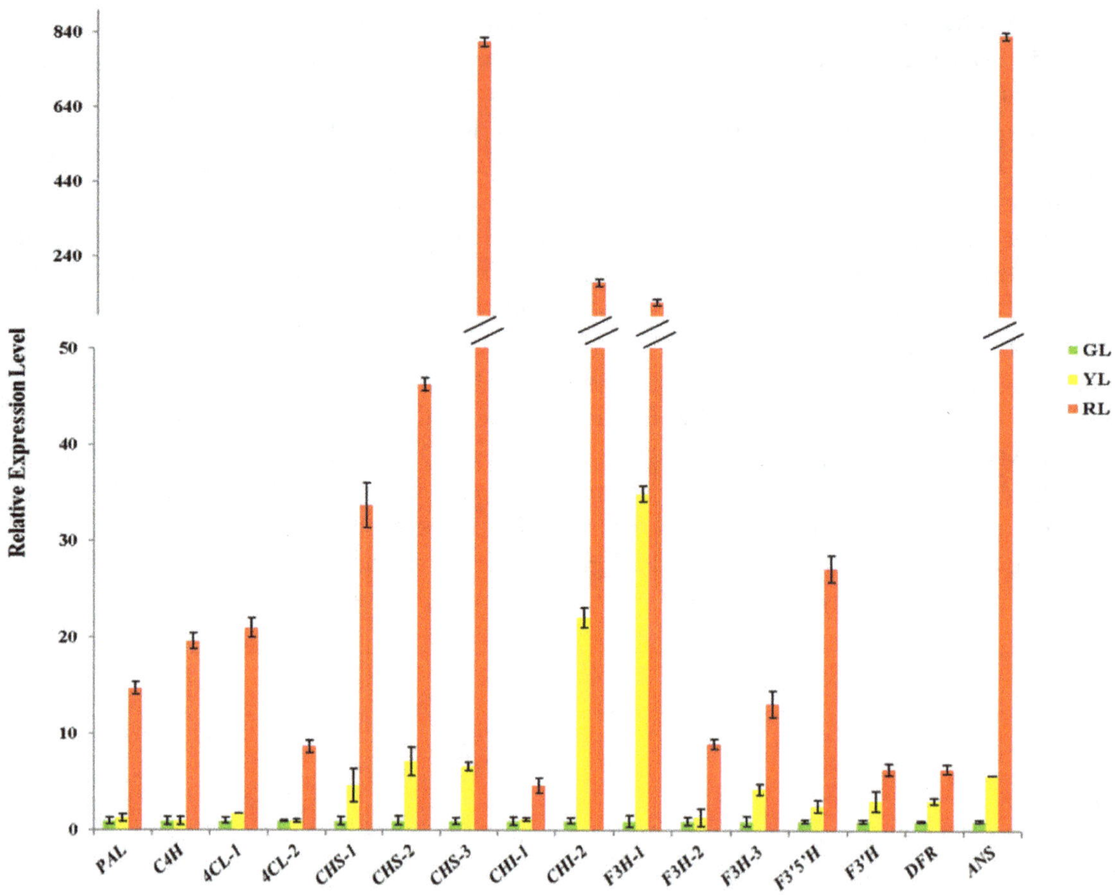

Figure 7. Expression pattern of genes coding enzymes related to anthocyanin synthesis in the process of *Cymbidium sinense* 'Red Sun' leaf color change.

3. Discussion

The regulation of plant leaf color is a complex process. Most of the previous studies analyzed the mechanism of leaf color formation by methods of physiology, cytology and molecular biology [1], but the mechanism of leaf color regulation from the perspective of small molecular metabolites needs to be studied further. In this study, based on UPLC-MS/MS, the changes of metabolites in the process of leaf color change of *C. sinense* 'Red Sun' were qualitatively and quantitatively analyzed, and the unique pattern of flavonoid-related metabolites in the process of leaf color change was constructed for the first time. At the same time, the regulation mechanism of leaf color was further analyzed by enzyme kinetics and gene expression analysis.

As far as we know, this is the first time to analyze the types of flavonoid-related metabolites in the leaves of *C. sinense* 'Red Sun' by UPLC-MS/MS (flavonoids-targeted) method. A total of 196 flavonoid-related metabolites were detected. These substances belong to anthocyanin, chalcone, dihydroflavonoid, dihydroflavonol, flavanols, flavone, flavonoid, flavonoid carbonoside, flavonoids, flavonols and isoflavones. Based on wide target metabolomics analysis, only 6 and 15 differential flavonoid-related metabolites were detected in tea leaves and *Ginkgo biloba* leaves, respectively [20,21]. Based on phenolic-targeted secondary metabolites analysis in purple fig peel, only 15 differential flavonoid-related metabolites (including four anthocyanins metabolites) were detected [18]. In this study, 119 kinds of differential flavonoid-related metabolites were found (including 10 kinds of anthocyanins). The above results show that Flavonoids-targeted metabolites method can identify more kinds of flavonoid-related metabolites, and has more advantages in mining the types and contents of flavonoid-related metabolites, especially anthocyanins metabolites.

Phalaenopsis and the materials of this study belong to Orchidaceae. In the metabolomic analysis between petals of white and purple *Phalaenopsis*, 142 differential flavonoid-related metabolites, including 17 anthocyanins metabolites, were detected by flavonoids-targeted metabolomic analysis [16]. In accordance with the results of this research, among the 119-differential flavonoid-related metabolites detected, there were 8 differential anthocyanins. The differential metabolites of eight anthocyanins identified in the leaves of *C. sinense* 'Red Sun' are Cyanidin 3-rutinoside (Keracyanin chloride), Cyanidin 3-O-galactoside, Peonidin 3-O-glucoside chloride, Cyanidin 3-O-malonylhexoside, Pelargonidin 3-O-malonylhexoside, Cyanidin O-acetylhexoside, Pelargonidin O-acetylhexoside and Anthocyanin 3-O-beta-D-glucoside. Compared with the 18 anthocyanin differential metabolites detected in the petals of *Phalaenopsis* [16], only Cyanidin 3-O-malonylhexoside and Cyanidin O-acetylhexoside are the same. Among the three main categories of pigmented glycosides, pelargonidin mainly shows orange/red, cyaniding mainly shows pink/magenta and delphinidin mainly shows purple/blue [22,23]. The metabolites with the highest proportion were cyanidin derivatives found in the petals of *Phalaenopsis* and leaves of *C. sinense* 'Red Sun'. Interestingly, during the process of *C. sinense* 'Red Sun' leaf color change, the two anthocyanins from existence to absence are Pelargonidin 3-O-malonylhexoside and Anthocyanin 3-O-beta-D-glucoside; the content of derivatives of six kinds of cyanidin and a kind of delphinidin was high in purple petals but zero in white petals of *Phalaenopsis* [16]. The above results are consistent with the phenotype of the corresponding materials.

As the upstream reaction of anthocyanin and other flavonoids, Phenylalanine was first converted to P-coumaroyl-CoA under the catalysis of PAL, C4H and 4CL [24]. The activities of PAL, C4H and 4CL were all up-regulated during the change of leaf color, while the coding genes expression of these three enzymes decreased significantly, which indicated that the activities of these three enzymes were also subject to post-transcriptional modification or post-translational modification [25]. Then P-coumaroyl-CoA was transformed into Naringenin under the action of CHS and CHI [24]. The activities of CHS and CHI were down-regulated in the process of leaf color change, which was consistent with the change pattern of gene expression. CHS is the first key enzyme in anthocyanin synthesis, its activity determines the formation of anthocyanin metabolic pathway, and the loss of its activity will lead to the loss of anthocyanin and other flavonoids [26,27]. From the results of qRT-PCR, the expression of CHS-3 in the red leaf stage was 813 times higher than that in the green leaf stage, indicating that CHS may play an important role in the process of leaf color change. Overall, similar correlations between gene expression and anthocyanin levels were also observed during the differential pigment deposition in crabapple cultivars with dark red, pink and white petal colors [28]. Next, Naringenin forms DHK under the catalysis of F3H. DHK then forms Dihydroquercetin (DHQ) and Dihydromyricetin (DHM) under the catalysis of F3′H and F3′5′H, respectively [29]. The enzyme activity of F3H and F3′H was down-regulated, while that of F3′5′H was up-regulated, and the coding gene expression of these three enzymes was significantly down-regulated. Both natural mutants and transgenic studies have proved that the competitions of three enzymes lead to different branching pathways at this critical point [30], and our results support this argument. DHK, DHQ and DHM formed unstable anthocyanins under the catalysis of DFR and ANS. The enzyme activity of DFR decreased at first and then increased, while the activity of ANS increased at first and then decreased, but the coding genes expression level of the two enzymes was significantly down-regulated. DFR from different plants has specific substrates biases for DHK, DHQ and DHM [6,31], and the downstream DFRs and ANSs is necessary for large sum of anthocyanin accumulation in *Phalaenopsis* [32,33]. These unstable anthocyanins eventually went through the action of UFGT to form stable anthocyanins [34]. During the period of color change, UFGT activity firstly decreased, and then increased, while the *UFGT* gene expression level was significantly down-regulated (Figure S5). Previous studies have found that overexpression of UFGT causes plants to show darker colors, such as crimson or purple, while overexpression of DFR or ANS only deepens the color to pink or lavender [34–36]. In this study, the results of qRT-PCR showed that the expression of ANS in red leaf stage was 833 times higher than that in green leaf stage, while the expression of UFGT in red leaf stage was 14 times higher than that in green leaf stage. The question of whether ANS or UFGT had a greater effect on the leaf color of *Cymbidium* remains to be further verified.

4. Materials and Methods

4.1. Plant Materials

The three-year-old *C. sinense* 'Red Sun' planted in the glass greenhouse located in Environmental Horticulture Research Institute of Guangdong Academy of Agricultural Sciences was used as the research material. According to the change pattern of leaf color, red leaf samples of three independent plants (RL1, RL2 and RL3) were taken on 15 January 2019, yellow leaf samples of three independent plants (YL1, YL2 and YL3) were taken on 25 February 2019 and green leaf samples of three independent plants (GL1, GL2 and GL3) were taken on 21 March 2019. The samples were quickly fixed with liquid nitrogen and stored at −80 °C.

4.2. Pigments Content Measurement

The contents of total chlorophyll and carotenoids were determined by spectrophotometric analyses. About 0.1 g leaves (accurately record the weight) were cut into pieces and rinsed with distilled water. Add 1 mL extracting solution (95% ethanol) and 50 mg calcium carbonate powder in mortar, the mixtures were grinded fully in low light conditions and transferred to 10 mL glass test tube. The mortar was rinsed with the extracting solution, all the washing solution was transferred into the glass tube and replenished to 10 mL with the extracting solution. Keep the glass tube in the dark until the tissue is completely whitened. Add 200 μL leach liquor in a 96-well plate, the automatic microplate reader (Sunrise, TECAN, Switzerland) was set to zero by extracting solution, then the absorbance values of 663 nm, 645 nm and 470 nm were determined and recorded as A663, A645 and A470, respectively. The calculation formula is as follows: total chlorophyll content (mg/g) = 0.02 × (20.21 × A6458.02 × A663) × N/M, total carotenoid concentration (mg/g) = 0.02 × [(1000A470–3.27Ca–104Cb)/229] × N/M (N represents dilution multiple, M represents sample fresh quantity).

The content of total anthocyanins was determined by colorimetry method. The sample was dried in the oven (37 °C) to a constant weight and crushed. After passing through a 40-mesh sieve, about 0.1 g sample was weighed (accurately recording the weight). Add 1 mL extracting solution (95% ethanol: 1.5 mol/L HCL = 85:15) in samples and the mixtures were extracted at 4 °C for 24 h. Then the mixtures were centrifuged at 8000× *g* for 10 min at room temperature, and the supernatant was taken to be tested. The automatic microplate reader (Sunrise, Switzerland) was preheated more than 30min, and reagent A (PH 1.0 buffer, 0.2mol/L KCL : 0.2mol HCL = 25:67) and reagent B (PH 4.5 buffer, 1 mol NaAC : 1 mol HCL: H_2O = 100:60:90) were preheated more than 10 min. Add 180 μL reagent A to 20 μL supernatant and let stand for 15 min, then the absorbance values at 530 nm and 700 nm were determined and were recorded as A1 and A2, respectively. Add 180 μL reagent B to 20 μL supernatant and let stand for 15 min, then the absorbance values at 530 nm and 700 nm were determined and were recorded as A3 and A4, respectively. The calculation formula is as follows: total anthocyanin concentration (μg/g) = 33.4 × [(A1 − A2) − (A3 − A4)] × N/M (N represents dilution multiple; M represents sample dry weight, g).

The content of total flavonoids was determined by colorimetry method. The sample was dried in the oven (37 °C) to a constant weight and crushed. After passing through a 40-mesh sieve, about 0.1 g sample was weighed (accurately recording the weight). Add 1 mL extracting solution (60% ethanol), the mixtures A were extracted by ultrasonic method (ultrasonic power: 300 W, crushing time: 5 s, interval time: 8 s) at 60 °C for 30 min. Then, the mixtures were 12,000 rpm for 10 min at room temperature, and the supernatant A was replenished to 1 mL by extracting solution and taken to be test. The automatic microplate reader (Sunrise, Switzerland) was preheated more than 30 min, set to 470 nm and zero by distilled water. Reagent A (5% nitrous acid), reagent B (10% aluminum nitrate), reagent C (4% sodium hydroxide solution) and standard (10 mg/mL tannic acid standard solution) were prepared. The supernatant was treated according to the instruction manual of the kit for the determination of total flavonoids in plants (Figure S6). The mixtures B was mixed well and placed in a water bath at 37 °C for 45 min. Then the mixtures B was centrifuged at 10,000× *g* for 10 min at room temperature, and 200 uL supernatant B was used to determine absorbance values at A470 in a 96-well plate. The calculation formula is as follows: the flavonoids concentration (mg/g) = C × N × V/M (C represents the content of flavonoids observed in the standard curve, mg/mL; N represents dilution multiple; V represents the total volume of extracting solution, ml; M represents the dry weight of sample, g).

Flavonoids-targeted metabolomics analysis was performed by liquid chromatography-mass spectrometry (LC-MS) at Metware Biotechnology Co.,Ltd (Wuhan, China) as described by [37], with small modifications. Metabolomics analysis includes two parts: metabolomics experiment and data analysis, based on the metabolite data obtained from experimental design, sample collection and processing, metabolite extraction and metabolite detection and analysis. It can carry out the identification of metabolites and the quality control analysis of sample data, and screen out some differential metabolites, so as to predict and analyze the related functions of the metabolites of the samples.

4.3. Analysis of Flavonoids-Targeted Metabolomics

4.3.1. Standards and Reagents

The standard (analytical purity) was purchased in BioBioPha (Kunming, China) or Sigma-Aldrich (St Louis, MO, USA), and dissolved in dimethyl sulfoxide (DMSO) or methanol as solvent and stored at −20 °C. Before mass spectrometry analysis, 70% methanol was diluted to different gradient concentrations. Methanol, acetonitrile and ethanol were all analytically pure and purchased in Merck (Darmstadt, Germany).

4.3.2. Sample Extraction Process

The leaves of *C. sinense* 'Red Sun' were vacuum freeze-dried and ground with a grinder (30 Hz for 1.5 min) to powder. The 0.1 g powder was dissolved in 70% methanol aqueous solution. The dissolved sample was swirled three times at 4 °C overnight to improve the extraction rate. After centrifugation (rotating speed 10,000× *g*, 10min), the supernatant was filtered with a microporous membrane (0.22 μ m pore size). The sample was stored in a sample injection bottle for UPLC-MS/MS analysis.

4.3.3. Acquisition Conditions of LC-MS

The data acquisition instrument system mainly includes Ultra Performance Liquid Chromatography (UPLC) (Shim-pack UFLC SHIMADZU CBM30A, http://www.shimadzu.com.cn/) and Tandem mass spectrometry (MS/MS) (Applied Biosystems 4500 QTRAP, http://www.appliedbiosystems.com.cn/).

UPLC was run under following conditions: (1) Waters ACQUITY UPLC HSS T3 C18 1.8 μm, 2.1 mm × 100 mm; (2) The aqueous phase was ultra-pure water (0.04% acetic acid is added) and the organic phase was acetonitrile (adding 0.04% acetic acid); (3) Elution gradients: 95:5 v/v at 0 min, 5:95 v/v at 11.0 min, 12 min, 12.1 min and 15 min; (4) The flow rate was 0.4 mL/min, the column temperature was 40 °C, and the injection volume was 5 μL. MS/MS worked with these conditions: (1) Curtain gas (CUR) was set to 25 psi; (2) Electrospray ionization (ESI) was set to 550 °C; (3) the MS voltage was set to 5500V; (4) Dclusteringotential (DP) was optimized; (5) Collision energy (CE) was optimized; (6) Collision-activated dissociation (CAD) was set to high.

4.3.4. Data Evaluation

Based on the self-built database MWDB (metware database) and the public database of metabolite information, the primary and secondary spectral data of mass spectrometry were qualitatively analyzed by software Analyst 1.6.3. For the qualitative analysis of some substances, the interference from isotope signals are removed, including duplicate signals of K^+, Na^+ and $NH_4{}^+$, as well as duplicate signals of fragment ions which were derived from other large molecules. The structure analysis of metabolites refers to the existing mass spectrometry public databases such as MassBank (http://www.massbank.jp/), KNAPSAcK (http://kanaya.naist.jp/KNApSAcK/), HMDB (http://www.hmdb.ca/) (Wishartetal.2013), MoToDB (http://www.ab.wur.nl/moto/) and METLIN (http://metlin.scripps.edu/index.php) [38].

The quantification of metabolites was carried out by using multiple reaction monitoring (MRM) mode. In the MRM model, the quadrupole first selected the precursor ions (parent ions) of the target substance. While screening the corresponding ions of other molecular weight substances to initially eliminate the interference, the precursor ions were ionized by the collision chamber to form a lot of fragment ions. Then, the fragment ions were filtered through the triple quadrupole to select a characteristic fragment ion, which eliminated the interference of non-target ions, making the quantitative inference more accurate and the better repeatability. After the metabolic substance spectrum analysis data of different samples were obtained, the mass spectrometry peaks of all substances were integrated, and the mass spectrometry peaks of the same metabolite in different samples were integrated and corrected [39].

4.3.5. Data Analysis

The original data obtained were preprocessed at first (noise filtering, peak matching and peak extraction) and the data were corrected [40]. Then, the data of quality control entered the stage of statistical analysis.

Statistical analysis used multivariate analysis. Data were log-transformed and mean-centred using SIMCA software (V14.1, MKS Data Analytics Solutions, Umea, Sweden) for PCA and OPLS-DA analysis. PCA analysis was followed by automated modelling analysis [41,42]. The first principal component (PC1) was subjected to OPLS-DA modelling, and the model quality was tested by 7-fold cross validation. After that, the resulting R2Y (the interpretability of the model on the categorical variable Y) and Q2 (the predictability of the model) were used to evaluate the validity of the model. The permutation test was performed multiple times to generate different random Q2 values, which were used to further test model validity [43].

R software (version 3.0.3) was used for Hierarchical clustering analysis (HCA) analysis. The data were log 2 transformed and similarity assessment for clustering was based on the Euclidean distance coefficient.

Based on OPLS-DA analysis, the differential metabolites were screened by the following criteria: (1) If the difference of metabolites content between the control group and the experimental group is more than 2 times or less than 0.5, the difference is considered to be significant; (2) On the basis of the above, the metabolites with VIP ≥ 1 are selected.

Finally, the Kyoto Encyclopaedia of Genes and Genomes (KEGG) Pathway database (http: //www.kegg.jp/kegg/pathway.html) [44] and MW-database (Metware Biotechnology Co., Ltd, Wuhan, China) were centred on metabolic reactions and concatenates possible metabolic pathways.

4.4. Enzyme Activity Determination

The enzyme activities of 10 enzymes were determined by enzyme-linked immunosorbent assay (ELISA). Leaves were ground to powder in liquid nitrogen and accurately weighed 0.1 g sample in the centrifuge tube. Then, added the same volume of 0.1 mol/L precooled PBS solution, centrifuged about 3000 rpm/min at low-temperature and low-speed for 10 min. 1 mL of the supernatant was collectedn to test. The ELISA detection kit manual of the corresponding enzyme (ProNets Biotechnology Co., Ltd., Wuhan, China) was used to determine the enzyme.

4.5. Total RNA Extraction and qRT-PCR Analysis

The total RNA of *C. sinense* 'Red Sun' was extracted by RNA extraction kit (TIANGEN, Biotech, Beijing, China). The content of RNA in three samples was determined by nanodrop2000 (Thermo

Fisher, Waltham, MA, USA). 500 ng RNA was taken from each sample for reverse transcription by HiScript Q RT SuperMix for qPCR kit (Vazyme Biotech Co., Nanjing, China) to obtain cDNA. cDNA dilution of 10-folds was used for qRT-PCRanalysis.

QRT-PCR used CFX96TM Real-Time System (Bio-Rad, Hercules, CA, USA) following the instructions based on ChamQ SYBR qPCR Master Mix kit (Vazyme Biotech Co., Nanjing, China). CsACTIN was used as normalization standard for gene expression. The gene expression was calculated by $2^{-\Delta\Delta CT}$. The primers for qRT-PCR are listed in Table S5.

5. Conclusions

In this study, a LC-MS-based metabolomics approach was used to evaluate the difference in metabolites during the change of leaf color of *C. sinense* 'Red Sun'. This is the first metabolomics study on *C. sinense*. A total of 196 flavonoid-related metabolites were detected. 42 metabolites were identified as differential metabolites during the process of leaf color change. In anthocyanin biosynthetic pathway, 15 metabolites were identified and the contents of them all showed decrease. Especially the contents of Naringenin chalcone, an important intermediate in anthocyanin synthesis, and two important anthocyanins metabolites, Pelargonidin O-acetylhexoside and Anthocyanin 3-O-beta-D-glucoside, decreased to zero in the green leaf stage. The enzyme activity of 10 enzymes related to anthocyanin synthesis showed different change patterns, while the expression of corresponding encoding genes was all down-regulated in the process of leaf color change. Overall, this study substantially contributes to the knowledge flavonoid-related metabolites composition in *C. sinense* and provides important reference values for breeders to improve the leaf color of *C. sinense*.

Author Contributions: Conceptualization, F.Y. and G.Z.; methodology, J.G. and R.R.; software, Y.W.; validation, J.J. and J.W.; formal analysis, J.G.; resources, C.L. and C.Z.; writing—original draft preparation, J.G.; writing—review and editing, S.A. and R.R. All authors have read and agreed to the published version of the manuscript.

Abbreviations

PAL	Phenylalanine ammonialyase
C4H	Cinnamate 4-hydroxylase
4CL	4-coumarateCoA ligase
CHS	Chalcone synthase
CHI	Chalcone isomerase
F3H	Flavonoid 3-hydroxylase
F3′H	Flavonoid 3′-hydroxylase
F3′5′H	Flavonoid 3′,5′-hydroxylase
DFR	Dihydroflavonolreductase
ANS	Anthocyanidin synthase
UFGT	anthocyanidin 3-O-glucosyltransferase
DHK	Dihydrokaempferol
DHQ	Dihydroquercetin
DHM	Dihydromyricetin
qRT-PCR	Quantitative real-time PCR

References

1. Zhu, G.; Yang, F.; Shi, S.; Li, D.; Wang, Z.; Liu, H.; Huang, D.; Wang, C. Transcriptome characterization of *Cymbidium sinense 'Dharma'* using 454 pyrosequencing and its application in the identification of genes associated with leaf color variation. *PLoS ONE* **2015**, *10*, e128592. [CrossRef] [PubMed]
2. Tsai, C.C.; Wu, Y.J.; Sheue, C.R.; Liao, P.C.; Chen, Y.H.; Li, S.J.; Liu, J.W.; Chang, H.T.; Liu, W.L.; Ko, Y.Z.; et al. Molecular Basis Underlying Leaf Variegation of a Moth Orchid Mutant (*Phalaenopsis aphrodite* subsp. *Formosana*). *Front. Plant Sci.* **2017**, *8*, 1333. [CrossRef] [PubMed]
3. Gould, K.S. Nature's swiss army knife: The diverse protective roles of anthocyanins in leaves. *J. Biomed. Biotechnol.* **2004**, *2004*, 314–320. [CrossRef] [PubMed]

4. Lee, D.W.; Gould, K.S. Why Leaves Turn Red: Pigments called anthocyanins probably protect leaves from light damage by direct shielding and by scavenging free radicals. *Am. Sci.* **2002**, *90*, 524–531. [CrossRef]
5. Albert, N.W.; Arathoon, S.; Collette, V.E.; Schwinn, K.E.; Jameson, P.E.; Lewis, D.H.; Zhang, H.; Davies, K.M. Activation of anthocyanin synthesis in *Cymbidium* orchids: Variability between known regulators. *Plant Cell Tissue Organ Cult. (PCTOC)* **2010**, *100*, 355–360. [CrossRef]
6. Saito, K.; Yonekura-Sakakibara, K.; Nakabayashi, R.; Higashi, Y.; Yamazaki, M.; Tohge, T.; Fernie, A.R. The flavonoid biosynthetic pathway in *Arabidopsis*: Structural and genetic diversity. *Plant Physiol. Biochem.* **2013**, *72*, 21–34. [CrossRef]
7. Dooner, H.K.; Robbins, T.P.; Jorgensen, R.A. Genetic and developmental control of anthocyanin biosynthesis. *Annu. Rev. Genet.* **1991**, *25*, 173–199. [CrossRef]
8. Pelletier, M.K.; Murrell, J.R.; Shirley, B.W. Characterization of flavonol synthase and leucoanthocyanidin dioxygenase genes in *Arabidopsis*. Further evidence for differential regulation of "early" and "late" genes. *Plant Physiol.* **1997**, *113*, 1437–1445. [CrossRef]
9. Kong, J.; Chia, L.; Goh, N.; Chia, T.; Brouillard, R. Analysis and biological activities of anthocyanins. *Phytochemistry* **2003**, *64*, 923–933. [CrossRef]
10. Castañeda-Ovando, A.; Pacheco-Hernández, M.D.L.; Páez-Hernández, M.E.; Rodríguez, J.A.; Galán-Vidal, C.A. Chemical studies of anthocyanins: A review. *Food Chem.* **2009**, *113*, 859–871. [CrossRef]
11. Fiehn, O. Metabolomics-the link between genotypes and phenotypes. *Plant Mol. Biol.* **2002**, *48*, 155–171. [CrossRef] [PubMed]
12. Mais, E.; Alolga, R.N.; Wang, S.; Linus, L.O.; Yin, X.; Qi, L. A comparative UPLC-Q/TOF-MS-based metabolomics approach for distinguishing Zingiber officinale Roscoe of two geographical origins. *Food Chem.* **2018**, *240*, 239–244. [CrossRef] [PubMed]
13. Rizzato, G.; Scalabrin, E.; Radaelli, M.; Capodaglio, G.; Piccolo, O. A new exploration of licorice metabolome. *Food Chem.* **2017**, *221*, 959–968. [CrossRef] [PubMed]
14. Zhou, F.; Peng, J.; Zhao, Y.; Huang, W.; Jiang, Y.; Li, M.; Wu, X.; Lu, B. Varietal classification and antioxidant activity prediction of Osmanthus fragrans Lour. Flowers using UPLC-PDA/QTOF-MS and multivariable analysis. *Food Chem.* **2017**, *217*, 490–497. [CrossRef]
15. Wang, S.; Tu, H.; Wan, J.; Chen, W.; Liu, X.; Luo, J.; Xu, J.; Zhang, H. Spatio-temporal distribution and natural variation of metabolites in citrus fruits. *Food Chem.* **2016**, *199*, 8–17. [CrossRef]
16. Meng, X.; Li, G.; Gu, L.; Sun, Y.; Li, Z.; Liu, J.; Wu, X.; Dong, T.; Zhu, M. Comparative metabolomic and transcriptome analysis reveal distinct flavonoid biosynthesis regulation between petals of white and purple *phalaenopsis amabilis*. *J. Plant Growth Regul.* **2019**. [CrossRef]
17. Dong, T.; Han, R.; Yu, J.; Zhu, M.; Zhang, Y.; Gong, Y.; Li, Z. Anthocyanins accumulation and molecular analysis of correlated genes by metabolome and transcriptome in green and purple asparaguses (*Asparagus officinalis*, L.). *Food Chem.* **2019**, *271*, 18–28. [CrossRef]
18. Wang, Z.; Cui, Y.; Vainstein, A.; Chen, S.; Ma, H. Regulation of Fig (*Ficus carica* L.) Fruit Color: Metabolomic and Transcriptomic Analyses of the Flavonoid Biosynthetic Pathway. *Front. Plant Sci.* **2017**, *8*, 1990. [CrossRef]
19. Holton, T.A.; Cornish, E.C. Genetics and biochemistry of anthocyanin biosynthesis. *Plant Cell* **1995**, *7*, 1071–1083. [CrossRef]
20. Shen, J.; Zou, Z.; Zhang, X.; Zhou, L.; Wang, Y.; Fang, W.; Zhu, X. Metabolic analyses reveal different mechanisms of leaf color change in two purple-leaf tea plant (*Camellia sinensis* L.) cultivars. *Hortic. Res.* **2018**, *5*, 7–14. [CrossRef]
21. Li, W.; Yang, S.; Lu, Z.; He, Z.; Ye, Y.; Zhao, B.; Wang, L.; Jin, B. Cytological, physiological, and transcriptomic analyses of golden leaf coloration in *Ginkgo biloba* L. *Hortic. Res.* **2018**, *5*. [CrossRef] [PubMed]
22. Ohmiya, A. Biosynthesis of plant pigments: Anthocyanins, betalains and carotenoids. *Plant. J.* **2010**, *54*, 733–749. [CrossRef]
23. Grotewold, E. The genetics and biochemistry of floral pigments. *Annu. Rev. Plant Biol.* **2006**, *57*, 761–780. [CrossRef] [PubMed]
24. Falcone Ferreyra, M.L.; Rius, S.; Casati, P. Flavonoids: Biosynthesis, biological functions, and biotechnological applications. *Front. Plant Sci.* **2012**, *3*, 222. [CrossRef] [PubMed]
25. Zhang, X.; Liu, C. Multifaceted regulations of gateway enzyme phenylalanine Ammonia-Lyase in the biosynthesis of phenylpropanoids. *Mol. Plant* **2015**, *8*, 17–27. [CrossRef]

26. Spribille, R.; Forkmann, G. Genetic control of chalcone synthase activity in flowers of *Antirrhinum majus*. *Phytochemistry* **1982**, *21*, 2231–2234. [CrossRef]
27. Hoshino, A.; Park, K.I.; Iida, S. Identification of r mutations conferring white flowers in the Japanese morning glory (*Ipomoea nil*). *J. Plant Res.* **2009**, *122*, 215–222. [CrossRef]
28. Tai, D.; Tian, J.; Zhang, J.; Song, T.; Yao, Y. A malus crabapple chalcone synthase gene, *McCHS*, regulates red petal color and flavonoid biosynthesis. *PLoS ONE* **2014**, *9*. [CrossRef]
29. Tanaka, Y.; Brugliera, F.; Kalc, G.; Senior, M.; Chandler, S. Flower color modification by engineering of the flavonoid biosynthetic pathway: Practical perspectives. *Biosci. Biotechnol. Biochem.* **2010**, *74*, 1760–1769. [CrossRef]
30. Nakatsuka, T.; Nishihara, M.; Mishiba, K.; Hirano, H.; Yamamura, S. Two different transposable elements inserted in flavonoid 3′,5′-hydroxylase gene contribute to pink flower coloration in *Gentiana scabra*. *Mol. Genet. Genom.* **2006**, *275*, 231–241. [CrossRef]
31. Hua, C.; Linling, L.; Shuiyuan, C.; Fuliang, C.; Feng, X.; Honghui, Y.; Conghua, W. Molecular cloning and characterization of three genes encoding dihydroflavonol-4-reductase from *Ginkgo biloba* in anthocyanin biosynthetic pathway. *PLoS ONE* **2013**, *8*, e72017. [CrossRef] [PubMed]
32. Yang, Y.; Wang, J.; Ma, Z.; Sun, G.; Zhang, C. De novo sequencing and comparative transcriptome analysis of white petals and red labella in *Phalaenopsis* for discovery of genes related to flower color and floral differentation. *Acta Soc. Bot. Pol.* **2014**, *83*, 191–199. [CrossRef]
33. Harborne, J.B.; Williams, C.A. Anthocyanins and other flavonoids. *Nat. Prod. Rep.* **2001**, *18*, 310–333. [CrossRef]
34. Zhao, Z.C.; Hu, G.B.; Hu, F.C.; Wang, H.C.; Yang, Z.Y.; Lai, B. The UDP glucose: Flavonoid-3-O-glucosyltransferase (UFGT) gene regulates anthocyanin biosynthesis in litchi (*Litchi chinesis* Sonn.) during fruit coloration. *Mol. Biol. Rep.* **2012**, *39*, 6409–6415. [CrossRef] [PubMed]
35. Hassani, D.; Liu, H.L.; Chen, Y.N.; Wan, Z.B.; Zhuge, Q.; Li, S.X. Analysis of biochemical compounds and differentially expressed genes of the anthocyanin biosynthetic pathway in variegated peach flowers. *Genet. Mol. Res.* **2015**, *14*, 13425–13436. [CrossRef] [PubMed]
36. Hu, C.; Gong, Y.; Jin, S.; Zhu, Q. Molecular analysis of a UDP-glucose: Flavonoid 3-O-glucosyltransferase (UFGT) gene from purple potato (*Solanum tuberosum*). *Mol. Biol. Rep.* **2011**, *38*, 561–567. [CrossRef] [PubMed]
37. Chen, W.; Gong, L.; Guo, Z.; Wang, W.; Zhang, H.; Liu, X.; Yu, S.; Xiong, L.; Luo, J. A novel integrated method for Large-Scale detection, identification, and quantification of widely targeted metabolites: Application in the study of rice metabolomics. *Mol. Plant* **2013**, *6*, 1769–1780. [CrossRef]
38. Zhu, Z.; Schultz, A.W.; Wang, J.; Johnson, C.H.; Yannone, S.M.; Patti, G.J.; Siuzdak, G. Liquid chromatography quadrupole time-of-flight mass spectrometry characterization of metabolites guided by the METLIN database. *Nat. Protoc.* **2013**, *8*, 451–460. [CrossRef]
39. Fraga, C.G.; Clowers, B.H.; Moore, R.J.; Zink, E.M. Signature-discovery approach for sample matching of a nerve-agent precursor using liquid chromatography-mass spectrometry, XCMS, and chemometrics. *Anal. Chem.* **2010**, *82*, 4165–4173. [CrossRef]
40. Dunn, W.B.; Broadhurst, D.; Begley, P.; Zelena, E.; Francis-McIntyre, S.; Anderson, N.; Brown, M.; Knowles, J.D.; Halsall, A.; Haselden, J.N.; et al. Procedures for large-scale metabolic profiling of serum and plasma using gas chromatography and liquid chromatography coupled to mass spectrometry. *Nat. Protoc.* **2011**, *6*, 1060–1083. [CrossRef]
41. Bro, R.; Smilde, A.K. Principal component analysis. *Anal. Methods* **2014**, *6*, 2812–2831. [CrossRef]
42. Wiklund, S.; Johansson, E.; Sjöström, L.; Mellerowicz, E.J.; Edlund, U.; Shockcor, J.P.; Gottfries, J.; Moritz, T.; Trygg, J. Visualization of GC/TOF-MS-Based metabolomics data for identification of biochemically interesting compounds using OPLS class models. *Anal. Chem.* **2008**, *80*, 115–122. [CrossRef] [PubMed]
43. Trygg, J.; Wold, S. Orthogonal projections to latent structures (O-PLS). *J Chemom.* **2002**, *16*, 119–128. [CrossRef]
44. Kanehisa, M.; Sato, Y.; Kawashima, M.; Furumichi, M.; Tanabe, M. KEGG as a reference resource for gene and protein annotation. *Nucleic Acids Res.* **2015**, *44*, D457–D462. [CrossRef]

7

Comparative Transcriptomics Provides Insight into Floral Color Polymorphism in a *Pleione limprichtii* Orchid Population

Yiyi Zhang [1], Tinghong Zhou [2], Zhongwu Dai [1], Xiaoyu Dai [1], Wei Li [1], Mengxia Cao [1], Chengru Li [1], Wen-Chieh Tsai [1,3], Xiaoqian Wu [1], Junwen Zhai [1], Zhongjian Liu [1,*] and Shasha Wu [1,*]

1 Key Laboratory of National Forestry and Grassland Administration for Orchid Conservation and Utilization, College of Landscape Architecture, Fujian Agriculture and Forestry University, Fuzhou 350002, China; zyy211802@126.com (Y.Z.); daizhongwu1995@163.com (Z.D.); dxy1020239286@126.com (X.D.); liwei961214@163.com (W.L.); mengxiacao0302@126.com (M.C.); lcr5060@126.com (C.L.); tsaiwc@mail.ncku.edu.tw (W.-C.T.); wu_xiaoqian@126.com (X.W.); zhai-jw@163.com (J.Z.)

2 Huanglong National Scenic Reserve, Songpan 623300, China; zth_17@tom.com

3 Institute of Tropical Plant Sciences and Microbiology, National Cheng Kung University, Tainan City 701, China

* Correspondence: zjliu@fafu.edu.cn (Z.L.); shashawu1984@126.com (S.W.);

Abstract: Floral color polymorphism can provide great insight into species evolution from a genetic and ecological standpoint. Color variations between species are often mediated by pollinators and are fixed characteristics, indicating their relevance to adaptive evolution, especially between plants within a single population or between similar species. The orchid genus *Pleione* has a wide variety of flower colors, from violet, rose-purple, pink, to white, but their color formation and its evolutionary mechanism are unclear. Here, we selected the *P. limprichtii* population in Huanglong, Sichuan Province, China, which displayed three color variations: Rose-purple, pink, and white, providing ideal material for exploring color variations with regard to species evolution. We investigated the distribution pattern of the different color morphs. The ratio of rose-purple:pink:white-flowered individuals was close to 6:3:1. We inferred that the distribution pattern may serve as a reproductive strategy to maintain the population size. Metabolome analysis was used to reveal that cyanindin derivatives and delphidin are the main color pigments involved. RNA sequencing was used to characterize anthocyanin biosynthetic pathway-related genes and reveal different color formation pathways and transcription factors in order to identify differentially-expressed genes and explore their relationship with color formation. In addition, qRT-PCR was used to validate the expression patterns of some of the genes. The results show that *PlFLS* serves as a crucial gene that contributes to white color formation and that *PlANS* and *PlUFGT* are related to the accumulation of anthocyanin which is responsible for color intensity, especially in pigmented flowers. Phylogenetic and co-expression analyses also identified a R2R3-MYB gene *PlMYB10*, which is predicted to combine with *PlbHLH20* or *PlbHLH26* along with *PlWD40-1* to form an MBW protein complex (MYB, bHLH, and WDR) that regulates *PlFLS* expression and may serve as a repressor of anthocyanin accumulation-controlled color variations. Our results not only explain the molecular mechanism of color variation in *P. limprichtii*, but also contribute to the exploration of a flower color evolutionary model in *Pleione*, as well as other flowering plants.

Keywords: *Pleione limprichtii*, flower color polymorphism; variation within populations; metabolome analysis; anthocyanin biosynthetic pathway; RNA sequencing; transcription factor

1. Introduction

Flower color is one of the most attractive characteristics of plants in nature. With such massive variation, flower color is regarded as an evolutionarily labile trait and has been shown to contribute to plant evolution [1–3]. In particular, flower color adaptive mutations mediated through pollinators are directly relevant to phenotypic evolution [4]. Color variation is considered a fixed difference between species and promotes the formation of population polymorphism [5,6]. Among angiosperms, sister species always display differences in flower hue and intensity, This pattern of flower color polymorphism is used as a model trait in the study of ecology, evolution, and gene regulation [7]. Color changes are related to flower pigment content. To date, the molecular mechanism of flower color transition has been investigated in several species, owing to the main floral pigments having been well characterized in many plants [8–13] providing sufficient information for studying floral color formation in non-model species and the opportunity to explore the relationship between phenotypic evolution and color variations.

Although flower color is influenced by many factors, flavonoids, especially anthocyanins which are produced by the anthocyanin biosynthesis pathway (ABP), are the primary components that contribute to floral pigments and they are produced by highly conservative structural and regulatory components [14]. The ABP involves multi-metabolic processes which mainly consist of seven core structural genes: *CHS*, *CHI*, *F3H*, *F3′H*, *F3′5′H*, *DFR*, and *ANS*, and several branch-enzyme genes [15]. Due to the instability of anthocyanidins, they exist mainly as anthocyanins, which are formed by anthocyanidins and various glycosides [16]. They play an irreplaceable role in the color development of plants and are primarily derived from three main anthocyanidins: Pelargonidin (brick red to scarlet), cyaniding (red to magenta), and delphinidin (purple to violet) [16]. Studies have shown that blocking the ABP can lead directly to variations in pigment production and affect flower color [17]. In addition to the structural genes in the ABP, transcription factors also contribute to flower color transition by regulating the spatial and temporal expression of the structural genes [18,19]. The ABP is regulated by three complex, interacting transcription factors: R2R3-MYB, basic helix–loop–helix (bHLH), and WD40-repeat (WDR) [20]. These transcription factors activate or suppress the transcription and expression of target genes, thereby regulating anthocyanin synthesis [21]. Generally, the structural and regulatory genes involved in the ABP have provided a number of targets to reveal the diversity of mutations that could block the ABP [22]. For flower polymorphism within populations, locating the blockage could elucidate the cause of flower color transition at the biochemical and molecular scales [8,10,11]. In addition, understanding their specific ABP is of benefit for predicting evolutionary influences from a genetic perspective.

The genus *Pleione* (Orchidaceae) comprises nearly 30 species of terrestrial, lithophytic, and epiphytic plants with high ornamental value [23]. There are 27 species in China, while Yunnan is the world biodiversity distribution center of this genus [24,25]. *Pleione* possesses different flower colors ranging between white, pink, lavender, magenta, light purple, and yellow [26]. In particular, populations of Huanglong, Sichuan Province, there remain a color polymorphism population, consisting of pink flowers of different intensities along with white mutant individuals, which can be considered an ideal situation to study the polymorphism formation mechanisms of *Pleione*, as well as benefit to explore potential correlation between color pattern and the species evolution.

The focus of our study was to understand the molecular mechanism of color polymorphism, including how the white individuals formed and the main reason caused pink flowers intensities, as well as summarize the general rules of the color distribution pattern. We aimed to investigate distribution of color monomorphic in the Huanglong *P. limprichtii* population and examine the transcriptome and biochemistry of their color polymorphic petals. RNA sequencing (RNA-seq) and ultra-performance liquid chromatography (UPLC) were used to identify the variation of related genes and the differences in flavonoid intermediates in the ABP that cause color transition, respectively.

2. Results

2.1. Color Differences and Quantity Distribution Pattern

After applying quantitative statistics to the individuals with the three distinct flowers at the full bloom stage randomly distributed in three rock populations, we concluded that the rose-purple individuals account for nearly 60%, the pink ones occupied about 30%, and the white ones not more than 10% (Table 1). The ratio of the number of the three differently flowered plants (rose-purple:pink:white) is roughly 6:3:1. Thus rose-purple is the main flower color in the Huanglong population, while white is the rarest color.

Table 1. Number of *P. limprichtii* individuals of each flower color polymorph in three selected populations at the opening stage.

Population	Rose-Purple	Pink	White
population1	409 (58%)	227 (32%)	70 (10%)
population2	31 (65%)	15 (31%)	2 (4%)
population3	92 (60%)	54 (35%)	8 (5%)

The percentage indicates the proportion of individuals with a given color account for the total number of individuals in the population.

According to principle component analysis, there were significant differences in the a*(redness and greenness) values of the three colors (Figure 1. A comparison of L* (lightness) and C* (chroma) values among the three color groups indicated that C* values can also be used as an indicator to distinguish these three colors, since the rose-purple group has the highest C* value, the white group the lowest, and the pink group an intermediate value (Table 2).

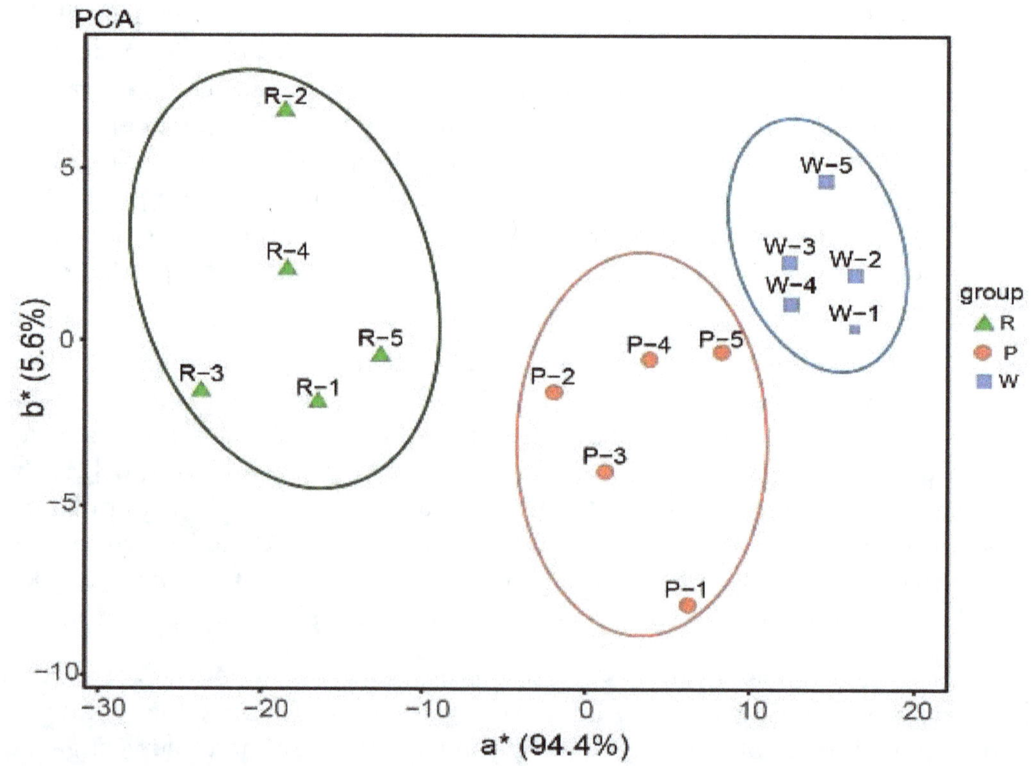

Figure 1. Principal component analysis of petal color of *P. limprichtii*. Distribution based on bivariate values of a* (redness and greenness) and b* (yellowness and blueness). a* means difference in red and green, b* means difference in yellow and blue.

Table 2. Color parameters of three distinct colors of *P. limprichtii* petal.

Flower Color	RHS	L*	C*	b*/a*
purple red	Purple-group N78B	67.92 ± 10.27	49.64 ± 9.41	−1.68 ± 0.61
		76.66 ± 4.51	52.88 ± 4.18	−2.35 ± 0.23
		67.35 ± 10.34	57.13 ± 6.24	−1.68 ± 0.61
		81.31 ± 8.72	51.89 ± 7.42	−1.91 ± 0.44
		79.68 ± 15.55	46.28 ± 11.76	−1.85 ± 0.79
Pink	Purple-group 75A	61.08 ± 5.29	26.71 ± 1.82	−1.20 ± 0.36
		66.05 ± 6.63	35.45 ± 9.60	−1.77 ± 0.71
		66.35 ± 2.73	31.72 ± 3.14	−1.57 ± 0.29
		58.80 ± 5.20	29.66 ± 6.05	−2.20 ± 1.07
		62.80 ± 5.78	25.43 ± 4.99	−2.56 ± 0.89
White	Purple-group 76D	93.08 ± 5.57	13.31 ± 9.67	−4.04 ± 2.06
		88.12 ± 8.13	18.53 ± 7.31	−13.60 ± 5.70
		95.85 ± 4.48	22.41 ± 6.77	−7.97 ± 3.88
		87.52 ± 11.45	25.27 ± 10.21	−6.11 ± 2.61
		87.07 ± 7.97	21.36 ± 6.12	−9.28 ± 4.73

RHS, Royal horticulture society color chart evaluation index; L*, lightness; a*, b*, chromatic components; C*, chroma (brightness).

2.2. *Major Anthocyanin Compounds in P. limprichtii*

UPLC analysis revealed that four anthocyanins and derivatives were detected in both the pigmented and white flower petals: Cyanidin 3-O-glucosyl-malonylglucoside, cyanidin 3-O-malonylhexoside, cyanidin, and delphinidin (Table 3). The cyanidin accumulation and delphinidin accumulation branches of the ABP in *P. limprichtii* indicate that anthocyanins and their derivatives are the main flower color pigments.

Table 3. Anthocyanin content of petal in *P. limprichtii*.

Index	Ion Mode	Molecular Weight	Substance
pme3609	positive	287.24	Cyanidin
pme0442	positive	303.24	Delphinidin
pmb0541	positive	697.1	Cyanidin 3-O-glucosyl-malonylglucoside
pmb0542	positive	535.1	Cyanidin 3-O-malonylhexoside

2.3. *RNA-Seq and Annotation of Unigenes*

To understand the molecular basis of flower polymorphism in *P. limprichtii*, three distinctly colored flower petals and lips were used for RNA-seq. A total of 80,525 unigenes with a mean length of 856 bp were obtained by de novo assembly. We assessed the quality of the unigenes in the transcriptome library. The length of N50 (sequence length of the shortest contig at 50%) was 1470 bp, the Q20 and Q30 percentages were 98% and 95%, the GC content was 40%, and the unigenes were generally distributed between 200 bp and 3000 bp in length. A total of 69,004,304 base pairs were aligned. These data show that the throughputs and sequencing quality were high enough to ensure further analysis.

According to the BLASTx results, a total of 33,724 unigenes were annotated, accounting for 41.88% of all the unigenes. Among these, 33,459 could be annotated using the Nr database (Non-redundant protein database, 41.55%), 21,177 using the Swiss-prot database (Swiss-Protein protein database, 26.30%), 11,067 using the KEGG database (Kyoto encyclopedia of genes and genomes, 24.68%), and 19,871 using the KOG database (Eukaryotic orthologous groups, 13.74%). In addition, there were 8396, 151, 42, and 23 unigenes annotated only using the Nr, Swiss-protein, KEGG, and KOG databases, respectively (Figure 2).

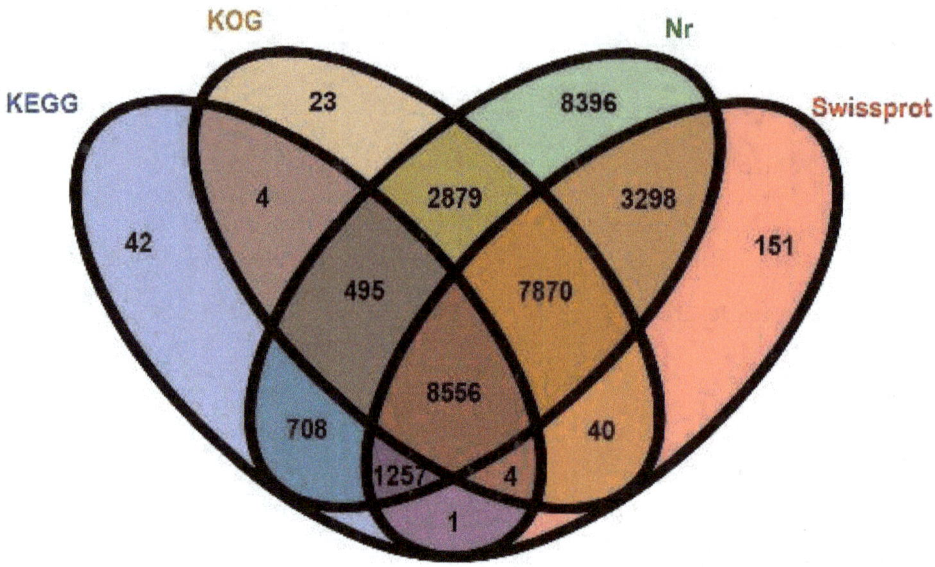

Figure 2. Venn diagram of the number of unigenes annotated by BLASTx with four protein databases.

Statistical analysis of the E-value (The probability due to chance) characteristics of the Nr annotations revealed that 32.05% of the mapped sequences showed strong homology (E-value < 1×10^{-3}), while 29.48% in the Swiss-protein database, 44.77% in the KEGG database, and 30.07% in the KOG database showed strong homology (Figure 3).

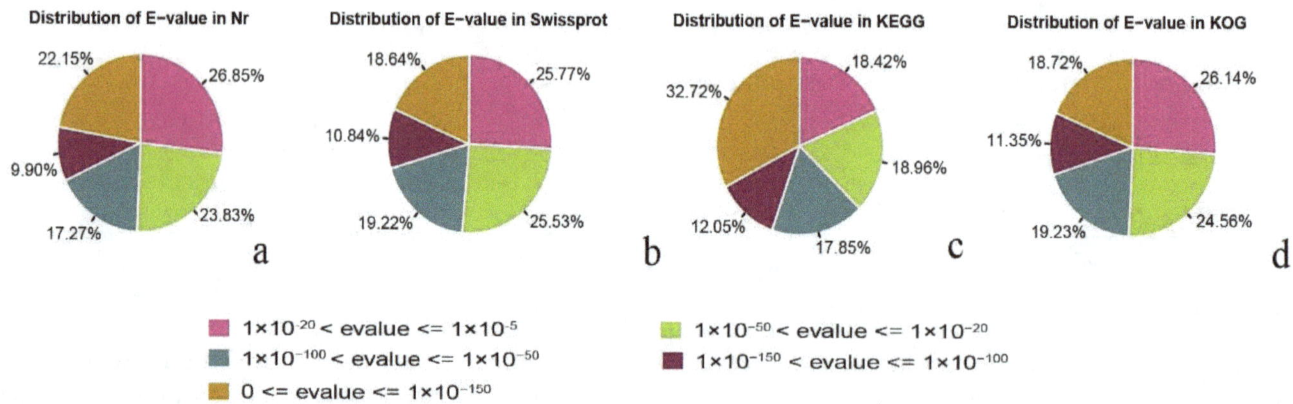

Figure 3. E-value distribution of top BLASTx hits against four protein databases for each unigene. (**a**) distribution E-values in non-redundant protein database; (**b**) distribution E-values in Swiss-protein database; (**c**) distribution E-values in the Kyoto Encyclopedia of Genes and Genomes database; (**d**) distribution E-values in the Eukaryotic Orthologous Group database.

Based on the Nr annotations, 6495 unigenes were classified into 53 functional categories, belonging to three functional terms: molecular function, cellular component, and biological process. The largest percentages of unigenes identified within each of the three functional terms were metabolic process (3602 unigenes), cell and cell part (2237 and 2233 unigenes), and binding and catalytic activity (2481 and 3409 unigenes), corresponding to biological process, cellular component, and molecular function, respectively (Figure 4).

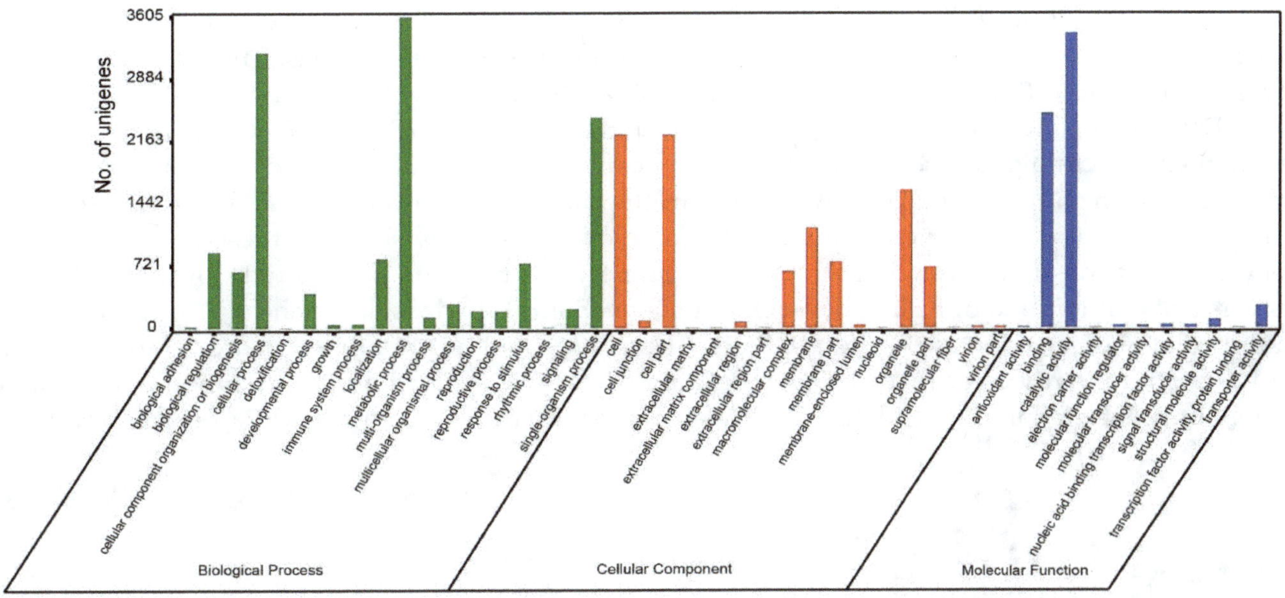

Figure 4. Function classification of the gene ontology of all unigenes based on Nr annotation.

To exhaustively explore the potential functions of the annotated unigenes, 11,067 unigenes were mapped onto 131 KEGG pathways, including metabolic pathways, biosynthesis of secondary metabolites, biosynthesis of antibiotics, and many other important metabolic pathways. Metabolic pathways and biosynthesis of secondary metabolites were enriched in the most unigenes, with 2450 and 1294 unigenes annotating to these two pathways, respectively, while the anthocyanin pathway was enriched in only one unigene. Among these 131 pathways, there are two color formation-related pathways, the flavonoid biosynthesis pathway (ko00941) to which 36 unigenes were mapped, and the ABP (ko00942) to which one unigene was mapped (the pathways are listed in Supplementary Table S3).

2.4. The Expression Patterns of ABP Genes

Genes involved in the flavonoid biosynthesis and ABPs that are related to color formation were isolated, including structural and regulatory genes, in order to identify those involved in the regulation of flower polymorphism in *P. limprichtii*. There were 21 ABP-related structural unigenes, which encoded 10 enzymes, expressed both in pigmented flowers and white flowers, with a total of 14 R2R3-MYB, 32 bHLH, and 23 WD40 regulatory unigenes. With regard to structural unigenes, six of the 10 belong to multigene families, with the exception of *F3H, F3′5′H, ANR*, and *FLS*, which were single copy genes (Table 4). All of these unigenes were used to analyze the expression pattern of the flower color polymorphism of *P. limprichtii*.

Table 4. Tentative anthocyanin biosynthesis pathway-related genes in *P. limprichtii*.

Pathway	Gene	Encoding Enzyme	Number
Flavonoid biosynthesis	*CHS*	Chalcone synthase	5
	CHI	Chalcone isomerase	2
	F3H	Flavanone 3-dioxygenase	1
	F3′H	Flavonoid 3′-hydroxylase	3
	F3′5′H	Flavonoid 3′,5′-hydroxylase	1
	DFR	Dihydroflavonol 4-reductase	3
	ANS	Anthocyanidin synthase	2
	ANR	Anthocyanidin reductase	1
	FLS	Flavonol synthase	1
Anthocyanin biosynthesis	*UFGT*	Anthocyanidin 3-O-glucosyltransferase	2

Since petal anthocyanins are detectable in different color petals, we inferred that floral color differences were caused by different expression patterns of ABP-related genes. The results show that the genes encoding flavonol synthese (*FLS, PlFLS*), anthocyanin synthese (*ANS, PlANS1, PlANS2*), and UDP-glucose anthocyanidin 3-O-glucosyltransferase (*UFGT, PlUFGT1, PlUFGT2*) were expressed at a higher level in pigmented flowers (rose-purple and pink) than in white flowers, and were more up-regulated in the rose-purple flowers than in the pink flowers (Figures 5a and 6). These genes were correlated with flower color intensity and color phenotypes. Besides these directly affected genes, other genes that have different expression patterns between pigmented flowers and white flowers also influence color formation, such as the flavanone 3′-hydroxylase gene (*F3′H; Pl F3′H3*) and dihydroflavonol 4- reductase (*DFR, PlDFR3*). These were both up-regulated in rose-purple flowers and contribute to red color formation. According to the ABP-related gene expression patterns and metabolites detected in the three distinct flower groups, we drew a putative ABP of *P. limprichtii*.

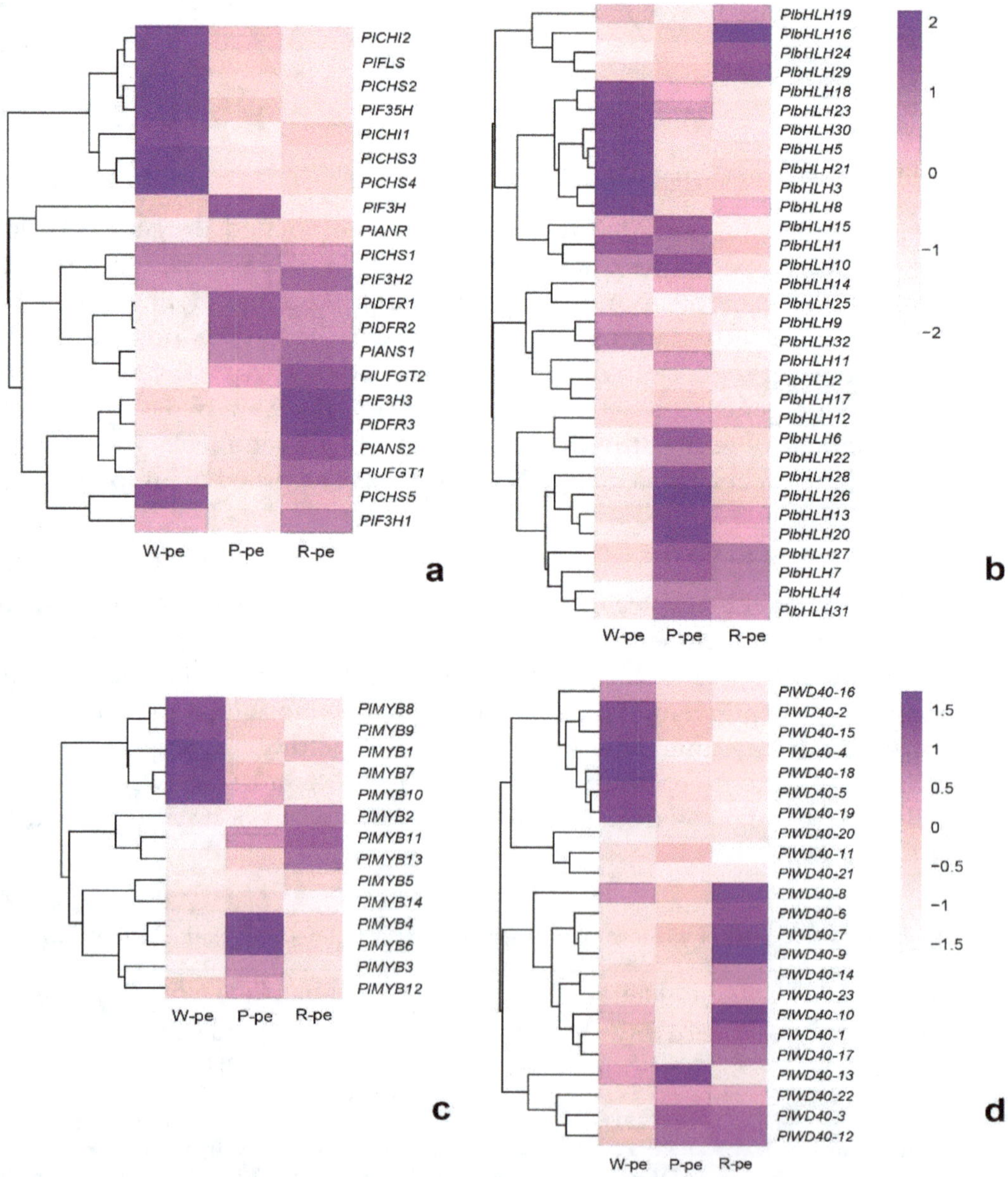

Figure 5. Expression pattern analysis base on RNA sequencing of anthocyanin biosynthesis pathway related genes and transcription factors in *P. limprichtii*. (**a**) Anthocyanin biosynthesis pathway-related unigenes; (**b**) *bHLH* unigenes; (**c**) *R2R3-MYB* unigenes; (**d**) *WD40* unigenes. W-pe (petal of white flower), P-petal (petal of pink flower), R-petal (petal of rose -purple flower).

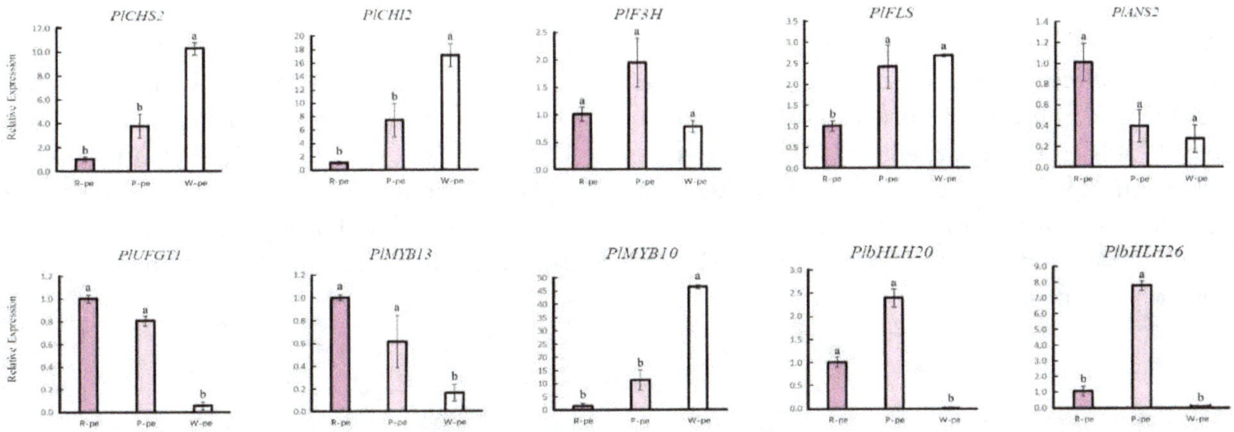

Figure 6. Real time quantitative reverse transcription-PCR of several genes in *P. limprichtii*. Each value is shown as average ± standard deviation from three biological replicate sampling.

The expression patterns of anthocyanin regulatory genes, including R2R3-MYB, bHLH, and WD40 were also investigated (Figure 5b–d). Phylogenetic analysis (Figure 7) shown that PlMYB13 (unigene0062421) and PlMYB4 (unigene0039181) were clustered with *AtMYB75*, *AtMYB90*, and *AtMYB113*, which belong to subgroup 6 of *A. thaliana* [27], and have been demonstrated to activate anthocyanin accumulation, while *PlMYB10* (Unigene0058559) was homologous to *AtMYB11*, *AtMYB12*, and *AtMYB111*, which belong to subgroup 7 in A. thaliana, and have been suggested to control flavonol biosynthesis [28]. The expression pattern of PlMYB13 was consistent with anthocyanin accumulation, while *PlMYB10* exhibited an inverse relationship between its expression and flower color intensity.

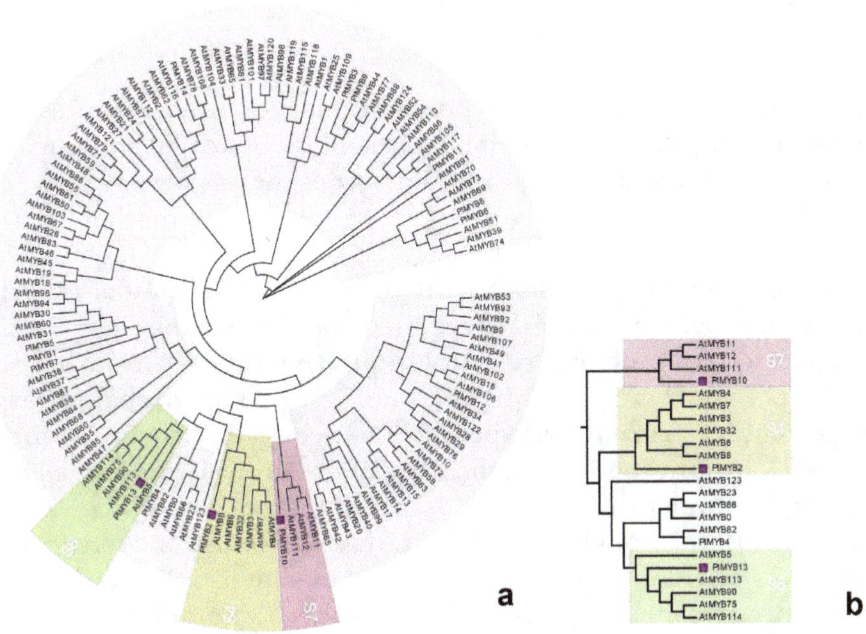

Figure 7. Phylogenetic analysis of R2R3-MYB DNA binding domains for *P. limprichtii* and Arabidopsis thaliana. (**a**) Circular phylogenetic tree; (**b**) amplification of S4, S6, and S7 branches. The R2R3 domains of the 14 MYBs identified in *P. limprichtii* petal transcriptome were aligned and analyzed using neighbor-joining phylogenetic methods.

2.5. The Relationship between Structure Genes and TFs

Through the analysis of the expression patterns of ABP-related genes and the phylogenetic tree of R2R3-MYB, we obtained several candidate genes that correlate with floral color intensity. We constructed a co-expression network (Figure 8) to identify the interactions between ABP-related genes and MBW complex proteins (MYB, bHLH, and WDR). The results show that ten structure unigenes and five MBW complex transcriptional unigenes composed of two R2R3-MYB, two bHLH and one WD40 unigene exhibited interaction relationships. The expression pattern of *PlMYB10* coincided with those of PlbHLH20 (Unigene0062784) and PlbHLH26 (Unigene00660), while *PlbHLH20* and *PlbHLH26* also correlated with *PlWD40-1* (Unigene0002153). PlMYB10 also showed a linear relationship with *PlFLS*. We therefore inferred that *PlMYB10* may interact with *PlbHLH20* or *PlbHLH26*, and *PlWD40-1* to form a MBW transcriptional complex, and regulate the expression pattern of *PlFLS*, finally affecting flower polymorphism in *P. limprichtii*.

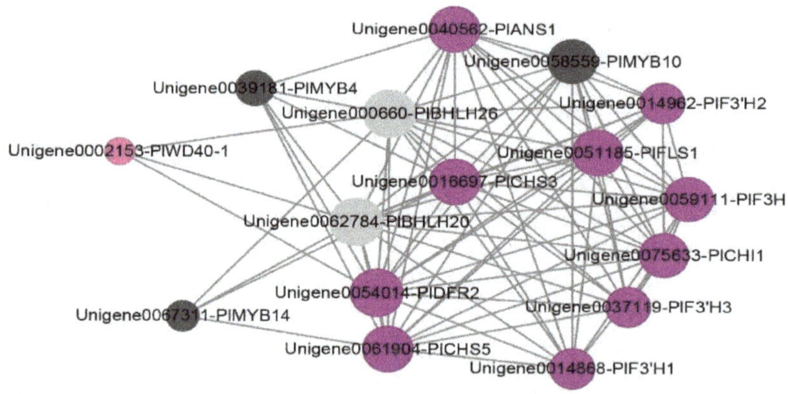

Figure 8. A co-expression network of anthocyanin biosynthesis pathway-related genes and transcription factors involved in pigmentation. Rose-purple circle represents structural genes; dark-gray circle represents R2R3-MYB; light-gray circle represents bHLH; pink circle represents WD40.

3. Discussion

We investigated the distribution of color polymorphic individuals in three rock populations of *P. limprichtii* within Huanglong District and combined chemical detection and transcriptomic analysis to isolate the main pigment compounds and candidate genes that determine flower color intensity. We present a putative biosynthesis pathway and discuss the regulatory mechanisms of color formation.

To explore color variation formation factors, we used the CIELAB evaluation system to distinguish rose-purple, pink, and white flower color, and then counted the number of individuals of each of these phenotypes in each rock population. In all three populations the color distribution pattern was nearly consistent, with a color ratio of rose-purple 6:pink 3:white 1. This distribution pattern is very rare in natural color polymorphic populations. Studies have shown that intraspecific flower color variation is often attributed to genetic drift, pollination-mediated selection, environmental conditions, or herbivory [13,29–31]. In our experiments, the populations grow across a small range, with the whole population occurring on similar rocks and exposed to the same climatic conditions. Environmental elements such as temperature, drought stress, and ultraviolet radiation were therefore not considered to be crucial promoters of color variation. Pollinator-mediated selection plays an important role in color variation, especially for deceptive pollination species in which competition for pollinators in sympatry promotes flower color divergence [32], and shifts in pollinators also contribute to the macro-evolution of flowers color [22]. Thus, we inferred that the flower color polymorphism within these populations might have been induced by pollinators. Color polymorphism may be a consequence of pollination competition or specific adaptations to pollinators, and pollinator behavior exerts strong selection stress on color variations. Some research also showed adaptive selection for pigmented flowers because colored flowers are less likely to be disrupted by herbivories than colorless ones [3]. According to our field observations, we found that white-flowered individuals were more susceptible to damage than individuals with pigmented flowers, and that white flower petals and cores were severely foraged when blooming. We thus inferred that the dominant pigmented color was beneficial to avoid herbivory, and reduce damage by herbivores to the population. When individuals were damaged they also suffered reduced attractiveness to pollinators, which is not conductive to the stability and development of the population. This explains the distribution pattern of the number of rose-purple flowers in the population. There is another view that flower color may not be the primary goal of natural selection, nor the initial choice of pollinators. Indeed, the biosynthetic precursors of pigments not only display color variations, but also serve other physiological functions [14]. Studies have shown that secondary metabolites associated with plant defense functions share the same biosynthetic pathway, the flavonoid synthesis pathway, correlating pigment with defense ability [33]. Therefore, colorful-flowered individuals were more resistant to some adversities. In summary, the phenomenon that rose-purple flowers were frequent and white ones rare within the population may be mediated by the pollinators and herbivores, and also related to survival adaptability of the *P. limprichtii*. The 6:3:1 distribution pattern of color polymorphism might be a reproductive strategy for the population to maintain the maximum population density, but further evidences should be investigated.

In theory, the transition from pigmented to white flowers could involve any mutations that block one or more steps in the anthocyanin pathway. This includes loss-of-function mutations in any pathway enzyme-coding genes, as well as the cis-regulatory genes that influence any of the pathway enzymes [34,35]. In our study, expression analysis identified several obvious differentially expressed genes in the petal which were down-regulated in white samples compared to pigmented samples, but metabolite detection found that Cy- and Del-derivatives existed in both white petals and pigmented petals, indicating that the color variations, especially the white petals, do not lacking any steps of the anthocyanin pathway. The cis-regulation of transcription factors is a crucial element to promote color divergence. This result is similar with the white color formation in *Primula vulgaris* which is caused by different genes expression pattern rather than loss- of-function mutations leading to the lack of anthocyanin [36].

For ABP-related genes, we isolated 21 transcripts which encode ten enzymes. Seven of the ten were flavonoid synthase genes, including *PlCHS, PlCHI, PlF3H, PlF3′H, PlF3′5′H, PlDFR*, and *PlANS*; one was a proanthocyanidin synthase gene, *PlANR*; one was a flavonol synthase gene, *PlFLS*; and one was an anthocyanin synthase gene, *PlUFGT*. Most of them are multi-gene families, only *PlF3H, PlF3′5′H, PlFLS*, and *PlANR* are single copy. To clearly illustrate the catalyzation steps of the ABP, we regarded Cy-related and Del-related biosynthesis processes as independent branches [8]. Thus we did not have to measure the content of each anthocyanin compound. Each branch of anthocyanin synthase was considered to make an equivalent contribution to pigmentation. Considering that both the Cy-related and Del-related branches may share the majority of enzymes, here, we selected ABP-related genes on the Cy-related branches to analyze their expression patterns.

The expression analysis of early step structural genes revealed a high level of *PlCHS* and *PlCHI* expression in white petals and a low level in pigmented petals, suggesting that white petals can produce a large amount of naringenin but cannot eventually flux this into anthocyanin synthase. Meanwhile, the expression levels of *PlF3′H, PlF3′5′H, PlDFR, PlANS*, and *PlUFGT* in pigmented flower petals was high, compared with white ones, and we inferred that downstream structural genes make a large contribution to coloration. Analysis also showed that *PlFLS* was significantly more upregulated in white petals than in pigmented petals. *FLS* encoding enzymes lead substrate into the flavones and flavonols pathway [37]. It has been suggested that the competition between the anthocyanin synthesis pathway and the flavone and flavonols pathways mainly results in substrate competition between *FLS* with *DFR*, while the *FLS* enzyme strengthens the metabolic flux toward the flavonols and limits anthocyanin accumulation [38]. This situation has also been reported in other species, such as in *Paeonia ostii*, a higher expression of *PoFLS4* in the nearly white flowers promotes dihydroflavonols transition into flavonols [11]. In onions, enhanced *AcFLS* could maximize flavonol production in the sheath [39]. Finally, in *Muscari armeniacun*, the conversion of substrate between *FLS* and *DFR* facilitates the elimination of blue pigmentation [8]. Thus, we confirmed *PlFLS* as one of the candidate genes for white color formation in *P. limprichtii*. The up-regulation of *PlDFR* in pigmented flowers is closely accompanied by a decrease of *PlFLS*; hence, more dihydroflavonols flow into anthocyanin production in pigmented flowers. From our analysis, the expression patterns of *PlANS* and *PlUFGT* are correlated with color intensity, and they showed their highest expression levels in the rose-purple flowers and their lowest in the white flowers. The ANS (Anthocyanidin synthase) encoding enzyme catalyzes the conversion of colorless leucocyanidin into colored anthocyanin [40], and anthocyanin is further glycosylated by different UFGT (UDP flavonoid glucosyl transferase) encoding enzymes that convert the anthocyanidins to different anthocyanin derivatives, exhibiting the final color [41]. We therefore speculate that *PlANS* and *PlUFGT* are two crucial genes that determine color intensity in *P. limprichtii*.

It has been revealed that the MBW protein complex, a combination of R2R3-MYB and bHLH transcription factors, along with WD40 proteins, play an important role in regulating the transcription of structural genes [41–45]. The activities of R2R3-MYB factors have distinct roles in determining the action of the complexes, either to promote or inhibit the transcription of anthocyanin biosynthesis genes [46,47]. By combining R2R3-MYB phylogenetic and co-expression network analyses, we isolated *PlMYB10*, which was homologous with the *AtMYB11*, *AtMYB12*, and *AtMYB111* belonging to S7 in *Arabidopsis* and that have been demonstrated to contribute to the regulation of genes that account for anthocyanin accumulation in all tissues [27,28]. The expression pattern was high expression in nearly white flower petals, gradually reducing in flowers with increasing color intensity. It also corresponded with the color polymorphism phenotypes. In other species, overexpression of *AmMYB330*, a negative regulator of the flavonoid biosynthesis, has been proven to inhibit phenylpropanoid metabolism in transgenic tobacco (*Nicotiana tabacum*) plants [48]. The co-expression network showed that *PlbHLH20* and *PlbHLH26*, along with *PlWD40-1* have a strong relationship to *PlMYB10*. It is likely that this potential MBW complex *PlMYB10/PlbHLH20/PlWD40-1* or *PlMYB10/PlbHLH26/PlWD40-1* may serve as a repressor responsible for the variation in color intensity in *P. limprichtii*. Co-expression network revealed *PlFLS* is the most likely target gene interacting with *PlMYB10*. Previous studies have verified that R2R3-MYB can regulate the expression pattern of *FLS* through the overexpression of *PsMYB114L* (from *Paeonia suffruticosa*) in *Arabidopsis* [49], which is consistent with our results. Thus, we tentatively speculate about the ABP of *P. limprichtii* (Figure 9). Further studies examining sequencing differences in these candidate genes and *PlMYB10* are necessary to assess our speculations. Metabolic substance quantification is also necessary to confirm the leading pigments in *P. limprichtii*.

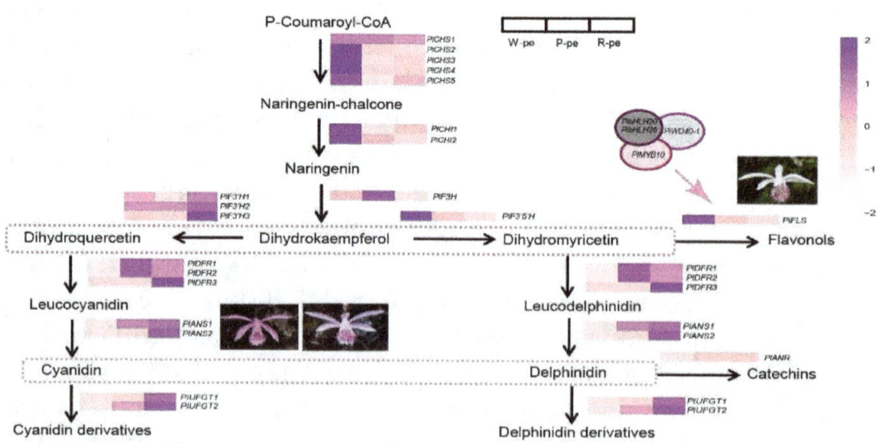

Figure 9. Tentative pathways for *P. limprichtii* color variations. The colored bar is the value of log2 (RPKM + 1), represented using the depth of color, with purple representing the up-regulated expression genes and pink representing the down-regulated expression genes. RPKM means the reads per kb per million reads mapped. CHS, chalcone synthase; CHI, chalcone-flavanone isomerase; F3H, flavanone-3-hydroxylase; F3′H, flavonoid 3′-hydroxylase; F3′5′H, flavonoid 3′5′-hydroxyla; DFR, dihydroflavonols 4-reductase; ANS, anthocyanidin synthase; UFGT, UDP flavonoid glucosyl transferase; FLS, flavonol synthase; The three gene complex consist of a MYB, bHLH and WD40 in most angiosperm.

4. Materials and Methods

4.1. Plant Materials

Pleione limprichtii, belongs to the Orchidaceae family, and is distributed in the Huanglong Nature Reverse in Sichuan Province, China, with three distinct color variations including rose-purple, pink, and white (Figure 10a–c). The study site was located at 32.68°N, 104.04°E, and 1851.61 m above sea level (Figure 10d). The plants are generally found living on rocks along streams. The sexual reproduction of *Pleione* is commonly dependent on deceptive pollination and each individual usually grows only one flower. The majority of individuals in the population have rose-purple flowers, while white flowers are relatively less common.

Figure 10. Flower polymorphism of *Pleione* limprichtii in Huanglong population. (**a**) rose-purple flower; (**b**) pink flower; (**c**) white flower; (**d**) one of the polymorphic populations on a rock.

4.2. Quantitative Statistics and Flower Colorimeter Analysis

Three rocks with *P. limprichtii* populations which contained individuals with all three color variations were selected, the number of each color individuals were counted. In addition, in order to evaluate the flower color objectively, a hand spectrophotometer (CS-280, Hangzhou Color Spectrum Technology Co., Ltd., Hangzhou, China) was used to measure the CIE L*a*b* color components with five technical repetitions using five different parts of the selected petals, in order to evaluate the flower color objectively. The formula $C^* = \sqrt{a*^2 + b*^2)}$, the L* (lightness) and C* (chroma) through the petals were measured using the lightness coefficient 'a*', which indicates greenness to redness as the value increases from negative to positive, and 'b*', which represents blueness to yellowness. Principle component analysis was performed using the a* and b* values as the principle components [32].

4.3. Sampling and RNA Extraction and cDNA Synthesis

Petals from rose-purple, pink, and white flowers at the full bloom stage were sampled, immediately frozen in liquid nitrogen, and preserved at −80 °C before pigment analysis and RNA extraction. RNA was isolated using a Quick RNA isolation Kit (Huayueyang Co., Ltd., Beijing, China), according to the manufacturer's instructions. The concentration and purity of the RNA was measured with a NanoDrop 2000 (Thermo Fisher Scientific Co., Ltd., Waltham, MA, USA) and Agilent 2100 (Agilent Technologies

Co., Ltd., Palo Alto, CA, USA) to verify RNA integrity. A total of nine samples, including three biological replicates, with high concentrations of RNA for each of the three color morphs were selected, then the strand cDNA synthesis was performed using a Revert Aid First Strand cDNA Synthesis Kit (Thermo Fisher, Foster City, CA, USA), according to the manufacturer's instructions, and was stored at −80 °C for RT-qPCR assays.

4.4. *Measurement of Flower Anthocyanin*

Sample preparation and extraction methods were as follow, the freeze dried tissues were crushed using a mixer mill (MM 400, Verder Shanghai Instruments and Equipment Co., Ltd., Shanghai, China) with a zirconia bead for 1.5 min at 30 Hz. A 100 mg sample of the powder was weighed and extracted overnight at 4 °C with 1.0 mL 70% aqueous methanol. Following centrifugation at 10, 000× *g* for 10 min, the extracts were absorbed and filtered. Then the sample extracts were analyzed using a UPLC system, the analytical conditions were as follow, HPLC: column, Waters ACQUITY UPLC HSS T3 C18 (1.8 μm, 2.1 mm × 100 mm); solvent system, water (0.04% acetic acid): acetonitrile (0.04% acetic acid); gradient program, 95:5 *v/v* at 0 min, 5:95 *v/v* at 11.0 min, 5:95 *v/v* at 12.0 min, 95:5 *v/v* at 12.1 min, 95:5 *v/v* at 15.0 min; flow rate,. 0.40 mL/min; temperature, 40 °C; injection volume: 2 μL [32,50]. The mass spectrometry data was analyzed using the software Analyst 1.6.1 (AB Sciex Pte. Ltd., Concord, Ontario, Canada) and based on the Metware Database (MWDB, Metware Biotechnology Co., Ltd., Wuhan, China), a local self-built database, and a public metabolite information database, to identify anthocyanin compounds. The experiments were repeated three times. 1.8 μm

4.5. *Library Preparation and Sequencing*

The construction of the libraries and RNA-seq were performed by the Omicshare Biotechnology Corporation (Guangzhou, China). The mRNA was enriched with Oligo (dT) beads, then broken into short fragments and reverse transcribed into cDNA as templates. First and second-strand cDNA were then synthesized. The cDNA fragments were purified with a QiaQuick PCR extraction kit (Qiagen, Valencia, CA, USA) and end repaired. Poly (A) was added and ligated to Illumina sequencing adapters. They were then sequenced using the Illumina HiSeqTM 4000 (Illumina Inc., Centre Drive, San Diego, CA, USA).

4.6. *De Novo Transcriptome Assembly Annotation*

Transcriptome de novo assembly was performed with clean data, filtered from the raw data by removing adaptors and unknown nucleotides (>10%), and those with low quality reads. The data were assembled using the Trinity platform [51] with the parameters 'K-mer = 31' and 'K-mer cover = 6'. First, short reads of a certain length were combined with overlap to form longer contigs. Then, based on their paired-end information, clean reads were mapped back to the corresponding contigs. Thus the sequences of the transcripts were finished, and defined as unigenes. All assembled unigenes were then annotated using BLASTx (E-value $\leq 1 \times 10^{-5}$) against protein databases, including the National Center for Biotechnology Information non-redundant (Nr, ftp://ftp.ncbi.nih.gov/blast/db/), Swiss-Protein (https://www.uniprot.org/), Kyoto Encyclopedia of Genes and Genomes (KEGG, https: //www.genome.jp/kegg/), and Gene Ontology (GO, http://geneontology.org/) databases. while Nr and Swiss-Protein annotate gene function, KEGG is used to understand biological systems and GO divides genes into different categories. BLASTx was used to search for the unigenes against the public databases with the following order of priority: Nr, Swiss-protein, KEGG, and COG. When a unigene could not be aligned to any of these protein databases, the protein code sequence and sequence direction was confirmed using the ESTscan program.

4.7. *Expression Profile and RT qPCR*

To compare color-related unigenes expression divergence, they were first normalized to RPKM (reads per kb per million reads). After that, different sample ratios of RPKM values were calculated.

The FDR (false discovery rate) value was used to identify the threshold of the *p*-value in multiple tests in order to compute the significance of the differences among unigenes. Here, only FDR significance scores < 0.05, and log2 ratios > 1 were regarded as differentially expressed genes and used in subsequent analysis. In order to validate the expression pattern of the RNA-Seq results, ten important different expression genes (the primers used are listed in Table S1) were selected and measured using RT-qPCR on a Quant Studio 5 Real-Time PCR System (Thermo Fisher, Foster City, CA, USA) using the PowerUp^{TM} SYBR^{TM} Green Master Mix (Thermo Fisher, Foster City, CA, USA), according to the manufacturer's instructions. *PlUBC34* and *PlUBC35* (ubiquitin-conjugating enzyme; primers are also listed in Table S1) actin genes from *Pleione* were used as the internal control for the normalization of gene expression. Each sample (including three biological repetitions) was quantified in triplicate.

4.8. Genes Related to the ABP and Phylogenetic Analyses

The ABP-related structure genes and transcription factors including *CHS, CHI, F3H, F3′5′H, DFR, FLS, ANS, FNS, UFGT, MYB, bHLH,* and *WD40* were used as queries to retrieve the corresponding unigenes from the *P. limprichtii* libraries. Meanwhile, every reference gene from model plants obtained from KEGG were aligned with the libraries using BLASTx to search for more related unigenes.

For the MYB genes, to determine which of the genes belonged to which R2R3-MYB subfamily, the R2R3-MYB genes from *Arabidopsis thaliana* were used to conduct phylogenetic analyses (Amino acid sequence obtained from Gene Bank, accession number is listed in Supplementary Table S2), and MEGA7.0 software (Institute of Molecular Evolutionary Genetics, PA, USA) [52] was used to perform sequence aligning and construct a circular phylogenetic tree according to neighbor joining method with 1000 interactions.

4.9. Structure Genes and Tfs Co-Expression Network

The String online database (https://string-db.org/) was used to search for the interaction relationships between structure genes and transcription factors. Then we used Cytoscape software (National Institute of General Medical Sciences, Bethesda, MD, USA) to construct a co-expression network to identify their main impact factors [53].

5. Conclusions

The ratio of 6:3:1 distribution patterns of *P. limprichtii* within the population in Huanglong District seems to reveal that pollinators, herbivores, and survival adaptability could promote the development of such a reproductive strategy in the population to maintain the maximum population density. Function and expression patterns indicated that *PlFLS* probably is a crucial gene in the formation of white and pigmented flowers. *PlANS* and *PlUFGT* were found to determine the color intensity in pigmented flowers. In addition, a putative MBW complex, *PlMYB10/PlbHLH20/PlWD40-1* or *PlMYB10/PlbHLH26/PlWD40-1* may serve as a repressor of regulated *PlFLS* expression that is responsible for variation in the color intensity of *P. limprichtii.* Our results provide valuable molecular information on floral color variations in *Pleione,* and also provide inspiration to further explore the relationship between color polymorphism and species evolution and to study its contribution to color evolution.

Author Contributions: This study was conceived by S.W. and Z.L. Y.Z. performed the transcriptome and gene expression analysis and wrote the manuscript. T.Z. and Z.D. helped with the sampling and field work. X.D., W.L., and M.C. helped to perform the experiment. C.L. and X.W. revised the manuscript. W.-C.T. and J.Z. provided writing guidance. All authors contributed to revising the manuscript. All authors have read and agreed to the published version of the manuscript.

Acknowledgments: The authors thank Bai-Jun Li, Jie-Yu Wang, Yan-Qiong Chen, Guo-Qiang Zhang, Si-jin Zeng, Zhi-yao Ren, and Xueyan Yuan who helped with this research. We also thank Qi-Xuan Peng and other staff of Huanglong National Scenic Reserve for assisting sampling.

Abbreviations

UPLC	Ultra performance liquid chromatography
ABP	Anthocyanin biosynthesis pathway
CHS	Chalcone synthase
CHI	Chalcone isomerase
F3H	Flavanone 3-hydroxylase
F3′H	Flavonoid 3′-hydroxylase
F3′5′H	Flavonoid 3′5′-hydroxyla
FLS	Flavonol synthase
DFR	Dihydroflavonol 4-reduct
ANS	Anthocyanidin synthase
ANR	Anthocyanidin reductase
UFGT	UDP flavonoid glucosyl transferase
RHS	Royal horticulture society
RHSCC	The royal horticulture society color chart
NR	Non-redundant protein database
Swiss-Prot	Swiss-Protein protein database
KEGG	Kyoto encyclopedia of genes and genomes
KOG	Eukaryotic orthologous groups
GO	Gene ontology
RPKM	reads per kb per million reads
FDR	false discovery rate
qRT-PCR	Real-time reverse transcription-PCR

References

1. Cooley, A.M.; Carvallo, G.; Willis, J.H. Is floral diversification associated with pollinator divergence? flower shape, flower colour and pollinator preference in Chilean Mimulus. *Ann. Bot.* **2008**, *101*, 641–650. [CrossRef] [PubMed]
2. Davies, K.M.; Albert, N.W.; Schwinn, K.E. From landing lights to mimicry: The molecular regulation of flower colouration and mechanisms for pigmentation patterning. *Funct. Plant Biol.* **2012**, *39*, 619–638. [CrossRef]
3. Vaidya, P.; Ansley, M.; Emily, M.; Michaela, K.; Lauren, C.; Lee, C.R.; Bingham, R.A.; Anderson, J.T. Ecological causes and consequences of flower color polymorphism in a self-pollinating plant (Boechera Stricta). *New Phytol.* **2018**, *218*, 380–392. [CrossRef] [PubMed]
4. Streinzer, M.; Roth, N.; Paulus, H.F.; Spaethe, J. Color preference and spatial distribution of glaphyrid beetles suggest a key role in the maintenance of the color polymorphism in the peacock anemone (Anemone pavonina, Ranunculaceae) in Northern Greece. *J. Comp. Physiol.* **2019**, *205*, 735–743. [CrossRef] [PubMed]
5. Kellenberger, R.T.; Byers, K.J.R.P.; De Brito Francisco, R.M.; Staedler, Y.M.; LaFountain, A.M.; Schönenberger, J.; Schiestl, F.P.; Schlüter, P.M. Emergence of a floral colour polymorphism by pollinator-mediated overdominance. *Nat. Commun.* **2019**, *10*, 63. [CrossRef] [PubMed]
6. Sobel, J.M.; Matthew, A.S. Flower color as a model system for studies of plant evo-devo. *Front. Plant Sci.* **2013**, *4*, 321. [CrossRef] [PubMed]
7. Rausher, M.D. Evolutionary transitions in floral color. *Int. J. Plant Sci.* **2008**, *169*, 7–21. [CrossRef]
8. Lou, Q.; Liu, Y.L.; Qi, Y.Y.; Jiao, S.Z.; Tian, F.F.; Jiang, L.; Wang, Y.J. Transcriptome sequencing and metabolite analysis reveals the role of delphinidin metabolism in flower color in grape hyacinth. *J. Exp. Bot.* **2014**, *65*, 3157–3164. [CrossRef]
9. Butler, T.; Dick, C.; Carlson, M.L.; Whittall, J.B. Transcriptome analysis of a petal anthocyanin polymorphism in the Arctic Mustard, Parrya Nudicaulis. *PLoS ONE* **2014**, *9*, e101338. [CrossRef]

10. Casimiro-Soriguer, I.; Eduardo, N.; Buide, M.L.; Valle, J.C.; Whittall, J.B. Transcriptome and biochemical analysis of a flower color polymorphism in Silene Littorea (Caryophyllaceae). *Front. Plant Sci.* **2016**, *7*, 204. [CrossRef]
11. Gao, L.X.; Yang, H.X.; Liu, H.F.; Yang, J.; Hu, Y.H. Extensive transcriptome changes underlying the flower color intensity variation in Paeonia Ostii. *Front. Plant Sci.* **2016**, *6*, 1205. [CrossRef] [PubMed]
12. Wang, H.; Lucie, C.C.; Bessière, J.M.; Guillaume, C.; Bertrand, S.; Eric, I. Flower color polymorphism in Iris Lutescens (Iridaceae): Biochemical analyses in light of plant–insect interactions. *Phytochemistry* **2013**, *94*, 123–134. [CrossRef] [PubMed]
13. Wu, C.A.; Matthew, A.S.; Laura, I.N.; Kaitlyn, A.C. The genetic basis of a rare flower color polymorphism in Mimulus Lewisii provides insight into the repeatability of evolution. *PLoS ONE* **2013**, *8*, e81173. [CrossRef] [PubMed]
14. Grotewold, E. The genetic and biochemistry of floral pigments. *Annu. Rev. Plant Biol.* **2006**, *57*, 761–780. [CrossRef] [PubMed]
15. Holton, T.A.; Edwina, C.C. Genetics and biochemistry of anthocyanin biosynthesis. *Plant Cell* **1995**, *7*, 1071–1083. [CrossRef] [PubMed]
16. Zhao, D.; Tao, J. Recent advances on the development and regulation of flower color in ornamental plants. *Front. Plant Sci.* **2015**, *6*, 261. [CrossRef]
17. Tanaka, Y.; Akemi, O. Seeing is believing: Engineering anthocyanin and carotenoid biosynthetic pathways. *Curr. Opin. Biotechnol.* **2008**, *19*, 190–197. [CrossRef]
18. Xie, D.Y.; Shashi, B.S.; Elane, W.; Wang, Z.Y.; Richard, A.D. Metabolic engineering of proanthocyanidins through co-expression of anthocyanidin reductase and the PAP1 MYB transcription factor. *Plant J.* **2006**, *45*, 895–907. [CrossRef]
19. Twyford, A.D.; Aaron, M.C.; Pratibha, C.; Ramesh, R.; Jannice, F. Loss of color pigmentation is maintained at high frequency in a monkey flower population. *Am. Nat.* **2018**, *191*, 135–145. [CrossRef]
20. Hichri, I.; Barrieu, F.; Bogs, J.; Kappel, C.; Delrot, S.; Lauvergeat, V. Recent advances in the transcriptional regulation of the flavonoid biosynthetic pathway. *J. Exp. Bot.* **2011**, *62*, 2465–2483. [CrossRef]
21. Albert, N.W.; Kevin, M.D.; David, H.L.; Zhang, H.B.; Mirco, M.; Cyril, B.; Boase, M.R.; Hanh, N.; Jameson, P.E.; Schwinn, K.E. A conserved network of transcriptional activators and repressors regulates anthocyanin pigmentation in Eudicots. *Plant Cell* **2014**, *26*, 962–980. [CrossRef] [PubMed]
22. Wessinger, C.A.; Rausher, M.D. Lessons from flower color evolution on targets of selection. *J. Exp. Bot.* **2012**, *63*, 695–709. [CrossRef] [PubMed]
23. Chen, X.; Cribb, P.J.; Gale, S.W. "Pleione". In *the Flora of China 25 (Orchidaceae)*; Wu, Z.Y., Raven, P.H., Hong, D.Y., Eds.; Beijing Science Press and Missouri Botanical Garden Press: Beijing, China, 2009; pp. 325–333.
24. Cribb, P.; Butterfield, I. *The genus Pleione*, 2nd ed.; Royal Botanic Gardens/Kew Press: London, UK, 1999; p. 27.
25. Govaerts, R.; Campacci, M.A.; Baptista, D.H.; Cribb, J.; George, A.; Kreutz, K. *Data from: World Checklist of Orchidaceae*; The Board of Trustees of the Royal Botanic Gardens/Kew: London, UK, 2016; Available online: http://apps.kew.org/wcsp (accessed on 20 August 2019).
26. Gravendeel, B.; Eurlings, M.C.M.; Berg, C.V.D.; Cribb, P.J. Phylogeny of *Pleione* (Orchidaceae) and parentage analysis of its wild hybrids based on plastid and nuclear ribosomal its sequences and morphological data. *Syst. Bot.* **2004**, *29*, 50–63. [CrossRef]
27. Stracke, R.; Werber, M.; Weisshaar, B. The R2R3-MYB gene family in Arabidopsis Thaliana. *Curr. Opin. Biotechnol.* **2001**, *4*, 447–456. [CrossRef]
28. Dubos, C.; Ralf, S.; Grotewold, E.; Weisshaar, B.; Martin, C.; Loïc, L. MYB transcription factors in Arabidopsis. *Trends Plant Sci.* **2010**, *15*, 573–581. [CrossRef] [PubMed]
29. Arista, M.; Talavera, M.; Berjano, R.; Ortiz, P.L. Abiotic factors may explain the geographical distribution of flower colour morphs and the maintenance of colour polymorphism in the scarlet pimpernel. *J. Ecol.* **2013**, *101*, 1613–1622. [CrossRef]
30. Imbert, E.; Wang, H.; Conchou, L.; Vincent, H.; Talavera, M.; Schatz, B. Positive effect of the yellow morph on female reproductive success in the flower colour polymorphic Iris lutescens (Iridaceae), a deceptive species. *J. Evol. Biol.* **2014**, *27*, 1965–1974. [CrossRef]
31. Chalker-Scott, L. Environmental significance of anthocyanins in plant stress responses. *Photochem. Photobiol.* **1999**, *70*, 1–9. [CrossRef]

32. Muchhala, N.; Sönke, J.; Smith, S.D. Competition for hummingbird pollination shapes flower color variation in Andean solanaceae: Competition for pollination shapes flower color variation. *Evolution* **2014**, *68*, 2275–2286. [CrossRef]
33. Parachnowitsch, A.L.; Christina, M.C. Predispersal seed herbivores, not pollinators, exert selection on floral traits via female fitness. *Ecology* **2008**, *89*, 1802–1810. [CrossRef]
34. Jin, X.H.; Huang, H.; Wang, L.; Sun, Y.; Dai, S.L. Transcriptomics and metabolite analysis reveals the molecular mechanism of anthocyanin biosynthesis branch pathway in different senecio cruentus cultivars. *Front. Plant Sci.* **2016**, *7*, 1307. [CrossRef] [PubMed]
35. Le Maitre, N.C.; Pirie, M.D.; Bellstedt, D.U. Floral color, anthocyanin synthesis gene expression and control in cape erica species. *Front. Plant Sci.* **2019**, *10*, 1565. [CrossRef] [PubMed]
36. Li, L.; Zhai, Y.; Luo, X.; Zhang, Y.; Shi, Q.Q. Comparative transcriptome analyses reveal genes related to pigmentation in the petals of red and white Primula vulgaris cultivars. *Physiol. Mol. Biol. Plants* **2019**, *25*, 1029–1041. [CrossRef] [PubMed]
37. Yuan, Y.W.; Alexandra, B.R.; Janelle, M.S.; Lauren, E.S.; Harvey, D.B. Competition between anthocyanin and flavonol biosynthesis produces spatial pattern variation of floral pigments between Mimulus species. *Proc. Natl. Acad. Sci. USA* **2016**, *113*, 2448–2453. [CrossRef]
38. Davies, K.M.; Schwinn, K.E.; Deroles, S.C.; Manson, D.G.; Lewis, D.H.; Bloor, S.J.; Bradley, J.M. Enhancing anthocyanin production by altering competition for substrate between flavonol synthase and dihydroflavonol 4-Reductase. *Euphytica* **2003**, *131*, 259–268. [CrossRef]
39. Sangkyu, P.; Kim, D.H.; Lee, J.Y.; Ha, S.H.; Lim, S.H. Comparative analysis of two flavonol synthases from different-colored onions provides insight into flavonoid biosynthesis. *J. Agric. Food Chem.* **2017**, *65*, 5287–5298.
40. Shi, S.G.; Li, S.J.; Kang, Y.X.; Liu, J.J. Molecular characterization and expression analyses of an anthocyanin synthase gene from magnolia sprengeri Pamp. *Int. J. Appl. Biotechnol. Biochem.* **2015**, *175*, 477–488. [CrossRef]
41. Wu, X.X.; Gong, Q.H.; Ni, X.P.; Zhou, Y.; Gao, Z.H. UFGT: The key enzyme associated with the petals variegation in Japanese Apricot. *Front. Plant Sci.* **2017**, *8*, 108. [CrossRef]
42. Feller, A.; Machemer, K.; Edward, L.B.; Grotewold, E. Evolutionary and comparative analysis of MYB and BHLH plant transcription factors. *Plant J.* **2011**, *66*, 94–116. [CrossRef]
43. Heim, M.A. The basic Helix-Loop-Helix transcription factor family in plants: A genome-wide study of protein structure and functional diversity. *Mol. Biol. Evol.* **2003**, *20*, 735–747. [CrossRef]
44. Baudry, A.; Heim, M.A.; Dubreucq, B.; Caboche, M.; Weisshaar, B.; Lepiniec, L. TT2, TT8, and TTG1 synergistically specify the expression of BANYULS and proanthocyanidin biosynthesis in Arabidopsis Thaliana. *Plant J.* **2004**, *39*, 366–380. [CrossRef] [PubMed]
45. Zimmermann, I.; Heim, M.; Weisshaar, B.; Uhrig, J. Comprehensive identification of Arabidopsis thaliana MYB transcription factors interacting with R/B-like bHLH proteins: Systematic analysis of MYB/BHLH-Interaction. *Plant J.* **2004**, *40*, 22–34. [CrossRef] [PubMed]
46. Akagi, T.; Ikegami, A.; Tsujimoto, T.; Kobayashi, S.; Sato, A.; Kono, A.; Yonemori, K. DkMyb4 is a myb transcription factor involved in proanthocyanidin biosynthesis in persimmon fruit. *Plant Physiol.* **2009**, *151*, 2028–2045. [CrossRef] [PubMed]
47. Matsui, K.; Yoshimi, U.; Masaru, O.T. AtMYBL2, a protein with a single MYB domain, acts as a negative regulator of anthocyanin biosynthesis in Arabidops. *Plant J.* **2008**, *55*, 954–967. [CrossRef] [PubMed]
48. Lang, X.A.; Li, N.; Li, L.F.; Zhang, S.Z. Integrated metabolome and transcriptome analysis uncovers the role of anthocyanin metabolism in *Michelia maudiae*. *Int. J. Genom.* **2019**, *2019*, 1–14. [CrossRef] [PubMed]
49. Gu, Z.; Zhu, J.; Hao, Q.; Yuan, Y.; Duan, Y.; Men, S.; Wang, Q.; Hou, Q.; Liu, Z.; Shu, Q.; et al. A novel R2R3-MYB transcription factor contributes to petal blotch formation by regulating organ-Sspecific expression of PsCHS in tree peony (Paeonia Suffruticosa). *Plant Cell Physiol.* **2018**, *60*, 599–611. [CrossRef] [PubMed]
50. Chen, W.; Gong, L.; Guo, Z.L.; Wang, W.S.; Zhang, H.Y.; Liu, X.Q.; Yu, S.B.; Xiong, L.Z.; Luo, J. A novel integrated method for large-scale detection, identification, and quantification of widely targeted metabolites: Application in the study of rice metabolomics. *Mol. Plant* **2013**, *6*, 1769–1780. [CrossRef]
51. Grabherr, M.G.; Haas, B.J.; Yassour, M.; Levin, J.Z.; Thompson, D.A.; Amit, I.; Adiconis, X.; Fan, L.; Raychowdhury, R.; Zeng, Q.; et al. Full-length transcriptome assembly from RNA-Seq data without a reference genome. *Nat. Biotechnol.* **2011**, *29*, 644–652. [CrossRef]

8

mRNA and miRNA Expression Analysis Reveal the Regulation for Flower Spot Patterning in *Phalaenopsis* 'Panda'

Anjin Zhao [1], Zheng Cui [1], Tingge Li [1], Huiqin Pei [1], Yuhui Sheng [1], Xueqing Li [1], Ying Zhao [1,2], Yang Zhou [1,2], Wenjun Huang [3], Xiqiang Song [1,2], Ting Peng [4] and Jian Wang [1,2,*]

1 Key Laboratory of Ministry of Education for Genetics and Germplasm Innovation of Tropical Special Trees and Ornamental Plants, Hainan Key Laboratory for Biology of Tropical Ornamental Plants Germplasm, College of Forestry, Hainan University, Haikou 570228, China
2 Research Center for Terrestrial Biodiversity of the South China Sea, College of Forestry, Hainan University, Haikou 570228, China
3 Department of Development and Design, Hainan University, Haikou 570228, China
4 Key Laboratory of Germplasm Innovation on Protection and Conservation of Mountain Plant Resources, Ministry of Education/College of Agriculture, Guizhou University, Guiyang 550025, China
* Correspondence: wjhainu@hainanu.edu.cn

Abstract: *Phalaenopsis* cultivar 'Panda' is a beautiful and valuable ornamental for its big flower and unique big spots on the petals and sepals. Although anthocyanins are known as the main pigments responsible for flower colors in *Phalaenopsis*, and the anthocyanins biosynthetic pathway in *Phalaenopsis* is generally well known, the detailed knowledge of anthocynins regulation within the spot and non-spot parts in 'Panda' flower is limited. In this study, transcriptome and small RNA libraries analysis from spot and non-spot sepal tissues of 'Panda' were performed, and we found *PeMYB7*, *PeMYB11*, and miR156g, miR858 is associated with the purple spot patterning in its sepals. Transcriptome analyses showed a total 674 differentially expressed genes (DEGs), with 424 downregulated and 250 upregulated (Non-spot-VS-Spot), and 10 candidate DEGs involved in anthocyanin biosynthetic pathway. The qPCR analysis confirmed that seven candidate structure genes (*PeANS, PeF3'H, PeC4H, PeF3H, PeF3H1, Pe4CL2,* and *PeCHI*) have significantly higher expressing levels in spot tissues than non-spot tissues. A total 1552 differentially expressed miRNAs (DEMs) were detected with 676 downregulated and 876 upregulated. However, microRNA data showed no DEMs targeting on anthocyanin biosynthesis structure gene, while a total 40 DEMs target transcription factor (TF) genes, which expressed significantly different level in spot via non-spot sepal, including 2 key MYB regulator genes. These results indicated that the lack of anthocyanidins in non-spot sepal may not directly be caused by microRNA suppressing anthocyanidin synthesis genes rather than the MYB genes. Our findings will help in understanding the role of miRNA molecular mechanisms in the spot formation pattern of *Phalaenopsis*, and would be useful to provide a reference to similar research in other species.

Keywords: *Phalaenopsis*; transcriptome; microRNA; anthocyanin biosynthesis; molecular mechanism

---

1. Introduction

Flower spots are heterochromatic dots or streaks with a specific texture and pattern appearing on the corollas of plants, which can affect the behavior of pollinators and the ornamental value of flowers [1–4]. Previous studies have confirmed the flower spot is caused by the accumulation of anthocyanins in a specific area of corollas. For example, peonidin-3-O-glucoside, malvidin-3-O-glucoside, delphinidin-3-*O*-diglucoside, and cyanidin-3-*O*-glucoside are the main anthocyanins found in petal and sepal spots in *Oncidium* [5], and cyanidin and delphinidin are the main anthocyanidins in the spot of pansy (*Viola* × *wittrockiana* Gams.) petals [6].

Molecular mechanism of anthocyanins accumulation has been clearly studied in some plants, such as *Dianthus hybrida* [7], *Antirrhinum majus* [8], *Petunia hibrida* [9], *Platycodonis Radix* [10], *Gerbera jamesonii* [11], and *Phalaenopsis equestris* [12], and the anthocyanin pathway, which is a branch of the flavonoid pathway, has been elucidated as well [13,14]. Three kinds of transcription factor genes families—MYB transcription factor, bHLH transcription factor, and WD40 repeat protein family (MBW)—were also found to regulate the anthocyanin biosynthesis genes [15].

As for the accumulation of anthocyanin in specific regions of colloras, the direct cause attributes to the specific expression of the biosynthesis genes involved in the anthocyanin pathway. For example, high level expression of the genes of OgCHI and OgDFR results in anthocyanin accumulation and pigmented spot formation in yellow lip in *Oncidium* [5,16]. Upregulation of *LhCHSA, LhCHSB*, and *LhDFR* is detected within the spots located in the center of the petals, in comparison to the low expression levels in the margin in *Lilium* 'Sorbonne' [17]. In *Dendrobium moniliforme*, pigment accumulation in the base of the column has been caused by a consequence of preferential expression of *DmF3′5′H* [18]. In *Clarkia gracilis*, precise spatiotemporal regulation of the expression of the anthocyanin genes *F3′H, F3′5′H, DFR1*, and *DFR2* produces spotted petals [19]. In pansy, *VwDFR, VwF3′5′H* and *VwANS* have more significantly higher level expression in cyanic flower areas [6]. Moreover, the MYB genes also play an important role for the production of flower spot by regulating the anthocyanin biosynthesis genes. For example, the large purple spots in *Phalaenopsis* 'Everspring Fairy' was mainly caused by the expression of *MYB* [12], and the *LhMYB6* and *LhMYB12* positively regulate anthocyanin biosynthesis and determine organ- and tissue-specific accumulation of anthocyanin in Asiatic hybrid lily 'Montreux' [20]. Hsu also conducted a detailed study on three *MYBs* in *Phalaenopsis*, and found that the color patterning of flower sepals, petals and lips is regulated by different MYB genes combinations, and the pigmented veins and spots on the petals are also regulated by these three *MYB* genes [21].

RNA interference, regulating gene expression by post-transcriptional mechanisms, has also been recognized to play an important role in the color special patterning model of some plants [22]. Koseki et al. [23] found that the star-type color pattern of *Petunia hybrida* Red star' flowers is induced by sequence-specific degradation of chalcone synthase RNA. Their further study found that the formation of bicolor flower types of petunia was due to RNA interference in two *CHSA* alleles (*PhCHS-A1* and *PhCHS-A2*) [23]. In *Arabidopsis thaliana*, miRNAs may function as regulators in anthocyanin biosynthesis by targeting on related transcription factors and lead to the different accumulation of anthocyanin [24].

Phalaenopsis spp. have become important ornamental plants worldwide for their long-lasting and various colorful flowers [25]. There are varieties of flower colors and corolla pigmentation patterning styles of *Phalaenopsis* spp., so it is very important to find more details about the flower color patterning for its breeding and production. However, there was little research in this field because of the complex anthocyanin synthesis pathways and the gene expression networks in *Phalaenopsis* spp. RNA sequencing can effectively identify the subtle differences in gene expression and find the targeted genes of small RNA in transcription level, so it has been used to study flower color patterning recently, such as in *Lilium* 'Tiny Padhye' [26], monkeyflowers (*Mimulus*) [27], and tree peony [28]. In this study, the transcriptome sequencing and small RNA sequencing in spot and non-spot sepal of *Phalaenopsis* 'Panda' were carried out by genomics Illumina sequencing platform, and the expression levels of candidate genes, as well as microRNAs, were verified by qPCR. Meanwhile, the model of flower spot formation pattern was predicted by the interactions of the structural genes, regulatory genes together with small RNA. This study created a joint research of transcriptome sequencing and small RNA sequencing to explain the spot pigmentation in *Phalaenopsis.* spp., and the results will be helpful for the breeding of new colorful cultivars of *Phalaenopsis* spp.

2. Results

2.1. *Anthocyanin Accumulation Patterns in Phalaenopsis 'Panda'*

The examination by light microscope showed that anthocyanin accumulated in the upper epidermal cells of spot area, while no anthocyanin accumulated in cells in non-spot area (Figure 1A–C). This result indicated that the visible spots resulted from accumulation of anthocyanin in cells of the upper epidermis. Furthermore, the anthocyanin content of the spot in *Phalaenopsis* 'panda' sepals was significantly different from that of the sepal according to the results by the UV-visible spectrophotometer scanning detection (Figure 1D,E). Anthocyanin content measured by the pH differential method revealed that high content of anthocyanin (1.0798 mg/g) was accumulated in spot area, while was barely detectable in the acyanic parts extraction.

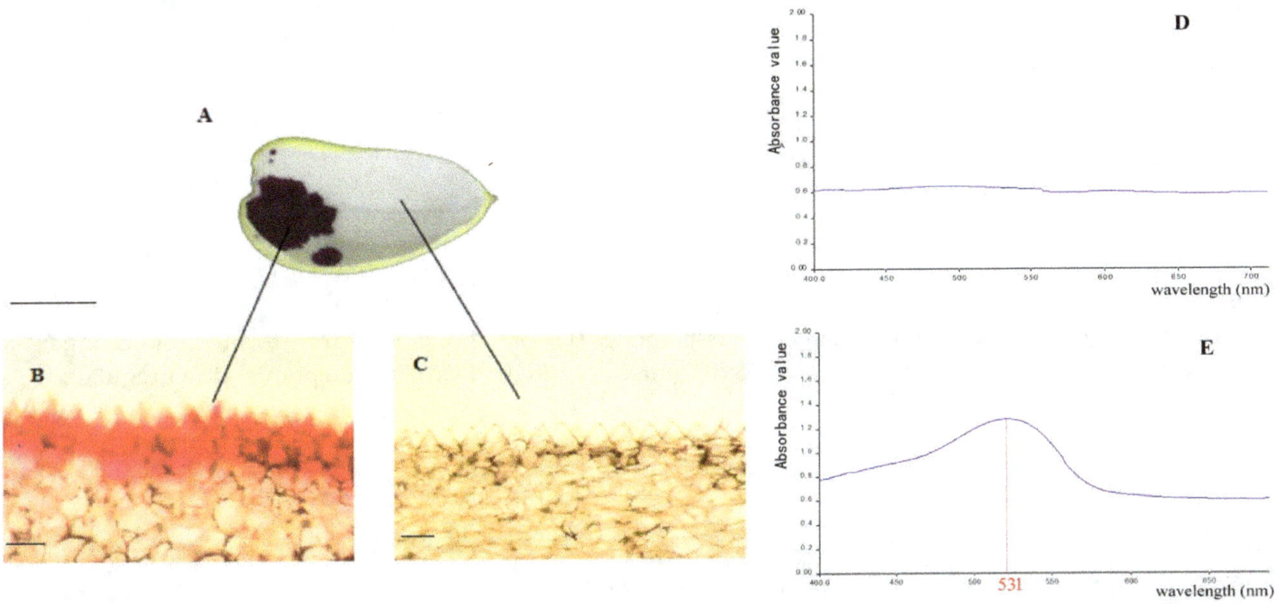

Figure 1. Anatomical structures and anthocyanin content of *Phalaenopsis* 'Panda' sepal. (**A**) Sepal (bar = 1 cm); (**B**) Upper epidermal cell in spot area (bar = 10 μm); (**C**) Upper epidermal cell in non-spot area (bar = 10 μm); (**D**) UV-visible spectrophotometer scanning of non-spot area; (**E**) UV-visible spectrophotometer scanning of spot area. Red number of 531 nm represents the absorption wavelength of anthocyanin.

2.2. *Construction of cDNA Library and Gene Mapping to the Reference Genomes*

Four cDNA libraries were constructed using Illumina Hiseq 2000 platform (Illumina, San Diego, CA, USA) and 44,632,028, 45,110,536, 45,380,086, 44,434,116 high-quality reads were obtained, respectively (Table S1 in suplememntary materials). The sequencing raw data has been deposited into the Short Reads Archive (SRA) database under the accession number SRP166213. These clean reads were mapped to reference genome of *Phalaenopsis equestris* and the average gene mapping ratio of each sample was 55.47%. We considered that the low homology between *Phalaenopsis* 'Panda' and *Phalaenopsis equestris* result to the low mapping ratio, but it has no effect on RNAseq quantitative analysis owing to the high clean reads quantity and enough sequencing data (Table S2).

2.3. *Functional Annotation and Classification*

To annotate the gene with putative functions, the assembled genes were searched against the public databases of NR. Among them, 17,871 genes were annotated to the NR database. To further illustrate the main biological functions of the transcripts, GO (15,588 genes) and KEGG pathway (19,864 genes) analyses were performed (Table S3).

2.4. Identification of Differentially Expressed Genes and KEGG Enrichment Analysis of DEGs

A total of 674 DEGs were detected in pairwise comparison with 424 downregulated and 250 upregulated genes (Non-spot-VS-Spot) (Figure 2). A total of 543 DEGs were mapped to all 106 pathways. Notably, the Flavonoid biosynthesis (ko00941, 12 DEGs of 105 gene with Q-value = 0.0005279107) and Phenylalanine metabolism (ko00360, 8 DEGs of 132 gene with Q-value = 1.900591×10^{-1}) were most significantly enriched in top20 pathways (Figure 3, Table S4).

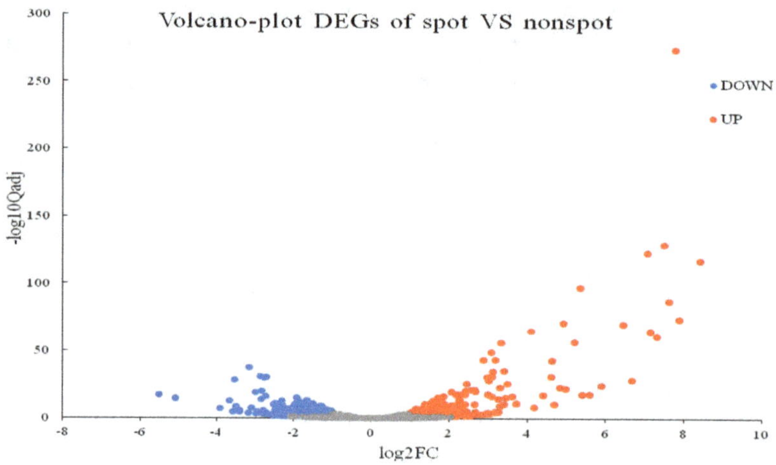

Figure 2. Volcano plot of differentially expressed genes (DEGs). X axis: log2 transformed fold change; Y axis: −log10 transformed significance; Red points: upregulated DEGs; Blue points: downregulated DEGs. Gray points: non-DEGs.

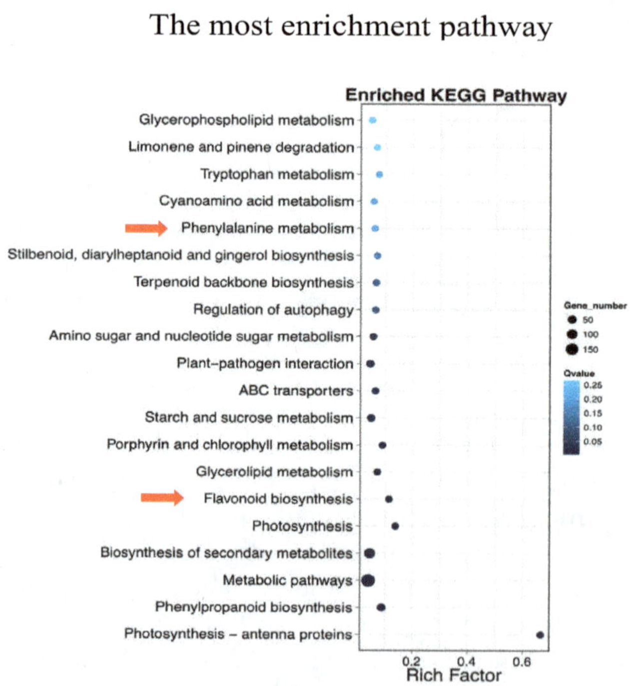

Figure 3. The most enrichment pathway of DEGs (TOP20). X axis: enrichment factor; Y axis: pathway name; The color: the q-value (high: white, low: blue), the lower q-value indicates the more significant enrichment; Point size: DEG number (The bigger dots refer to larger amount); Rich Factor: the value of enrichment factor, which is the quotient of foreground value (the number of DEGs) and background value (total Gene amount), the larger the value, the more significant enrichment. Red arrows represent the pathways related directly to the anthocyanin biosynthesis.

2.5. DEGs in Anthocyanin Biosynthesis and MBW Genes

There were 10 DEGs in flavonoid biosynthesis which are directly related to spot pattern, including *PeANS* (PEQU_25924), *PeF3′H* (PEQU_00400), *PeC4H* (PEQU_12025), *PeDFR* (PEQU_34933), *PeCHI* (PEQU_22606), *PeF3H1* (PEQU_38891), *PeF3H* (PEQU_22432), *PePAL* (PEQU_01877), *Pe4CL2* (PEQU_00756), Pe4CL (PEQU_07458), and most of them presented higher expression level in spot areas than non-spot areas (Table 1, Figure 4).

Table 1. Putative anthocyanin structural genes identified from differentially expressed genes (DEGs).

Gene ID	Annotation	FPKM (non-spot)	FPKM (spot)	Log2FC	Padj	Up/Downregulation
PEQU_25924	*PeANS*	119.9303758	21,612.25078	7.493508409	4.71×10^{-130}	Up
PEQU_00400	*PeF3′H*	31.02321	6117.42	7.623432	5.88×10^{-88}	Up
PEQU_12025	*PeC4H*	2272.365	8325.431	1.87333	1.79×10^{-8}	Up
PEQU_34933	*PeDFR*	17.40537	5973.723	8.422954	1.13×10^{-117}	Up
PEQU_22606	*PeCHI*	2467.193	707.8831	−1.80129	1.13×10^{-6}	Down
PEQU_07458	*Pe4CL*	11.87304	253.4331	4.415844	3.15×10^{-18}	Up
PEQU_38891	*PeF3H1*	52.59874	5401.439	6.682172	2.59×10^{-29}	Up
PEQU_22432	*PeF3H*	21.49244	3076.48	7.161307	2.17×10^{-65}	Up
PEQU_01877	*PePAL*	14783.45	6368.878	−1.21487	8.83×10^{-8}	Down
PEQU_00756	*Pe4CL2*	383.56	1290.653	1.750578	0.000153	Up

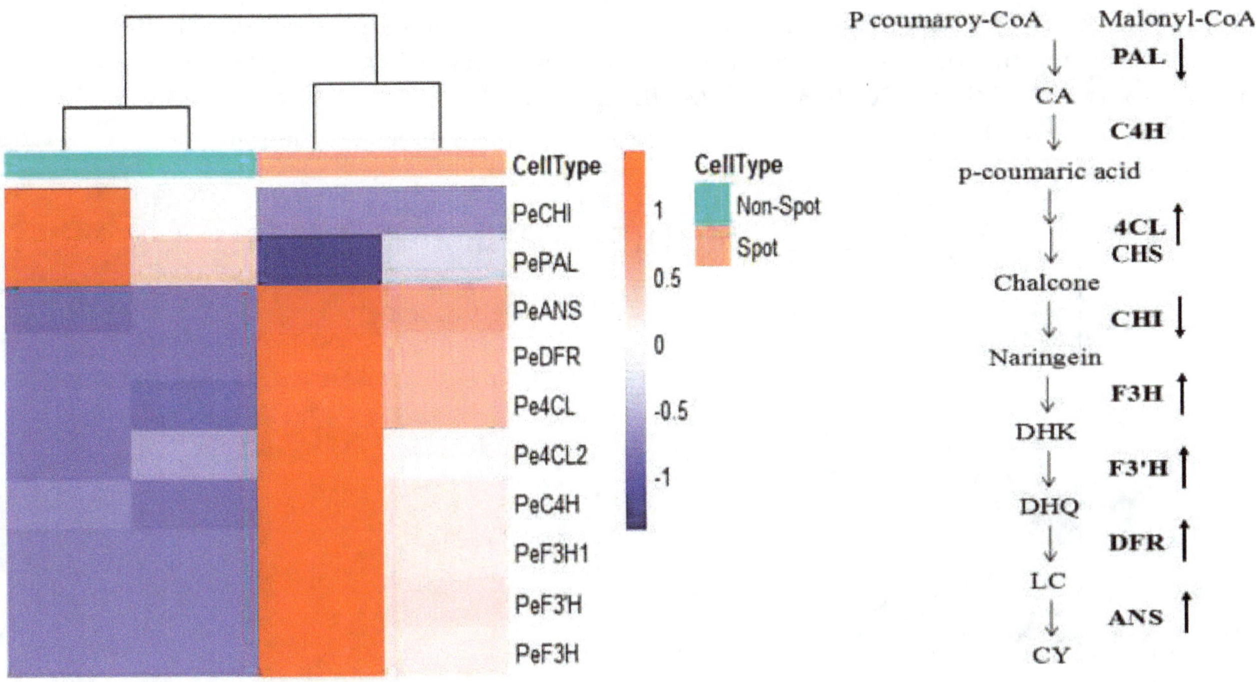

Figure 4. Heatmap of differentially expressed structure genes (DEGs) related to anthocyanin biosynthesis in *Phalaenopsis*. Bold arrow means up/downregulations of genes. (PAL, phenylalanine ammonia lyase; C4H, cinnamate-4-hydroxylase; 4CL, 4-coumarate–CoA ligase; CHS, chalcone synthase; CHI, chalcone isomerase; F3H, flavanone 3-hydroxylase; F3′H, Flavonoid 3′-hydroxylase; DFR, dihydroflavonol 4-reductase; ANS, Anthocyanidin synthase; DHK, dihydrokaempferol; DHQ, dihydroquercetin; DHM, dihydromyricetin; LC, leucocyanidin; Cy, cyaniding).

As for the MBW transcriptor genes related to the anthocyanin synthesis, 4 unigenes were found having significant expression difference including 3 MYB uingenes and 1 bHLH unigene (Table 2). In these transcriptor unigenes, 2 MYB unigenes and 1 bHLH unigene were upregulated and 1 MYB unigene was downregulated.

Table 2. Putative MBW genes identified from differentially expressed genes (DEGs).

Gene ID	Annotation	FPKM (non-spot)	FPKM (spot)	Log2FC	Padj	Up/Downregulation
PEQU_03393	*PeMYB7*	29.41538126	113.6604505	0.04965	0.00303	UP
PEQU_10361	*PeMYB11*	10.58835723	1693.628622	7.32149	8.71×10^{-62}	UP
PEQU_10362	*PeMYB11*	4.49564678	216.34208328	5.58864183	2.35×10^{-18}	UP
PEQU_09064	*PeMYB16*	1022.11354	508.0418102	−1.00853	0.03605	DOWN
PEQU_19747	*PebHLH1*	149.2074009	801.9464953	2.42618	2.57×10^{-18}	UP

2.6. Data Analysis of Small RNA Sequencing

Four sRNA libraries were constructed and 23,352,936, 24,896,769, 26,605,438, 25,926,370 high-quality tag were obtained, respectively (Table S5). The sequencing raw data has been deposited into the Short Reads Archive (SRA) database under the accession number SRP161646. These clean tags were mapped to reference genome of *Phalaenopsis equestris* and other sRNA databases. The average gene mapping ratio of each sample was 77.38% (Table S6).

A total of 1552 DEMs were detected in pairwise comparison with 676 downregulated and 876 upregulated (Non-spot-VS-Spot) (Figure 5). miRNA target gene prediction resulted in 1396 common unigenes between TargetFinder (http://http://targetfinder.org/) and psRobot (http://http://omicslab.genetics.ac.cn/psRobot/), 1648 by TargetFinder and 7652 by psRobot, and totally 1307 target unigenes were aligned to KEGG database. However, there were no DEMs were found targeting on anthocyanin biosynthesis structure gene. As for the regulator genes, we found totally 12 microRNAs targeting on 10 significant different expressed MYB, *bHLH* or *WRKY* unigenes (Table 3). However, only *PeMYB7* (PEQU_03393), *PeMYB11* (PEQU_10361, PEQU_10362) were significantly upregulated in these TF families according to the transcriptomes data, and these microRNAs belonged to miR156 or miR858 families.

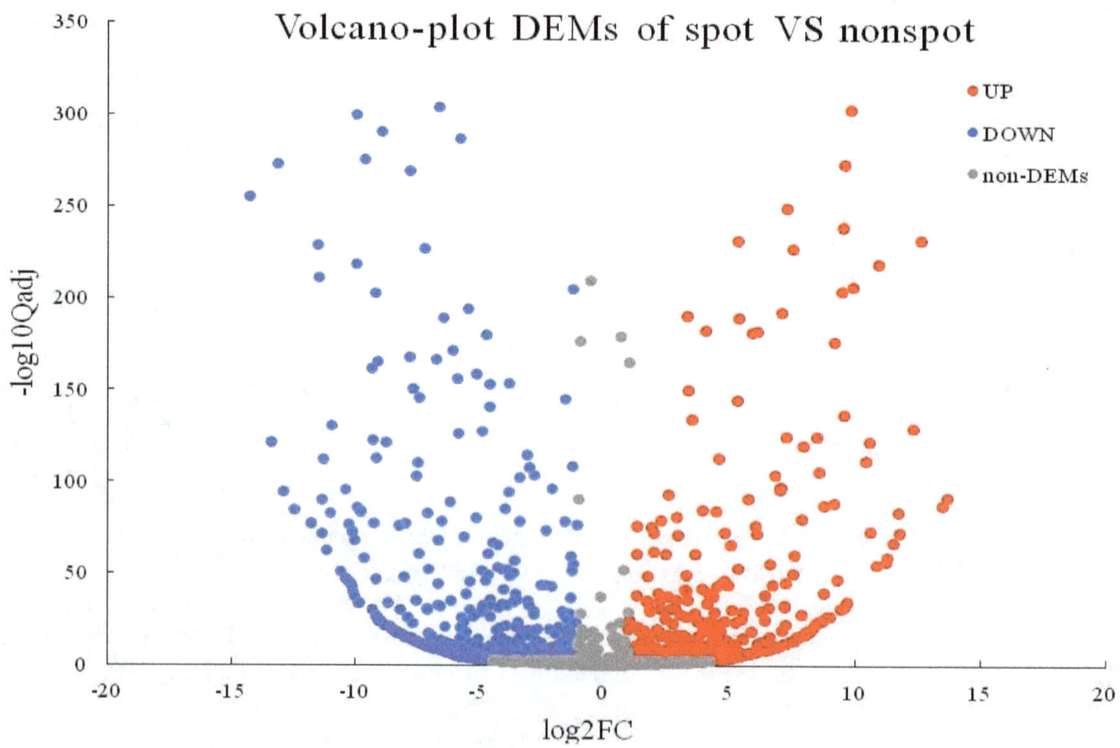

Figure 5. Volcano plot of DEMs. X axis represents log2 transformed fold change. Y axis represents −log10 transformed significance. Red points represent upregulated DEMs. Blue points represent downregulated DEMs. Gray points represent non-DEMs.

Table 3. Putative miRNAs targeting on *MYB*, *bHLH* or *WRKY* unigenes.

miRNA id	miRNA Expression in Non-spot	miRNA Expression in Spot	Up/Down Regulation of miRNA	*p*-Value	Target Gene	Target Unigene ID	Up/Downregulation of Target Unigene in Spot
Novel-m1700-5p	12	0	UP	0.000818887	PeMYB39	PEQU_22029	NA
mtr-miR156g-3p	49	1	UP	2.66×10^{-12}	PeMYB7	PEQU_03393	UP
Novel-m0210-3p	82	10	UP	1.69×10^{-13}	PeMYB7	PEQU_03393	UP
cme-miR858	52	0	UP	1.57×10^{-14}	PeMYB11	PEQU_10361	UP
					PeMYB11	PEQU_10362	UP
					PeMYB8	PEQU_10866	NA
					PeMYB12	PEQU_20333	NA
ath-miR858	336	116	UP	4.67×10^{-34}	PeMYB11	PEQU_10361	UP
					PeMYB11	PEQU_10362	UP
					PeMYB12	PEQU_20333	NA
ata-miR528-3p	12	0	UP	0.000818887	PebHLH086	PEQU_08299	NA
zma-miR528a-3p	49	1	UP	2.66×10^{-12}	PebHLH086	PEQU_08299	NA
osa-miR162b	82	10	UP	1.69×10^{-13}	PebHLH13	PEQU_33912	NA
Novel-m0112-3p	0	52	DOWN	1.57×10^{-14}	PebHLH	PEQU_26133	NA
gma-miR169v	54	0	UP	2.83×10^{-13}	katanin p80 WD40	PEQU_05516	NA
Novel-m0290-5p	24	0	UP	9.92×10^{-7}	katanin p80 WD40	PEQU_05516	NA

2.7. qPCR of Key Structural Genes, Regulate Genes and miRNA

The expression levels of 10 anthocyanin genes and 4 transcriptor genes between the non-spot and spot areas were verified by qPCR. The results showed the genes *Pe4CL2*, *PeANS*, *PeF3H*, *PeF3H1*, *PeF3'H* and *PeMYB7*, *PeMYB11* presented significantly higher expression levels in spot areas than the non-spot areas (Figure 6), which was generally consistent with the results of the transcriptome data (Table 1).

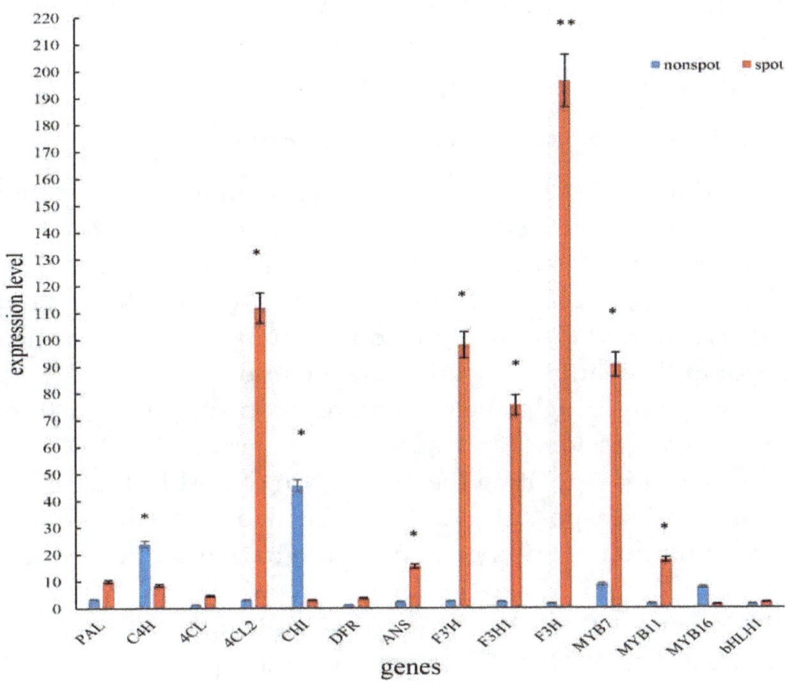

Figure 6. The expression level of structure gene and regulate gene Associated with Anthocyanin Biosynthesis, * represent there are significant differences between spot and non-spot area; ** represent there are extremely significant differences; red: spot area, blue: non-spot area.

The expression of microRNA of RNA miR858, miR156g were also analyzed by qPCR between the non-spot and spot areas (Figure 7). The results showed that they presented higher expression levels in non-spot areas, which means their targeting genes, *PeMYB7* and *PeMYB11*, having less expression levels in the non-spot areas.

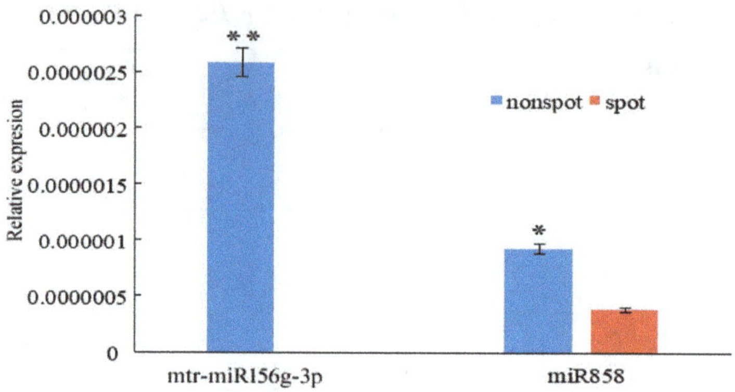

Figure 7. The expression level of mtr-miR156g-3p and miR858. ** represent an extremly significant difference ($p < 0.01$); * represent an significant difference ($p < 0.05$).

3. Discussion

3.1. Low Expression of Anthocyanin Genes Causing the Lack of Pigments in Non-Spot Areas

In this study, we found anthocyanin accumulated in the upper epidermal cells in spot area, while no anthocyanin accumulated in cells in non-spot areas of 'Panda' (Figure 1). Transcriptome data and qPCR indicated 7 anthocyanin genes were low expression in the non-spot areas, especially for the very different expression levels of *4CL2, F3'H, F3H* and *F3H1* (Table 2, Figure 6). In *Phalaenopsis* 'Everspring Fairy', which also has purple spot on the white petal and sepal, DFR is the main gene which differently expressed between the purple and white part of the petals and sepals [12]. However, in *Phalaenopsis* 'Panda' expression of *DFR* in spot tissue was not significantly changed in comparison to non-spot area. These results showed that the same phenotype may be caused by different genes in the pigmentation in *Phalaenopsis.*

3.2. PeMYB7 and PeMYB11 Are Important Genes in Spot Formation

MYBs have been found to be very important in the floral pigmentation patterning in *Phalaenopsis* spp. In the sepals/petals, silencing of *PeMYB2, PeMYB11,* and *PeMYB12* resulted in the loss of the full-red pigmentation, red spots, and venation patterns, respectively; *PeMYB11* was responsive to the red spots in the callus of the lip, and *PeMYB12* participated in the full pigmentation in the central lobe of the lip [21]. In *P. amabilis* and *P. schilleriana*, anthocyanin-specific Myc and Wd were expressed, however, Myb specific for anthocyanin biosynthesis were undetectable in *P. amabilis*; in *Phalaenopsis* 'Everspring Fairy' petals and sepals, high levels of anthocyanin-specific Myb transcripts were present in the purple, but not in the white sectors [12]. Hsu, et al. [29] verified *PeMYB11* as the major regulatory R2R3-MYB transcription factor for regulating the production of the black color, and a retrotransposon, named Harlequin Orchid RetroTransposon 1 (HORT1), resulted in strong expression of *PeMYB11* and thus extremely high accumulation of anthocyanins in the harlequin flowers of the *P.* Yushan Little Pearl variety.

In our study, *PeMYB7* and *PeMYB11* expressed significantly different between the spot and non-spot areas, while *PeMYB2* and *PeMYB12* had not different expression levels (Table 2 and Figure 6). This result indicated that the gene of *PeMYB11* may also play an important role in spot pigmentation in *Phalaenopsis* 'Panda', which was similar to these previous researches [12,21,29,30]. *PeMYB7* had different expression levels between the spot and non-spot areas of sepals in this research, which indicated this gene may also associate with purple spot patterning in *Phalaenopsis* 'Panda'. The function of *PeMYB7* is not very clear presently, however, in the phylogenetic tree inferred from MYB genes of *Phalaenopsis equestris* and *Oryza sativa* (Figure 8A), *PeMYB7* was in the same clad of *PeMYB2* and *PeMYB8*, suggesting *PeMYB7* may have similar function as *PeMYB2* which can activate anthocyanin synthesis [30].

3.3. *miR156g, miR858 Silence PeMYB7, and PeMYB11*

MiRNAs can influence tissue anthocyanin formation in previous studies. In petunia, the sequence-specific degradation of CHS RNA is the primary cause of the formation of white sectors in 'Red Star' flowers [23]. In potato (*Solanum tuberosum* L.), small RNAs of miR828 and TAS4 D4(−) can decrease the expression levels of MYB12 and R2R3-MYB genes in purple skin and flesh, which caused to anthocyanin accumulation [31]. In our research, miR156 and miR858 were predicted as the main interference RNAs for PeMYB7 and PeMYB11 (Table 3). In order to confirm *PeMYB7* and *PeMYB11* are the direct targets of the small RNAs, we download the complete cDNAs of *PeMYB7* (GenBank: KF769472.1) and *PeMYB11* (GenBank: MH675649.1), and analyzed the target sites of miR156 and miR858 in these two genes (Figure 8B). This finding confirms that miR156g and miR858 can silence *PeMYB7* and *PeMYB11*.

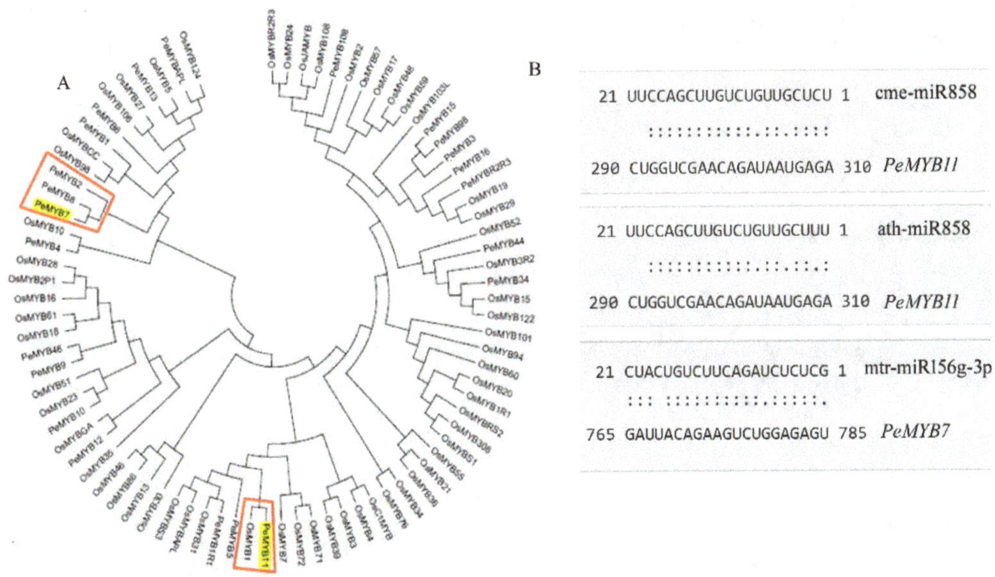

Figure 8. Identification of *PeMYB7*, *PeMYB11* and their target sites by miR156g, miR858. (**A**) Phylogenetic tree inferred from the MYB sequences of *Phalaenopsis equestris* and *Oryza sativa*; (**B**) Identification of target sites of miR156 and miR858 in *PeMYB7* and *PeMYB11*.

In *Arabidopsis thaliana*, miR156 negatively regulates the anthocyanin accumulation by regulating the expression level of target gene *MYB113* indirectly [32]. High miR156 level promotes anthocyanin biosynthesis through the negative regulation of *SPL9* gene because SPL9 can destabilize the MYB-bHLH-WD40 transcriptional activation complex. However, in this study, *SPL9* gene in *Phalaenopsis* doesn't express differently, which indicated that mtr-miR156g-3p may be able to directly silence *PeMYB7* and caused the anthocyanin lacking on the white sector.

The biological functions of miR858 has not been fully explored, however, small RNA sequencing in *Arabidopsis thaliana* [24] and *Vitis vinifera* L. [33] verified that its target gene is *MYB12*, which is a positive transcript factor for the anthocyanin synthesis. Chitwood et al. transferred the constructed binary vector into wild-type tomato plants to reduced MIR858 expression level, and found the expression levels of *SLY-Myb12* in transgenic plants were lower than wild type, with upregulations of the related structure gene *PAL*, *DFR*, *ANS* and *CHS* involved in anthocyanin biosynthesis, and increasing of anthocyanin content [34]. Ballester et al. [35] used VIGS technology silenced *MYB12* and then found the tomato fruits turned pink. In our study the targeted genes of miR858 were predicted to be both *MYB12* and *MYB11* (Table 3), however, the unigene of *PeMYB12* didn't show significantly different expression levels between the spot and non-spot areas, suggesting that miR858 explored higher effect on the unigene of *PeMYB11* in 'Panda'.

3.4. A Proposed Modelsummarizing of Spot Formation Pattern in Phalaenopsis 'Panda'

Our research confirmed *Pe4CL2*, *PeF3H*, *PeF3'H*, and *PeANS* expression were low or repressed in non-spot areas in 'Panda' sepals, and MIR156g and MIR858 may silence the expression of *PeMYB11* and *PeMYB7* in non-spots parts. In conclusion, we illustrated a proposed molecular mechanism in spots formation in 'Panda' (Figure 9). The diagram revealed the unique spot formation in 'panda' may result from the higher expression of MIR156g and MIR858 clusters in non-spot tissue, which targeted and suppressed the expression of key regulate genes *PeMYB7*, *PeMYB11* and then caused very low expression of *Pe4CL2*, *PeF3H*, *PeF3'H*, and *PeANS* in non-spot tissue. These results finally led to the lack of synthesis and accumulation of anthocyanin in non-spot area.

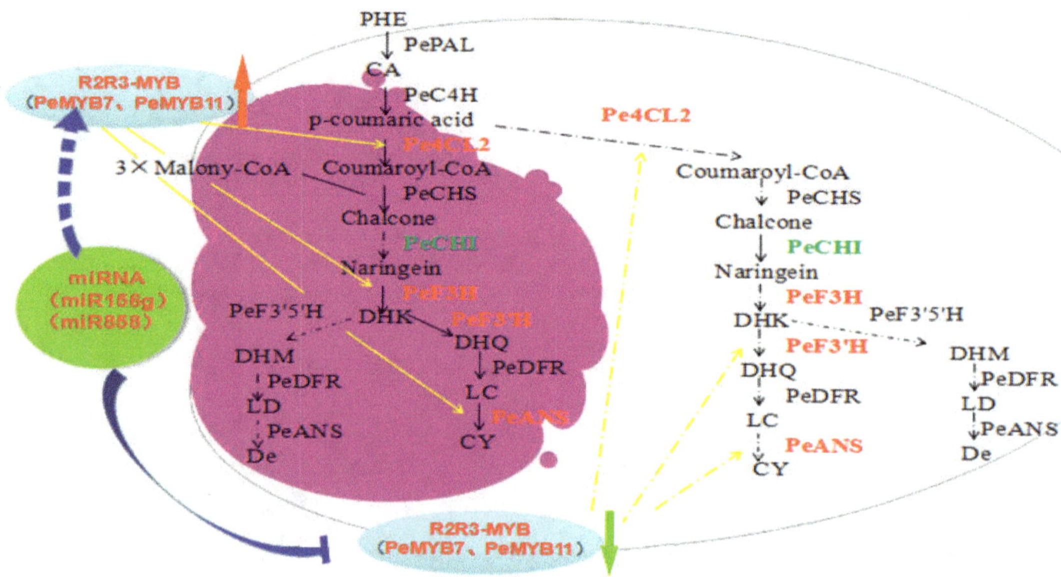

Figure 9. A proposed model summarizing of spot formation pattern in *Phalaenopsis* 'Panda' sepal. The purple segment illustrates the accumulation of anthocyanidin in spot area of 'Panda'. MIR156g, MIR858 target on key regulate genes *PeMYB7* and *PeMYB11*, suppressed the transcript level of *Pe4CL2, PeF3H, PeF3'H, PeANS* and resulted in reduced anthocyanidin production in the non-spot area, while relatively lower expression levels of MIR156g and MIR858 and high levels of transcription of these genes in spot area cause accumulation of anthocyanidin. The black dashed arrows represent low levels of transcription, while the black solid-line arrows represent high levels of transcription. The yellow solid-line arrows represent promoting transcription, while yellow dashed arrow represent lack of promotion function. The blue bold solid-line T-arrow represents interfering transcription, while blue bold dashed arrow represents lack of interference function. The red bold arrow represents upregulation, and the green bold arrow represents downregulation. DHK, dihydrokaempferol; DHQ, dihydroquercetin; DHM, dihydromyricetin; LC, leucocyanidin; LD, leucodelphinidin; Cy, cyanidin; De, delphinidin.

4. Materials and Methods

4.1. Plant Materials

Phalaenopsis 'Panda' (Figure 10) were grown at the horticultural farm of Hainan University (Latitude: 20.03N, longitude: 110.33E), Haikou, Hainan Province, China. The sepal tissues were divided into two parts, spots area and non-spots area for total anthocyanin analyses and totally RNA extraction. Three independent biological replicates were collected for each sample. All samples were immediately frozen in liquid nitrogen and stored at −80 °C until use.

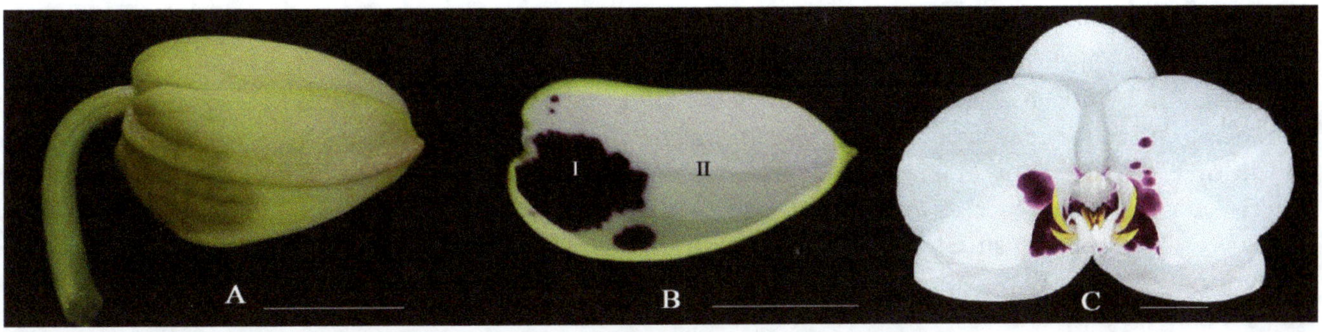

Figure 10. The flower tissue of *Phalaenopsis* 'Panda'. Bars = 1 cm. (**A**) Full bud; (**B**) Sepal; I. spot area; II. non-spot area. (**C**) Full bloom stage.

4.2. Observations of Sepal Anatomy and Determination of Total Anthocyanin Content

Spot and non-spot sepal tissue of full bud were cut into segments longitudinally (approx.10 mm × 5 mm), and the upper epidermis and dorsal epidermis were separated with forceps. The sections were observed and photographed with a light microscope (Primo Vert, Zeiss, Germany).

The total anthocyanin content of *Phalaenopsis* sepal in spot and non-spot areas were determined by the pH differential method [36].

4.3. RNA Extraction, cDNA Library Construction, and mRNA Sequencing

RNAs were extracted from spot and non-spot parts of sepals of full bud using the modified Trizol method [6]. Two biological replicates were used for spot sepal and non-spot sepal. A total of 4 RNA-seq libraries were constructed from these sepals. RNA-Seq libraries were constructed using the RNA Library Prep Kit for Illumina according to the manufacturer's instructions (NEB, Ipswich, MA, USA). Library quality was assessed on the Agilent Bioanalyzer 2100 system. The mRNA libraries were sequenced on the Illumina HiSeq 2000 platform (Illumina, San Diego, CA, USA) based on sequencing by synthesis with 150 bp paired-end reads (Biomarker Technologies, Wuhan, China).

4.4. mRNA Transcriptome Data Analysis

The reads with adaptors were removed firstly. The reads in which unknown bases comprised more than 5% of the total and low quality reads (the percentage of the low quality bases of quality value $\leq$ 15 is more than 20% in a read) were also removed. The clean reads were aligned to *Phalaenopsis equestris* genome accessed from (https://www.ncbi.nlm.nih.gov/bioproject/192198) by HISAT2 (http://www.ccb.jhu.edu/software/hisat/index.shtml) [37].

Gene alignment rate was counted with Bowtie2 v2.2.5 (http://bowtie-bio.sourceforge.net/bowtie2/index.shtml), gene and transcript expression levels were calculated with RSEM v1.2.12 (http://deweylab.biostat.wisc.edu/RSEM). The differentially expressed genes (DEGs) between spot and non-spot tissues were identified and filtered with the R package DESeq2 based on $|\log2\ (\text{foldchange})| \geq 1$ and Padj < 0.05 [38]. The heatmap displays of the Trimmed Mean of M-values (TMM) normalized against the Fragment Per Kilobase of transcripts per Million (FPKM) mapped reads were performed using the R package pheatmap (https://cran.r-project.org/src/contrib/Archive/pheatmap/). The DGEs were imported into Blast2GO software v2.5 [39] and in-house perl scripts for gene ontology (GO) term analysis, while the KEGG pathways were assigned to the assembled genes using the online KEGG Automatic Annotation Server (KAAS, http://www.genome.jp/kegg/), and then the phyper function in R software was used for enrichment analysis with $FDR \leq 0.01$. To identify transcription factors, the assembled transcriptome were searched against the Plant Transcription Factor Database PlnTFDB (http://plntfdb.bio.uni-potsdam.de/v3.0/downloads.php) using hmmseach v3.0 (http://hmmer.org).

4.5. *qRT-PCR Analyses of mRNA*

10 DEGs in anthocyanin biosynthesis and 4 DEGs of transcription factor were chosen for qRT-PCR analyses. qRT-PCR analyses were performed using SYBR Premix Ex TaqTM II (Tli RNaseH Plus) (Takara, Dalian, China) according to the manufacturer's instructions with the following reaction conditions: denaturation at 95 °C for 30 s and 40 cycles of amplification (95 °C for 5 s, 60 °C for 30 s). The β-Actin gene was used as an internal control for normalization. Relative expression levels of target genes were calculated using the $2^{-\Delta\Delta Ct}$ [40] against the internal control. The gene-specific primers are shown in Table S7. Experiments were performed with three independent biological replicates and three technical replicates.

4.6. *Small RNA Library Construction and Sequencing*

RNA was extracted from spot and non-spot sepal using the modified Trizol method [6], and 4 small RNA libraries from sepal spot and non-spot samples with 2 biological replicates were constructed and sequenced by SBQ500 sequencing method (BGI, Wuhan, China). Then the low-quality, 5′ primer contaminants, without 3′ primers and insert tags, and sequences fewer than 18 nucleotides (nt) were filtered out to obtain clean reads from raw data reads. The final clean reads of the 4 libraries were mapped to *Phalaenopsis* genome and other smallRNA databases using Bowtie2 [41]. All the remaining clean sequences were annotated into different classes to remove rRNA, scRNA, snoRNA, snRNA, and tRNA using miRBase, siRNA, piRNA, snoRNA database with Bowtie2 [42] and Rfam database with cmsearch [43]. The novel miRNAs were predicted by Mireap software (https://sourceforge.net/projects/mireap/).

4.7. *Data Analysis of Small RNA Sequencing*

The expression level of miRNAs was compared between spot and non-spot tissues to identify differentially expressed miRNAs (DEMs). The differentially expressed miRNA between spot and non-spot tissues were identified and filtered with the R package DESeq2 and Poisson Distribution [38,44]. The DEMs were determined based on FDR ≤ 0.001 and the absolute value of |Log2FC| ≥ 1. The heatmap displays of the Trimmed Mean of M-values (TMM) normalized against the Fragment Per Kilobase of transcripts per Million mapped reads (FPKM) were performed created using the R package pheatmap (https://cran.r-project.org/src/contrib/Archive/pheatmap/).

The potential target genes by differentially expressed miRNAs were predicted and analyzed with 2 different software, including psRobot [45] and Targetfinder [46]. In order to increase the level of confidence and get more reliable results, we selected only those binding sites that were predicted by both of two softwares.

All protein-coding target genes regulated by DEMs were annotated against the KEGG databases. The KEGG enrichment analysis for the target genes of the DEMs was performed by Fisher's exact test (P-values) in Blast2GO pipeline, and P-values were used to conduct multiple test correction by FDR. KEGG terms with FDR< 0.05 were considered to be significantly enriched.

4.8. *Verification of miRNA Expression Levels by qPCR*

Two miRNA targets on key regulate gene transcription factor *PeMYB11* and *PeMYB7* were chosen for qRT-PCR analyses. The qRT-PCR analyses were performed using SYBR Premix Ex TaqTM II (Tli RNaseH Plus) (Takara, Dalian, China) according to the manufacturer's instructions with the following reaction conditions: Denaturation at 95 °C for 30 s and 40 cycles of amplification (95 °C for 5 s, 60 °C for 30 s). The U6 gene was used as an internal control for normalization. Relative expression levels of target genes were calculated using the $2^{-\Delta\Delta Ct}$ [40] against the internal control. The specific stem-ring primers were designed according to the miRNA qPCR primer design method [47], showed in Table S8. Experiments were performed with three independent biological replicates and three technical replicates.

Author Contributions: Conceptualization, J.W. and X.S.; methodology, A.Z., Z.C. and T.L.; software, H.P.; materies culture, Y.S. and X.L.; validation, A.Z., Y.Z. (Ying Zhao) and Y.Z. (Yang Zhou); writing—original draft preparation, A.Z.; writing—review and editing, J.W., W.H. and T.P.

References

1. Eckhart, V.M.; Rushing, N.S.; Hart, G.M.; Hansen, J.D. Frequency-dependent pollinator foraging in polymorphic Clarkia xantiana ssp. xantiana populations: Implications for flower colour evolution and pollinator interactions. *Oikos* **2010**, *112*, 412–421. [CrossRef]
2. Miller, R.; Owens, S.J.; Rørslett, B. Plants and colour: Flowers and pollination. *Opt. Laser Technol.* **2011**, *43*, 282–294. [CrossRef]
3. Moeller, D.A. Pollinator community structure and sources of spatial variation in plant–pollinator interactions in Clarkia xantiana ssp. xantiana. *Oecologia* **2005**, *142*, 28–37. [CrossRef] [PubMed]
4. Shang, Y.; Venail, J.; Mackay, S.; Bailey, P.C.; Schwinn, K.E.; Jameson, P.E.; Martin, C.R.; Davies, K.M. The molecular basis for venation patterning of pigmentation and its effect on pollinator attraction in flowers of Antirrhinum. *New Phytol.* **2011**, *189*, 602–615. [CrossRef] [PubMed]
5. Chiou, C.Y.; Wu, K.; Yeh, K.W. Characterization and promoter activity of chromoplast specific carotenoid associated gene (CHRC) from Oncidium Gower Ramsey. *Biotechnol. Lett.* **2008**, *30*, 1861–1866. [CrossRef] [PubMed]
6. Li, Q.; Wang, J.; Sun, H.Y.; Shang, X. Flower color patterning in pansy (Viola x wittrockiana Gams.) is caused by the differential expression of three genes from the anthocyanin pathway in acyanic and cyanic flower areas. *Plant Physiol. Biochem.* **2014**, *84*, 134–141. [CrossRef] [PubMed]
7. Forkmann, G.; Dangelmayr, B. Genetic control of chalcone isomerase activity in flowers of Dianthus caryophyllus. *Biochem. Genet.* **1979**, *18*, 519–527. [CrossRef]
8. Matin, C.; Prescott, A.; Mackay, S.; Bartlett, J.; Vrijlandt, E. Control of anthocyanin biosynthesis in flowers of Antirrhinum majus. *Plant J.* **1991**, *1*, 37–49.
9. Holton, T.A.; Brugliera, F.; Tanaka, Y. Cloning and expression of flavonal synthase from Petunia hybrida. *Plant J.* **1993**, *4*, 1003–1010. [CrossRef]
10. Kawabata, S.K.Y.; Kusuhara, Y.K.; Li, Y.H.; Sakiyama, R.Z. The regulation of anthocyanin Biosythesis in Eustoma grandiflorum under low light condition. *J. Jpn. Soc. Hort. Sci.* **1999**, *68*, 519–526. (In Japanese) [CrossRef]
11. Bashandy, H.; Pietiainen, M.; Carvalho, E.; Lim, K.J.; Elomaa, P.; Martens, S.; Teeri, T.H. Anthocyanin biosynthesis in gerbera cultivar 'Estelle' and its acyanic sport 'Ivory'. *Planta* **2015**, *242*, 601–611. [CrossRef] [PubMed]
12. Ma, H.M.; Pooler, M.; Griesbach, R. Anthocyanin Regulatory/Structural Gene Expression in Phalaenopsis. *J. Am. Soc. Hortic. Sci.* **2009**, *134*, 88–96. [CrossRef]
13. Katsumoto, Y.; Fukuchi-Mizutani, M.; Fukui, Y.; Brugliera, F.; Holton, T.A.; Karan, M.; Nakamura, N.; Yonekura-Sakakibara, K.; Togami, J.; Pigeaire, A. Engineering of the rose flavonoid biosynthetic pathway successfully generated blue-hued flowers accumulating delphinidin. *Plant Cell Physiol.* **2007**, *48*, 1589. [CrossRef] [PubMed]
14. Saito, K.; Yonekura-Sakakibara, K.; Nakabayashi, R.; Higashi, Y.; Yamazaki, M.; Tohge, T.; Fernie, A.R. The flavonoid biosynthetic pathway in Arabidopsis: Structural and genetic diversity. *Plant Physiol. Biochem.* **2013**, *72*, 21–34. [CrossRef] [PubMed]
15. Zhao, L.; Gao, L.; Wang, H.; Chen, X.; Wang, Y.; Yang, H.; Wei, C.; Wan, X.; Xia, T. The R2R3-MYB, bHLH, WD40, and related transcription factors in flavonoid biosynthesis. *Funct. Integr. Genom.* **2013**, *13*, 75–98. [CrossRef] [PubMed]
16. Chiou, C.Y.; Yeh, K.W. Differential expression of MYB gene (OgMYB1) determines color patterning in floral tissue of Oncidium Gower Ramsey. *Plant Mol. Biol.* **2008**, *66*, 379–388. [CrossRef] [PubMed]
17. Yamagishi, M. Oriental hybrid lily Sorbonne homologue of LhMYB12 regulates anthocyanin biosyntheses in flower tepals and tepal spots. *Mol. Breed.* **2011**, *28*, 381–389. [CrossRef]
18. Whang, S.S.; Wan, S.U.; Song, I.J.; Lim, P.O.; Choi, K.; Park, K.W.; Kang, K.W.; Mi, S.C.; Koo, J.C. Molecular Analysis of Anthocyanin Biosynthetic Genes and Control of Flower Coloration by Flavonoid 3′,5′-Hydroxylase (F3′5′H) in Dendrobium moniliforme. *J. Plant Biol.* **2011**, *54*, 209–218. [CrossRef]

19. Martins, T.R.; Berg, J.J.; Blinka, S.; Rausher, M.D.; Baum, D.A. Precise spatio-temporal regulation of the anthocyanin biosynthetic pathway leads to petal spot formation in Clarkia gracilis (Onagraceae). *New Phytol.* **2013**, *197*, 958–969. [CrossRef]
20. Yamagishi, M.; Shimoyamada, Y.; Nakatsuka, T.; Masuda, K. Two R2R3-MYB genes, homologs of petunia AN2, regulate anthocyanin biosyntheses in flower tepals, tepal spots and leaves of Asiatic hybrid lily. *Plant Cell Physiol.* **2010**, *51*, 463–474. [CrossRef]
21. Hsu, C.C.; Chen, Y.Y.; Tsai, W.C.; Chen, W.H.; Chen, H.H. Three R2R3-MYB transcription factors regulate distinct floral pigmentation patterning in Phalaenopsis spp. *Plant Physiol.* **2015**, *168*, 175–191. [CrossRef]
22. Kaucsár, T.; Rácz, Z.; Hamar, P. Post-transcriptional gene-expression regulation by micro RNA (miRNA) network in renal disease. *Adv. Drug Deliv. Rev.* **2010**, *62*, 1390–1401. [CrossRef]
23. Koseki, M.; Goto, K.; Masuta, C.; Kanazawa, A. The star-type color pattern in Petunia hybrida 'red Star' flowers is induced by sequence-specific degradation of chalcone synthase RNA. *Plant Cell Physiol.* **2005**, *46*, 1879–1883. [CrossRef]
24. Guan, X.; Pang, M.; Nah, G.; Shi, X.; Ye, W.; Stelly, D.M.; Chen, Z.J. miR828 and miR858 regulate homoeologous MYB2 gene functions in Arabidopsis trichome and cotton fibre development. *Nat. Commun.* **2014**, *5*, 3050. [CrossRef]
25. Cai, J.; Liu, X.; Vanneste, K.; Proost, S.; Tsai, W.C.; Liu, K.W.; Chen, L.J. The genome sequence of the orchid Phalaenopsis equestris. *Nat. Genet.* **2015**, *47*, 65–72. [CrossRef]
26. Xu, L.; Yang, P.; Feng, Y.; Xu, H.; Cao, Y.; Tang, Y.; Yuan, S.; Liu, X.; Ming, J. Spatiotemporal Transcriptome Analysis Provides Insights into Bicolor Tepal Development in Lilium "Tiny Padhye". *Front. Plant Sci.* **2017**, *8*, 398. [CrossRef]
27. Li, C.Q.; Shen, W.; Yan, P.; Li, X.Y.; Zhou, P. Bioinformatic Prediction of microRNA and Their Target Gene in Mimulus gutatus. *Guizhou Agric. Sci.* **2015**, *1*, 8–12.
28. Zhang, Y.; Cheng, Y.; Ya, H.; Xu, S.; Han, J. Transcriptome sequencing of purple petal spot region in tree peony reveals differentially expressed anthocyanin structural genes. *Front. Plant Sci.* **2015**, *6*, 964. [CrossRef]
29. Hsu, C.C.; Su, C.J.; Jeng, M.F.; Chen, W.H.; Chen, H.H. A HORT1 retrotransposon insertion in the PeMYB11 promoter causes harlequin/black flowers in Phalaenopsis orchids. *Plant Physiol.* **2019**, *180*, 1535–1548. [CrossRef]
30. Hsu, C.C.; Chen, H.H. *Flower Color and Pigmentation Patterns in Phalaenopsis Orchids. Orchid Biotechnology III*; World Scientific Publishing: Singapore, 2017; pp. 393–420.
31. Bonar, N.; Liney, M.; Zhang, R.; Austin, C.; Dessoly, J.; Davidson, D.; Stephens, J.; McDougall, G.; Taylor, M.; Bryan, G.J.; et al. Potato miR828 Is Associated With Purple Tuber Skin and Flesh Color. *Front. Plant Sci.* **2018**, *9*, 1742. [CrossRef]
32. Gou, J.Y.; Felippes, F.F.; Liu, C.J.; Weigel, D.; Wang, J.W. Negative regulation of anthocyanin biosynthesis in Arabidopsis by a miR156-targeted SPL transcription factor. *Plant Cell* **2011**, *23*, 1512–1522. [CrossRef]
33. Chitwood, D.H.; Klein, L.L.; O'Hanlon, R.; Chacko, S.; Greg, M.; Kitchen, C.; Miller, A.; Londo, J.P. Latent developmental and evolutionary shapes embedded within the grapevine leaf. *New Phytol.* **2016**, *210*, 343–355. [CrossRef]
34. Jia, X.; Shen, J.; Liu, H.; Li, F.; Ding, N.; Gao, C.; Pattanaik, S.; Patra, B.; Li, R.; Yuan, L. Small tandem target mimic-mediated blockage of microRNA858 induces anthocyanin accumulation in tomato. *Planta* **2015**, *242*, 283–293. [CrossRef]
35. Ballester, A.R.; Molthoff, J.; Vos, R.D.; Hekkert, B.T.L.; Orzaez, D.; Fernándezmoreno, J.P.; Tripodi, P.; Grandillo, S.; Martin, C.; Heldens, J. Biochemical and molecular analysis of pink tomatoes: Deregulated expression of the gene encoding transcription factor SlMYB12 leads to pink tomato fruit color. *Plant Physiol.* **2010**, *152*, 71–84. [CrossRef]
36. Giusti, M.M.; Wrolstad, R.E. Charaterization and Measurement of Anthocyanins by UV-Visible Spectroscopy. *Food Anal. Chem.* **2001**. [CrossRef]
37. Kim, D.; Langmead, B.; Salzberg, S.L. HISAT: A fast spliced aligner with low memory requirements. *Nat. Method* **2015**, *12*, 357–360. [CrossRef]
38. Anders, S.; Huber, W. Differential expression analysis for sequence count data. *Genome Biol.* **2010**, *11*, R106. [CrossRef]

39. Conesa, A.; Götz, S. Blast2GO: A comprehensive suite for functional analysis in plant genomics. *Int. J. Plant Genom.* **2008**, *2008*, 619832. [CrossRef]
40. Livak, K.J.; Schmittgen, T.D. Analysis of Relative Gene Expression Data Using Real-Time Quantitative PCR and the $2^{-\Delta\Delta Ct}$ Method. *Methods* **2001**, *25*, 402–408. [CrossRef]
41. Langmead, B.; Salzberg, S.L. Fast gapped-read alignment with Bowtie 2. *Nat. Methods* **2012**, *9*, 357–359. [CrossRef]
42. Langmead, B.; Trapnell, C.; Pop, M.; Salzberg, S.L. Ultrafast and memory-efficient alignment of short DNA sequences to the human genome. *Genome Biol.* **2009**, *10*, R25. [CrossRef]
43. Nawrocki, E.P.; Eddy, S.R. Infernal 1.1: 100-fold faster RNA homology searches. *Bioinformatics* **2013**, *29*, 2933–2935. [CrossRef]
44. Audic, S.; Claverie, J.M. The significance of digital gene expression profiles. *Genome Res.* **1997**, *7*, 986–995. [CrossRef]
45. Wu, H.J.; Ma, Y.K.; Chen, T.; Wang, M.; Wang, X.J. PsRobot: A web-based plant small RNA meta-analysis toolbox. *Nucleic acids Res.* **2012**, *40*, W22–W28. [CrossRef]
46. Fahlgren, N.; Carrington, J.C. miRNA Target Prediction in Plants. *Methods Mol. Biol.* **2010**, *592*, 51–57.
47. Varkonyi-Gasic, E.; Hellens, R.P. *qRT-PCR of Small RNAs. Methods in Molecular Biology*; Humana Press: Totowa, NJ, USA, 2010; pp. 109–122.

9

Volatile Organic Compounds from Orchids: From Synthesis and Function to Gene Regulation

Mummadireddy Ramya [1], Seonghoe Jang [2], Hye-Ryun An [1], Su-Young Lee [1], Pil-Man Park [1] and Pue Hee Park [1,3,*]

1 Floriculture Research Division, National Institute of Horticultural and Herbal Science, RDA, Wanju-gun, Jellabuk-do 55365, Korea; ramya87.4u@gmail.com (M.R.); hryun@korea.kr (H.-R.A.); lsy8542224@korea.kr (S.-Y.L.); pmpark@korea.kr (P.-M.P.)

2 World Vegetable Center Korea Office (WKO), Wanju-gun, Jellabuk-do 55365, Korea; seonghoe.jang@worldveg.org

3 Department of Horticultural Science and Biotechnology, Seoul National University (SNU), Seoul 08826, Korea

* Correspondence: puehee@korea.kr or puehee@gmail.com

Abstract: Orchids are one of the most significant plants that have ecologically adapted to every habitat on earth. Orchids show a high level of variation in their floral morphologies, which makes them popular as ornamental plants in the global market. Floral scent and color are key traits for many floricultural crops. Volatile organic compounds (VOCs) play vital roles in pollinator attraction, defense, and interaction with the environment. Recent progress in omics technology has led to the isolation of genes encoding candidate enzymes responsible for the biosynthesis and regulatory circuits of plant VOCs. Uncovering the biosynthetic pathways and regulatory mechanisms underlying the production of floral scents is necessary not only for a better understanding of the function of relevant genes but also for the generation of new cultivars with desirable traits through molecular breeding approaches. However, little is known about the pathways responsible for floral scents in orchids because of their long life cycle as well as the complex and large genome; only partial terpenoid pathways have been reported in orchids. Here, we review the biosynthesis and regulation of floral volatile compounds in orchids. In particular, we focused on the genes responsible for volatile compounds in various tissues and developmental stages in *Cymbidium* orchids. We also described the emission of orchid floral volatiles and their function in pollination ecology. Taken together, this review will provide a broad scope for the study of orchid floral scents.

Keywords: *Cymbidium*; floral scents; Orchidaceae; pollination; volatile organic compounds

1. Introduction

The Orchidaceae family is one of the largest and widely diverse families of flowering plants, with more than 28,000 accepted species spanning 763 genera [1]. These plants are absent only in polar and desert regions, but are particularly abundant in the wet tropics worldwide [2]. However, a majority of orchids are distributed locally and generally rare [3]. Associated with the massive number of species in Orchidaceae, orchids display extraordinary floral diversification and represent a highly advanced and terminal line of floral evolution in the monocotyledons. As fascinating and highly popular plants, orchids are valued because of their exquisite flowers and long floral lifespan. These plants consist of great diversity in floral form, size, color, fragrance, and texture. A specific interaction between a pollinator and orchid flower may be one of the factors that promote orchid species richness [4]. The Orchidaceae family can be categorized into four subfamilies (Cypripedioideae,

Epidendroideae, Orchidoideae, and Vanilloideae) [5] and comprises a considerable diversity in life forms, with approximately 30% of species being terrestrial and mainly growing as epiphytes and lithophytes [6]. Furthermore, commercial production of orchids has greatly expanded and become a very profitable industry. Dominant species, such as those of *Cymbidium*, *Paphiopedilum*, and *Phalaenopsis*, are cultivated based on consumer flower preferences [7].

Orchids have complex life histories and diversified adaptation strategies; consequently, researchers have focused on orchid flower development and orchid pollination interactions. Flower color and scent are main traits for many floricultural crops. Floral scents emit various types of volatile organic compounds (VOCs). Orchids currently account for a prominent share of the world's flower trade, with annual sales of more than $4 billion (USD). It is widely used in perfumes, cosmetics, florivores, and medicinal applications. Some are also used as food and traditional medicines in many countries [8]. For example, dried vanilla seed pods (especially *Vanilla planifolia*) are commercially important as a flavoring used in baking, as well as for perfume manufacturing and aromatherapy [9]. *Gastrodia elata* is one of three orchids listed in the earliest known Chinese Materia Medica, and is used for treating headaches, dizziness, tetanus, and epilepsy [10]. However, because of its economic value in floral and pharmaceutical industries, *G. elata* has suffered great losses in habitat, resulting in a rare species [11,12].

Flower color and volatile compounds are key characteristics for many floricultural crops. Synthesis of VOCs occur in all plant organs, including roots, stems, leaves, seeds, fruits, as well as flowers, which are reported to emit the highest amounts and diversity of VOCs [13,14]. To date, more than 1700 floral VOCs have been identified in around 1000 seed plants [14]. In general, VOCs formed in other organs, apart from flowers, are involved in defense mechanisms. Although floral volatiles play a crucial role in reproductive process by attracting pollinators, they also have other adaptive roles [15,16], such as repellents [17–19] and physiological protectors against abiotic stresses [14,20]. In addition, floral volatiles are widely used as components of perfumes, cosmetics, flavorings, and even for therapeutic applications. Together with floral color, volatiles emitted by flowers represent key floral signals used by insects to detect and select rewarding flower species [21,22]. Floral scents emit different types of VOCs. VOCs are generally lipophilic and have low molecular weights and high melting points. Based on their origin, function, and biosynthesis, floral scents are grouped into three major clusters: terpenoids, phenylpropanoids, and fatty acid derivatives. Floral volatiles with terpene synthases (TPSs) have been identified in orchids [23,24].

Various species with large genomes are observed in monocots, such as species in Alliaceae, Asparagaceae, Liliaceae, Melanthiaceae, and Orchidaceae [25]. Among these, Orchidaceae, with genome sizes in a 168-fold range (1C = 0.33–55.4 pg), are perhaps the most diverse angiosperm families [25]. Epidendroideae, in Orchidaceae, contain variable genome sizes, with genome sizes in a range of over 60-fold (1C = 0.3–19.8 pg). Orchidoideae, with the largest descending/offspringing from species in subtribe Orchidinae, are pictured by a more restricted range of genomes (1C = 2.9–16.4 pg). Cypripedioideae show genome sizes in only a 10-fold range (1C = 4.1–43.1 pg). Cypripedioideae contain the largest mean genome size (1C = 25.8 pg) among all the subfamilies. Some species in Vanilloideae have been estimated, ranging from 1C = 7.3 to 55.4 pg. *Pogonia ophioglossoides* presents the largest genome size (1C = 55.4 pg) in this family [25]. Apostasioideae, the primitive subfamilies, contain calculated 1C-values ranging from 0.38 to 5.96 pg in a close to 16-fold range [26].

Orchids are one of the most diversified angiosperms and have mesmerized botanists for centuries. For orchids, floral color, shape, and fragrance are primary key determinants of consumer preferences. Many floricultural crops have lost their scents, following traditional breeding. However, only a few genomic resources are available for these non-model plants. Despite its economic as well as biological importance, metabolic engineering approaches on floral scents are still at the stage of infancy in orchids. In this review, we give an overview of orchid floral volatiles with a focus on *Cymbidium* orchids; we review their importance in pollination ecology, genes encoding enzymes and transcription factors (TFs) responsible for the biosynthesis, and the regulation of orchid floral volatiles. We hope that our information will provide guidance for future studies on orchid floral scents.

2. Orchid Volatile Compounds and Biosynthetic Pathways

Plant volatile compounds are a complex mixture of low molecular weight lipophilic molecules that have low melting points [27]. Biosynthesis of VOCs depends on the availability of carbon, nitrogen, and sulfur together with energy provided from the primary metabolism. Flower color and volatile compounds are key characteristics in many floricultural crops. Depending on their origin and functions, floral volatile compounds are categorized into one of three groups: terpenoids, phenylpropanoids/benzenoids, and fatty acid derivatives [20] (Figure 1).

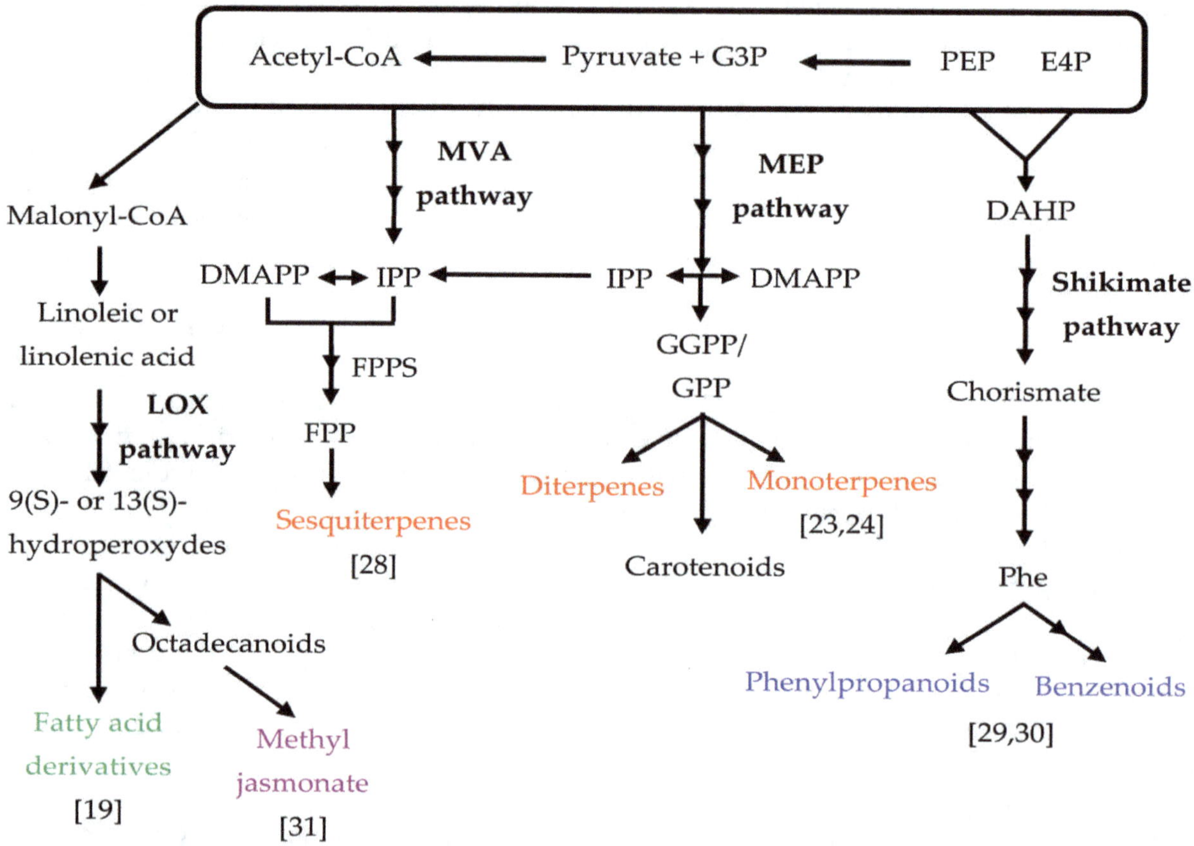

Figure 1. Floral volatile compound responsible pathways in orchid flowers. Major orchid floral volatile compounds are highlighted in colors (sesquiterpenes [28], monoterpenes [23,24], phenylpropanoids/benzenoids [29,30] and fatty acid derivatives/methyl jasmonate [19,31]). Abbreviations: MVA: mevalonic acid; MEP: methyl erythritol phosphate; LOX: lipoxygenase; PEP: phosphoenolpyruvate; G3P: glyceraldehyde-3-phosphate; E4P: erythrose 4-phosphate; DMAPP: dimethylallyl pyrophosphate; FPPS: farnesyl pyrophosphate synthase; FPP: farnesyl pyrophosphate; GGPP, geranylgeranyl pyrophosphate; GPP, geranyl pyrophosphate; IPP: isopentenyl pyrophosphate; DAHP: 3-deoxy-D-arabinoheptulosonate-7phosphate; Phe: phenylalanine.

2.1. *Terpenoids*

Terpenoids, or terpenes, represent the largest group of plant floral volatiles [27]. They play key roles in attracting pollinators for successful reproduction [32,33] and in defense against pathogens and florivores [34,35]. Moreover, from their natural roles, terpenoids are widely used in the cosmetic and perfume industries and as food additives because of their distinctive aromas and flavors [36,37]. Studies on floral scents have mainly focused on isolation and characterization of *terpene synthase* (*TPS*) genes encoding the key enzymes responsible for the synthesis of terpenes. All terpenoids are produced from isopentenyl diphosphate (IPP) and dimethyl allyl diphosphate (DMAPP), which are five-carbon (C5) precursors [38]. IPP and DMAPP are derived from two alternative biosynthetic pathways localized in different cellular compartments. The classical mevalonic-acid (MVA) pathway, which is localized in the cytosol, gives rise to IPP from three molecules of acetyl-CoA. In contrast, the methylerythritol phosphate (MEP) pathway takes place in plastids and produces IPP from pyruvate and glyceraldehyde 3-phosphate. In plants, monoterpenes, diterpenes, carotenoids, ubiquinones, and phytols are produced in the plastid via the MEP pathway, while all other plant terpenoids (sesquiterpenes, triterpenes, and polyterpenes) are produced using the MVA pathway. Floral volatiles with TPSs have been identified in such orchids as *P. bellina* [24] and *C. goeringii* [28].

Terpenoids are dominant in floral volatiles including those emitted by orchids. Geraniol, linalool, and their derivatives are major compounds of scented *P. bellina* flowers. Monoterpenes (Table 1) play a key role in the volatile profile [23,24]; in *C. goeringii*, floral volatile organic compounds include farnesol, methyl epi-jasmonate, (*E*)-β-farnesene, and nerolidol. Sesquiterpenes play a key role in the scent profile [28]. In the *Cymbidium* hybrid "Sunny Bell," linalool is the major compound found in the petal [39]. The volatile floral scents inside species and cultivars of *Cymbidium* have been reported [28,39]. Among the volatiles α-pinene, eucalyptol, trans-β-ocimene, α-copaene, and β-caryophyllene terpenoid were leading components in the volatile mixture [40,41]. The Vanda Mimi Palmer flower mainly contains cimene, linalool oxide, and linalool, which are classified as monoterpenes [42]. In addition, nerolidol is a sesquiterpene [40,41]. Compared with model plants, there are few reports on floral scent terpenoids in orchids.

Table 1. Major volatile organic components in orchid flowers.

Compound	Structure	Species	Reference
Terpenoids			
Linalool	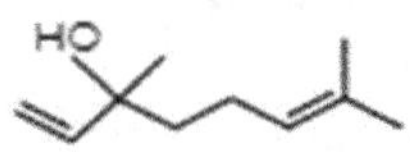	*P. bellina* C. cv. Sunny Bell	[23] [39]
Geraniol	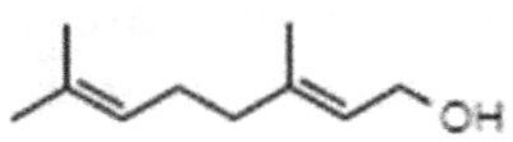	*P. bellina*	[24]
Ocimene	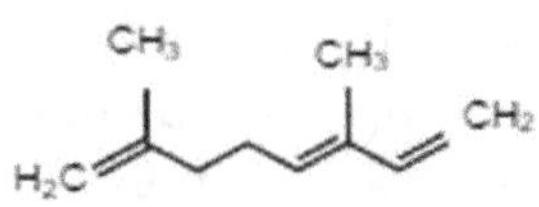	*Vanda* Mimi Palmer	[42]
Farnesol	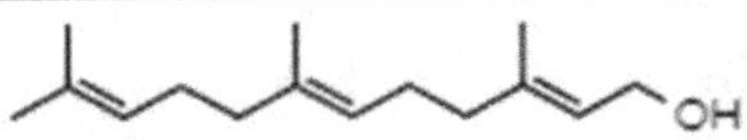	*C. goeringii*	[28]
β-Caryophyllene	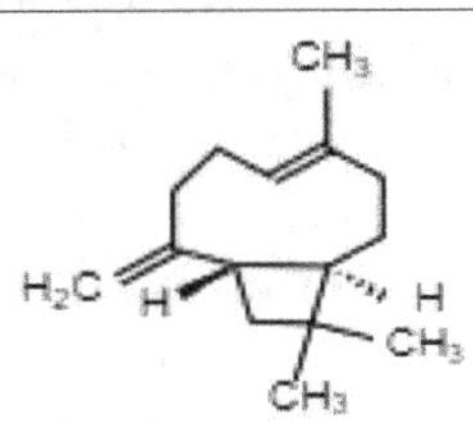	*M. tenufolia*	[43]
Phenylproponoids/Benzenoids			
Eugenol	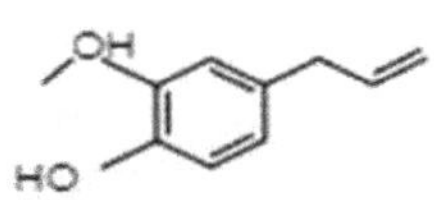	*Gymnadenia* Species	[30]
2-methyl butanal	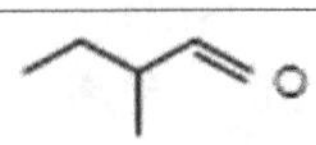	C. cv. Sael Bit	[29]

Table 1. *Cont.*

Compound	Structure	Species	Reference
Benzyl acetate	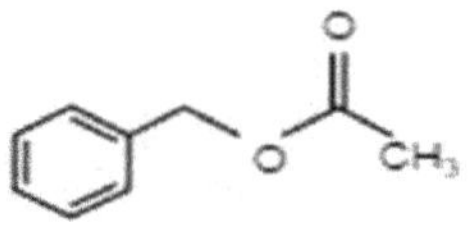	*Vanda* Mimi Palmer	[42]
Fatty Acid Derivatives			
Methyl jasmonate	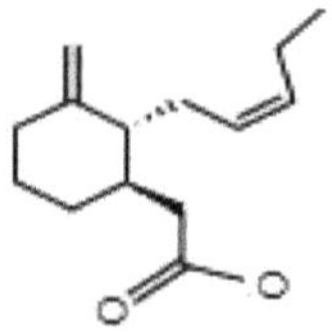	*C. ensifolium* *C. faberi*	[31] [44]

2.2. Phenylproponoids and Benzenoids

Phenylpropanoids and benzenoids are the second most dominant group of volatile compounds [14]. Phenylpropanoids/benzenoids are produced from the aromatic amino acid phenylalanine, which is produced in plastids via the shikimate pathway and arogenate pathway (Figure 1) through seven and three enzymatic steps [20]. Phenylalanine ammonia lyase (PAL) catalyzes the conversion of phenylalanine to trans-cinnamic acid, as a step in the phenylpropanoid pathway of plants [45]. The conversion of cinnamic acid to phenylpropanoids/benzenoids is followed by a shortening of the propyl chain via either the β-oxidative pathway or the non-β-oxidative pathway [45]. Recently, it was reported that the β-oxidative pathway for the formation of benzoic acid (BA) and benzenoids contributes to the production of volatile benzenoids in petunia flowers [46–49]. Moreover, in the fatty acid metabolism starting from cinnamic acid activation to its CoA thioester, hydration, oxidation, and thiolysis occurs in the peroxisome.

The formation of benzenoids (C6-C1) from cinnamic acid requires a shortening of the propyl side chain by two carbons and has been shown to proceed via a β-oxidative pathway, a non-β-oxidative pathway, or a combination of these pathways. Benzaldehyde formation, through the non-β-oxidative pathway, is oxidized by NAD+-dependent benzaldehyde dehydrogenase to benzoic acid, which is isolated from snapdragon flowers [50]. However, the formation of benzaldehyde in the enzymatic steps of the non-β-oxidative pathway remains unknown. Furthermore, floral phenylpropanoid and benzenoid compounds play a crucial role in scent production via two super families, SABATH methyl transferases and BAHD acyltransferases, in several plants [20]. In contrast, the formation of floral volatile phenyl propenes such as eugenol and isoeugenol starts with lignin and takes two enzymatic steps; the oxygen functionality at the C9 position is removed, and coniferyl alcohol is produced [20]. Finally, eugenol and isoeugenol are formed through the conversion of coniferyl acetate. From the *Cymbidium* cultivar Sael Bit, benzenes were reported in the full blooming stage [29]. Floral volatiles such as phenylpropanoids and benzenoids are also emitted from scentless flowers of *P. equestris*. In addition, *Gymnadenia* species quantitatively and qualitatively release diverse blends of nearly 50 volatile compounds and attract different suits of pollinators. Eugenol and benzyl acetate are two predominant compounds among the scents of these species [51].

2.3. Fatty Acid Derivatives

Among the floral volatile compounds, fatty acid derivatives are the smallest group of volatiles. They mainly consist of floral fatty acids synthesized from C18 polyunsaturated fatty acids and linolenic and linoleic acids. Methyl jasmonate is an important volatile fatty acid compound found in orchids. The biosynthetic process of fatty acid or its derivatives begins from the stereo-specific oxygenation catalyzed by lipoxygenases (LOXs) to produce 9-hydroxy and 13-hydroperoxy intermediates. These intermediates can enter two different batches of the LOX pathway to produce volatile compounds. Allene oxide synthase (AOS) catalyzes the first step in the biosynthesis of jasmonic acid from lipoxygenase-derived hydroperoxides of free fatty acids. In addition, the AOS pathway generates the C6 and C9 aldehydes through condensation of hydroperoxide derivatives by hydroperoxide lyase (HPLS). Limited data are available regarding the synthesis/pathways of fatty acids and/or their derivatives in flowers. In *Antirrhinum majus* flowers, 20 fatty acid derivatives have been identified [52]. Furthermore, methyl jasmonate and jasmonic acid involved in the floral scent pathway in *C. ensifolium* and *C. faberi* have been identified. Various volatile fatty acids were also found synthesized in the orchid genus *Ophrys*; among them, *alkenes* have an important function *in attracting pollinators* [17]. Two genes encoding stearoyl-acyl carrier protein desaturase (SAD) isoforms, SAD1 and SAD2, were reported to be flower-specific, and these genes broadly parallel alkene production during flower development of *Ophrys sphegodes* and *O. exalanta*; in particular, SAD2 showed a tight association with alkene production [19]. Further study is required to better understand the floral scent pathways in orchids.

3. Transcriptional Factors in Floral Volatile Regulation

Transcription factors (TFs) are sequence-specific DNA-binding proteins that interact with the regulatory regions of the target genes and modulate the transcription initiation rate by RNA polymerases [53]. Although several types of transcription factors have been reported to be involved in the biosynthesis of volatile compounds and secondary metabolites in plants, a limited number of TFs involved in the formation of molecules responsible for floral scent have been identified. Recently, a growing number of research results have reported that several types of TFs, including basic helix-loop-helix (bHLH), basic leucine zipper (bZIP), ethylene response factor (ERF), NAC, MYB, and WRKY family members, are involved in regulation for terpene biosynthesis [53].

TFs play a key role in controlling the expression level of genes involved in various developmental processes and physiological pathways, including plant secondary metabolism [54]. *ODORANT1* (*ODO1*) is the first *MYB* TF identified in flowers. It is a member of the R2R3-MYB TF family, which regulates genes involved in floral scent production through the shikimate and phenylpropanoid pathways [43]. Terpene biosynthesis is regulated by genes encoding required enzymes and their regulators. *Phalaenopsis* TFs regulate the terpenoid pathway; *PbbHLH4* regulates the *geranyl diphosphate synthase* (*GDPS*) gene for the synthesis of monoterpenoids in *P. bellina* [55]. Moreover, a higher expression of five genes encoding TFs (*PbbHLH4, PbbHLH6, PbbZIP4, PbERF1*, and *PbNAC1*) was reported in the scented orchid. Especially, 10-fold higher levels of α-terpineol (a monoterpenoid) were detected in *PbbZIP4*-overexpressing flowers of *P. aphrodite* compared to the control [56]. *HY5*, an *bZIP* TF, is known to play a critical role in mastering both the light and circadian signaling pathway. Identification of HY5-interacting motifs on the upstream regulatory fragments of *PbNAC1* implies that the light and circadian clock signals are likely to manage monoterpene biosynthesis in *P. bellina* [56].

Transcriptomic analyses of *C. goeringii* flowers led to identification of 1179 genes that were clustered into 64 groups encoding putative TFs with the three largest being bHLH (73 members), ERF (71 members), and C2H2 zinc finger proteins (65 members). *CgbHLH1* and *CgbZIP3* are homologs of *AabHLH1* and *AabZIP1*, respectively. *CgbZIP7* is a homolog of *PbbZIP4*, which regulates monoterpene biosynthesis in *P. bellina* (Table 2) [57], while *CgERF2* is a homolog of *CitAP2.10*, which is associated with sesquiterpene (+)-valencene synthesis in sweet orange [34]. *CgNAC5*, a homolog of *AaNAC4*, controls monoterpene synthesis in kiwifruit [35], and *CgWRKY1* and CgWRKY2, which are homologs of GaWRKY1, regulate sesquiterpene (+)-δ-cadinene synthesis in cotton [36]. Cymbidium Sael Bit MYB1 expression is detected at various flower developmental stages and is highest in petals and columns of the fully open flower. *Cymbidium* Sael Bit *MYB1* is regarded as a regulator of phenylpropanoid/benzenoid genes in floral scent profiles. The key component of flower scent in *C. faberi* is MeJA, which is regulated by the crosstalk of many plant hormones. A total of 379 TFs identified as belonging to 37 TF families are differentially expressed between blooming and withered flowers of *C. faberi* [58], and the top 10 groups belong to the *MYB, AP2-EREBP, bHLH, NAC, GRAS, C2H2, C2C2-Dof, MADS, WRKY*, and

ABI3VP1 families. Furthermore, an increase in the levels of floral volatiles in tissues resulted from a large increase of various transcription factors in orchids.

Table 2. Representative genes responsible for floral scents in orchids.

Floral Scent Gene	Metabolism Pathway	Species	Reference
Genes			
PbGDS	Terpenoid pathway	*Phalaenopsis bellina*	[24]
VMPAAT	Terpenoid pathway	*Vanda* species	[59]
VMDXS	Terpenoid pathway	*Vanda Mimi Palmer*	[42]
GdEGS	Benzenoid pathway	*Gymnadenia* species	[30]
OsSAD1	Benzenoid pathway	*Ophrys sphegodes*	[19]
Transcription Factors (TFs)			
CsMYB1	Phenylprponid/benzenoid	*Cymbidium* cv. Sael Bit	[29]
PbbZIP4	Monoterpene pathway	*Phalaenopsis aphrodite*	[60]
PbBHLH2	Monoterpene pathway	*Phalaenopsis bellina*	[23]

4. Spatial and Temporal Emission of Volatile Organic Compounds

Scent is an important property of flowers and plays a vital role in the ecological, economic, and aesthetic properties of flowering plants. Each plant possesses a distinct and unique floral scent. Floral scent is composed of all the VOCs, including terpenoids, phenylpropanoids, benzenoids, fatty-acids, and their derivatives, which are emitted by floral tissues (Table 2). In *Phalaenopsis bellina*, expression analysis of the *PbGDS* (*geranyl diphosphate synthase*) gene encoding a homodimeric GDS showed that its expression is flower-specific and that maximal expression is concomitant with maximal emission of monoterpenes on Day 5 post-anthesis [23]. In *P. abies*, expression of *PaIDS1* (*isoprenyl diphosphate synthase 1*) encoding a bifunctional GDS and GERANYLGERANYL DIPHOSPHATE SYNTHASE (GGDS) exhibits a peak in wood where oleoresin, comprising monoterpenes and diterpenes, is accumulated [24].

Plants emit a large variety of VOCs that are actively involved in plant growth and protection. VOCs are defined as any organic compound with vapor pressures high enough under normal conditions to be vaporized into the atmosphere [53]. VOC emissions are strongly dependent on environmental conditions and developmental stages of the plant tissue. In plants, emission of VOCs is spatiotemporally regulated; a majority of VOCs are emitted from flowers compared to other plant tissues/organs, and the level of emission increases when the floral bud is close to opening and decreases as it moves to the senescence stage [61,62].

In *Vanda* Mimi Palmer orchids, various types of sesquiterpenes and benzenoids were highly expressed at the full blooming stage with the expression of floral scent genes [42,59,63]. Floral volatile emission increased bud to flowering stages in *Cymbidium goeringii*. Different types of *Maxillaria* orchids emit strong vanilla or coffee-like scents, which are responsible for pollinator attraction [64]. *M. tenuifolia* Lindl is called a "coconut orchid" due to its strong coconut-like scent, and was recognized as the best scented orchid in the 18th World Orchid Conference [65]. The sepal is a source of floral volatiles in *Maxillaria* species, and the highest level of floral volatiles separated through electronic nose and GC-MS analyses was detected at the initial flowering stage [43,66]. In addition, methyl jasmonate (mJA) emission predominantly occurs in sepals and floral parts of *C. ensifolium* [31]. *Phalaenopsis* is undoubtedly the most widely grown orchid in the world, and in *P. bellina*, various types of monoterpenes are produced in the full flowering stage [23,24]. Furthermore, in a comparison of fragrant and non-fragrant *Phalaenopsis* flowers, terpene compounds were found to be much more abundant with increased levels of relevant gene expression in flowers of fragrant orchids [67].

Developmental regulation of scent emission occurs at several levels, including orchestrated expression of scent biosynthetic genes [68], enzyme activities, and substrate availability [69]. Based on an evolutionary study of floral scent genes in three closely related orchid species of the genus

Gymnadenia, it is likely that the switch from the production of one to two scent compounds evolved under relaxed purifying selection [51]. Two major volatile compounds, α-copaene and β-caryophyllene, have been identified in all floral organs of *M. tenuifolia*, with the highest levels in the petal. α-copaene and β-caryophyllene were found to be emitted in all flower developmental stages except the floral bud stage I [43]. In fact, volatile compounds of *M. tenuifolia* include α-copaene, β-caryophyllene, 1,8-cineole, limonene, β-myrcene, α-pinene, β-pinene, sabinene, and δ-decalactone, which is responsible for the typical coconut aroma. The majority of studies on *Maxillaria* fragrance reported only the chemical composition of the floral scent; however, little data are available on the spatiotemporal emission of the floral volatiles. In addition, sulfur- and nitrogen-containing volatile compounds contribute to the attraction of pollinators to flowers by mimicking food or brood sources such as carrion or dung. Besides the importance of floral scents in plant ecology, identification and functional validation of relevant genes responsible for biosynthetic and/or regulatory pathways of floral volatiles are required for a better understanding of floral scent production and for the development of novel cultivars with desirable characteristics. Transcriptomic and metabolic analyses together with genetic engineering approaches will be of great help in driving towards this goal.

5. Gene Evolution for VOCs

The evolution of orchids has resulted in an immense diversity of flower traits such as color and scent. Orchidaceae consist of extraordinary adaptations that may have guaranteed its evolutionary success. To date, most of the examined plant gene families originated through gene duplication. Gene duplication plays a key role in species evolution because it provides raw materials for the evolution of new genes and new genetic functions. Multiple mechanisms contribute to gene duplication, including tandem duplication, segmental duplication, transposon-mediated duplication, and retro duplication. Studies of floral scent gene duplications in orchids have been limited. Orchid TPSs are the key enzymes that generate the structure diversity of terpenes. Through the analysis of plant genome, researchers have shown that the plant TPS gene family have their gene numbers ranging from 20 to 150 and thus belong to a mid-size family [70]. *Phalaenopsis equestris* genome has 23 TPSs belonging to TPS-a, -b, -c, e/f, and -g. Twenty-three TPSs found and predicted as having mono-, di- and sesqui-terpene synthase evolutionary relationships among orchids and experiencing duplication and then sub- or neo-functionalization, have occurred during evolution [71]. It has been proposed [72] that diterpene synthases are the origin of mono- and sesqui-terpene synthases during evolution. *P. aphrodite*, *PaCHS3*, *PaCHS4*, and *PaCHS5* formed a tandemly arrayed gene cluster, and the intervals between the three *CHS* genes were approximately 13.3 kb (*PaCHS3* and *PaCHS4*) and 7.7 kb (*PaCHS4 and PaCHS5*). This arrangement was also observed in a closely related orchid, *P. equestris*, in which the three CHS genes were all positioned on Scaffold 000036. Thus, the tandem array of three *CHS* genes was probably present in a common ancestor before speciation within *Phalaenopsis*. Tandem gene duplications represent a substantial proportion of all plant genes [73].

6. Functions of Orchid Volatile Compounds

Previous reviews provide a good summary of floral emissions and the involvement of biochemical processes in the interactions of flowers with their flower visitors [20], their action over pollinator behavior [21], and the ecological processes that drive their evolution [22], wherein they mediate intra- and interspecific interactions. The role of vegetative VOCs has been extensively reviewed [20]. Here, we present the functions of the floral VOCs, which are especially involved in the attraction of pollinators [20]. Floral volatiles have a role in many multifaceted functions that contribute to pollinator attraction, plant defense, plant reproduction, and plant diversity (Figure 2).

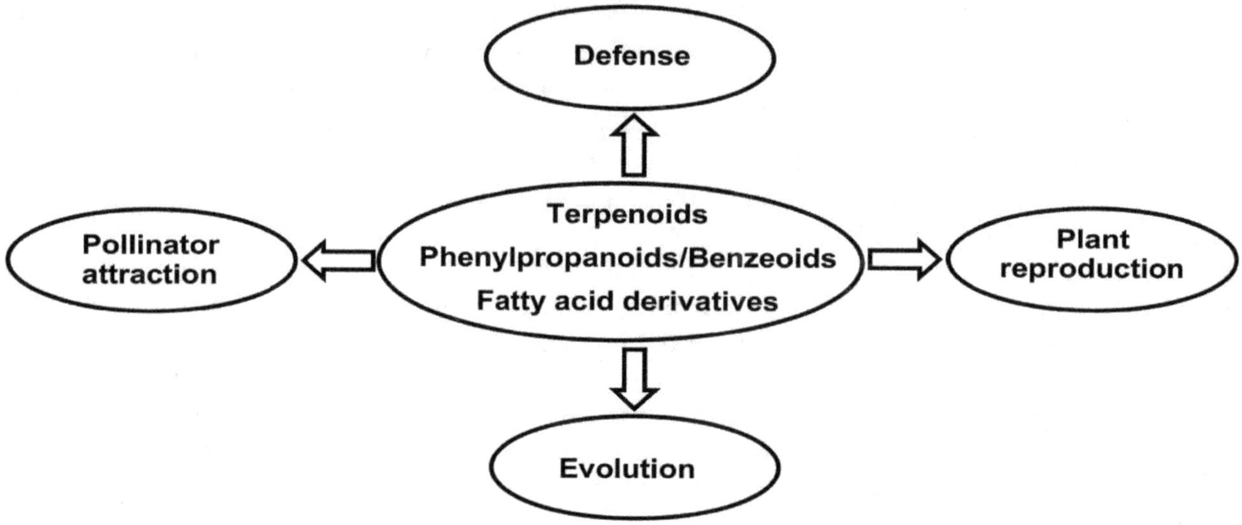

Figure 2. Functions of floral volatiles in orchid flowers.

6.1. *Flower Defense*

Generally, flowers have effective physical barriers comprising highly lignified cell walls, although the generation of these cell walls renders flowers highly vulnerable to pathogens and florivores. Plants constitutively emit VOCs from flowers, leaves, and roots. Emission usually increases when plants are attacked by antagonists such as insect herbivores or pathogens [51].

Many VOCs were shown to exhibit antimicrobial and antifungal activities in vitro [74] or inferred to have these antimicrobial activities based on tissue-specific expression patterns [75]. However, only a few VOCs have been explored for their role in defense against pathogens. (*E*)-β-caryophyllene, emitted from stigmas of *Arabidopsis* flowers, was shown to limit bacterial growth; furthermore, *Arabidopsis* plants lacking (*E*)-β-caryophyllene emission displayed denser bacterial populations on their stigmas and reduced seed weight than wild-type plants, indicating that (*E*)-β-caryophyllene acts in the defense against pathogenic bacteria and is also important for plant fitness [76]. VOCs emitted by petals of *Saponaria officinalis* were also shown to inhibit bacterial growth, supporting their roles in controlling bacterial community diversity in petals [77].

6.2. *Pollinator Attraction*

In many flowering plant species, the emission of volatile scents from the flower is important for attracting insect pollinators. Orchid flowers exhibit visual, chemical, and morphological advertisements to guide their pollinators, and may offer rewards such as nectar, pollen, fragrance, or oil [78]. Over the past few years, evidence has supported the role of floral volatiles in pollinator attraction. In fact, a high occurrence of non-rewarding flowers has been noted in orchids compared to other plant families [79]. Floral volatile profiles are specific to each species depending on the type of pollinator [80]. However, the selection of pollinator has played a key role affecting the pattern of floral VOC profile across angiosperms.

Approximately one-third of all orchid species reach pollination over food deception, whereby flowers contain no nectar or other rewards but resemble or mimic floral signals of rewarding plants to attract pollinators [78]. Subsequently, intraspecific variation in floral traits is estimated to be high in food-deceptive orchids, since flowers must delay the avoidance learning of pollinators [79]. Flowers of the fly-pollinated *Satyrium pumilum* orchids emit a cocktail of six compounds containing sulfurous oligo sulfides such as dimethyl disulfide (DMDS) and dimethyl trisulfide (DMTS). Secretion of these volatiles is also tissue-specific, which is anticipated to be the key olfactory cue for attracting flesh-eating fly pollinators [80] and lepidopteran pollinators [81]. Floral scent can be distinguished even among closely related taxa when species differ in pollination systems, such as lepidopteran vs. bee fly pollination in

Narcissus species [80] and bee vs. hummingbird pollination in two *Mimulus* species [80], suggesting that differences in the dominating functional group of pollinators drive divergence in floral scent.

6.3. Plant Reproduction

Pollinator attraction is often intermediated by multimodal signaling mechanisms including floral morphology, color, and scent [81]. In deceptive species, attractiveness is very important for ensuring reproductive success. For instance, *Ophrys* and *Neotinea* species are known to produce complex bouquets of volatiles typically consisting of more than 100 chemical compounds [82]. The species belonging to these genera are all deceptive, but *Ophrys* species use a sexual deception strategy, while *Neotinea* is a food-deceptive genus [83]. Various orchid volatiles play a key role in plant reproduction. In the Dracula orchid, *Dracula lafleurii*, the labellum acts as both a visual and an olfactory mimic of mushrooms that often grow alongside these orchids [84]. The labellum emits an unusual floral volatile blend of mushroom alcohols, especially (R)-1-octen-3-ol [85]. *Cypripedium calceolus* is pollinated by bees. Scent profile consists mainly of aliphatics, terpenoids, and aromatics. In this context, orchids are highly pertinent models for studying plant reproduction, as they present a great variety of floral traits and trait associations. This is mirrored by the great diversity of reproductive strategies in orchids. One of the most intriguing strategies is deceptive pollination (i.e., nectarless flowers), which is found in about one-third of orchid species. Floral scent analyses in *Ophrys orchids* [81] showed that their flowers emit attractive blends of VOCs. Moreover, different *Ophrys* species, which mainly use alkenes with certain double-bond positions as key signals for plant reproduction (e.g., non-hydrocarbons with low molecular weight), were found as "long-range" attractants [85].

Though by no means exclusive to orchids, deceptive pollination approaches are particularly well-developed in the Orchidaceae, with an estimated one-third of the family (around 10,000 species) using such strategies [85].

6.4. Evolution

Evolution of angiosperms has resulted in an immense diversity of flower traits such as shape, size, color, and scent. Remarkably, the quality and quantity of emitted volatiles are species-specific and vary among different populations within a species [20]. While much effort has so far been invested in describing scent composition in various flowering species, the mechanisms driving the evolution and diversification of floral scent remain underexplored. Analysis of the genetic basis for differences in scent profiles between these two species revealed that only two quantitative trait loci are responsible for the distinct scent phenotypes [86]. One of these locus maps to the MYB TF ODO1, which controls flux over the shikimate pathway and, therefore, the amount of precursors available for benzenoid biosynthesis [87], while the genetic identity of the second locus is presently unknown. *Ophrys* may rely on species-specific alkene emission profiles that are distinct in enzyme activity and on the gene expression of a few stearoyl acyl carrier protein desaturases of the *Ophrys* genus, and only limited genetic variation among species and populations was observed with microsatellite markers. These findings suggest that divergent pollinator-mediated selection rather than genetic drift explains the strong differences in volatile profiles. Taken together, the above examples demonstrate that small genetic variations can have large effects on floral scent chemistry and interactions with pollinators.

7. Case Studies of *Cymbidium* Floral Volatiles

Orchids are the largest, most highly diverse flowering plants, and form an extremely peculiar group of plants. *Cymbidium* is one of the most important genera of orchids for the cut-flower and potted plant markets. *Cymbidium* spp. have great horticultural value as ornamental plants because of their beautiful and fragrant flowers. The *Cymbidium* genus consists of nearly 55 species that are distributed mainly in tropical and subtropical Asia, reaching as far south as Papua New Guinea and Australia [88]. The *Cymbidium* genus can be divided into three subgenera (*Cymbidium*, *Cyperorchis*, and

Jensoa) [89,90] and includes *C. sinense*, *C. goeringii*, *C. forrestii*, *C. faberi*, *C. ensifolium*, and *C. kanran*. *C. sinense* is a winter blooming epiphytic orchid usually regarded as a "Spring Festival" flower.

Great efforts have been made to better understand the flowering of orchids such as *Cymbidium*, *Phalaenopsis*, *Dendrobium* and *Cattleya* through biotechnological approaches including tissue culture and transgenic technologies [91–94]. Moreover, while *Cymbidium* orchid species are not all widely cultivated, hybrids of *Cymbidium* orchids lend themselves to cultivation. Some commercially important hybrids have been created for over 100 years. Because of their ornamental and commercial value, *Cymbidium* orchids have been the subject of taxonomic studies and, particularly, species identification [95–97]. In the past few decades, the application of diverse molecular techniques have contributed to widening our knowledge in the flowering/flower development, species identification, and volatile compounds of orchids.

7.1. Floral Volatile Research on Cymbidium

Floral VOCs are important compounds derived from flowers. Floral scent is a key trait for many floricultural crops. The molecular mechanisms underlying the regulation of biosynthesis and emission of volatiles from orchids remain largely elusive. For adding commercial value, studies on floral scents in orchids aim not only to understand the molecular and genetic mechanisms of the biosynthesis and emission of floral scents but also to assist in *Cymbidium* breeding programs. Currently, many orchid researchers are focusing on the development of cultivars with desirable floral scents. Recent reports on the propagation of sterile seedlings [98–100], leaf and flower morphogenesis [101–103], and the characterization of volatiles as floral scents [104,105] in *Cymbidium* spp. also reflect this trend. In fact, the biosynthesis of widespread VOCs in plant tissues is involved in multiple biological functions such as defense against pathogens, parasites, and herbivores [106–108]. The development of floral scents is likely to be a vital event in biological evolution, providing olfactory signals that plants can utilize to attract pollinators. We are presented major floral scent *Cymbidiums* (Figure 3).

Figure 3. *Cymbidium* flowers described to the floral scent. (**A**) *C. goeringii*, (**B**) *C. faberi*, (**C**) *C. ensiforium*, (**D**) *C.* "Sael Bit," (**E**) *C.* "Sunny Bell.".

7.1.1. Cymbidium goeringii

C. goeringii is one of the most popular terrestrial species indigenous to temperate Eastern Asia, cultivated as an ornamental, and whose flowers are used as an ingredient in soup, alcoholic drinks, and tea [88]. Floral fragrance is determined by a mixture of volatile compounds. In *C. goeringii*, floral scent pathways have been studied in various developmental stages during flowering [28]. Sesquiterpenes are the major compounds in the *C. goeringii* floral scent profile. The dominant floral scent compounds were identified as farnesol, methyl epi-jasmonate, (*E*)-β-farrnesene, and nerolidol. In particular, examination of farnesol emission from the day of anthesis (D0) to the fifth day after anthesis (D+5) demonstrated that emission had a peak at the D+2 stage. Transcriptomic analyses focusing on floral scent pathways have been performed using three different stages of flowers in *C. goeringii*. Most terpenoid pathway genes, including *1-deoxy-D-xylulose-5-phosphate reductoisomerase* (*DXR*), *1-deoxy-D-xylulose-5-phosphate synthase* (*DXS*), and *farnesyl diphosphate synthase* (*FDPS*), were expressed at the initial flowering stage compared to bud and/or full flowering stages [28]. Furthermore, 32 and 38 unigenes known to be associated with MVA and MEP pathways, respectively, were reported in *C. goeringii.* Several TPSs and TFs were also identified in the floral transcriptome of *C. goeringii* including *CgTPS7*, which encodes a key enzyme involved in sesquiterpene synthesis. A putative terpenoid pathway responsible for the volatile profile in *C. goeringii* has also been reported.

7.1.2. Cymbidium faberi

C. faberi Rolfe is one of the most significant species with elegant flower scents [44], and is one of the oriental orchids that has been longest cultivated. There are more than 100 compounds in the flower scent of blooming *C. faberi* flowers. Among these, MeJA is the most abundant, but is almost untraceable in the volatile emission of withered flowers [44]. The major pathways include α-linolenic acid metabolism, pyruvate metabolism, and fatty acid degradation, which contribute to the conversion of α-linolenic acid to MeJA. One of the differentially expressed genes (DEGs), *jasmonic acid carboxyl methyltransferase* (*CfJMT*), was highly regulated in the blooming flower of *C. faberi.* Consequently, the full-length coding sequence and genomic sequence of *CfAOS* from *C. faberi*, which is localized to the chloroplasts, has no introns, and is one of the most important enzymes in the MeJA biosynthetic pathways in *C. faberi* [58], was cloned. *CfAOS* has numerous roles including insect attraction and mediation of anti-microbial and stress tolerance. AOS and allene oxide cyclase (AOC) are crucial enzymes in the MeJA biosynthetic pathway in *C. faberi.*

7.1.3. Cymbidium ensifolium

C. ensifolium is a popular miniature terrestrial orchid that produces fragrant flowers and is often marketed as a potted specimen. MeJA was one of the key scent compounds found in *C. ensifolium* [31]. It is known that MeJA is synthesized via the octadecanoid pathway. MeJA is primarily recognized as a floral scent compound in flowers of *Jasminum grandiflorum*, and is also universally distributed among the plant kingdom including many *Cymbidium* orchids [31]. MeJA emission was at very low levels in unopened or half-opened *C. ensifolium* flowers, and scent emission reached maximal levels between Days 4 and 6 and decreased from Days 7 to 10 post-anthesis. *C. ensifolium* tissue-specific manner and high MeJA emission was found in sepals and petals.

7.1.4. *Cymbidium* Cultivar Sael Bit

The small green and light yellow colored flowers of *Cymbidium* cultivar Sael Bit are highly fragrant [109]. The strong emission of volatiles was detected in petals at the blooming stage. The dominant floral VOCs in Sael Bit were identified to be alkenes, benzenes, and estaric compounds; i.e., 2-methyl butanal, 2-methyl pentanal and (Z)-2-octenal at the blooming stage.

Full-length cDNA of *Cymbidium* Sael Bit *MYB1* from the petal has been isolated; fragrance genes such as *MYB1*, *OOMT*, *AOS*, *LOX*, and *SAMS* were expressed during the entire floral developmental stages, and all genes were highly expressed in the full opening flowers [29]. Sael Bit *MYB1* regulates the floral scent phenylproponoid- and benzenoid-responsible genes during scent emission.

7.1.5. *Cymbidium* Cultivar Sunny Bell

A *Cymbidium* variety Sunny Bell (*C. karan* x *C. eburneum*) was developed at the National Institute of Horticultural & Herbal Science, Rural Development Administration, Suwon, Korea in 2013 [39]. Monoterpenes, sesquiterpenes, and aliphatics have been recognized as the major volatile compounds in *Cymbidium* Sunny Bell [39]. Twenty-four volatile components were identified in the Sunny Bell flowers; among the total volatiles, petals produced dominant volatile compounds compared to other floral tissues such as sepal, labellum, and column. Linalool is the major compound responsible for the floral volatile profile in Sunny Bell.

Some species in the genus *Cymbidium*, including *C. floribundum*, *C. pumilum*, and *C. suavissimum*, release identical volatiles for pollinator attraction. Various types of alkenes, esters, and fatty acid derivative compounds are released for pollinator attraction. It has been reported that *Cymbidium* flowers are rich in volatile compounds including cineole, isoeugenol, and (-) selinene [104]. Floral scent and color are major traits for floriculture crops in developing new cultivars of *Cymbidium*. Furthermore, 21–28 floral scent compounds have been identified as major volatile components in the flowers of three *Cymbidium* varieties [105]. The volatiles mainly comprise monoterpenes, aliphatics, and sesquiterpenes, and their content values have exceeded 90% [105]. Their aromatic characteristics can be determined by the profiles of each VOC that may vary depending on each genotype [105].

8. Final Remarks and Future Directions for Overcome the Challenges

At present, orchid industries worldwide are facing various difficulties. For developing new cultivars, physiological and genomic maps have been needed to produce markers. RNA-illumina sequencing technology has been extensively used for identifying gene expression at a genome-wide scale in many organisms, including non-model plants. The adoption of this technique, especially mRNA sequencing from floral tissues and de novo transcriptome construction (Figure 4), has been performed in several orchid species, with a goal of identifying genes involved in the biosynthesis and/or biosynthetic pathways of floral volatiles [28,110].

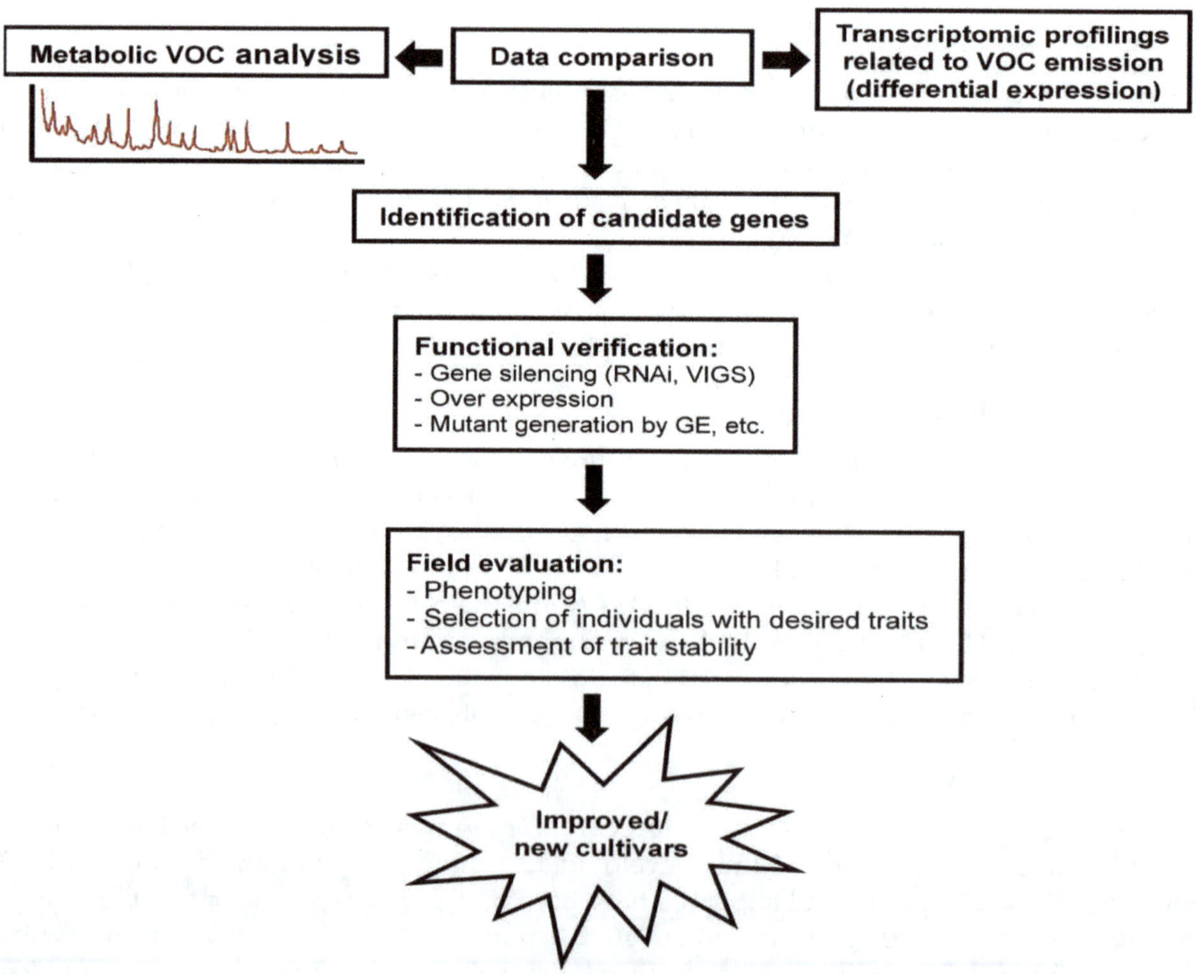

Figure 4. Schematic representation of functional studies for orchid breeding to develop floral scent trait.

Moreover, other strategies include molecular evolutionary analysis tools. For example, testing for gene duplication and selection signatures in hypothesized pathway genes, from a phylogenetic perspective, is frequently used. For the identification of significant candidate genes and pathways, targeted and strategic transcriptome analyses of fragrant and non-fragrant flower organs and tissues are often the first key step. Regulation of DEGs between fragrant and non-fragrant tissues and developmental stages can be investigated. The latter provides a key baseline for identifying DEGs in fragrant tissues. This approach was utilized for the breakthrough discovery on volatile biosynthesis [58,111] and may hold potential for elucidating other biosynthetic pathways. RNA-sequencing analysis across species, with the goal of identifying shared gene expression and metabolic pathways, may also prove informative. Transformation technology has been developed for orchids; a few successful methods using virus induced gene-silencing (VIGS) approaches have recently been demonstrated as efficient strategies for functional studies of genes in orchids (Figure 4). Furthermore, transgenic approaches, such as the overexpression of floral scent genes and/or genome editing, have also been recently developed for orchids.

9. Conclusions

Over the last few decades, studies on plant volatile compounds and their biosynthetic processes have markedly increased. In orchids, volatile compounds play a key role in pollination, which ensures fertilization. To date, the biosynthesis of orchid floral fragrance is not well understood, with only some terpenoid pathways reported. Plant volatiles are generally produced at very low concentrations with low quantity, even in floral tissues. Thus, isolation of each component in the volatile compounds is inefficient and expensive. Despite the presence of studies of floral VOCs, many aspects of their biosynthesis together with transcriptional regulation and function require further studies. Further advances in functional studies on key genes for floral scent may rely on a breakthrough in orchid transformation technology that may lead to more efficient results. The genome sequences of several orchids have now been determined [112,113]. Overall, it is clear that genetic manipulation of orchid volatile compounds may be possible, but requires the selection of the appropriate species. In the future, studies in scent research may focus on orchid floral traits and on increasing phytochemical compounds, flavor, and aroma through the regulation of genes by transcription factors in floriculture crops. This review provides an important theoretical reference for aromatic volatile compound studies in orchids.

Author Contributions: M.R., S.J., and P.H.P., drafted the manuscript. M.R., H.-R.A., P.-M.P., and S.J. collected the background information. S.J. and S.-Y.L. revised the manuscript. All authors have read and agreed to the published version of the manuscript.

References

1. Christenhusz, M.J.M.; Byng, J.W. The number of known plant species in the world and its annual increase. *Phytotaxa* **2016**, *261*, 201–217. [CrossRef]
2. Chase, M.W. Classification of Orchidaceae in the age of DNA data. *Curtis's Bot. Mag.* **2005**, *22*, 2–7. [CrossRef]
3. Waterman, R.J.; Bidartondo, M.I. Deception above, deception below: Linking pollination and mycorrhizal biology of orchids. *J. Exp. Bot.* **2008**, *59*, 1085–1096. [CrossRef] [PubMed]
4. Cozzolino, S.; Widmer, A. Orchid diversity: An evolutionary consequence of deception? *Trends Ecol. Evol.* **2005**, *20*, 487–494. [CrossRef] [PubMed]
5. Phillips, R.D.; Brown, A.P.; Dixon, K.W.; Hopper, S.D. Orchid biogeography and factors associated with rarity in a biodiversity hotspot, the Southwest Australian Floristic Region. *J. Biogeogr.* **2011**, *38*, 487–501. [CrossRef]

6. Gravendeel, B.; Smithson, A.; Slik, F.J.W.; Schuiteman, A. Epiphytism and pollinator specialization: Drivers for orchid diversity? *Philos. Trans. R. Soc. Lond. Ser. B Biol. Sci.* **2004**, *359*, 1523–1535. [CrossRef]
7. Hew, C.S.; Yong, J.W.H. *The Physiology of Tropical Orchids in Relation to the Industry*, 2rd ed.; World Scientific Publishing: Singapore, 2004.
8. Arditti, J. *Fundamentals of Orchid Biology*; John Wiley Sons: New York, NY, USA, 1992.
9. Lubinsky, P.; Bory, S.; Hernandez, J.H. Origins and dispersal of cultivated vanilla (*Vanilla planifolia* Jacks. [Orchidaceae]). *Econ. Bot.* **2008**, *62*, 127–138. [CrossRef]
10. Tsaia, C.F.; Huang, C.L.; Lind, Y.L. The neuroprotective effects of an extract of *Gastrodia elata. J. Ethnopharmacol.* **2011**, *138*, 119–125. [CrossRef]
11. Luo, Y.B.; Jia, J.S.; Wang, C.L. A general review of the conservation status of Chinese orchids. *Biodivers. Sci.* **2003**, *11*, 70–77.
12. Liu, Q.; Chen, J.; Corlett, R.T. Orchid conservation in the biodiversity hotspot of southwestern China. *Conserv. Biol.* **2015**, *29*, 1563–1572. [CrossRef]
13. Yue, Y.; Yu, R.; Fan, Y. Transcriptome profiling provides new insights into the formation of floral scent in *Hedychium coronarium. BMC Genom.* **2015**, *16*, 470. [CrossRef] [PubMed]
14. Knudsen, J.T.; Eriksson, R.; Gershenzon, J.; Stahl, B. Diversity and distribution of floral scent. *Bot. Rev.* **2006**, *72*, 1–120. [CrossRef]
15. Piechulla, B.; Pott, M.B. Plant scents-mediators of inter-and intraorganismic communication. *Planta* **2003**, *217*, 687–689. [CrossRef] [PubMed]
16. Raguso, R.A. Start making scents: The challenge of integrating chemistry into pollination ecology. *Entomol. Exp. Appl.* **2008**, *128*, 196–207. [CrossRef]
17. Borg-Karlson, A.K.; Groth, I.; Agren, L.; Kullenberg, B. Form-specific fragrances from *Ophrys insectifera* L. (Orchidaceae) attract species of different pollinator genera: Evidence of sympatric speciation? *Chemoecology* **1993**, *4*, 39–45. [CrossRef]
18. Kessler, D.; Gase, K.; Baldwin, I.T. Field experiments with transformed plants reveal the sense of floral scents. *Science* **2008**, *321*, 1200–1202. [CrossRef]
19. Paulus, H.F.; Gack, C. Pollination of *Ophrys* (Orchidaceae) in Cyprus. *Plant Syst. Evol.* **1990**, *169*, 177–207. [CrossRef]
20. Dudareva, N.; Klempien, A.; Muhlemann, J.K.; Kaplan, I. Biosynthesis, function and metabolic engineering of plant volatile organic compounds. *New Phytol.* **2013**, *198*, 16–32. [CrossRef]
21. Chittka, L.; Raine, N.E. Recognition of flowers by pollinators. *Curr. Opin. Plant Biol.* **2006**, *9*, 428–435. [CrossRef]
22. Delle-Vedove, R.; Juillet, N.; Bessière, J.M.; Dormont, L.; Pailler, T.; Schatz, B. Colour-scent associations in a tropical orchid: Three colours but two odours. *Phytochemistry* **2011**, *72*, 735–742. [CrossRef]
23. Hsiao, Y.; Tsai, W.; Kuoh, C.; Huang, T.; Wang, H.; Wu, T.; Leu, Y.; Chen, W.; Chen, H. Comparison of transcripts in Phalaenopsis bellina and Phalaenopsis equestris (Orchidaceae) flowers to deduce the monoterpene biosynthesis pathway. *BMC Plant Biol.* **2006**, *6*, 14. [CrossRef]
24. Hsiao, Y.; Jeng, M.; Tsai, W.; Chung, Y.; Li, C.; Wu, T.; Kuoh, C.; Chen, W.; Chen, H. A novel homodimeric geranyl diphosphate synthase from the orchid *Phalaenopsis bellina* lacking a DD(X)2-4D motif. *Plant J.* **2008**, *55*, 719–733. [CrossRef] [PubMed]
25. Leitch, I.J.; Kahandawala, I.; Suda, J.; Hanson, L.; Ingrouille, M.J.; Chase, M.W.; Fay, M.F. Genome size diversity in orchids: Consequences and evolution. *Ann. Bot.* **2009**, *104*, 469–481. [CrossRef] [PubMed]
26. Jersáková, J.; Trávníček, P.; Kubátová, B.; Krejčíková, J.; Urfus, T.; Liu, Z.J.; Lamb, A.; Ponert, J.; Schulte, Z.K.; Čurn, V.; et al. Genome size varia- tion in Orchidaceae subfamily Apostasioideae: Filling the phylogenetic gap. *Bot. J. Linn. Soc.* **2013**, *172*, 95–105.
27. Dudareva, N.; Pichersky, E. *Biology of Floral Scent*; CRC Press: Boca Raton, FL, USA, 2006.
28. Ramya, M.; Park, P.H.; Chen, Y.C.; Kwon, O.K.; An, H.R.; Park, P.M.; Baek, Y.S.; Kang, B.C.; Tsai, W.C.; Chen, H.H. RNA sequencing analysis of *Cymbidium goeringii* identifies floral scent biosynthesis related genes. *BMC Plant Biol.* **2019**, *337*, 1–15. [CrossRef]
29. Ramya, M.; Lee, S.Y.; An, H.R.; Park, P.M.; Kim, N.S.; Park, P.H. MYB1 transcription factor regulation through floral scent in *Cymbidium* cultivar 'Sael Bit'. *Phytochem. Lett.* **2019**, *32*, 181–187. [CrossRef]
30. Gupta, A.K.; Schauvinhold, I.; Pichersky, E.; Schiestl, F.P. Eugenol synthase genes in floral scent variation in *Gymnadenia* species. *Funct. Integr. Genom.* **2014**, *14*, 779. [CrossRef]

31. Huanga, M.; Maa, C.; Yub, R.; Mub, L.; Houa, J.; Yua, Y.; Fana, Y. Concurrent changes in methyl jasmonate emission and the expression of its biosynthesis-related genes in *Cymbidium ensifolium* flowers. *Physiol. Plant.* **2015**, *153*, 503–512. [CrossRef]
32. Blight, M.M.; Le Metayer, M.; Delegue, M.H.P.; Pickett, J.A.; Marion-Poll, F.; Wadhams, L.J. Identification of floral volatiles involved in recognition of oilseed rape flowers, *Brassica napus* by honeybees, *Apis mellifera*. *J. Chem. Ecol.* **1997**, *23*, 1715–1727. [CrossRef]
33. Byers, K.J.; Bradshaw, H.D.; Riffell, J.A. Three floral volatiles contribute to differential pollinator attraction in monkeyflowers (*Mimulus*). *J. Exp. Biol.* **2014**, *217*, 614–623. [CrossRef]
34. Junker, R.R.; Gershenzon, J.; Unsicker, S.B. Floral odor bouquet loses its ant repellent properties after inhibition of terpene biosynthesis. *J. Chem. Ecol.* **2011**, *37*, 1323–1331. [CrossRef] [PubMed]
35. Huang, M.; Sanchez-Moreiras, A.M.; Abel, C.; Sohrabi, R.; Lee, S.; Gershenzon, J.; Tholl, D. The major volatile organic compound emitted from *Arabidopsis thaliana* flowers, the sesquiterpene (*E*)-β-caryophyllene, is a defense against a bacterial pathogen. *New Phytol.* **2012**, *193*, 997–1008. [CrossRef] [PubMed]
36. Caputi, L.; Aprea, E. Use of terpenoids as natural flavouring compounds in food industry. *Recent Pat. Food Nutr. Agric.* **2011**, *3*, 9–16. [CrossRef] [PubMed]
37. Schwab, W.; Davidovich-Rikanati, R.; Lewinsohn, E. Biosynthesis of plant-derived flavor compounds. *Plant J.* **2008**, *54*, 712–732. [CrossRef] [PubMed]
38. McGarvey, D.J.; Croteau, R. Terpenoid metabolism. *Plant Cell.* **1995**, *7*, 1015–1026.
39. Baek, Y.S.; Ramya, M.; An, H.R.; Park, P.M.; Lee, S.Y.; Baek, N.I.; Park, P.H. Volatiles Profile of the Floral Organs of a New Hybrid *Cymbidium*, 'Sunny Bell' Using Headspace Solid-Phase Microextraction Gas Chromatography-Mass Spectrometry Analysis. *Plants* **2019**, *8*, 251. [CrossRef]
40. Zhang, Y.; Wang, Y.; Tian, M.; Zhou, W.W. Analysis of aroma components in different orchid varieties. *J. Anal. Sci.* **2012**, *28*, 502–506.
41. Zhang, Y.; Tian, M.; Wang, C.X.; Chen, S. Component analysis and sensory evaluation of flower aroma of Oncidium Sharry Baby 'Sweet Fragrance' under different temperature conditions. *J. Plant Resour. Environ.* **2015**, *24*, 112–114.
42. Mohd-Hairul, A.R.; Parameswari, N.; Gwendoline Ee, C.L.; Janna, O.A. Terpenoid, benzenoid and phenylpropanoid compounds in the floral scent of Vanda Mimi Palmer. *J. Plant Biol.* **2010**, *53*, 358–366. [CrossRef]
43. Kim, S.Y.; An, H.Y.; Park, P.M.; Baek, Y.S.; Kwon, O.K.; Park, S.Y.; Park, P.H. Analysis of floral scent patterns in flowering stages and floral organs of *maxillaria* using an electronic nose. *Flower Res. J.* **2016**, *24*, 171–180. [CrossRef]
44. Omata, A.; Nakamura, S.; Yomogia, K.; Moriai, K.; Ichikawa, Y.; Watanabe, I. Volatile components of To-Yo-Ran flowers (*Cymbidium faberi* and *Cymbidium virescens*). *Agric. Biol. Chem.* **1990**, *54*, 1029–1033. [CrossRef]
45. Verdonk, J.C.; Haring, M.A.; van Tunen, A.J.; Schuurink, R.C. ODORANT1 regulates fragrance biosynthesis in petunia flowers. *Plant Cell* **2005**, *17*, 1612–1624. [CrossRef]
46. Colquhoun, T.A.; Marciniak, D.M.; Wedde, A.E.; Kim, J.Y.; Schwieterman, M.L.; Levin, L.A.; Moerkercke, A.V.; Schuurink, R.C.; Clark, D.G. A peroxisomally localized acyl-activating enzyme is required for volatile benzenoid formation in a Petunia × hybrida cv. 'Mitchell Diploid' flower. *J. Exp. Bot.* **2012**, *63*, 4821–4833. [CrossRef] [PubMed]
47. Schlüter, P.M.; Xu, S.; Gagliardini, V.; Whittle, E.; Shanklin, J.; Grossniklaus, U.; Schiestl, F.P. Stearoyl-acyl carrier protein desaturases are associated with floral isolation in sexually deceptive orchids. *Proc. Natl. Acad. Sci. USA* **2011**, *108*, 5696–56701. [CrossRef] [PubMed]
48. Wildermuth, M.C. Variations on a theme: Synthesis and modification of plant benzoic acids. *Curr. Opin. Plant Biol.* **2006**, *9*, 288–296. [CrossRef] [PubMed]
49. Klempien, A.; Kaminaga, Y.; Qualley, A.; Nagegowda, D.A.; Widhalm, J.R.; Orlova, I.; Shasany, A.K.; Taguchi, G.; Kish, C.M.; Cooper, B.R.; et al. Contribution of CoA ligases to benzenoid biosynthesis in petunia flowers. *Plant Cell* **2012**, *24*, 2015–2030. [CrossRef] [PubMed]
50. Qualley, A.V.; Widhalm, J.R.; Adebesin, F.; Kish, C.M.; Dudareva, N. Completion of the core β-oxidative pathway of benzoic acid biosynthesis in plants. *Proc. Natl. Acad. Sci. USA* **2012**, *109*, 16383–16388. [CrossRef]
51. Huber, F.; Kaiser, R.; Sauter, W.; Schiestl, F. Floral scent emission and pollinator attraction in two species of *Gymnadenia* (Orchidaceae). *Oecologia* **2005**, *142*, 564–575. [CrossRef]

52. Suchet, C.; Dormont, L.; Schatz, B.; Giurfa, M.; Simon, V.; Raynaud, C.; Chave, J. Floral scent variation in two *Antirrhinum majus* subspecies influences the choice of naïve bumblebees. *Behav. Ecol. Sociobiol.* **2011**, *65*, 1015–1027. [CrossRef]
53. Ramya, M.; Kwon, O.K.; An, H.R.; Park, P.M.; Baek, Y.S.; Park, P.H. Floral scent: Regulation and role of MYB transcription factors. *Phytochem. Lett.* **2017**, *19*, 114–120. [CrossRef]
54. Xie, D.Y.; Sharma, S.B.; Wright, E.; Wang, Z.Y.; Dixon, R.A. Metabolic engineering of pro anthocyanidins through co-expression of anthocyanidin reductase and the PAP1 MYB transcription factor. *Plant J.* **2006**, *45*, 895–907. [CrossRef] [PubMed]
55. Zvi, M.M.; Shklarman, E.; Masci, T.; Kalev, H.; Debener, T.; Shafir, S. PAP1 transcription factor enhances production of phenylpropanoid and terpenoid scent compounds in rose flowers. *New Phytol.* **2012**, *195*, 335–345. [CrossRef] [PubMed]
56. Ben Zvi, M.M.; Negre-Zakharov, F.; Masci, T.; Ovadis, M.; Shklarman, E. Interlinking showy traits: Co-engineering of scent and colour biosynthesis in flowers. *Plant Biotechnol. J.* **2008**, *6*, 403–415.
57. Yang, C.Q.; Fang, X.; Wu, X.M.; Mao, Y.B.; Wang, L.J.; Chen, X.Y. Transcriptional regulation of plant secondary metabolism. *J. Integr. Plant Biol.* **2012**, *54*, 703–712. [CrossRef] [PubMed]
58. Xu, Q.; Wang, S.; Hong, H.; Jhou, Y. Transcriptomic profiling of the flower scent biosynthesis pathway of *Cymbidium faberi* Rolfe and functional characterization of its jasmonic acid carboxyl methyltransferase gene. *BMC Genom.* **2019**, *20*, 125. [CrossRef]
59. Chan, W.S.; Abdullah, J.O.; Namasivayam, P. Isolation, cloning, and characterization of fragrance-related transcripts from Vanda Mimi Palmer. *Sci. Hortic.* **2011**, *127*, 388–397. [CrossRef]
60. An, F.M.; Chan, M.T. Transcriptome-wide characterization of miRNA-directed and non-miRNA-directed endonucleolytic cleavage using Degradome analysis under low ambient temperature in *Phalaenopsis aphrodite* subsp. *formosana. Plant Cell Physiol.* **2012**, *53*, 1737–1750. [CrossRef]
61. Zhang, Y.; Li, X.L.; Wang, Y.; Tian, M.; Fan, M.H. Changes of aroma components in Oncidium Sharry Baby in different florescence and flower parts. *Sci. Agric. Sin.* **2011**, *44*, 110–117.
62. Zheng, J.A.; Hu, Z.H.; Guan, X.L.; Dou, D.Q.; Bai, G.; Wang, Y.; Guo, Y.; Li, W.; Leng, P. Transcriptome ssanalysis of *Syringa oblata* lindl: Inflorescence identifies genes associated with pigment biosynthesis and scent metabolism. *PLoS ONE* **2015**, *10*, e142542. [CrossRef]
63. Teh, S.L.; Chan, W.S.; Janna, O.A.; Parameswari, N. Development of expressed sequence tag resources for Vanda Mimi Palmer and data mining for EST-SSR. *Mol. Biol. Rep.* **2011**, *38*, 3903–3909. [CrossRef]
64. Flach, A.; Donon, R.C.; Singer, R.B.; Koehler, S.; Amaral, M.D.E.; Marsaioli, A.J. The chemistry of pollination in selected brazilian Maxillariinae orchids: Floral rewards and fragrance. *J. Chem. Ecol.* **2004**, *30*, 1045–1056. [CrossRef] [PubMed]
65. Perraudin, F.; Popovici, J.; Bertrand, C. Analysis of headspace-solid micro extracts from flowers of *Maxillaria tenuifolia* Lindl. by GC-MS. *Electron. J. Nat. Subst.* **2006**, *1*, 1–5.
66. Baek, Y.S.; Kim, S.K.; Park, P.H.; An, H.R.; Park, P.M.; Baek, N.I.; Kwon, O.K. Analysis of volatile floral scents in *maxillaria* species and cultivars. *Flower Res. J.* **2016**, *24*, 282–289. [CrossRef]
67. Been, C.G.; Kang, S.B.; Kim, D.G.; Cha, Y.J. Analysis of fragrant compounds and gene expression in fragrant *Phalaenopsis. Flower Res. J.* **2014**, *22*, 255–263. [CrossRef]
68. Colquhoun, T.A.; Verdonk, J.C.; Schimmel, B.C.; Tieman, D.M.; Underwood, B.A.; Clark, D.G. Petunia floral volatile benzenoid/phenylpropanoid genes are regulated in a similar manner. *Phytochemistry* **2010**, *71*, 158–167. [CrossRef] [PubMed]
69. Pichersky, E.; Lewinsohn, E.; Croteau, R. Purification and characterization of Slinalool synthase: An enzyme involved in the production of floral scent in *Clarkia breweri. Arch. Biochem. Biophys.* **1995**, *316*, 803–807. [CrossRef]
70. Chen, F.; Tholl, D.; Bohlmann, J.; Pichersky, E. The family of terpene synthases in plants: A mid-size family of genes for specialized metabolism that is highly diversified throughout the kingdom. *Plant J.* **2011**, *66*, 212–229. [CrossRef]
71. Tsai, W.C.; Dievar, A.; Hsu, C.C.; Hsiao, Y.Y.; Chiou, S.Y.; Huang, H.; Chen, H.H. Post genomics era for orchid research. *Bot. Stud.* **2017**, 58–61. [CrossRef]
72. Cao, R.; Zhang, Y.; Mann, F.M.; Huang, C.; Mukkamala, D.; Hudock, M.P.; Mead, M.E.; Prisic, S.; Wang, K.; Lin, F.Y.; et al. Diterpene cyclases and the nature of the isoprene fold. *Proteins* **2010**, *78*, 2417–2432. [CrossRef] [PubMed]

73. Kuo, Y.T.; Chao, Y.T.; Chen, W.C.; Shih, M.C.; Chang1, S.B. Segmental and tandem chromosome duplications led to divergent evolution of the chalcone synthase gene family in *Phalaenopsis* orchids. *Ann. Bot.* **2019**, *123*, 69–77. [CrossRef]
74. Schiestl, F.P. The evolution of floral scent and insect chemical communication. *Ecol. Lett.* **2010**, *13*, 643–656. [CrossRef] [PubMed]
75. Johnson, K.B.; Stockwell, V.O. Management of fire blight: A case study in microbial ecology. *Annu. Rev. Phytopathol.* **1998**, *36*, 227–248. [CrossRef] [PubMed]
76. Chen, F.; Tholl, D.; Auria, J.C.; Farooq, A.; Pichersky, E.; Gershenzon, J. Biosynthesis and emission of terpenoid volatiles from *Arabidopsis* flowers. *Plant Cell* **2003**, *15*, 481–494. [CrossRef] [PubMed]
77. Bakkali, F.; Averbeck, S.; Averbeck, D.; Idaomar, M. Biological effects of essential oils-a review. *Food Chem.* **2008**, *46*, 446–475. [CrossRef]
78. Schiestl, F.P.; Huber, F.; Gomez, J. Phenotypic selection on floral scent: Trade-off between attraction and deterrence? *Evol. Ecol.* **2011**, *25*, 237–248. [CrossRef]
79. Pellmyr, O.; Thien, L.B. Insect reproduction and floral fragrances: Keys to the evolution of the angiosperms. *Taxon* **1986**, *35*, 76–85. [CrossRef]
80. Schiestl, F.P.; Ayasse, M. Do changes in floral odor cause speciation in sexually deceptive orchids? *Plant Syst. Evol.* **2002**, *234*, 111–119. [CrossRef]
81. Vereecken, N.J.; Cozzolino, S.; Schiestl, F.P. Hybrid floral scent novelty drives pollinator shift in sexually deceptive orchids. *BMC Evol. Biol.* **2010**, *10*, 103. [CrossRef]
82. Dobson, H.E.M.; Danielson, E.M.; Van Wesep, I.D. Pollen odor chemicals as modulators of bumble bee foraging on *Rosa rugosa* Thunb. (Rosaceae). *Plant Species Biol.* **1999**, *14*, 153–166. [CrossRef]
83. Muhlemann, J.K.; Waelti, M.O.; Widmer, A.; Schiestl, F.P. Postpollination changes in floral odor in *Silene latifolia*: Adaptive mechanisms for seed-predator avoidance? *J. Chem. Ecol.* **2006**, *32*, 1855–1860. [CrossRef]
84. Muhlemann, J.K.; Maeda, H.; Chang, C.Y.; San Miguel, P.; Baxter, I.; Cooper, B.; Dudareva, N. Developmental changes in the metabolic network of snapdragon flowers. *PLoS ONE* **2012**, *7*, e40381. [CrossRef] [PubMed]
85. Chen, H.; Fan, Y.P. Analysis of aroma components of *Oncidium*. *Acta Agric. Univ. Jiangxiensis* **2012**, *34*, 692–698.
86. Hong, G.J.; Xue, X.Y.; Mao, Y.B.; Wang, L.J.; Chen, X.Y. *Arabidopsis* MYC2 interacts with DELLA proteins in regulating sesquiterpene synthase gene expression. *Plant Cell* **2012**, *24*, 2635–2648. [CrossRef] [PubMed]
87. Tsai, W.C.; Fu, C.H.; Hsiao, Y.Y.; Huang, Y.M.; Chen, L.J.; Wang, M.; Liu, Z.J.; Chen, H.H. OrchidBase 2.0: Comprehensive collection of Orchidaceae floral transcriptomes. *Plant Cell Physiol.* **2013**, *54*, 7. [CrossRef]
88. Chung, M.Y.; Nason, J.D. Spatial demographic and genetic consequences of harvesting within papulations of terrestrial orchid *Cymbidium goeringii*. *Biol. Conserv.* **2007**, *137*, 125–137. [CrossRef]
89. Chen, X.Q.; Liu, Z.J.; Zhu, G.H.; Lang, K.Y.; Ji, Z.H.; Luo, Y.B.; Jin, X.H.; Cribb, P.J.; Wood, J.J.; Gale, S.W. Orchidaceae. In *Flora of China*; Science Press & Missouri Botanical Garden Press: Beijing, China, 2009; p. 260.
90. Du Puy, D.; Cribb, P. *The Genus Cymbidium*, 2nd ed.; Surrey, Royal Botanic Gardens, Kew Publishing: London, UK, 2007.
91. Da Silva, J.A.T.; Chan, M.T.; Sanjaya; Chai, M.L.; Tanaka, M. Priming abiotic factors for optimal hybrid *Cymbidium* (Orchidaceae) callus induction, plantlet formation, and their subsequent cytogenetic stability analysis. *Sci. Hortic.* **2006**, *109*, 368–378. [CrossRef]
92. Da Silva, J.A.T.; Singh, N.; Tanaka, M. Priming biotic factors for optimal protocorm-like body in hybrid *Cymbidium* (Orchidaceae), and assessment of cytogenetic stability in regenerated plants. *Plant Cell Tissue Organ Cult.* **2006**, *84*, 119–128. [CrossRef]
93. Da Silva, J.A.T.; Chin, D.P.; Van, P.T.; Mii, M. Transgenic orchids. *Sci. Hortic.* **2011**, *130*, 673–680. [CrossRef]
94. Huan, L.V.T.; Takamura, T.; Tanaka, M. Callus formation and plant regeneration from callus through somatic embryo structures in *Cymbidium* orchid. *Plant Sci.* **2004**, *166*, 1443–1449. [CrossRef]
95. Van den Berg, C.; Ryan, A.; Cribb, P.J.; Chase, M.W. Molecular phylogenetics of *Cymbidium* (Orchidaceae: Maxillarieae): Sequence data from internal transcribed spacers (ITS) of nuclear ribosomal DNA and plastid matK. *Lindleyana* **2002**, *17*, 102–111.
96. Sharma, S.K.; Dkhar, J.; Kumaria, S.; Tandon, P.; Rao, S.R. Assessment of phylogenetic inter-relationships in the genus *Cymbidium* (Orchidaceae) based on internal transcribed spacer region of rDNA. *Gene* **2012**, *495*, 10–15. [CrossRef]

97. Pornarong, S. DNA barcoding of the *Cymbidium* species (Orchidaceae) in Thailand. *Afr. J. Agric. Res.* **2012**, *7*, 393–404. [CrossRef]
98. Tao, J.; Yu, L.; Kong, F.; Zhao, D. Effects of plant growth regulators on in vitro propagation of *Cymbidium faberi* Rolfe. *Afr. J. Biotechnol.* **2011**, *10*, 15639–15646. [CrossRef]
99. Pindel, A. Optimization of isolation conditions of *Cymbidium* protoplasts. *Folia Hortic.* **2007**, *19*, 79–88.
100. Chen, Y.; Liu, X.; Liu, Y. In vitro plant regeneration from the immature seeds of *Cymbidium faberi*. *Plant Cell Tissue Organ Cult.* **2005**, *81*, 247–251. [CrossRef]
101. Huang, W.; Fang, Z.; Zeng, S.; Zhang, J.; Wu, K.; Chen, Z.; da Silva, J.A.T.; Duan, J. Molecular cloning and functional analysis of three FLOWERING LOCUS T (FT) homologous genes from Chinese Cymbidium. *Int. J. Mol. Sci.* **2012**, *13*, 11385–11398. [CrossRef]
102. Li, X.; Luo, J.; Yan, T.; Xiang, L.; Jin, F.; Qin, D.; Sun, C.; Xie, M. Deep sequencing based analysis of the *Cymbidium ensifolium* floral transcriptome. *PLoS ONE* **2013**, *8*, e85480. [CrossRef]
103. Ning, H.; Zhang, C.; Fu, J.; Fan, Y. Comparative transcriptome analysis of differentially expressed genes between the curly and normal leaves of *Cymbidium goeringii* var. longibracteatum. *Genes Genom.* **2016**, *38*, 985–998. [CrossRef]
104. Peng, H. Study on the Volatile, Characteristic Floral Fragrance Components of Chinese Cymbidium. Ph.D. Thesis, Chinese Academy of Forestry, Beijing, China, 2009.
105. Kim, S.M.; Jang, E.J.; Hong, J.W.; Song, S.H.; Pak, C.H. A comparison of functional fragrant components of Cymbidium (Oriental Ochid) species. *Korean J. Hortic. Technol.* **2016**, *34*, 331–341.
106. Christenson, E.A. *Phalaenopsis: A Monograph*; Timber Press: Portland, OR, USA, 2001; pp. 24–25.
107. Kaiser, R. *The Scent of Orchids-Olfactory and Chemical Investigations*; Elsevier: Amsterdam, The Netherlands, 1993; pp. 239–240.
108. Schlossman, M.L. *The Chemistry and Manufacture of Cosmetics*; Allured Publishing Corporation: Carol Stream, IL, USA, 2009; Volume 2, p. 851.
109. Park, P.H.; Ramya, M.; An, H.R.; Park, P.M.; Lee, S.Y. Breeding of Cymbidium 'Sale Bit' with Bright Yellow Flowers and Floral Scent. *Korean J. Breed. Sci.* **2019**, *51*, 258–262. [CrossRef]
110. Tsai, C.C.; Wu, K.M.; Chiang, T.Y.; Huang, C.Y.; Chou, C.H.; Li, S.J.; Chiang, Y.C. Comparative transcriptome analysis of *Gastrodia elata* (Orchidaceae) in response to fungus symbiosis to identify gastrodin biosynthesis-related genes. *BMC Genom.* **2016**, *17*, 212. [CrossRef]
111. Hossain, M.M.; Kant, R.; Van, P.T.; Winarto, B.; Zeng, S.; da Silva, J.A.T. The application of biotechnology to orchids. *Crit. Rev. Plant Sci.* **2013**, *32*, 69–139. [CrossRef]
112. Cai, J.; Liu, X.; Vanneste, K.; Proost, S.; Tsai, W.C.; Liu, K.W.; Chen, L.J.; He, Y.; Xu, Q.; Bian, C.; et al. The genome sequence of the orchid *Phalaenopsis equestris*. *Nat. Genet.* **2015**, *47*, 65–72. [CrossRef] [PubMed]
113. Chen, W.H.; Kao, Y.K.; Tang, C.Y.; Tsai, C.C.; Lin, T.Y. Estimating nuclear DNA content within 50 species of the genus *Phalaenopsis* Blume (Orchidaceae). *Sci. Hortic.* **2013**, *161*, 70–75. [CrossRef]

10

Identification of (Z)-8-Heptadecene and *n*-Pentadecane as Electrophysiologically Active Compounds in *Ophrys insectifera* and its *Argogorytes* Pollinator

Björn Bohman [1,2,*], Alyssa M. Weinstein [3], Raimondas Mozuraitis [4], Gavin R. Flematti [1] and Anna-Karin Borg-Karlson [5]

1 School of Molecular Sciences, the University of Western Australia, Crawley, WA 6009, Australia
2 Department of Plant Protection Biology, Swedish University of Agricultural Sciences, 23053 Alnarp, Sweden
3 Research School of Biology, the Australian National University, Acton, ACT 2601, Australia
4 Department of Zoology, Stockholm University, 106 91, Stockholm, Sweden
5 Department of Chemical Engineering, Mid Sweden University, 85170 Sundsvall, Sweden
* Correspondence: bjorn.bohman@uwa.edu.au

Abstract: Sexually deceptive orchids typically depend on specific insect species for pollination, which are lured by sex pheromone mimicry. European *Ophrys* orchids often exploit specific species of wasps or bees with carboxylic acid derivatives. Here, we identify the specific semiochemicals present in *O. insectifera*, and in females of one of its pollinator species, *Argogorytes fargeii*. Headspace volatile samples and solvent extracts were analysed by GC-MS and semiochemicals were structurally elucidated by microderivatisation experiments and synthesis. (Z)-8-Heptadecene and *n*-pentadecane were confirmed as present in both *O. insectifera* and *A. fargeii* female extracts, with both compounds being found to be electrophysiologically active to pollinators. The identified semiochemicals were compared with previously identified *Ophrys* pollinator attractants, such as (Z)-9 and (Z)-12-C_{27}-C_{29} alkenes in *O. sphegodes* and (Z)-9-octadecenal, octadecanal, ethyl linoleate and ethyl oleate in *O. speculum*, to provide further insights into the biosynthesis of semiochemicals in this genus. We propose that all these currently identified *Ophrys* semiochemicals can be formed biosynthetically from the same activated carboxylic acid precursors, after a sequence of elongation and decarbonylation reactions in *O. sphegodes* and *O. speculum*, while in *O. insectifera*, possibly by decarbonylation without preceding elongation.

Keywords: *Ophrys*; sexual deception; semiochemicals; fly orchid; pollination

---

1. Introduction

Pseudocopulation as a means of pollination was first reported over 100 years ago, in two parallel systems [1,2]. Correvon and Pouyanne made observations of European *Ophrys* orchids [1], while in Australia, *Cryptostylis* orchids were reported to use the same sexually deceptive strategy, in which insect pollinators attempt copulatory or courtship behaviour with the flower, thereby transferring pollinia [2,3]. Insect sexual attraction is induced through chemical and physical mimicry of female insects. Pollination by sexual deception is now known to be a phenomenon that has evolved independently multiple times on different continents. There are several hundred confirmed cases in the Orchidaceae, with many more likely to be discovered with future studies [4–6]. There are also single reports of sexual deception in the Asteraceae [7] and Iridaceae [8], indicating that this pollination strategy may be more common than is currently known.

Following the initial observations of pollination via sexual deception in *Ophrys* and *Cryptostylis* orchids, an intensive Swedish research program was launched in 1948 to investigate the chemical cues underlying this bizarre pollination strategy. *Ophrys insectifera* and some southern European *Ophrys* and their solitary bee pollinator species were the main study species [9]. In these early studies, field experiments demonstrated that floral volatiles were the key to pollinator attraction [9,10]. With the use of electroantennography (EAG), it was later shown that two species of male sphecid wasp pollinator, *Argogorytes mystaceus* and *A. fargeii*, unlike their conspecific females, responded to tentatively identified alkanes, alkenes, and terpenes in sorption headspace extracts of *O. insectifera* flowers [11]. A few years later, the first evidence of chemical mimicry of several species of *Andrena* bee pollinators by *O. fusca* and *O. lutea*, was found: aliphatic alcohols, monoterpene- and sesquiterpene alcohols, and aldehydes attracted the patrolling males to varying degrees [12,13].

The first identification of pollinator sexual attractants in the genus *Ophrys* did not occur until the late 1990s, with the successful structural elucidation of attractants from *O. sphegodes* [14,15]. A key to the detection and identification of the semiochemicals from this species was the use of gas chromatography coupled with electroantennogram detection (GC-EAD), which revealed a set of 14 electrophysiologically active compounds to be shared among the orchid and the female of its bee pollinator, *Andrena nigroaenea*. Before being confirmed as attractants in field bioassays, these compounds were identified by GC-MS, including microderivatisation experiments, as a series of long-chained alkanes and alkenes. Furthermore, three (Z)-7 alkenes were discovered to be responsible for the attraction of male *Colletes cunicularius* bees to *O. exaltata* [16]. The chemical stimuli for the sexual attraction of various *Ophrys* pollinators also include other types of structures, as shown when a mixture of hydroxy- and keto acids, together with aldehydes and esters, were identified as the attractants in *O. speculum*, which is pollinated by male *Campsoscolia ciliata* scoliid wasps [17].

In Australian sexually deceptive orchids, 1,3-cyclohexanediones (chiloglottones) have been identified as pollinator attractants in *Chiloglottis* [18], as have hydroxymethylpyrazines and a β-hydroxylactone (drakolide) in *Drakaea* [19–22], (methylthio)phenols, acetophenones and monoterpenes in *Caladenia* [23–25], and tetrahydrofuran acid derivatives in *Cryptostylis* [26].

Besides the discovery of a broad range of compounds pivotal for pollination in sexually deceptive orchids, there has also been interest in the biosynthesis of these compounds, with the aim to link biosynthesis to the evolution and speciation of orchids. Schlüter and Schiestl [27] predicted that, in *Ophrys*, the biosynthesis of alkenes would follow the biosynthetic pathway for alkanes [28], but with the addition of an extra desaturation step, potentially achieved by stearoyl-acyl carrier protein desaturases (SAD). Later, three putative SAD genes (SAD1-SAD3) were isolated [29]. Transgenic expression and in vitro enzyme assays revealed SAD2 to be a functional desaturase capable of introducing 18:1 Δ^9 and 16:1 Δ^4 fatty acid intermediates, from which it was hypothesized that (Z)-9 alkenes and (Z)-12 alkenes are built. Three additional putative SAD genes (SAD4-SAD6) were also identified from an *O. sphegodes* transcriptome [30].

In *O. sphegodes* and *O. exaltata*, SAD1 and SAD2 expression levels were shown to be significantly correlated with (Z)-9 and (Z)-12-alkene production, while high SAD5 expression was correlated with the (Z)-7-alkene production unique to *O. exaltata* [31]. In vitro enzyme activity studies further showed that a putative housekeeping desaturase, SAD3, catalyses the general reactions of stearate to oleate (18:0-ACP to 18:1 Δ^9-ACP), and palmitate to palmitoleate (16:0-ACP to 16:1 Δ^9-ACP), whereas SAD5 is a specialized 16:0 Δ^9-ACP enzyme [32]. Subsequent elongation of a 16:1 Δ^9-ACP to a 26:1 Δ^{19}-coenzyme A precursor, followed by decarbonylation, would yield the (Z)-7 alkene (25:1 Δ^7) that characterizes *O. exaltata*.

In *O. speculum*, the pollinator attractants were also identified as carboxylic acid derivatives [17]. The most attractive compounds from both floral extracts and females of the scoliid wasp pollinator

Campsoscolia ciliata were (ω-1)-hydroxy- and -oxo acids. However, it is noteworthy that the pseudo-copulation rates in field bioassay experiments more than doubled when aldehydes such as (Z)-9-octadecenal and octadecanal, together with the esters ethyl linoleate and ethyl oleate, were added to the dummy female [17].

The phylogenetic relationships within *Ophrys* are currently under debate [33–37], with some phylogenetic analyses indicating that the *Argogorytes*-pollinated *O. insectifera* group represents a basal taxon, while the latest studies place the *O. fusca* complex, including *O. iricolor*, as ancestral [36,37]. All studies agree that wasp pollination is ancestral to bee pollination in *Ophrys*.

To obtain a broader understanding of the chemical details of semiochemicals in the wasp-pollinated *O. insectifera*, and sex pheromone candidates in its pollinator *A. fargeii*, we used GC-EAD, GC-MS, microderivatisation reactions, and organic synthesis to identify EAD-active compounds. These semiochemicals were compared with previously identified pollinator attractants from the bee-pollinated *O. sphegodes* and wasp-pollinated *O. speculum*, and biosynthetic relationships within *Ophrys* were proposed.

2. Results and Discussion

To identify semiochemicals in *O. insectifera*, and sex pheromone candidates in *Argogorytes fargeii* pollinators, solvent extractions of flowers and insects, and floral headspace sampling, were conducted. Samples of *O. insectifera* labella were extracted in solvents of increasing polarity, from *n*-hexane, to dichloromethane, to methanol. Headspace volatile sampling was performed using solid phase extraction (SPME). Furthermore, whole females of *A. fargeii* were extracted in dichloromethane. Due to the very limited number of pollinators available, we were restricted to evaluating biological activity using gas chromatography coupled with electroantennography (GC-EAD). Since we were unable to locate males of *A. fargeii*, GC-EAD was used to detect which components of the various extracts were detected by *A. mystaceus*, a closely related species that is the second main pollinator of *O. insectifera* [9]. Two compounds from the floral extracts were repeatedly EAD-active (elicited responses in six out of 10, and two out of 10 EAD experiments). These two compounds were tentatively identified by mass spectrometry (GC-MS) as a C17 alkene and *n*-pentadecane. In previous studies on *O. insectifera*, *n*-pentadecane (**2**, Figure 1a) was indeed found to be active in EAG experiments, while no alkenes were isolated or identified [11]. Here, we found that *n*-pentadecane and the C17 alkene were present in the female *A. fargeii* (six extracts of individual insects) and were also present in only minor amounts in floral solvent extracts (three extracts of 10 flowers). We investigated the double bond location by dimethyldisulfide (DMDS) microderivatisation of a semi-preparative GC purified compound that was extracted from the wasp. The observation of identical retention times and mass spectra between the semiochemical isolated from the wasp and the synthesized (Z)-8-heptadecene (**1**), before and after treatment with DMDS, meant that the double bond position and configuration of the natural product could be confirmed. Furthermore, a floral extract was treated analogously, and was confirmed to contain identical mass fragments at the same relative intensity and retention time, confirming that the compound detected by *A. mystaceus* was shared between *O. insectifera* and female *A. fargeii*. In addition to the semiochemicals identified from flowers, another two C15-alkenes and one C17-diene were identified from females of *A. fargeii*. These compounds were also isolated by semi-preparative GC and treated with DMDS. Candidate compounds were synthesized and co-injected with natural extracts (on two GC columns) and tested with GC-EAD. The monoenes were subsequently confirmed as (Z)-6-pentadecene (**3**) and (Z)-7-pentadecene (**4**), while the diene was identified as (Z,Z)-6,9-heptadecadiene (**5**) (Figure 1).

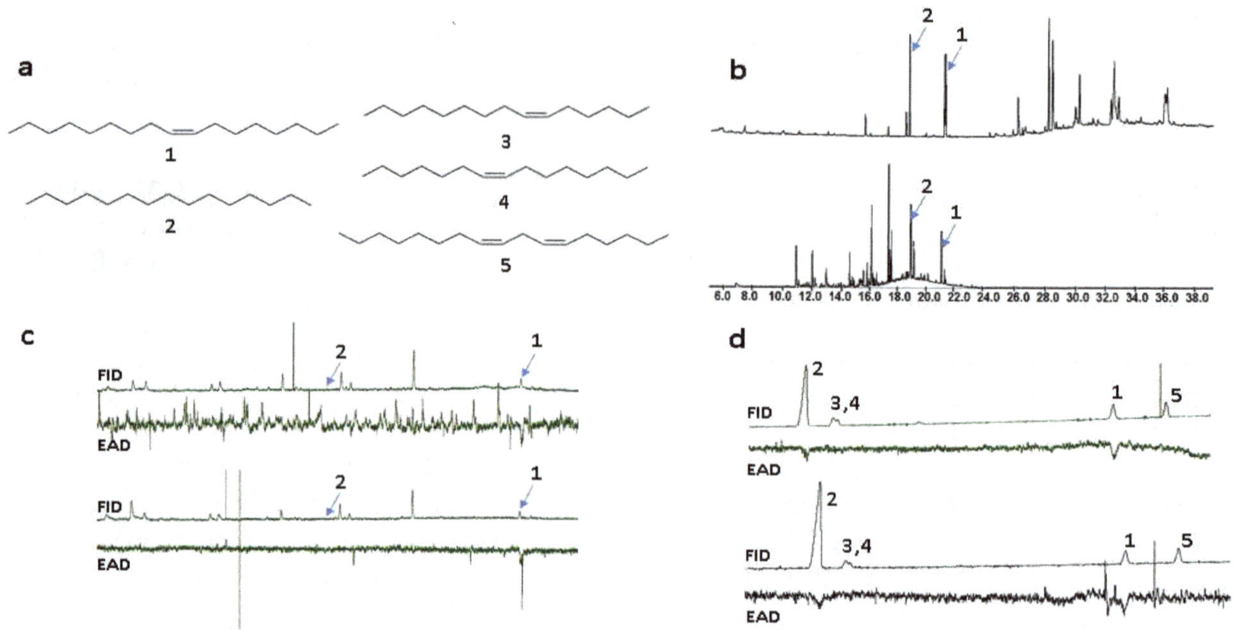

Figure 1. (**a**) Semiochemicals from *Ophrys insectifera* (**1–2**; **1** = (Z)-8-heptadecene, **2** = *n*-pentadecane) and female *Argogorytes mystaceus* (**1–5**; **3** = (Z)-6-pentadecene, **4** = (Z)-7-pentadecene, **5** = (Z,Z)-6,9-heptadecadiene). (**b**) GC-MS total ion chromatograms of female *A. fargeii* (upper trace) and *O. insectifera* (lower trace). (**c**) GC-EAD of SPME extracts of *O. insectifera* to antenna of *A. mystaceus* males. Two replicated analyses are shown. (**d**) GC-EAD of synthetic standards **1–5** to antenna of *A. mystaceus*. Two replicated analyses are shown.

The GC-EAD and GC-MS analyses of the floral extracts showed that *n*-pentadecane (**2**) was of low abundance and was electrophysiologically active in only two experiments, while (Z)-8-heptadecene (**1**) was active in six experiments. When tested as synthetics at higher concentrations (100 ng to 1 μg), both compounds were strongly EAD-active in replicated experiments. However, the additional alkenes **3–5** from *A. fargeii*, when tested as synthetic samples at the higher concentration, elicited consistently less frequent and/or weaker EAD responses compared to the orchid-produced **1** and **2** (Figure 1, Table 1).

Table 1. Occurrence of semiochemicals in *Ophrys insectifera* (SPME extracts) and *Argogorytes fargeii* females (solvent extracts), with electroantennographic responses in *A. mystaceus* males.

	Compound	Abundance in *O. insectifera*	Abundance in *A. fargeii* (Female)	EAD-Activity
	1	✔✔	✔✔	✔✔
	2	✔	✔✔	✔
	3	–	✔	(✔)
Argogorytes mystaceus visiting	4	–	✔	(✔)
Ophrys insectifera	5	–	✔✔	(✔)

✔✔= very abundant compound (>20% of base peak area); repeated (6 extracts, >6 synthetic samples) strong EAD-responses. ✔ = abundant compound (>10% of base peak area); repeated EAD-responses (2 extracts, >6 synthetic samples). (✔) = occasional weaker EAD-response (generally less than 50% of response of orchid semiochemicals, >3 synthetic samples). Photo A.M. Weinstein.

By analysing the GC-MS traces of floral extracts, it was observed that larger amounts of compounds **1** and **2** were present in headspace samples of flowers compared with solvent extracts. Although headspace extractions and solvent extractions are not directly comparable, our findings indicate that the flowers likely continuously produce compounds (indicated by increasing quantity with an increase in SPME sampling time), rather than depend on stored compounds (indicated by very low amounts in solvent extracts) in the floral tissue. This observation is in agreement with earlier studies of *O. insectifera* and *O. sphegodes*, favouring headspace sorption extraction over solvent extraction [14,38].

In addition to comparing observations between various *Ophrys* systems, it is of further interest to extend this comparison to other sexually deceptive orchids with known semiochemistry. Such cases are predominantly Australian, where the pollinator attractants in hammer orchids and spider orchids, unlike in *Ophrys*, have been found to be stored in relatively large amounts within the floral tissue [21,23,25].

The discovery of (Z)-8-heptadecene (**1**) in *O. insectifera*, detected by males of *A. mystaceus*, provides important insights about the chemistry of *Ophrys* orchids. In earlier studies of the biosynthetic pathways for the longer chained C_{25} and C_{27} alkenes from *O. exaltata* and *O. sphegodes*, C_{16}- and C_{18} activated carboxylic acids have been proposed as intermediates [32] (Figure 2). In fact, it has been proposed that in the plastid of the lip epidermis cell of the labellum of *O. exaltata* and *O. sphegodes*, 16:0-ACP and 18:0-ACP are transformed to 16:1 Δ^4-ACP and 18: 1 Δ^9-ACP by SAD2, before being elongated in the cuticle [29]. If instead, 16:0-ACP and 18:1 Δ^9-ACP are decarbonylated, the exact compounds found to be EAD-active in *O. insectifera*, *n*-pentadecane (**2**) and (Z)-8-heptadecane (**1**), would be formed (Figure 2).

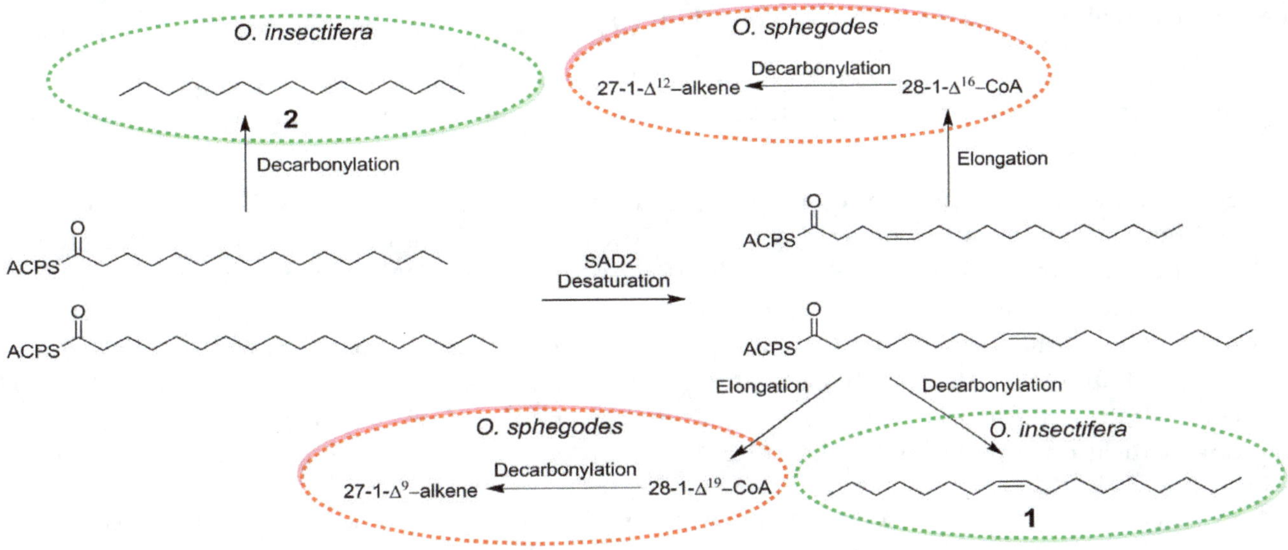

Figure 2. Proposed biosynthesis of bioactive alkenes in *Ophrys sphegodes* (from [32]) and *O. insectifera*.

In a similar manner, our results can be compared to the pollinator attractants previously identified in *O. speculum*. Out of the blend of eight electrophysiologically active compounds that showed the highest pollinator attraction in field bioassays, three compounds: hexadecanal, (Z)-9-octadecenal, and ethyl oleate, show strong structural similarity with the hydrocarbons that we identified in *O. insectifera*. In fact, decarbonylation of these semiochemicals, in a similar way as proposed in the case of *O. sphegodes* (Figure 2), would yield pentadecane (**2**) from hexadecanal and (Z)-8-heptadecene (**1**) from (Z)-9-octadecenal and ethyl oleate.

Compared to the recent studies of Australian *Drakaea* and *Caladenia* orchids, where multiple, structurally diverse pollinator attractants have been identified in multiple species [21–23,25], the structural similarities between the semiochemicals of *O. insectifera*, *O. sphegodes*, and *O. speculum* are evident, all being clearly biosynthetically closely related carboxylic acid derived compounds. It is also interesting to note the difference in volatility compared to the widely studied Australian systems, where "traditional" volatiles are used as long-range attractants, while the European systems utilise less volatile cuticular hydrocarbons, such as the C_{27}–C_{29} alkenes in *O. sphegodes*, which have been proven sufficiently volatile to lure pollinators from a distance as attractants [15]. Furthermore, it is relevant to note that in the case of *O. insectifera* and *A. fargeii*, the orchid and pollinator share the exact same semiochemicals, which is in agreement with other investigated *Ophrys* systems, including *O. sphegodes* [39] and *O. speculum* [17], as well as with most Australian systems [4] (but see [40]).

In conclusion, we have identified (*Z*)-8-heptadecene (**1**) and pentadecane (**2**) as shared semiochemicals from *O. insectifera* and *A. fargeii*. Access to denser populations of *A. fargeii* or *A. mystaceus* would be required to undertake bioassays testing the field activity of these compounds as pollinator attractants. Nevertheless, this study provides an important first step in the identification of key compounds that, once pollinator populations have been located, are available to be tested in field behavioural bioassays. Furthermore, the identification of these semiochemicals and comparison with related species within the genus shows strong commonalities in structures and suggests a conserved biosynthetic pathway for semiochemical production within *Ophrys*.

3. Materials and Methods

3.1. General Experimental Procedures

NMR spectra (Supplementary Materials) were acquired on a Bruker Avance (Bruker, Billerica, MA, USA) 500 or 600 MHz spectrometer with $CDCl_3$ as solvent. Chemical shifts were calibrated to resonances attributed to the residual solvent signal.

EIMS (70 eV) were recorded on an Agilent 5973 mass detector connected to an Agilent 6890 GC equipped with a DB5-MS column (Agilent, Santa Clara, CA, USA, 50 m × 0.2 mm × 0.33 μm) using helium as the carrier gas.

Semi-preparative gas chromatography was performed on an HP 5890 GC (Agilent, USA), equipped with a three-way glass splitter separating the gas flow post column into the FID and the collector. An RTX-5 column, 30 m × 0.53 mm id × 5 μm film (Restek, Bellefonte, PA, USA) was used. Samples of 3 μL were injected in splitless mode (1 min) and helium was used as the carrier gas. A custom-made manual fraction collector was used, with samples collected in glass capillaries (100 × 1.55 mm id, Hirschmann Laborgeräte, Eberstadt, Germany) positioned in an aluminium holder submerged in a dry ice/acetone bath. All fractions were eluted with dichloromethane and stored at −20 °C until used for microderivatisation experiments [26].

GC-EAD data were recorded using an Agilent 6890 GC equipped with an identical column as the GC-MS and a flame ionization detector (FID) using helium as carrier gas. A GC effluent splitter (split ratio 1:1) was used to split the flow to the FID and EAD. The split for EAD was passed through a Syntech effluent conditioner (Syntech, Buchenbach, Germany) containing a heated transfer line, with the outlet placed in a purified and humidified airstream, where the electrodes holding the antenna were presented. For each EAD run, an excised antenna with the tip cut off, was mounted on a holder consisting of two electrodes using electrode gel. The electrode was connected to a PC via a serial Syntech intelligent data acquisition controller (IDAC) interface for simultaneous recording of the FID and EAD signals in the Syntech software package.

Solvents for extractions and purifications were of HPLC grade.

3.2. Plant Material and Insects

All plants and insects were collected in June over four years (2016–2019) at various field locations in Sweden. *Ophrys insectifera* were sourced from several populations across the central parts of Öland. Flowers were kept in cooler boxes (ca. 4 °C) while transported to the laboratory, where they were either sampled with solid-phase microextraction (SPME), extracted with hexane, dichloromethane or methanol, or kept as baiting flowers to collect male insects. Male *Argogorytes mystaceus* were collected from *O. insectifera* flowers on stems (20 flowers) near Torslunda, Öland or near Södertälje, Södermanland. Female *Argogorytes fargeii* were collected from food plants, *Pastinaca sativa*, near Långöre, Öland.

3.3. Extraction and Isolation

Flowers for SPME were enclosed in oven bags (Multix 25 cm × 38 cm, McPherson's Limited, Kingsgrove, NSW, Australia) and sampled for 24 h (DVB/CAR/PDMS, Supelco, Bellefonte, PA, USA) at room temperature. For solvent extractions, labella were removed and batches of 20 were extracted in each solvent (ca. 1 mL) for 48 h. The extracts were concentrated under a gentle stream of nitrogen at room temperature to a final volume of ca. 0.1 mL, stored at −20 °C and subsequently used for GC-MS and GC-EAD analyses. Female *A. fargeii* were extracted in dichloromethane (ca. 0.5 mL) for 48 h. The extracts were concentrated and treated as for the floral extracts.

3.4. Structure Elucidations of Alkenes

From a concentrated dichloromethane extract of a female *A. fargeii*, the two fractions containing C15-alkenes and C17-alkenes were isolated by semi-preparative GC. Each fraction, in hexane (30 μL), was treated with DMDS (50 μL) and iodine in diethyl ether (5 μL, 60 mg/mL). The reaction mixtures were left at 40 °C in vials over night before being washed with sodium bisulphite (5%) and concentrated to ca. 20 μL under nitrogen before being analysed by GC-MS [41]. The fraction containing C15-alkenes contained two compounds, with characteristic ions for 6-pentadecene (M = 304, fragments m/z = 131, 173) and 7-pentadecene (M = 304, fragments m/z = 145, 159). The fraction containing C17-alkenes contained one monoene and one diene, with characteristic ions for 8-heptadecene (M = 332, fragments m/z = 159, 173, also present in *O. insectifera*) and 6,9-heptadecene (M = 362, fragments m/z = 131, **155**, 159, **183**, 203 and 231) [42].

3.5. Preparation of Alkenes

All alkenes apart from 7-pentadecene were prepared from the corresponding C16- and C18-carboxylic acids (oleic acid, linoleic acid, palmitoleic acid) via a modified Barton reductive decarboxylation [43]. 7-Pentadecene was synthesised via 7-pentadecyne, prepared from 1-bromoheptyne and octylmagnesium chloride [44]. The alkyne was partially reduced to the *cis*-alkene in a low yield by the method of Obora et al. [45], although in amounts sufficient to our needs.

Author Contributions: Conceptualization, B.B. and A.-K.B.-K; methodology, B.B.; R.M. and G.R.F.; investigation, B.B., A.M.W., A.-K.B.-K.; resources, B.B., R.M. and G.R.F.; writing—original draft preparation, B.B.; writing—review and editing, all authors.; funding acquisition, B.B. All authors have read and agreed to the published version of the manuscript.

Acknowledgments: B.B. acknowledges the Australian Research Council for funding (DE160101313). D. Bainbridge is gratefully acknowledged for designing and fabricating the preparative GC collector used in this study. The authors acknowledge the facilities, and the scientific and technical assistance of the Australian Microscopy & Microanalysis Research Facility at the Centre for Microscopy, Characterization & Analysis, The University of Western Australia, a facility funded by the University, State and Commonwealth Governments. The authors thank Kerstin Persson for allowing the use of her land for this study, Bo G. Svenson for assistance with the identification of insects, and Monica Ruibal and Rod Peakall for DNA sequencing of several of our wasp samples, which provided confirmation of our species identification.

Abbreviations

EAG	Electroantennography
GC	Gas chromatography
EAD	Electroantennographic detection
SAD	Stearoyl-acyl carrier protein desaturase
ACP	Acyl-carrier protein
EIMS	Electron impact mass spectrometry
GCMS	Gas chromatography mass spectrometry
NMR	Nuclear magnetic resonance

FID	Flame ionization detector
HPLC	High-pressure liquid chromatography
SPME	Solid-phase microextraction
DVB	Divinyl benzene
CAR	Carboxen
PDMS	Polydimethylsiloxane
DMDS	Dimethyl disulphide

References

1. Correvon, H.; Pouyanne, M.A. Un curieux cas de mimétisme chez les Ophrydées. *Journal de la Société Naturelle Horticole de France* **1916**, *4*, 29–47.
2. Coleman, E. Pollination of the orchid *Cryptostylis leptochila*. *Vic. Nat.* **1927**, *44*, 20–23.
3. Coleman, E. Pollination of an Australian orchid by the male ichneumonid *Lissopimpla semipunctata* Kirby. *Trans. R. Entomol. Soc. Lond.* **1929**, *76*, 533–539. [CrossRef]
4. Bohman, B.; Flematti, G.R.; Barrow, R.A.; Pichersky, E.; Peakall, R. Pollination by sexual deception—It takes chemistry to work. *Curr. Opin. Plant Biol.* **2016**, *32*, 37–46. [CrossRef]
5. Gaskett, A.C. Orchid pollination by sexual deception: Pollinator perspectives. *Biol. Rev.* **2011**, *86*, 33–75. [CrossRef]
6. Phillips, R.D.; Scaccabarozzi, D.; Retter, B.A.; Hayes, C.; Brown, G.R.; Dixon, K.W.; Peakall, R. Caught in the act: Pollination of sexually deceptive trap-flowers by fungus gnats in *Pterostylis* (Orchidaceae). *Ann. Bot.* **2014**, *113*, 629–641. [CrossRef]
7. Ellis, A.G.; Johnson, S.D. Floral mimicry enhances pollen export: The evolution of pollination by sexual deceit outside of the Orchidaceae. *Am. Nat.* **2010**, *176*, E143–E151. [CrossRef]
8. Vereecken, N.J.; Wilson, C.A.; Hötling, S.; Schulz, S.; Banketov, S.A.; Mardulyn, P. Pre-adaptations and the evolution of pollination by sexual deception: Cope's rule of specialization revisited. *Proc. R. Soc. B Biol. Sci.* **2012**, *279*, 4786–4794. [CrossRef]
9. Kullenberg, B. On the scents and colours of *Ophrys* flowers and their specific pollinators among the aculeate Hymenoptera. *Svensk Botanisk Tidskrift* **1956**, *50*, 25–46.
10. Kullenberg, B. Studies in *Ophrys* pollination. *Zoologiska Bidrag Från Uppsala* **1961**, *34*, 1–340.
11. Ågren, L.; Borg-Karlson, A.-K. Responses of *Argogorytes* (Hymenoptera: Sphecidae) males to odor signals from *Ophrys insectifera* (Orchidadeae). Preliminary EAG and chemical investigation. *Nova Acta Regiae Soc. Sci. Ups. C Bot. Gen. Geol. Phys. Geogr. Palaeontol. Zool.* **1984**, *3*, 111–117.
12. Borg-Karlson, A.-K. Chemical basis for the relationship between *Ophrys* orchids and their pollinators I. Volatile compounds of *Ophrys lutea* and *O. fusca* as insect mimetic attractants/excitants. *Chem. Scr.* **1985**, *25*, 283–294.
13. Borg-Karlson, A.K.; Tengö, J. Odor mimetism? *J. Chem. Ecol.* **1986**, *12*, 1927–1941. [CrossRef] [PubMed]
14. Schiestl, F.P.; Ayasse, M.; Paulus, H.F.; Erdmann, D.; Francke, W. Variation of floral scent emission and postpollination changes in individual flowers of *Ophrys sphegodes* subsp. *Sphegodes*. *J. Chem. Ecol.* **1997**, *23*, 2881–2895. [CrossRef]
15. Schiestl, F.P.; Ayasse, M.; Paulus, H.F.; Löfstedt, C.; Hansson, B.S.; Ibarra, F.; Francke, W. Orchid pollination by sexual swindle. *Nature* **1999**, *399*, 421. [CrossRef]
16. Mant, J.; Brändli, C.; Vereecken, N.J.; Schulz, C.M.; Francke, W.; Schiestl, F.P. Cuticular hydrocarbons as sex pheromone of the bee *Colletes cunicularius* and the key to its mimicry by the sexually deceptive orchid, *Ophrys exaltata*. *J. Chem. Ecol.* **2005**, *31*, 1765–1787. [CrossRef] [PubMed]
17. Ayasse, M.; Schiestl, F.P.; Paulus, H.F.; Ibarra, F.; Francke, W. Pollinator attraction in a sexually deceptive orchid by means of unconventional chemicals. *Proc. R. Soc. Lond. Ser. B Biol. Sci.* **2003**, *270*, 517–522. [CrossRef]
18. Schiestl, F.P.; Peakall, R.; Mant, J.G.; Ibarra, F.; Schulz, C.; Franke, S.; Francke, W. The chemistry of sexual deception in an orchid-wasp pollination system. *Science* **2003**, *302*, 437–438. [CrossRef]
19. Bohman, B.; Jeffares, L.; Flematti, G.; Byrne, L.T.; Skelton, B.W.; Philips, R.D.; Dixon, K.W.; Peakall, R.; Barrow, R.A. Discovery of tetrasubstituted pyrazines as semiochemicals in a sexually deceptive orchid. *J. Nat. Prod.* **2012**, *75*, 1589–1594. [CrossRef]

20. Bohman, B.; Jeffares, L.; Flematti, G.; Phillips, R.D.; Dixon, K.W.; Peakall, R.; Barrow, R.A. The discovery of 2-hydroxymethyl-3-(3-methylbutyl)-5-methylpyrazine: A semiochemical in orchid pollination. *Org. Lett.* **2012**, *14*, 2576–2578. [CrossRef]
21. Bohman, B.; Phillips, R.D.; Menz, M.H.M.; Berntsson, B.W.; Flematti, G.R.; Barrow, R.A.; Dixon, K.W.; Peakall, R. Discovery of pyrazines as pollinator sex pheromones and orchid semiochemicals: Implications for the evolution of sexual deception. *New Phytol.* **2014**, *203*, 939–952. [CrossRef] [PubMed]
22. Bohman, B.; Tan, M.; Phillips, R.; Scaffidi, A.; Sobolev, A.; Moggach, S.; Flematti, G.; Peakall, R. A specific blend of drakolide and hydroxymethylpyrazines—An unusual pollinator sexual attractant used by the endangered orchid *Drakaea micrantha*. *Angew. Chem. Int. Ed.* **2020**, *59*, 1124–1128. [CrossRef] [PubMed]
23. Bohman, B.; Phillips, R.D.; Flematti, G.R.; Barrow, R.A.; Peakall, R. The spider orchid *Caladenia crebra* produces sulfurous pheromone mimics to attract its male wasp pollinator. *Angew. Chem.* **2017**, *56*, 8455–8458. [CrossRef] [PubMed]
24. Bohman, B.; Phillips, R.D.; Flematti, G.R.; Peakall, R. (Methylthio)phenol semiochemicals are exploited by deceptive orchids as sexual attractants for *Campylothynnus* thynnine wasps. *Fitoterapia* **2018**, *126*, 78–82. [CrossRef] [PubMed]
25. Xu, H.; Bohman, B.; Wong, D.C.J.; Rodriguez-Delgado, C.; Scaffidi, A.; Flematti, G.R.; Phillips, R.D.; Pichersky, E.; Peakall, R. Complex sexual deception in an orchid is achieved by co-opting two independent biosynthetic pathways for pollinator attraction. *Curr. Biol.* **2017**, *27*, 1867–1877.e5. [CrossRef]
26. Bohman, B.; Weinstein, A.M.; Phillips, R.D.; Peakall, R.; Flematti, G.R. 2-(Tetrahydrofuran-2-yl)acetic acid and ester derivatives as long-range pollinator attractants in the sexually deceptive orchid *Cryptostylis ovata*. *J. Nat. Prod.* **2019**, *82*, 1107–1113. [CrossRef]
27. Schlüter, P.M.; Schiestl, F.P. Molecular mechanisms of floral mimicry in orchids. *Trends Plant Sci.* **2008**, *13*, 228–235. [CrossRef]
28. Samuels, L.; Kunst, L.; Jetter, R. Sealing plant surfaces: Cuticular wax formation by epidermal cells. *Annu. Rev. Plant Biol.* **2008**, *59*, 683–707. [CrossRef]
29. Schlüter, P.M.; Xu, S.; Gagliardini, V.; Whittle, E.; Shanklin, J.; Grossniklaus, U.; Schiestl, F.P. Stearoyl-acyl carrier protein desaturases are associated with floral isolation in sexually deceptive orchids. *Proc. Natl. Acad. Sci. USA* **2011**, *108*, 5696–5701. [CrossRef]
30. Sedeek, K.E.M.; Qi, W.; Schauer, M.A.; Gupta, A.K.; Poveda, L.; Xu, S.; Liu, Z.-J.; Grossniklaus, U.; Schiestl, F.P.; Schlüter, P.M. Transcriptome and proteome data reveal candidate genes for pollinator attraction in sexually deceptive orchids. *PLoS ONE* **2013**, *8*, e64621. [CrossRef]
31. Xu, S.; Schlüter, P.M.; Grossniklaus, U.; Schiestl, F.P. The genetic basis of pollinator adaptation in a sexually deceptive orchid. *PLOS Genet.* **2012**, *8*, e1002889. [CrossRef] [PubMed]
32. Sedeek, K.E.; Whittle, E.; Guthörl, D.; Grossniklaus, U.; Shanklin, J.; Schlüter, P.M. Amino acid change in an orchid desaturase enables mimicry of the pollinator's sex pheromone. *Curr. Biol.* **2016**, *26*, 1505–1511. [CrossRef] [PubMed]
33. Breitkopf, H.; Onstein, R.E.; Cafasso, D.; Schlüter, P.M.; Cozzolino, S. Multiple shifts to different pollinators fuelled rapid diversification in sexually deceptive *Ophrys* orchids. *New Phytol.* **2015**, *207*, 377–389. [CrossRef] [PubMed]
34. Devey, D.S.; Bateman, R.M.; Fay, M.F.; Hawkins, J.A. Friends or relatives? Phylogenetics and species delimitation in the controversial European orchid genus *Ophrys*. *Ann. Bot.* **2008**, *101*, 385–402. [CrossRef] [PubMed]
35. Tyteca, D.; Baguette, M. *Ophrys* (Orchidaceae) systematics—When molecular phylogenetics, morphology and biology reconcile. *Berichte aus den Arbeitskreisen Heimische Ochideen* **2017**, *34*, 37–103.
36. Bateman, R.M. Two bees or not two bees? An overview of *Ophrys* systematics. *Berichte aus den Arbeitskreisen Heimische Ochideen* **2018**, *35*, 5–46.
37. Piñeiro Fernández, L.; Byers, K.J.; Cai, J.; Sedeek, K.E.; Kellenberger, R.T.; Russo, A.; Qi, W.; Aquino Fournier, C.; Schlüter, P.M. A phylogenomic analysis of the floral transcriptomes of sexually deceptive and rewarding european orchids, *Ophrys* and *Gymnadenia*. *Front. Plant Sci.* **2019**, *10*, 1553. [CrossRef]
38. Borg-Karlson, A.-K. Chemical basis for the relationship between *Ophrys* orchids and their pollinators III. Volatile compounds of species in the *Ophrys* sections Fuciflorae and Bombyliflorae as insect mimetic attractants/excitants. *Chem. Scr.* **1987**, *27*, 313–325.

39. Schiestl, F.P.; Ayasse, M.; Paulus, H.F.; Löfstedt, C.; Hansson, B.S.; Ibarra, F.; Francke, W. Sex pheromone mimicry in the early spider orchid (*Ophrys sphegodes*): Patterns of hydrocarbons as the key mechanism for pollination by sexual deception. *J. Comp. Physiol. A* **2000**, *186*, 567–574. [CrossRef]
40. Bohman, B.; Peakall, R. Pyrazines attract *Catocheilus* thynnine wasps. *Insects* **2014**, *5*, 474–487. [CrossRef]
41. Buser, H.R.; Arn, H.; Guerin, P.; Rauscher, S. Determination of double bond position in mono-unsaturated acetates by mass spectrometry of dimethyl disulfide adducts. *Anal. Chem.* **1983**, *55*, 818–822. [CrossRef]
42. Kuwahara, Y.; Ohshima, M.; Sato, M.; Kurosa, K.; Matsuyama, S.; Suzuki, T. Chemical ecology of astigmatid mites Xl Identification of the alarm pheromone and new C17 hydrocarbons from *Tortonia* sp., a pest attacking the nest of *Osmia cornifrons*. *Appl. Entomol. Zool.* **1995**, *30*, 177–184. [CrossRef]
43. Ko, E.J.; Savage, G.P.; Williams, C.M.; Tsanaktsidis, J. Reducing the cost, smell, and toxicity of the Barton reductive decarboxylation: Chloroform as the hydrogen atom source. *Org. Lett.* **2011**, *13*, 1944–1947. [CrossRef] [PubMed]
44. Cahiez, G.; Gager, O.; Buendia, J. Copper-catalyzed cross-coupling of alkyl and aryl grignard reagents with alkynyl halides. *Angew. Chem. Int. Ed.* **2010**, *49*, 1278–1281. [CrossRef] [PubMed]
45. Obora, Y.; Moriya, H.; Tokunaga, M.; Tsuji, Y. Cross-coupling reaction of thermally stable titanium (ii)-alkyne complexes with aryl halides catalysed by a nickel complex. *Chem. Commun.* **2003**, 2820–2821. [CrossRef] [PubMed]

11

An Overview of Orchid Protocorm-Like Bodies: Mass Propagation, Biotechnology, Molecular Aspects and Breeding

Jean Carlos Cardoso [1], Cesar Augusto Zanello [2] and Jen-Tsung Chen [3,*]

[1] Laboratory of Plant Physiology and Tissue Culture, Department of Biotechnology, Plant and Animal Production, Centro de Ciências Agrárias, Universidade Federal de São Carlos, Rodovia Anhanguera, km 174, CEP 13600-970 Araras, SP, Brazil; jeancardctv@gmail.com

[2] Masterscience degree by Programa de Pós Graduação em Produção Vegetal e Bioprocessos Associados, Centro de Ciências Agrárias, Universidade Federal de São Carlos, CEP 13600-970 Araras, SP, Brazil; cesarzanello1@gmail.com

[3] Department of Life Sciences, National University of Kaohsiung, Kaohsiung 811, Taiwan

* Correspondence: jentsung@nuk.edu.tw

Abstract: The process through induction, proliferation and regeneration of protocorm-like bodies (PLBs) is one of the most advantageous methods for mass propagation of orchids which applied to the world floricultural market. In addition, this method has been used as a tool to identify genes of interest associated with the production of PLBs, and also in breeding techniques that use biotechnology to produce new cultivars, such as to obtain transgenic plants. Most of the molecular studies developed have used model plants as species of *Phalaenopsis*, and interestingly, despite similarities to somatic embryogenesis, some molecular differences do not yet allow to characterize that PLB induction is in fact a type of somatic embryogenesis. Despite the importance of species for conservation and collection purposes, the flower market is supported by hybrid cultivars, usually polyploid, which makes more detailed molecular evaluations difficult. Studies on the effect of plant growth regulators on induction, proliferation, and regeneration of PLBs are the most numerous. However, studies of other factors and new technologies affecting PLB production such as the use of temporary immersion bioreactors and the use of lighting-emitting diodes have emerged as new tools for advancing the technique with increasing PLB production efficiency. In addition, recent studies on *Phalaenopsis equestris* genome sequencing have enabled more detailed molecular studies and the molecular characterization of plantlets obtained from this technique currently allow the technique to be evaluated in a more comprehensive way regarding its real applications and main limitations aiming at mass propagation, such as somaclonal variation.

Keywords: biotechnology; breeding; mass propagation; Orchidaceae; protocorm-like bodies; somaclonal variation; somatic embryogenesis

---

1. Introduction

Orchids (Family Orchidaceae) represent one of the two largest plant families, including from 736 [1] to 899 genera and 27,800 accepted species names [2] and over 100,000 hybrids produced by artificial pollination [3]. In addition to their unquestionable botanical and ecological importance, orchids participate in current cultivation systems using high-tech horticulture, grown in environments with good climate control, especially temperature, which allows the induction of flowering regardless of the time of year, especially aiming at the scheduled supply of potted and cut flowers in the competitive world flower market. Some species of orchids, such as the genera *Dendrobium*, *Gastrodia*, and *Bletilla*, have also been used for medicinal purposes, using the basis of traditional Chinese medicine [4] and some *Vanilla* species is also used for food purposes [5].

In this economic context, family Orchidaceae currently represents one of the most important in the world commercial floriculture, with emphasis on the genus *Phalaenopsis* as well as its interspecific hybrids, which is currently the main potted flower marketed in the main world flower markets. To have an idea of the importance of this genus in the expansion of world floriculture, only in the Dutch market, the largest in the world, in 2014, 121 million pots of *Phalaenopsis* were sold generating approximately US$ 500 million [6]. In addition to *Phalaenopsis*, other genera of economic importance to floriculture include the genera *Cattleya*, *Dendrobium*, and *Oncidium* and their hybrids [7–9] as well as *Cymbidium* and *Vanda* used for production of potted or even cut flowers.

Despite the individual importance of these genera, a commercial classification for orchids must be set separately from the botanical classification. This is because although genera have a greater genetic and morphological contribution to commercial plants, most commercial flower production of these genera occurs through the production of hybrids from interspecific crosses, which include the use of crosses between species of the same genus, but also species of different genera (intergeneric hybrids) [9]. An example of this case is the very frequent use of *Doritis* in crossings with *Phalaenopsis*, generating the hybrid genus known as *Doritaenopsis* [10,11]. Nevertheless, commercially these hybrids are all called *Phalaenopsis* because considering the morphological similarity and commercialization value, there is no commercial justification for separation into two classes.

Another justification for the separation of botanical and commercial classification is the recent changes of genera in many species, including those of commercial importance and resulting from the advancement of available molecular techniques that allow genetic rather than just morphological comparisons [1]. An example would be the genera *Laelia* and *Sophronitis*, commonly used in crossings with the genus *Cattleya* to incorporate hybrids with red, yellow and orange flowers, little present in *Cattleya*. Both *Laelia* and *Sophronitis* have undergone more than one change in their names in the last decade, with new changes possibly still remaining due to advances in molecular markers and phylogenetic aspects related to this complex and diverse plant family [12,13].

Thus, it is important to highlight this botanical difference from the commercial one, due to the complexity of the family and its high hybridization capacity. Thus, using as an example the commercial classification encompassing these genera includes not only the genus, but its many hybrids used for the genetic improvement and development of new cultivars for the world floriculture. When mentioning *Cattleya*, this includes genera such as *Laelia*, *Sophronitis*, *Broughtonia*, *Epidendrum*, *Encyclia*, *Caularthron*, among other correlates and with possible hybridization with *Cattleya*. The same occurs in *Oncidium*, in which plants of different genera such as *Brassia*, *Ionopsis*, *Odontoglossum*, *Miltonia*, among others [14] are used for breeding intergeneric hybrids and many commercial hybrids are the result of combinations of more than two genera.

In few plant families it is possible to obtain so many viable and fertile combinations of progenies from very different morphologically species and genera. This allows breeders to incorporate numerous traits of interest into a single plant, which brings the innovative aspect of flower production as well as the advance in breeding, using these same mostly fertile hybrids for the advancement of generations of crosses and obtaining new hybrids. This high hybridization capacity may be a result of the specific process of embryogenic development and later protocorm development that occur in orchids [15]. In other species, it has been reported that lack of hybridization and hybrid seed abortion is associated with disruption of proper endosperm development or mismatch between endosperm development and embryo [16]; and zygotic embryogenesis in family Orchidaceae, embryo development occurs in the absence of endosperm [15].

After obtaining the hybrid of commercial interest, propagation is the factor that defines the time for this hybrid to be available in the market for clonal propagation, which ensures the maintenance of the selected characteristics in propagated plants, quickly, on a large scale and allowing the production of plantlets throughout the year. These propagation characteristics, in addition to ensuring the quality of the plantlets produced, also aim to maintain the commercial scale necessary to meet the target market. The only viable technique that combines all these characteristics has been in vitro micropropagation of orchids [17].

Among the in vitro cultivation techniques used for the in vitro seedling or plantlets production of orchids, it can be used the in vitro asymbiotic germination and micropropagation techniques aiming at the large-scale production of clonal plantlets.

Asymbiotic germination involves the in vitro inoculation and germination of orchid seeds with the aid of a sucrose-containing culture medium [18,19], under conditions free of microorganisms; including those symbionts that assist in germination, especially under natural conditions, a technique known as symbiotic germination, which can be done in vitro [19,20], ex vitro, or in situ and which, unlike asymbiotic, considers the use of symbiotic microorganisms to assist in the germination and early development of newly germinated seedlings, and lacking nutritional reserves to support early seedling development [20,21].

Techniques involving the germination of orchid seeds under in vitro conditions are especially used in: Conservation and production of seedlings of native species; germination of seedlings from crosses aiming at genetic improvement and production of new orchid cultivars [8]; aiming at the production of protocorms in order to study somatic embryogenesis in vitro, also known as protocorm-like bodies or simply PLBs [17,22]. They can also be used for commercial propagation and seedlings production, but with high genetic variability inherent in the family Orchidaceae, including commercial groups used for flower production [8].

In vitro germination of orchids makes it possible to increase the efficiency of conservation and breeding programs, since in vitro germination rates higher than 70% are commonly reported [23], while in ex vitro conditions under natural environmental conditions, these rates hardly exceed 5% germinated seeds [24]. This is especially due to the fact that orchid seeds do not contain nutritional reserves [25], and the embryo and seedlings at early germination are highly dependent on symbiosis with microorganisms known as mycorrhizae, which nutritionally supply these plants during a long time until the complete establishment of the seedling in the natural environment [26]. In *Serapias vomeracea* orchid, in symbiosis with *Tulasnella calospora* there was observed a differential gene expression related to organic nitrogen transport and metabolism, showing the nutritionally supply of fungus to orchids in early development of protocorms [27].

A characteristic of the in vitro asymbiotic germination of orchids is the formation of the so-called protocorms, prior to budding, mainly containing the first leaves and undeveloped stem, followed by the roots [25] and later on with the development of the leaf and pseudobulb.

The term protocorm-like bodies (PLBs) is used as a reference to this type of protocorm-producing germination, characteristic of orchids. The main difference between the germination and the sexual reproduction process, which includes the fertilization process, zygotic embryogenesis, followed by the germination and formation of protocorms, is that PLBs comes from somatic tissues, therefore being considered a type of vegetative propagation.

The production of PLBs, therefore, can be compared to a specific type of somatic embryogenesis that occurs in orchids, and the anatomy, development and characteristics of cells and some cell wall markers at the beginning of PLB formation are similar to those in the development of protocorms in orchids [28]. These authors observed that in non-embryogenic callus of *Phalaenopsis* orchids, the inability to synthesize some cell wall components such as the JIM11 and JIM20 epitopes resulted in loss of morphogenic capacity of these calli, and the correct formation of the cell wall is directly associated with the ability of cell division and elongation in these cell types. In contrast, embryogenic calli synthesized these components, similar to what occurred in zygotic embryogenesis [28].

Despite these anatomical and cellular similarities between PLB induction and zygotic embryogenesis, molecularly, zygotic embryogenesis in *Phalaenopsis aphrodite* is considered different from PLB formation, and that induction of PLBs follows a different route from the embryogenic program [29]. One explanation for these differences is a consequence of the degree of speciation for the development of the embryogenic program in orchids, which follows a very specific pattern and different from the conventional embryogenic program occurring in species of other families, such as the absence of endosperm development and gene expression for establishing symbiotic relationships during seed germination process [15].

Due to these still-present doubts regarding comparisons of zygotic embryogenesis with induction of PLBs in orchids, we have adopted the term IPR–PLB (induction, proliferation, and regeneration of PLBs) as the standard to describe this technique in this paper. IPR–PLBs in orchids have different applications in the world flower industry. Undoubtedly the one with the largest commercial application is aimed at the mass propagation of clonal plants to meet the world's demanding flower production market, in which orchids play a significant part in both the pot and cut flower market [6,30]. However, other applications such as for species conservation purposes [31] and obtaining transgenic plants [32] can be found in the literature.

Despite a significant amount of studies with IPR-PLB in different orchid species and hybrids, such as *Coelogyne cristata and C. flaccida* [33,34], *Cyrtopodium paludicolum* [35], *Grammatophyllum speciosum* [36] among others, this review has as its main objective to compile the recent studies and advances found in the induction, proliferation and regeneration of PLBs from the two most important genera in the world flower market, especially *Phalaenopsis* and *Oncidium* hybrid groups.

2. Genus *Phalaenopsis* and Related

The limited efficiency of clonal multiplication by the induction of shoots from floral stems cultivated in vitro has been one of the main difficulties faced in micropropagation of *Phalaenopsis*, resulting in an increase in the production cost of micropropagated plantlets [37] and associated with falling prices in the international market [6] place in vitro plantlets as the current major cost of producing *Phalaenopsis*. In this sense, the IPR–PLBs can be an important tool in the micropropagation of commercial hybrids of this genus aiming to increase the production efficiency, being necessary to know the main factors involved in each phase of plantlets from PLBs production, e.g., induction, proliferation, and regeneration, which result in efficient clonal and mass propagation techniques for *Phalaenopsis*.

The first studies involving clonal micropropagation of *Phalaenopsis* were conducted by [38–40] using *Phalaenopsis amabilis* as a model. Soon after, [41] concluded that leaf segments obtained from inflorescence buds grown in vitro when grown in New Dogashima Medium (NDM) [41] medium supplemented with 0.1 mg L^{-1} NAA (Naphthaleneacetic Acid) and 1.0 mg L^{-1} BA (6-Benzyladenine) could generate up to 10,000 PLBs within a year. Ref. [42] also reported PLB regeneration from a callus induction phase (indirect somatic embryogenesis) using Vacin Went medium [43] supplemented with 20% coconut water and 4% sucrose with the hybrid *Phalaenopsis* Richard Shaffer 'Santa Cruz'.

In orchids, PLBs are suggested to be somatic embryos due to the morphological similarity and developmental pattern observed between them and the zygotic embryos [42,44]. Besides that, ontogenetic studies based on histological and histochemical methods developed by [28] compared the early developmental pattern of zygotic embryos and PLBs, which led to the conclusion that cytological characteristics and cell wall markers were similar in the early developmental stages of both zygotic embryos and PLBs, which would justify saying that PLBs are somatic embryos. Still, histological analyses made by [45] also showed that the formation of PLBs occurs directly on the epidermal surface of the leaf segment with a cluster of meristem cells in constant division and without connection with the leaf vascular system, which is interesting from a commercial point of view, since it ensures the health of plants obtained through PLBs [46–49] and enable success of genetic transformation [50,51].

In several plant species, some genes that are involved in somatic embryogenesis, known as *SERK* (somatic embryogenesis receptor-like kinase), are described. Ref. [52] characterized and analyzed the expression of 5 of these genes in *Phalaenopsis* and which were described by the authors as *PhSERK*. According to this study, the expression of these 5 genes was observed in various parts of plants (root, leaf, apical bud, and flower meristem) as well as during seed germination and PLB induction. According to the authors, PLBs segmented and grown in secondary PLB-inducing medium showed high *PhSERK5* expression during the third week, when secondary PLBs became visible, suggesting that this SERK transcription may be closely associated with the acquisition of embryogenic competence during formation of PLBs. It is noteworthy that transformed *Arabidopsis* plants with overexpression of the *AtSERK1* gene showed high capacity for induction of somatic embryos in in vitro culture [53], showing that this gene is indeed involved in somatic embryogenesis, at least in *Arabidopsis*.

Although cytological features indicate that a PLB is a somatic embryo and studies have shown *PhSERK* gene expression during PLB induction [52], transcriptome studies developed by [29] analyzing gene expression in *Phalaenopsis aphrodite* concluded that PLBs are molecularly distinct from zygotic embryos. According to the authors, PLBs share different transcriptomic signatures from zygotic embryos, and early processes of PLB development show a distinct regeneration program, not following the embryogenesis program. In addition, the authors report that the SHOOT MERISTEMLESS gene, a class I KNOTTED-LIKE HOMEOBOX gene, probably plays an important role in PLB regeneration and should be further investigated.

The genetic transformation with *AtRKD4* gene, which encode proteins with RWP-RK transcription factor and is associated to early embryogenic pattern in *Arabidopsis thaliana* [54], also increases the number of PLBs produced in leaves of this *Phalaenopsis* 'Sogo vivien' [55] and *Dendrobium phalaenopsis* [56] transgenic plants.

Recent studies with *Phalaenopsis equestris* genome sequencing [57], with 2n = 2x = 38 and 29,431 predicted protein-coding genes and *Phalaenopsis* Brother Spring Dancer 'KHM190' [58], 2n = 2x = 38 and 41,153 protein coding genes, make room for further detailed studies on the identification and expression of genes involved in the production of PLBs from different types of somatic tissue in orchids, which can be compared with other model species and in which the embryogenic pathway is already better elucidated, similar to the studies already carried out that brought new discoveries about flowering and the development of floral organs [58].

Among the several factors that regulate somatic embryogenesis in *Phalaenopsis*, the absence of light is described as responsible for the PLB induction step [59]. After maintaining the leaf segments for 60 days in the dark, it is possible to observe at the ends of the segments the formation of embryo-like structures, still with a yellowish-white color (Figure 1A). After about 15 days under 14 h light photoperiod, PLBs change color to light green and dark green (Figure 1B) and after 90 days subjected to light there is the onset of differentiation of PLBs with leaf primordia to their complete differentiation with leaf and root formation. The PLBs also could be induced from shoots and proliferate in solid (Figure 1C) or liquid medium under shake agitation (Figure 1D).

From these observations, it is possible to infer that the absence of light plays an important role in the induction of PLBs, just as light influences the differentiation of PLBs into plantlets. Also, according to [60], the type of light used can also optimize the regeneration of PLBs, with the use of red and white LED combined with sucrose as a carbohydrate source, or blue and white LED with trehalose as the carbohydrate source, which had the best response for the regeneration of PLBs. However, only 17.5% of papers described a dark-period to induce PLBs, while 67.5% used light period (12-16-h photoperiod) to induce and regeneration of PLBs in *Phalaenopsis* (Table 1).

Figure 1. Induction, proliferation and regeneration of protocorm-like bodies in *Dendrobium* and *Phalaenopsis* orchids. Protocorm-like bodies (PLBs)-directly induced from leaf segments of *Phalaenopsis* hybrid '501' **(A)** obtained from young in vitro shoots from inflorescence nodal segments and details of secondary PLBs (**B**) obtained in New Dogashima Medium (NDM) culture medium. Proliferation of PLBs in agar (**C**) and liquid (**D**) MS½ culture medium of *Dendrobium* 'Hybrid 3'. Bars = 1 cm. Unpublished photos of Cesar A. Zanello (A,B) and Jean C. Cardoso (C,D).

Besides the influence of light, another admittedly important factor in the induction of PLBs in *Phalaenopsis* and orchids in general is the genotype [61]. This means that under the same cultivation condition, the induction responses of PLBs may be significantly different [62], which is still considered a limitation of the technique. Ref. [30] evaluated the induction of PLBs in two commercial hybrids (Ph908—red-painted yellow flowers and RP3—dark red) of *Phalaenopsis* and reported significant differences in both percentage of PLB leaf segments (45% and 10%, respectively) as in the number of PLBs per leaf segment (25 and 2 PLBs, respectively).

Regarding the type of explant, leaf segments of plants grown in vitro have been the most suitable for induction of PLBs in *Phalaenopsis* (45% of papers; Table 1), but there are reports of protocols that used in vitro roots of *P.* 'Join Angle × Sogo Musadian' cultivated in MS½ medium supplemented with NAA, BAP, and IAA (0.5 ppm, 5 ppm, and 0.5 ppm, respectively) and up to 49.33 PLBs/explant [63].

Table 1. Compliance of studies with induction, proliferation and regeneration of PLBs (IPR-PLBs) with *Phalaenopsis* and *Doritaenopsis*.

Species or Hybrids	Origin and Age of Explants	Culture Media	Growth Conditions	Main Results	Evaluation and Detection of SV	Reference
12 cultivars of *Phalaenopsis*	Shoot tips derived from flower stalk buds	NDM added 10 g L^{-1} sucrose, 2 g L^{-1} Gelrite, 0.1 mg L^{-1} NAA and 1–5 mg $^{L-1}$ BA	23 ± 1 °C, 14-h photoperiod, 33 μmol m^{-2} s^{-1}	93–100% survival rate of explants, 33–40% PLB formation, green color of PLBs showed multiplication, 27–28% PLBs formed shoots	Non-evaluated	[41]
Phalaenopsis Nebula	Calluses derived from 1–2 months protocorms	MS½ + 100 mg L^{-1} myo-inositol + 0.5 mg L^{-1} niacin and pyridoxine + 0.1 mg L^{-1} thiamine + 2.0 mg L^{-1} glycine + 170 mg L^{-1} NaH_2PO_4 + 20 g L^{-1} sucrose + 2.2 g L^{-1} Gelrite, pH 5.2	26 ± 2 °C, 16-h photoperiod, PPFD 28–36 μmol m^{-2} s^{-1}	Both TDZ and BA were able to induce PLBs in calluses, but interestingly equal number of PLBs per callus (74) was obtained when callus was transferred to free-PGR medium	Not observed any phenotypic abnormality and no chromosome number alterations were observed in 2–3 months plantlets	[64]
Phalaenopsis Hybrid with pink striped flowers	Section transversely cutted from apical meristems (2-mm in size) of PLBs obtained from leaf segments	Liquid Hyponex modified medium (Kano, 1965—1 g L^{-1} of 6.5N – 4.5P – 19K + 1 g L^{-1} 20N – 20P – 20K + 1% potato homogenate)	25 ± 2 °C, 16-h photoperiod, PPFD 60 μmol m^{-2} s^{-1}, white fluorescent light, under shaker at 100 rpm or temporary our continuous immersion bioreactor system	100 ml medium per 0.5 g inoculum under agitation (9.2 PLBs/PLB section) or air-lift balloon with 10.0 g inoculum (12.6 PLBs/PLB section); charcoal filter attached to bioreactor increased to 17 PLBs/PLB section; Hyponex medium increased percentage of PLB regeneration, rooting and fresh weight of plantlets	Non-evaluated	[65]
9 genotypes of *Phalaenopsis*	Shoot tips from flower stalk buds and callus from cell suspension cultures	Shoot tips to PLBS, NDM + 2 g L^{-1} Gellan gum, pH 5.4; Cell suspension, liquid NDM + 58.4 mM sucrose; Induction of PLBs from calluses, NDM + 29.2 μM sucrose + 2 g L^{-1} Gellan gum	23 ± 1 °C, 14-h photoperiod, 33 μmol m^{-2} s^{-1}, cell suspension culture were obtained in liquid medium under agitation 0f 80 rpm	44.4% PLB formation from shoot tips were obtained with 0.5 μM NAA and 4.44 μM BA and 29.2 mM sucrose; increases in sucrose concentration (58.4 mM increased callus formation); calluses could induced to PLBs with 29.2 mM sucrose	The type and frequency of morphological variants were large dependent on genotype: in *P.* Snow Parade and *P.* Little Steve any variants was reported, while 47.9% variants were observed in *P.* Reichentea	[66]
Phalaenopsis Tinny Sunshine 'Annie'; 'Taisuco Hatarot'; Teipei Gold 'Golden Star'; Tinny Galaxy 'Annie'	Young leaf segments (10 × 5 mm) derived in vitro shoots from flower stalk nodes	induction of PLBs: MS½ + 10% coconut water/Proliferation of PLBs: different saline formulation + 2 g L^{-1} peptone + 3% potato homogenate + 0.05% activated charcoal + 30 g L^{-1} sucrose	Temp 25 ± 1 °C, 16-h photoperiod by cool white fluorescent lamps, PPFD 30 μmol m^{-2} s^{-1}; liquid media in shaker at 50 rpm	70–90% of explants with PLBs depending on cultivar; 85% explants with PLBs and 12 PLBs/explant with 88.8 μM BA + 5.4 μM NAA; 45 g L^{-1} sucrose showed highest number PLBs per explant (6) and low light intensity (10 μmol m^{-2} s^{-1}) resulted in best PLBs induction (90%) and number of PLBs/explant (12); liquid with cotton raft support Hyponex medium increased PLBs proliferation (20.5 PLBs)	Non-evaluated	[67]

Table 1. *Cont.*

Species or Hybrids	Origin and Age of Explants	Culture Media	Growth Conditions	Main Results	Evaluation and Detection of SV	Reference
Doritaenopsis 'New Candy' × (*D.* 'Mary Anes' × *D.* 'Ever Spring'	Leaf segments 1 mm thick from three months old leaves from in vitro plantlets	MS½ + 20% coconut water + 10 mg L^{-1} adenine sulphate + 2.3 g L^{-1} Gelrite, pH 5.5	1 week in dark at 27 °C followed by 25 ± 1 °C, 16-h photoperiod by cool white fluorescent lamps, PPFD 10 μmol m^{-2} s^{-1}	9.0 μM TDZ resulted in best PLB formation (72.3%); Thin leaf segments—1 mm— resulted in best PLB formation (>50%) than thick leaf sections—5 mm (10%) and are correlated with ethylene content (ppm)	Irregular shaped bodies (CLBs) increased with increases in concentrations of TDZ (0.57% at free-PGR to 11.56% at 22.5 μM) and BA (32.14% at 4.4 μM); However, no phenotypic variations were observed in vegetative growth in greenhouse	[68]
Doritaenopsis 'New Candy' × (*D.* 'Mary Anes' × *D.* 'Ever Spring'	Root tips (<0.5 cm) from 3-months old in vitro plantlets	MS + 20% coconut water + 10 mg L^{-1} adenine sulphate + 2.3 g L^{-1} Gelrite, pH 5.5	Temp 25 °C, cool white fluorescent lamps, PPFD 30 μmol m^{-2} s^{-1}, 16-h photoperiod	TDZ at 2.3 μM showed best PLB formation (47.2% of root tips with 2–6 PLBs each) compared to BA and Zea; most of PLBs originated from cortex tissues of root	Non-evaluated	[69]
Phalaenopsis Snow Parade and Wedding Promenade, *Doritaenopsis* New Toyohashi	Cell suspension from calluses	NDM + 2 g L^{-1} gellan gum, pH 5.4	23 ± 1 °C, 14-h photoperiod, 33 μmol m^{-2} s^{-1}, cell suspension culture were obtained in liquid medium under agitation 0f 80 rpm	The response were genotype-dependent: Glucose at 58.4 mM and sucrose at 29.2 mM showed several increases in number (>2000) and fresh weight of PLBs for P. Snow Parade, while glucose at 14.6–29.2 mM showed highest number of PLBs in *P.* Wedding Promenade	Non-evaluated	[70]
Phalaenopsis 'Little Steve'	Leaf explants (1cm2) derived from flower stalk buds eighteen-month-old in vitro plants	MS½ added 4.54 μM TDZ, 100 mg L^{-1} myo-inositol + 0.5 mg L^{-1} niacin + 0.5 mg L^{-1} pyridoxine + 0.1 mg L^{-1} thiamine + 2.0 mg L^{-1} glycine + 1000 mg L^{-1} peptone + 2.2 g L^{-1} Gelrite + 20 g L^{-1} sucrose, pH 5.2	Dark for 2 months followed by 16-h photoperiod	40% explants with PLBs; not reported the number of PLBs per explant	Non-evaluated	[71]
Phalaenopsis amabilis var. formosa	Leaf tip segments obtained from in vitro germinated seedlings and leaf-derived nodular masses	M½ S added 3 mg L^{-1} TDZ, 100 mg L^{-1} myo-inositol + 0.5 mg L^{-1} niacin + 0.5 mg L^{-1} pyridoxine + 0.1 mg L^{-1} thiamine + 2.0 mg L^{-1} glycine + 1000 mg L^{-1} peptone + 2.2 g L^{-1} Gelrite + 20 g L^{-1} sucrose, pH 5.2	Temp 26 ± 1 °C; 16-h photoperiod	93.8% explants with PLBs and 19.4 PLBs per explant for leaf tip segments; 5.4 proliferation rate and 13.8 PLBs per explant for leaf-derived embryogenic masses	Non-evaluated	[72]
Phalaenopsis gigantea	Trimmer base protocorms 1 mm from 150-d in vitro germinated protocorms	XER medium (Ernst, 1994) + 20 g L^{-1} fructose + 1% agar, pH 5.7	Temp 25 ± 2 °C, under continuous illumination from cool fluorescent lamps, PPFD 20–50 μmol m^{-2} s^{-1}	Trimmed protocorms increased PLBs proliferation (56.8%) and number of PLBs/protocorm (4.24) using 15% coconut water and 2.5 g L^{-1} activated charcoal, compared to untrimmed (4.56% and 0.56 PLB/protocorm) and shoot regeneration from PLBs were increased using only coconut water at 10% (33.56% shoot formation)		[73]

Table 1. *Cont.*

Species or Hybrids	Origin and Age of Explants	Culture Media	Growth Conditions	Main Results	Evaluation and Detection of SV	Reference
Alba flower hybrid' of *Phalaenopsis*	Nodular masses	NDM culture medium added 1.0 mg L^{-1} BA and 0.1 mg L^{-1} NAA, 100 mg L^{-1} myo-inositol + 1.0 mg L^{-1} (niacin, pyridoxine, thiamine, cysteine, calcium pantothenate) + 0.1 0 mg L^{-1} biotin + 20 g L^{-1} sucrose + 2.0 g L^{-1} Phytagel, pH 5.8	Not reported growth conditions	8.5 PLBs per explant; not reported percentage of explants with PLBs	Non-evaluated	[74]
Phalaenopsis amabilis cv. 'Cool Breeze'	Inflorescence axis thin sections	MS½ added 2,0 mg L^{-1} BA, 0,5 mg L^{-1} NAA, 2% sucrose, 10% coconut water, 2 g L^{-1} peptone and 1 g L^{-1} activated charcoal		20 PLBs/explant after 12 weeks	Non-evaluated	[75]
Phalaenopsis var. Hawaiian Clouds × *Phalaenopsis* Carmela's Dream	Clumps of callus (8 mm diameter)	NDM culture medium added 1 mg L^{-1} TDZ, 10 g L^{-1} maltose, 2.8 g L^{-1} Gelrite	Temp 25 ± 2 °C, in the dark	52.5% callus with PLBs	Non-evaluated	[76]
Phal. amabilis; Phal. 'Nebula'	Cut end of leaf explants (1.0 cm length); clonal plantlets of *P. amabilis* and in vitro germinated seedlings for *P.* 'Nebula'	MS½ added 3 mg L^{-1} TDZ, 100 mg L^{-1} myo-inositol + 0.5 mg L^{-1} niacin + 0.5 mg L^{-1} pyridoxine + 0.1 mg L^{-1} thiamine + 2.0 mg L^{-1} glycine + 1000 mg L^{-1} peptone + 2.2 g L^{-1} Gelrite + 20 g L^{-1} sucrose, pH 5.2	Temp 26 ± 1 °C; dark for 60-d (induction) 45-d for subculture period;	50% explants with PLBS and 8.2 PLBs/explant for *P. amabilis*; 80% explants with PLBs and 3.5 PLBs for *P.* 'Nebula'	Non-evaluated	[59,77,78]
10 genotypes of *Phalaenopsis*	Basal portion of sectioned horizontally protocorms (3–5 mm) were placed upward in contact with the culture medium	3.5 g L^{-1} HyponexTM #1 + 1 g L^{-1} tryptone + 0.1 g L^{-1} citric acid + 1 g L^{-1} activated charcoal + 20 g L^{-1} sucrose + 20 g L^{-1} homogenized potato + 25 g L^{-1} homogenized banana + 7.5 g L^{-1} agar, pH 5.5	Temp 25 ± 2 °C, 16-h photoperiod with PPFD 10 µmol m^{-2} s^{-1}	22% of sectioned protocorms induced PLBs and 17.5 PLBs per responsive protocorms were obtained	High endopolyploidy were observed in *Phalaenopsis* protocorms; from 22 diploid protocorms used as explant, 34.1% of derived-PLBs were polyploidy at first cycle and 51.7% at second cycle of proliferation	[79]
Phalaenopsis violacea	Leaf segments (1 × 1 cm) from in vitro shoots derived from flower stalks	MS½ + 5% banana extract	Temp 25 °C, 16-h photoperiod, PPFD 40 µmol m^{-2} s^{-1} by white fluorescent tubes	70% of leaf segments formed PLBs with 0.8 µM BAP, while TDZ were able to induce PLBs only in 40% of explants and BAP (0.6 µM) was more effective to PLBs proliferation than TDZ and Zea	Non-evaluated	[80]

Table 1. *Cont.*

Species or Hybrids	Origin and Age of Explants	Culture Media	Growth Conditions	Main Results	Evaluation and Detection of SV	Reference
Phal. amabilis cv. Lovely (purple flowers)	Young emerging leaves from in vivo plants	MS1/2 + 2% sucrose + 10% coconut water + 2 g L^{-1} peptone + 1 g L^{-1} activated charcoal + 2.2 g L^{-1} Gelrite, pH 5.6;	Temp 24 ± 1 °C, cool white fluorescent light, PPFD 30 µmol m^{-2} s^{-1}, 16-h photoperiod	2.0 mg L^{-1} BA and 0.5 mg L^{-1} NAA resulted in 75.5% explants formed PLBs and 10 PLBs/explant; MS½ + 10% coconut water + 150 mg L^{-1} glutamine showed best proliferation rate of PLBs (200.5 PLBs/explant)	Non-evaluated	[81]
Phalaenopsis bellina	*In vivo* leaf	MS½ + 100 mg L^{-1} myo-inositol + 0.5 mg L^{-1} niacin and pyridoxine + 0.1 mg L^{-1} thiamine + 2.0 mg L^{-1} glycine + 3.0 mg L^{-1} TDZ + 2% sucrose + 3.0 g L^{-1} Gelrite + 10% fresh banana extract, pH 5.6	PLBs from leaf (S0), PLBs proliferation after 3 months (S1) and after six months (S2)	Efficiency of induction and regeneration of PLBs not presented by authors	Minimal dissimilarity in *P. bellina* by RAPD markers; S0 presented 96% similarity, S1 87% and S2 80% similarity to the mother plant	[82]
Phalaenopsis bellina	Young leaves (1.5 cm^2) of a nursery plant	MS½ + 2% sucrose + 100 mg L^{-1} myo-inositol + 0.5 mg L^{-1} niacin and pyridoxine + 0.1 mg L^{-1} thiamine + 2.0 mg L^{-1} glycine + 10% fresh ripen banana extract + 3.0 mg L^{-1} TDZ + 3.0 g L^{-1} Gelrite, pH 5.6	Temp 25 ± 2 °C, 14-h photoperiod for 12–16 weeks	71.9–78.1% explants with PLBs; 14.3–14.8 PLBs per flask; MS1/2 was the best for PLB proliferation compared to VW	Non-evaluated	[83]
Phalaenopsis gigantea	Leaf tip segments (1.0 cm length) from in vitro germinated seedlings	NDM culture medium, sucrose 20 g L^{-1} + 1.0 mg L^{-1} NAA and 0.1 mg L^{-1} TDZ	Temp 25 ± 2 °C, 12-h photoperiod for 6 weeks	The authors only report that NAA and TDZ treatment was the best for callus induction and PLBs after 6 weeks of culture.	Non-evaluated	[84]
Phalaenopsis amabilis cv. 'Golden Horizon'	Young emerging leaves from in vivo plants	MS½ + 2% sucrose + 10% coconut water + 2 g L^{-1} peptone + 1 g L^{-1} activated charcoal + 2.2 g L^{-1} Gelrite, pH 5.6	Temp 24 ± 1 °C, cool white fluorescent light, PPFD 30 µmol m^{-2} s^{-1}, 16-h photoperiod	BA at 2.0 mg L^{-1} combined with NAA 0.5 mg L^{-1} resulted in 80.5% explants with PLBs and 15 PLBs/explant; MS½ + 10% coconut water + 150 mg L^{-1} glutamine showed best proliferation rate of PLBs (250.5 PLBs/explant)	Non-evaluated	[85]
Phalaenopsis Gallant Beau 'George Vasquez'	Longitudinally bisected PLBs (2–3 mm in diameter) and 2-months old	Miracle Pack®culture system with liquid VW + 20% coconut water without sucrose, pH 5.3	Temp 25 °C, 16-h photoperiod, PPFD 45 µmol m^{-2} s^{-1}, plant growth fluorescent lamps, under magnetic fields	Although higher Fresh weight of PLBs was obtained with 0.1 Tesla–South (237.4 g), best number of PLBs was obtained in control without magnetic fields; 0.15 Tesla for 7 weeks (South) also increased PLB fresh weight, control treatment not differed from the best results using magnetic fields	Non-evaluated	[86]

Table 1. *Cont.*

Species or Hybrids	Origin and Age of Explants	Culture Media	Growth Conditions	Main Results	Evaluation and Detection of SV	Reference
Phalaenopsis cornu-cervi	Leaf segments from in vitro germinated seedlings with 2-months	MS½ added 0.1 mg L^{-1} NAA, 0.1 mg L^{-1} TDZ and 15% coconut water	Temp 25 ± 1 °C; 16-h photoperiod for 45 days	100% explants with PLBs; 35 PLBs per explant	Non-evaluated	[87]
Phalaenopsis gigantea	PLBs obtained from leaf tip segments (1.5 cm length) from young leaves	Liquid medium with 20% coconut water, pH 5.4.	Temp 25 ± 2 °C, under 16-h photoperiod using fluorescent lighting 30 μmol m^{-2} s^{-1}, 60 rpm rotary shaker	VW medium with 10 mg L^{-1} chitosan resulted in higher number of PLBs (177) and fresh weight of PLBs (8.4 g)	ISSR, non-detected somaclonal variations in *P. gigantea* related to mother plants	[88]
Phalaenopsis 'R11 × R10'	Leaves, root tips and stem explants from eight months (plantlets or seedlings?)	MS½ + 15% coconut water + 0.01% activated charcoal + 0.03% polyvinylpyrrolidone (PVP) + 88.8 μM BA + 5.37 μM NAA + 0.025% Phytagel, pH 5.6–5.8	Temp. 25 °C, 16-h photoperiod	Stem segments were interesting explant for PLB induction; sucrose at 3% (71.2 PLBs) was more effective than maltose (39 PLBs) in PLBs proliferation	Non-evaluated	[89]
Phalaenopsis Tropican Lady	Young etiolated shoots leaves segments (5 × 10 mm) from flower stalk nodes for induction and PLBs for proliferation	PLB induction: $\frac{1}{4}$ macroelements and full-strength microelements, glycine and vitamins of MS + 30 g L^{-1} sucrose + 0.5 mg L^{-1} TDZ + 7 g L^{-1} agar / PLB Proliferation: 3 g L^{-1} Hyponex (7-6-19) + 1 g L^{-1} tryptone + 50 g L^{-1} potato homogenate + 50 g L^{-1} banana homogenate + 30 g L^{-1} sucrose + 2 g L^{-1} activated charcoal + 7.5 g L^{-1} agar, pH 5.6	Temp 25 ± 2 °C, under 12-h photoperiod by cool white fluorescent lamps, PPFD 23.2 μmol m^{-2} s^{-1},	Basal part of sectioned of bi or trisectioned PLBs resulted in highest explants with PLB formation (46.8–96.3%) and number of PLBs/explant (15.4–22.9); wounding stimulate ethylene production and gene expression for stimulation of cell division	Non-evaluated	[90]
Phalaenopsis cornu-cervi	Whole leaves and leaf-segments (proximal, middle and distal regions) from 120-d old seedlings	MS½ + 3% sucrose + 15% coconut water + 0.23% Gelrite, pH 5.6	Temp 25 ± 1 °C, under 16-h photoperiod, cool white fluorescent lamps, PPFD 20 μmol m^{-2} s^{-1} or pre-treated with 1 week in the dark before photoperiod	Highest percentage of explants with PLBs (30%) and number of PLBs per leaf segment (5.3) were obtained with 9 μM of TDZ under without dark period. Dark period reduced number of PLBs/explant	Non-evaluated	[91]
Phalaenopsis gigantea	Leaf tip segments from young leaves of in vitro seedlings	NDM medium added 0.1 mg L^{-1} TDZ, 10 mg L^{-1} chitosan, 0.2% Gelrite and pH 5.7	Temp 25 ± 2 °C, 16-h photoperiod, 33 μmol m^{-2} s^{-1}	353 PLBs per explant and 4.8 g PLBs fresh weight	ISSR, SV detected after the subculture four (5 to 20%)	[92]
Phalaenopsis hybrids	Intact and transversally divided protocorms (two or four divisions) 1.0–1.5 mm width	MS + 15% coconut water + 7.0 g L^{-1} agar	Temp 25 ± 2 °C, 16-h photoperiod, 25 μmol m^{-2} s^{-1}	No PLBs formed in intact protocorms; Middle and Basal part of sectioned protocorms showed 40 and 44% PLB formation and 11.7 and 13.3 PLBs per explant in Free-PGR culture medium, respectively; Four division of protocorms increased PLBs formation and number of PLBs	Non-evaluated	[93]

Table 1. *Cont.*

Species or Hybrids	Origin and Age of Explants	Culture Media	Growth Conditions	Main Results	Evaluation and Detection of SV	Reference
P. aphrodite subsp. *formosana*	in vitro germinated seedlings with 2-months	Using 2-step method: Liquid MS½ for 2 months and then transferred to solid MS (half strength) with 1 cm of medium Liquid MS (half strength) for a further 2 months. All media with 1 mg L^{-1} TDZ.	Temp 25 ± 2 °C; followed by 16-h photoperiod	44 PLBs per seedling	Non-evaluated	[94]
Phalaenopsis amabilis cv. 'Surabaya'	Leaf segments from in vitro shoots obtained from inflorescence stalk segments	-	Temp 25 ± 1 °C; 16-h photoperiod, subcultured each 14-d	5 mg L^{-1} BA + 2 mg L^{-1} NAA produced 8.7 number of PLBs and TDZ at 3.0 mg L^{-1} showed 22.45 PLBs	non reported by authors that acclimatized and cultivated regenerated plantlets until flowering stage	[95]
Phalaenopsis 'Fmk02010'	Single PLBs	MS with 412.5 mg L^{-1} NH_4NO_3 and 950 mg L^{-1} of KNO_3 + 20;0 g L^{-1} sucrose + 2.0 g L^{-1} Phytagel, pH 5.5–5.8	-	Hyaluronic acid 9 and 12, at 0.1 mg L^{-1}, increased percentage of explants with PLBs (100%), PLB number (18.2 to 23.3) and fresh weight of PLBs (0.291 to 0.596 g) compared to control (86.7%, 12.9 and 0.198 g)	no malformation was observed in regenerated plantlets	[96]
P. 'Join Angle × Sogo Musadian'	*In vitro* roots	MS½ added NAA (0.5 ppm), BA (5 ppm) and IAA (0.5 ppm)	Temp 26 ± 1 °C; dark for 1 month (induction) followed by 16-h photoperiod (4 weeks)	49.33 PLBs per explant; not reported percentage of explants with PLBs	Non-evaluated	[63]
Phalaenopsis Classic Spoted Pink	leaf segments (1.0 cm^2) with 90-d obtained from in vitro shoots	MS½ added NAA (0,537 µM) and TDZ (13,621 µM)	Temp 25 ± 2 °C, dark for 90-d (induction) followed by 16-h photoperiod	The percentage of explants in regeneration and the number of PLBs/explant were not described	Non-evaluated	[45]
Phalaenopsis amabilis var. 'Manila'	Leaf segments (1 cm × 0.5 cm) obtained from in vitro flower stalk nodes	MS added 15 mg L^{-1} BA and 3 mg L^{-1} NAA	Temp 25 ± 1 °C; 16-h photoperiod	50.65 PLBs per explant after 6 weeks	Non-evaluated	[97]
Phalaenopsis amabilis	Protocorms (4 weeks-old), roots, leaves and stems (6-month-old) cut transversely	NP (New Phalaenopsis) medium added 3 mg L^{-1} TDZ	25 ± 1 °C with 1000 lux intensity of continuous light; 8 weeks	Protocorm: 100% explants with PLBs and 23.3 PLBs/explant; Leaf: 100% explants with PLBs and 7.75 PLBs/explant; Root: 80% explants with PLBs and 8.25 PLBs/explant; Stem: 100% explants with PLBs and 28.25 PLBs/explant	Non-evaluated	[98]
Phalaenopsis 'Fmk02010'	Single PLBs	MS with 412.5 mg L^{-1} NH_4NO_3 and 950 mg L^{-1} of KNO_3 + 2.2 g L^{-1} Phytagel, pH 5.5–5.8	Temp 25 ± 2 °C, 16-h photoperiod, PPFD 54 µmol m^{-2} s^{-1}	Highest number of PLBs (54.13) were obtained with Red-White LEDs and with sucrose at 20 g L^{-1} and highest fresh weight of PLBs (0.167 g) was obtained with Red-Blue-White LEDs and trehalose (20 g L^{-1})	Non-evaluated	[60]
Phalaenopsis 'RP3' and '908'	Leaf segments (0.4–0.5 cm^2) obtained from in vitro shoots	NDM culture medium added 0.25 mg L^{-1} TDZ (908) or 1.0 mg L^{-1} NAA, 20.0 mg L^{-1} BA and 0.125 mg L^{-1} TDZ (RP3)	Temp 25 ± 2 °C, dark for 60-d (induction) followed by 14-h photoperiod	45% (908) and 10% (RP3) explants with PLBs; 25 and 2 PLBs/explant respectively	Non-evaluated	[30]

NDM: New Dogashima Medium [41]; MS: Murashige and Skoog Medium [99]; Hyponex medium: [100]; XER medium: [101]; VW: Vacin Went medium [43]; NP: New *Phalaenopsis* medium [102]. 2,4-D, 2-4-Dichlorofenoxiacetic acid; BA, 6-Benzyladenine; IAA, 3-Indoleacetic acid; IBA, Indole-3-butyric acid; NAA, Naphtaleneacetic acid; PPFD: Photosynthetically Photon Flux Density; Temp, Temperature; TDZ, Thidiazuron.

Segmentation made in leaf segments of *Phalaenopsis* to induce PLBs results in a process called phenolic oxidation, which is the release of polyphenol oxidase (PPO) [103] and other compounds toxic to plant tissue, which may cause its death [74], consequently reducing the induction of PLBs. The immersion of leaf segments in solution of cystine and ascorbic acid during the leaf segmentation stage is reported as a way to reduce the release of these compounds capable of impairing the formation of PLBs [74].

One of the influential factors in the induction of PLBs that has been widely evaluated is the concentrations and possible combinations of plant growth regulators (PGRs). Based on the current literature, successful induction of PLBs seems to be mainly influenced by cytokinin BA (6-benzyladenine) and cytokinin-like compound TDZ (thidiazuron), and in some cases the combination of these cytokinins with an auxin [30,45] also proved beneficial. Protocols citing the use of cytokinin BA recommend concentrations between 0.5 mg L^{-1} [78] and 20 mg L^{-1} [67]. For the induction of PLBs with the use of TDZ, the recommended concentrations range from 0.25 mg L^{-1} [30] to 3.0 mg L^{-1} [72]. With the combined use of cytokinins and auxins, the most commonly used auxin is NAA, which varies in concentration from 0.1 mg L^{-1} [45,74] to 1.0 mg. L^{-1} [30,82].

Ref. [104] reviewed the influence of auxins in orchids, including in PLBs and concluded that auxins is important for callus induction and PLB formation and proliferation, while is inhibitory for PLB regeneration into shoots.

As already described, the addition of PGRs is critical to the success of the PLB induction and regeneration technique in *Phalaenopsis*. Cytokinin-like compound such as TDZ (47.5%) and BA (35%) was the most PGRs used to IPR-PLB technique (Table 1). Nevertheless, the use of these regulators may also result in somaclonal variation. This variation can be assessed by morphological, physiological, biochemical traits or molecular markers [105]. Using Random Amplified Polymorphic DNA (RAPD) markers, [82] reported 17% dissimilarity between PLBs and the parent plant in *P. bellina*. Ref. [89] observed 20% dissimilarity after 20 weeks of cultivation in *P. gigantea* using ISSR (Inter Simple Sequence Repeats) markers, leading to the conclusion that PLB proliferation should be done for up to 16 weeks to reduce somaclonal variations and morphological changes. It should be noted that changes in alleles will not always result in phenotypic changes [106], so the variations observed by the markers will not always cause some kind of morphological change in plants.

According to [107], the combination of red light and far red contributes to decrease endoreduplication rates during PLB induction and regeneration, and consequently may reduce somaclonal variations during mass propagation processes.

Bioreactors could be used to improve the proliferation of PLBs in *Phalaenopsis*. The authors of [108] obtained 18,000 PLBs from 1000 PLBs sections using 0.5 or 2.0 L volume of air per volume of medium min^{-1}.

3. *Oncidium* Hybrids Group

According to the World Checklist of Selected Plant Families of the Kew Botanical Garden, in December 2019, there are 374 accepted names of *Oncidium* species with more than 90% of accepted names allocated in Southern America and the last in Northern America. In addition to the species, thousands more interspecific and intergeneric hybrids have been registered with the Royal Horticultural Society and are used in the commercial production of cut and pot flowers worldwide [9,109]. Different chemical and physical factors alter the response to PLB induction in *Oncidium*. Using *Oncidium* 'Gower Rampsey' shoot tips, [109] observed a higher percentage of shoot tips induced to produce PLBs (96.7%) in monochromatic red-light emitting diodes (RR), compared to blue LED (83.3%) and fluorescent white light (76.7%) used as control. However, the use of RR, as well as green LEDs, increased in inhibition of differentiation of PLBs into green buds, while blue LEDs enhanced differentiation. Associated with this response, the authors also observed that in blue light, PLBs contained higher contents of carotenoids, chlorophyll, soluble proteins, lower amounts of soluble sugars and carbohydrates. The authors further argue that in red LEDs, where a higher PLB induction response was obtained, there was a greater accumulation of soluble sugars, starch and carbohydrates, while in blue light, where there was a greater differentiation of PLBs, there was a greater accumulation of proteins and pigments such as chlorophylls and carotenoids.

PGRs are one of the most tested factors in IPR–PLBs in *Oncidium* (Table 2). Benzyladenine (BA) at 2.0 mg L^{-1} + 0.2 mg L^{-1} Naphthaleneacetic Acid (NAA) has been shown to be the most efficient treatment for inducing PLBs in *Oncidium* 'Sweet Sugar' apical and axillary buds [110] and the combination of 0.1 mg L^{-1} BA + 0.2 mg L^{-1} ANA resulted in better response for *Oncidium* Aloha 'Iwanaga' [111]. In this context, BA can be used efficiently to obtain PLBs in *Oncidium* in 31.8% of the papers, and auxin NAA is the one most used along with BAP (Table 2).

Interestingly, [112] reported the individual and combined effects of BA and NAA PGRs at different stages of in vitro induction, proliferation and regeneration of PLBs on *Oncidium* sp. These authors identified that previous callus production in culture medium containing 2,4-D at 1.0 mg L^{-1}, prior to induction, was beneficial for the production of PLBs from in vitro shoots, and from callus it was possible to observe up to 98 PLBs/callus cluster using 0.75 mg L^{-1} NAA, while only 28.2 PLBs/shoot cluster were directly obtained using the combination of 0.5 + 0.5 mg L^{-1} NAA and BA, respectively. The use of 1.0 mg L^{-1} NAA alone allowed PLB proliferation (up to 79.2 PLBs/sample), while the addition of 1.0 mg L^{-1} BA resulted in shoot bud formation (up to 12.4 shoots/PLB). Similarly, [113] observed that the concentration of 2.0 mg L^{-1} BA resulted in the highest number of shoot buds obtained from PLBs (4.3/PLB) in *Oncidium* 'Sweet Sugar'.

Thidiazuron (TDZ) also appears to have a pronounced effect on direct induction of PLBs in *Oncidium* leaf segments and were reported in 54.5% of the papers (Table 2), being higher for the percentage of explants directly forming PLBs (60–75%) and number of PLBs per explant (10.3–10.7) compared to other cytokinins such as kinetin, zeatin, 2-isopentenyladenine and BA itself [114]. Ref. [115] reported direct regeneration of PLBs from mainly the epidermis and cut regions of young leaf segments of *Oncidium* 'Gower Ramsey' using TDZ alone (0.3–3.0 mg L^{-1}), rather than BA in the culture medium, while the combination 2,4-D and TDZ was not beneficial for induction of PLBs. The production of PLBs from tissue damaged regions of inflorescence segments (65%) of *Oncidium* 'Gower Ramsey' using 3 mg L^{-1} TDZ [116] has also been reported. A similar experiment using the same cultivar observed that calli from root apexes and stem segments produced PLBs in medium containing 0.3–3.0 mg L^{-1} TDZ, being beneficial the addition of NAA for the formation of embryos n root and leaf calli [117], being a tissue-specific response.

Other PGRs as GA_3 is reported as an inhibitor of PLB induction in *Oncidium*, while the use of antigibberellins, as ancymidol and Paclobutrazol, increased the percentage of leaf explants with PLBs and the number of PLBs obtained [118].

The use of liquid medium, rather than semi-solidified with Agar, is also an alternative for in vitro PLB proliferation (Figure 2). Ref. [113] used 5 L balloon-type air-lift bioreactor to provide mass propagation of *Oncidium* 'Sweet Sugar', and show that this system provides 326.3 g PLBs and growth ratio of 10.2, and is more efficient than semi-solid (2.7 g PLBs and Growth ratio of 3.4) and liquid-agitated flask culture (3.5 g PLBs and growth ratio of 4.4). In bioreactor, the lag phase was observed in the first 10-d culture, accompanied by a sharp drop in pH (5.7 to 4.7) and EC (3.2 to 1.5 mS cm^{-1}) in the first 20-d of cultivation, followed by an intense mass growth from 10 to 40 days of cultivation, when the pH increased again to 5.9. An interesting fact was the dynamics of sugars in the culture medium, and a fast and drastic reduction of sucrose in the medium was observed, from 27 (day zero) to 5.5 (day five), 1.2 (day 10) and zero (day 20), associated with a substantial increase in glucose and fructose in the first 10 days of cultivation, with the exhaustion of these sugars at 40 days of cultivation, when the PLBs entered the stationary phase, demonstrating that during a certain period the PLBs release invertases in the culture medium to reduce sugars, and these are metabolized during the exponential phase of production of PLBs [113].

Table 2. Compliance of studies with induction, proliferation and regeneration of PLBs (IPR-PLBs) technique used with *Oncidium* species and hybrids.

Species or Hybrids	Origin and Age of Explants	Culture Media	Growth Conditions	Main Results	Evaluation and Detection of SV	Reference
Oncidium varicosum	Root tips 1.5 mm long from seedlings	Modified VW (replace $Fe_2(C_4H_4O_6)_3$ by 27.8 mg L^{-1} Fe-EDTA + 15% coconut water (PLBs proliferation from PLB) + 1.25 mg L^{-1} NAA (callus and PLB induction), pH 5.5	25 ± 1 °C, Gro-lux bulbs with 16-h photoperiod and 500 lux	Only one callus formed PLB and proliferation of PLBs occurred only in liquid medium with 15% coconut water	Non-evaluated	[119]
Oncidium 'Gower Ramsey'	Leaf segments 5 mm in length from in vitro plantlets leaves of 2–4 cm and 5–7 cm	MS½ + 100 mg L^{-1} inositol + niacin and pyridoxine (0.5 mg L^{-1}) + thiamine (0.1 mg L^{-1}) + glycine (2.0 mg L^{-1}), peptone (1000 mg L^{-1}), NaH_2PO_4 (170 mg L^{-1}), sucrose (20,000 mg L^{-1}) + Gelrite (2,500 mg L^{-1}), pH 5.2	Temp 26 ± 2 °C, PPFD 28–36 µmol m^{-2} s^{-1}, daylight fluorescent tubes, 16-h photoperiod	Donor leaves with 5–7 cm long showed higher percentage formed PLBs (25–35%) and number of PLBs/leaf segment (17–24.4) than 2–4 cm donor leaves (15–25% and 5.3–13.0) using 1–3 mg L^{-1} TDZ; proliferation of PLBs was highest with 0.3 mg L^{-1} TDZ, and regeneration of PLBs showed best response in absence of PGRs	Non-evaluated	[115]
Oncidium 'Gower Ramsey'	Leaves 2–4 and 5–7 cm, stem internodes 5mm and root tips 1 cm	MS½ + 100 mg L^{-1} inositol + niacin and pyridoxine (0,5 mg L^{-1}) + thiamine (0.1 mg L^{-1})+ glycine (2.0 mg L^{-1}), peptone (1000 mg L^{-1}), NaH_2PO_4 (170 mg L^{-1}), sucrose (20,000 mg L^{-1}) + Gelrite (2200 mg L^{-1}), pH 5.2: callus phase, 3.0 mg L^{-1} 2,4-D + 3.0 mg L^{-1} TDZ; PLBs, 0.1 NAA + 3.0 mg L^{-1} TDZ	Temp 26 ± 1 °C, PPFD 28–36 µmol m^{-2} s^{-1} white cool fluorescente, 16-h photoperiod	10% and 25% callusing from stem and root tips, 3.38 and 3.86 callus proliferation rate from stem and root tips, until 93.8 callus forming embryos and 29.1 embryos/callus from roots	Different callus lines showed large differential response to PLBs induction (0% to 93.8%) and number of PLBs/explant (0 to 29.1)	[117]
Oncidium 'Gower Ramsey' and *O.* 'Sweet Sugar'	Internodes 5 mm length from 15–20 cm inflorescence length	MS½ + 100 mg L^{-1} inositol + niacin and pyridoxine (0,5 mg L^{-1}) + thiamine (0.1 mg L^{-1})+ glycine (2.0 mg L^{-1}), peptone (1000 mg L^{-1}), NaH_2PO_4 (170 mg L^{-1}), sucrose (20,000 mg L^{-1}) + Gelrite (2200 mg L^{-1}), pH 5.2	Temp 26 ± 1 °C, PPFD 28–36 µmol m^{-2} s^{-1}, daylight fluorescent tubes, 16-h photoperiod	TDZ 1–3 mg L^{-1} increased explants produced PLBs directly in *O.* Sweet Sugar, but not in *O.* Gower Ramsey. Callus from explants on NAA + TDZ both at 1.0 mg L^{-1} showed 19 PLBs/callus. PLBs regeneration into shoots occurred in free-PGR MS½	Non-evaluated	[116]
Oncidium 'Gower Ramsey'	Leaf explants 1 cm in length from two-month old donor in vitro plantlets	MS½ + 100 mg L^{-1} inositol + niacin and pyridoxine (0,5 mg L^{-1}) + thiamine (0.1 mg L^{-1})+ glycine (2.0 mg L^{-1}), peptone (1000 mg L^{-1}), NaH_2PO_4 (170 mg L^{-1}), sucrose (20,000 mg L^{-1}) + Gelrite (2200 mg L^{-1}), pH 5.2	Temp 26 ± 1 °C, PPFD 28–36 µmol m^{-2} s^{-1}, daylight fluorescent tubes, 16-h photoperiod	Auxins IAA, NAA, IBA and 2,4-D inhibited direct PLB induction, while cytokinins promoted; TDZ 0.3–3.0 mg L^{-1} increased percentage of explants formed PLBs (60–75% in leaf tips and 25–40% in adaxial surfaces, with 9.5–10.7 PLBs/explant	Non-evaluated	[114]

Table 2. *Cont.*

Species or Hybrids	Origin and Age of Explants	Culture Media	Growth Conditions	Main Results	Evaluation and Detection of SV	Reference
Oncidium bifolium	Leaf segments 4 × 4 mm from germinated seedlings	MS½ + 2% sucrose + 2 g L^{-1} Phytagel + 1.0 mg L^{-1} TDZ, pH 5.5	27 ± 2 °C, 14-h photoperiod	25.5% of leaf segments formed PLBs and 12 PLBs/explant	Non-evaluated	[120]
Oncidium 'Gower Ramsey'	Leaf explants 1-cm length from two month old in vitro donor plantlets	MS½ + 1.0 mg L^{-1} TDZ, pH 5.2	Temp 26 ± 1 °C, PPFD 28–36 μmol m^{-2} s^{-1}, daylight fluorescent tubes, 16-h photoperiod	Leaf tips and leaves with adaxial surface in contact with culture medium was the best region for PLB induction, sucrose at 10–20 g L^{-1}, NaH_2PO_4 170 mg L^{-1}, peptone 1.0 g L^{-1} (65–80% explants with PLBs and 10.7 to 11.2 PLBs/explant);	Non-evaluated	[121]
Oncidium 'Gower Ramsey'	Leaf tips 1-cm length from two month old in vitro donor plantlets	MS½ + 100 mg L^{-1} inositol + niacin and pyridoxine (0,5 mg L^{-1}) + thiamine (0.1 mg L^{-1})+ glycine (2.0 mg L^{-1}), peptone (1000 mg L^{-1}), NaH_2PO_4 (170 mg L^{-1}), sucrose (20,000 mg L^{-1}) + Gelrite (2200 mg L^{-1}), pH 5.2	Temp 26 ± 1 °C, PPFD 28–36 μmol m^{-2} s^{-1}, daylight fluorescent tubes, 16-h photoperiod	GA_3 inhibited PLB formation, while anti-gibberellins Ancymidol (2.5 mg L^{-1}) and paclobutrazol (10 mg L^{-1}) increased explants formed PLBs (80–87.5% leaf tips formed PLBs and 154.8–193.2 PLBs/petri dish)	Non-evaluated	[118]
Oncidium taka	Axillary buds 0.5–1.0 cm lenght	MS + 3% sucrose + 0.7% agar, pH 5.7–5.8. PLBs induction at 1.0 mg L^{-1} BA + 0.5 mg L^{-1} NAA; PLBs regeneration at 2.0 mg L^{-1} BA + 1.0 mg L^{-1} BA	26 ± 2 °C, 12-h photoperiod, 3000 lux cool white fluorescent light	90% explants with PLBs and 9.4 shoots per culture	Non-evaluated	[122]
Oncidium 'Gower Ramsey'	Shoot tips 2–3 mm length for callus induction and 9-months age callus line for PLBs induction	MS½ + thiamine (1.0 mg L^{-1}) + nicotinic acid and pyridoxine (0.5 mg L^{-1}) + glycine (2.0 mg L^{-1}) + inositol (100 mg L^{-1}) + 2% sucrose + 7.5 g L^{-1} Agar, pH 5.7: callus proliferation, 1.0 mg L^{-1} 2,4-D + 0.5–1.0 mg L^{-1} TDZ / PLBs induction, 0.1 mg L^{-1} NAA and 0.4 mg L^{-1} BA with sucrose, maltose or trehalose	callus induction and proliferation in dark for 60-d (induction), subcultured every 2-weeks; PLBs induction, Temp 26 ± 2 °C, PPFD 57 μmol m^{-2} s^{-1}, 16-h photoperiod	680–732 g callus FW (1.0 mg L^{-1} 2,4-D and 0.5–1.0 mg L^{-1} TDZ); 1478 PLBs/0.25 g callus (Sucrose 10–20 g L^{-1}); 24 to 52.9 efficiency of plantlet conversion from PLBs (trehalose at 20 g L^{-1})	Non-evaluated	[123]
Oncidium 'Gower Ramsey' and *O.* 'Sweet Sugar'	Leave tips 1-cm long from in vitro plantlets	MS½ + 1.0 mg L^{-1} TDZ	Temp 26 ± 1 °C, PPFD 28–36 μmol m^{-2} s^{-1} daylight fluorescent tubes, 16-h photoperiod	Leaf tips and Adaxial region of leaves showed most response to PLB formation; 95% explants with PLBs with 20 g L^{-1} fructose in two cultivars; 31.1 (*O.* Sweet Sugar) to 33.7 (*O.* Gower Ramsey) PLBs/explant with 20 and 30 g L^{-1} sucrose, respectively	non evaluated	[124]

Table 2. *Cont.*

Species or Hybrids	Origin and Age of Explants	Culture Media	Growth Conditions	Main Results	Evaluation and Detection of SV	Reference
Cut flower varieties of *Oncidium*	New lateral buds	MS + 25 g L^{-1} sucrose + 10% coconut water + 7.5 g L^{-1} agar + 3.0 mg L^{-1} BA + 0.3 mg L^{-1} NAA, pH 5.6	25 ± 2 °C, 10–12-h photoperiod, 2000–2500 lux cool white fluorescent light	proliferation of 2.96	Non-evaluated	[125]
Oncidium 'Gower Ramsey'	PLBs from callus	Method described by ref. [123] using 10 g L^{-1} maltose	Callus at 26 ± 2 °C in the darkness; PLBs from callus in 50 µmol m^{-2} s^{-1} for 16-h photoperiod, under blue (455 nm), red (660 nm) and Far-red (730 nm)	2986 PLBs under fluorescent lamps statistically equal to red + blue + far red LEDs (3114 PLBs)	Non-evaluated	[126]
Oncidium flexuosum	Leaf apices 0.5 cm in length from 4-m seedlings	MS½ + 30 g L^{-1} sucrose + myo-inositol 100 mg L^{-1} + 5 g L^{-1} agar + nicotinic acid and Pyridoxine (0.5 mg L^{-1}) + thiamine (0.1 mg L^{-1}) + glycine (2.0 mg L^{-1}), pH 5.8	Temp 25 ± 2 °C, PPFD 40 µmol m^{-2} s^{-1}, 16-h photoperiod	Darkness for 90-d before photoperiod increased explants regenerating PLBs from 5 (Light) to 80% (Dark) and 10.8 PLBs per explant using 1.5 mg L^{-1} TDZ. Until 29.3 PLBs/explant 60-d after transfer PLBs to free-PGR MS	Non-evaluated	[127]
Oncidium 'Sugar Sweet'	Shoot tips 0.5 mm length for callus induction and PLBs obtained from callus	Callus: MS½ +2.0 mg L^{-1} BA + 0.3 mg L^{-1} NAA + 30 g L^{-1} sucrose + 7.0 g L^{-1} agar, pH 5.7; PLBs proliferation in MS + 30 g L^{-1} sucrose + 1.0 mg L^{-1} BA + 0.2 mg L^{-1} NAA, pH 5.8; PLBs regeneration, MS + 2.0 mg L^{-1} BA + 0.1 mg L^{-1} NAA + 30 g L^{-1} sucrose + 7.0 g L^{-1} agar	PLBs proliferation: 25 °C, 16-h photoperiod, white fluorescent light at 30 µmol m^{-2} s^{-1} at 5 l balloon type air lift bioreactor, 20 g fresh weight PLBs per bioreactor	3335.5 g fresh weight PLBs per vessel and 16.8 growth ratio; until 4.3 shoots/PLB and 1.17 g fresh weight per explant	Non-evaluated	[113]
Oncidium 'Gower Ramsey'	Shoot tips 5 mm for PLB induction and PLBs sections 3–4 mm diameter	PLBs induction: MS½ + 30 g L^{-1} sucrose + 6.0 g L^{-1} agar + 1.0 mg L^{-1} BA / PLBs proliferation: MS + 30 g L^{-1} sucrose + 6.0 g L^{-1} agar + 1.0 mg L^{-1} BA + 0.5 mg L^{-1} NAA	Temp 25 ± 2 °C, 16-h photoperiod	Red LEDs (660 nm) resulted in best induction rate (83.3% explants), Fresh weight (≡ 20 g) and propagation rate (>6) of PLBs, while Blue LEDs showed 90% of differentiation rate of PLBs into shoots	Non-evaluated	[109]
Oncidium 'Gower Ramsey'	Root tips segments 1 cm in length from 6-months old in vitro plantlets	Callus induction: MS½, pH 5.2 / PLB induction: MS½ + 0.1 mg L^{-1} NAA + 3.0 mg L^{-1} TDZ	Temp 25 ± 1 °C, darkness	Age of callus from 0.5 to 2 years resulted in best percentage (80–100%) of callus produced PLBs and number of PLBs/callus (6.2–6.6); the increase in age of callus reduced it embryogenesis capacity	Different callus lines showed large differential response to PLBs induction. However, 3-years old plantlets greenhouse cultivated showed same color, size and morphology of O. Gower Ramsey	[128]

Table 2. *Cont.*

Species or Hybrids	Origin and Age of Explants	Culture Media	Growth Conditions	Main Results	Evaluation and Detection of SV	Reference
Oncidium forbesii (*Brasilidium forbesii*)	Transverse and lateral Thin cell layers 1mm thickness from in vitro germinated protocorms	WPM + 3% sucrose + 0.6% agar, pH 5.8	Temp 25 ± 1 °C/19 ± 1 °C (day/night), 16-h photoperiod, white fluorescent tubes 40 µmol m^{-2} s^{-1}	Lateral thin cell layers in culture medium with BA at 2.0 µM increased PLB induction in 64 to 82% explants and both from lateral and transversal TCL at 1.0 µM promoted the number of PLBs obtained/explant (17.1–24.6)	Non-evaluated	[129]
Oncidium 'Gower Ramsey'	PLBs sections obtained from nodal explants from inflorescences	MS½ (full strength MS vitamins) + 1 g L^{-1} tryptone + 20 g L^{-1} sucrose + 1 g L^{-1} activated charcoal + 65 g L^{-1} potato tuber + 8 g L^{-1} agar + 5 µM TDZ (TDZ, vitamins and glycine were filter sterilized)	Temp 22 ± 2 °C, 16-h photoperiod	PLBs regeneration from PLBs section increased with addition of chloro or methyl or nitro derivatives (compounds 5a–5c) using 2–5 µM, from 41 (control) until 95 plantlets per culture bottle using 5 µM of 5c compound	Non-evaluated	[130]
Oncidium sp. (Vu Nu Orchids)	In vitro shoots	MS½ + 20 g L^{-1} sucrose + 10% coconut water + agar, pH 5.8	Temp 26 ± 2 °C, PPFD 22.2 µmol m^{-2} s^{-1}, 12-h photoperiod	NAA 0.75 mg L^{-1} produced highest number of PLBs/callus (98) and 1 mg L^{-1} BA promoted PLBs regeneration into shoots (12.42/PLB)	Non-evaluated	[112]
Tolumnia Snow Fairy	Leaf segments from different in vitro plantlets height and leaf positions	MS½ (with Fe-NaEDTA, vitamins and glycine at full-strength MS) + 100 mg L^{-1} myo-inositol + NaH_2PO_4 (170 mg L^{-1}), 30 g L^{-1} sucrose + 8.0 g L^{-1} agar, pH 5.2	Temp 25 ± 2 °C, 8-weeks in dark and transferred to dim light, PPFD 5 µmol m^{-2} s^{-1}, cool white fluorescent tubes, 12-h photoperiod	Leaves from 1–2 cm plantlet height showed highest explants induced PLBs using 2.0 mg L^{-1} BA (16.7%), but highest number of embryos was obtained with 4.0 mg L^{-1} BA and from plantlets with 2–3 cm (41 PLBs/explant), upper wounding region of bigger PLBs improved PLBs proliferation and number of PLBs per explant	Plants were transferred to plastic pots and flowered after one-year without reports of somaclonal variations in vegetative and reproductive phase	[31]

MS: Murashige and Skoog Medium [99]; VW: Vacin Went medium [43]; WPM: Wood Plant Medium [131]. 2,4-D, 2-4-Dichlorofenoxiacetic acid; BA, 6-Benzyladenine; IAA, 3-Indoleacetic acid; IBA, Indole-3-butyric acid; NAA, Naphtaleneacetic acid; PPFD: Photosynthetically Photon Flux Density; Temp, Temperature; TDZ, Thidiazuron.

Another study conducted in a gelled medium by [124] observed that the use of 2% fructose resulted in 95% explants containing PLBs in *Oncidium* Gower Ramsey or 2% glucose resulted in 85% explants containing PLBs in *Oncidium* Sweet Sugar [124]. However, for the number of PLBs per explant, the best results were obtained with 2–3% sucrose (31.1–33.7 PLBs/explants), demonstrating that sucrose is the most suitable sugar for IPR–PLB. The use of other types of sugars, cellobiose, maltose and trehalose do not result in benefits for number of PLBs from callus in *Oncidium* Gower Ramsey [122] or for direct production of PLBs from young leaves [124].

There are no doubt about the application of PLBs in mass clonal production of *Oncidium* [132], but recent studies also showed and confirmed the presence of somaclonal variation in *Oncidium* obtained from IPR–PLBs [133], similar to observed with *Phalaenopsis* genus.

4. Some News with *Cymbidium*, *Dendrobium*, and Others

The most of results obtained with *Phalaenopsis* and *Oncidium* were similar to reported with other species of orchids of importance in floriculture, as *Cymbidium* and *Dendrobium* genera, such as the main PGRs used for IPR–PLBs. As example, the combination of cytokinin BA (5.0 mg L^{-1}) and auxin NAA (2.5 mg L^{-1}) were used to induce PLBs (20.55 PLB per primary protocorm) in *Cymbidium mastersii* protocorms [134]. Thin cell layers (TCL) from different types of tissues was a technique used to improve the production of PLBs in *Cymbidium* [135], *Dendrobium* [136,137], *Oncidium* [129], and *Phalaenopsis* [93].

In *Dendrobium*, a wide and complete study about molecular research was exhaustively carried out by [138], and considered especially the identification, classification and breeding of *Dendrobium*. Similarly, other study with micropropagation of *Dendrobium* was realized by [17] and concluded that PLBs were used as explants in 21.8% of studies, and together with nodal or nodal segments explants is one of the major method used for *Dendrobium* micropropagation.

Thidiazuron was also an important PGR for induction of PLBs in *Dendrobium* orchids, but the response to different cytokinins depends on genotype. In *Dendrobium aqueum*, only the cytokinin 2iP [*N*-6-(2-isopentyl) adenine] at 1.5 mg L^{-1} proved it efficiency in production of PLBs (42.7 PLBs per explants) from callus, compared to other cytokinins BA, Kin and Zea, and cytokinin-like compound TDZ. These authors also observed that arginine at 25 mg L^{-1} increased direct somatic embryogenesis, instead of callus derived PLBs [137]. Meta-Topolins, a natural aromatic type of cytokinin, were also reported used in induction and regeneration of PLBs in *D. nobile*, which combined with 0.5 mg L^{-1} NAA resulted in best PLBs formation (92%) and shoots/explants (9.2) [139]. These same authors observed that addition of polyamines, such as spermidine and putrescine increased regeneration of shoots from PLBs and secondary PLB formation.

In our laboratory, PLBs of *Dendrobium* Hybrid 'H3', could be induced and proliferated in one-step, and obtained from in vitro shoots, using liquid MS½ medium with 1.0 mg L^{-1} BA, and under agitation of 80 rpm (Figure 1D).

5. Applications of IPR–PLB Technique on Orchid Propagation and Breeding and Main Limitations of the Technique

Induction, proliferation, and regeneration of PLBs in orchids have many advantages to conventional micropropagation by shoot proliferation or use of shoots from inflorescence stalk segments as in *Phalaenopsis* [140], as increased rate of proliferation/multiplication [141] and single-cell derived PLBs [123], which could be used for propagation, but also for breeding purposes and to obtain disease free plantlets.

In breeding programs using in vitro techniques, PLBs could be used to obtain autotetraploid plants with use of anti-mytotic agents as oryzalin [142] and colchicine [143], and to obtain mutants by the use of chemical mutagens as sodium azide [144] or physical mutagens as gamma-irradiation [145].

PLBs can be also used for transformation protocols and successful protocols were developed and obtained stable transgenics with target characteristics for floriculture [146,147]. In genetic transformation of orchids, the use of PLBs derived directly from individual epidermal cells resulted in solid transgenic plants with clonal identity of *Oncidium* Sharry Baby 'OM8' [32], an exceptional advantage over PLBs from callus and with multicellular origin [126], which may result in the emergence of somaclonal variants [42] and chimeric tissues when used for genetic transformation, which are difficult to characterize and separate [32]. Using this technique, these authors reported 33–43% PLBs expressing the β-glucuronidase gene (GUS) and obtained six lineages that amplified the transgenes pepper ferredoxin-like protein (pflp) and hygromycin phosphotransferase (hpt) using the particle bombardment technique. *Agrobacterium tumefasciens*-mediated transformation has also been successfully used in the production of transgenic plants of *Oncidium* 'Sharry Baby OM8' and *Oncidium* Gower Ramsey using the induction of secondary PLBs from in vitro-maintained PLBs [148,149].

From a phytosanitary point of view, it is known that the use of seeds for in vitro asymbiotic sowing of orchids is a real way to obtain virus-free seedlings in orchids from contaminated mother plants, as observed for *Cymbidium* species [150,151]. Ref. [152] confirmed on a large scale (1000 plants) that in vitro plants from seeds are free of *Cymbidium Mosaic Virus* (CyMV) and *Ondontoglossum Ringspot Virus* (ORSV).

The technique of culturing apical meristems may also be effective in eliminating viral diseases in orchids, but it requires great manual skill for excision of tiny meristems leading to contamination-free tissue [153]. These requirements and the individual characteristics of viral diseases may lead to breakthroughs in the technique, which may result in in vitro plantlets containing viral diseases, as reported in *Brassolaeliocattleya, Cattleya, Dendrobium, Epicattleya, Oncidium,* and *Mokara* grown in vitro, for which CyMV virus was reported to be present in 27.6% of 880 plantlets evaluated, while ORSV was not detected in these samples [152].

Furthermore, in genera such as *Phalaenopsis*, the most commercially important in the world, only stem apex culture may not be effective in completely eliminating important viral diseases in the crop [140], and may still result in the need to kill the mother plant to obtain the apical meristem, since these plants are monopodial and have poorly developed stem [150]. In this sense, in vitro IPR–PLBs is an alternative to the production of virus-free clonal plants in orchids. In *Phalaenopsis* hybrid 'V3', Ref. [140] obtained PLBs from stem apexes of donor plants contaminated with *Ondontoglossum* ringspot virus and *Cymbidium* mosaic virus, and observed that the first PLBs produced directly from the stem apex had 31.25% PLBs with viruses, identified by the enzyme-linked immunosorbent assay (ELISA) and RT-PCR and were only eliminated in the process after some subcultures. The PLBs identified as virus-free were subcultured in PLB proliferation medium, and in the second subculture 18.18% positive PLBs were identified for both viruses. Only in the third subculture of PLB proliferation, it was possible to obtain 100% virus-free PLBs, which remained until the end of the experiment.

PLBs can also be used for orchid propagation using the synthetic seed technique and for cryopreservation. In *Dendrobium* 'Sonia', the use of PLBs stored at 4 °C for 15 days in the pro-meristematic and leaf primordium stages and encapsulated with 3–4% sodium alginate + 75–100 mM $CaCl_2{*}2H_2O$ resulted in 100% germinated PLBs, with the appearance of the first leaf at 22–27 days and the first root at 30–35.8 days, and the technique can be replicated with similar results for *Oncidium* 'Gower Ramsay' and *Cattleya leopoldii* [154].

In *Dendrobium candidum* and *Dendrobium nobile*, PLBs have also been used to increase the production of bioactive compounds. In *D. nobile*, an increase was observed in the production of secondary metabolites such as phenols, flavonoids and alkaloids extracted from PLB-micropropagated plants, when compared to the mother plant [139]. In *D. candidum*, the increase in methyl-jasmonate elicitor concentrations, although resulting in a proportional reduction in PLBs mass gain, increased the concentrations of alkaloids, polysaccharides, phenols and flavonoids when used between 75 and 100 μM [155].

Although the IPR–PLB technique is widely used for large scale plantlet production, breeding and conservation, some difficulties still limit the wider use of the technique on a commercial scale. Among the main limitations are the high genotype-dependence of PLB induction and proliferation responses in vitro, and the occurrence of undesirable somaclonal variations, which greatly hinder the proliferation of clonal propagation of PLBs for a wide range of commercial cultivars available and required by the market.

Ref. [30] used NDM culture medium plus TDZ (0.25 mg L^{-1}) and NAA (1.0 mg L^{-1}) and observed distinct responses between '908' genotype (45% explants with PLBs and up to 25 PLBs/leaf segment) and 'RP3' genotype (10% explants with PLBs and only 2 PLBs/leaf segment), the latter being highly recalcitrant to the induction and proliferation of PLBs from leaf segments of plants grown in vitro. A study by [59] also noted important differences between the PLBs induction responses between *P. amabilis* (up to 50% explants with PLBs and 15.6 PLBs/explant) and the commercial cultivar *P. nebula* (80% explants with PLBs and up to 5.3 PLBs/explant). The same occurred in another study with the same cultivars, in which the cytokinin types and concentrations that resulted in the highest percentage of explants with PLBs were 13.32 μM BAP in *P. amabilis* (80%) and 13.62 μM TDZ in *P. nebula* (65%). The largest number of PLBs per explant was obtained with 13.62 μM TDZ in *P. amabilis* (7.8 PLBs/explant) and 4.65 μM Kin in *P. nebula* (16 PLBs/explant) [77].

Ref. [156] point out that one of the biggest difficulties in *Phalaenopsis* micropropagation by PLBs is that not all genotypes respond to a single protocol and the same cultivation conditions, and often result in plants with undesirable characteristics. Ref. [41] compared eight cultivars of *Phalaenopsis* and *Doritaenopsis* to obtain PLBs from shoot tips of inflorescence stalk buds with best percentage of PLB formation in four genotypes using 1.0 mg L^{-1} BAP (26.9–71.4% depending on genotype), while two respond better with 2.0 mg L^{-1} (60–75% explants with PLBs) and one produced 50% PLBs independently of the concentration of BAP (1, 2, or 5.0 mg L^{-1}). Testing other four genotypes authors reported ranges from 7.1% to 40% of PLBs formation only in NDM culture medium, while in ½MS only two cultivars produced PLBs [41].

Ref. [156] have been associated undesirable characteristics observed in some plantlets with the identification of somaclonal variants from PLBs, which can be morphologically identified even at the shoot bud regeneration and in vitro plantlet production stage. According to [157], the occurrence of SV in the IPR–PLBs technique is higher than that observed from adventitious bud propagation, and that most commercial laboratories use a maximum of three generations of PLBs subcultures to avoid high frequencies of somaclonal variations in this type of propagation.

In our laboratory conditions, using leaf segments from in vitro plantlets to obtain PLBs (Figure 1A,B) somaclonal variations are observed in rooting phase of PLB-derived plantlets of *Phalaenopsis* 'Ph908', while were not observed in plantlets derived from shoot-proliferation using inflorescence stem nodal segments (Figure 2A). The main symptoms were the limited development of plantlets that remains in acclimatized plantlets, with morphological abnormalities in leaves (Figure 2B), also observed and called as 'creased leaves' by [66] and flowers deformities as absence of lip in some flowers of the inflorescence (Figure 2C,D) possibly associated with mutations rather than epigenetic variations.

Ref. [139] used induction of PLBs from pseudostems from in vitro germinated *Dendrobium nobile* plants in MS + 1.5 mg L^{-1} TDZ and 0.25% activated charcoal medium and verified 94% explants producing PLBs and up to 11.6 PLBs/explant. These authors observed a somaclonal variation rate close to 6% in the obtained plants, being the main cause of the somaclonal variations detected by molecular markers Random amplified polymorphic DNA (RAPD) and Start codon targeted (SCoT), attributed by the authors to the use and exposure time to TDZ.

Although the cytokinin-like compound TDZ is appointed as one of the major causes of SV in orchid PLB induction, there were some contradictory reports.

As example, the cytokinin Kinetin at 1.5 mg L^{-1} resulted in increases of somaclonal variations frequency of PLBs in *Dendrobium* Sabin Blue, detected by ISSR and DAMD molecular markers, when compared with use of TDZ at 4.0 mg L^{-1} added activated charcoal [158].

Figure 2. Somaclonal variations observed in *Phalaenopsis* induction, proliferation and regeneration of protocorm-like Bodies in *Phalaenopsis* Hybrid "908". Normal vegetative developed plant (**A**) and somaclonal variation observed in vegetative development with "creased leaves" (red arrow) (**B**); (**C**,**D**), Normal vegetative developed plants with somaclonal variations in flower development, with first and last flower without of labellum (red arrow, wl) in the same inflorescence with normal flowers (nf). All figures are unpublished photos from J.C.C.

In addition, [159] observed somaclonal variants in *Phalaenopsis* True Lady 'B79-19', obtained from the induction of PLBs and from young leaves obtained from in vitro plants in VW culture medium containing only BA and NAA as phytoregulators, i.e., without using TDZ. These authors also reported that variant plants were discarded during in vitro subcultures (not quantified), and out of the plants obtained and without morphological variations in the leaves, only 20 out of a total of 1360 obtained (1.5%) were somaclonal variants, indicated by the different flowers of the original clone.

Also the use of topolins *meta*-Topolins (*m*T) and *meta*-Topolins Riboside (*m*TR), a natural aromatic cytokinin reported as reducing phytotoxic effects in micropropagation, it use not solved the problem of somaclonal variation obtained in vitro [160] and, although was reported increasing efficiency of PLB induction it use not resulted in absence of somaclonal variation in orchids [139].

These observations with other cytokinins PGRs diminish the importance of TDZ as the unique or main factor for VS inducing in orchid IPR–PLBs, and include other causes, such as the differential susceptibility of genotype and the number of subcultures under proliferation stage of PLB production.

Genotype susceptibility is appointed one of the main factors lead to VS in *Phalaenopsis* and *Doritaenopsis* orchids micropropagation, ranging from zero to 100% SV depending on genotype and is not exclusive of the PLB technique [72,161]. Similarly, [70] also observed that some genotypes of *Phalaenopsis* not presented any variants, while others showed until 47.9% of variants. Among them, most of SV in this genus were reported in flowering stage [161], by modification of inflorescence and flower characteristics, such as the perloric and semi-perloric mutants observed in *Phalaenopsis* Zuma Pixie '#1', *P.* Little Mary and *Doritaenopsis* Minho Diamond 'F607' [162]. Lose of part of flowers were also reported, such as pollinia [162] and absence of labellum (Figure 2C,D).

Ref. [161] evaluated until the flowering stage (1.0–1.5 years after acclimatization) plants of 10 genotypes of *Phalaenopsis* and *Doritaenopsis* hybrids micropropagated by the PLB technique, and subcultured in vitro for 5 to 10x and identified the presence of seven types of VS, possible to be identified only at the flowering stage. The plants had deficiencies or divergences in the petals and sepals or in the development of the inflorescence, but with similar vegetative development in relation to the mother plant. These authors observed that the produced VS were not polyploid mutants, maintaining the same amount of genetic material as the mother plants.

Although most of SV was reported in flowering stage, transcript analysis by Real-Time PCR demonstrated that mutants has also many other alterations in factors of transcription and transcripts were detailed reported in *Phalaenopsis* and *Doritaenopsis* by [162]. In *Oncidium* 'Milliongolds' were also observed chlorophyll SV (whole yellow or with streaked leaves) in vegetative development of in vitro plantlets [133].

Another factor related to the origin of VS in PLBs in orchids is the phase in which VS occurs. It has been reported that in the proliferation phase, undesirable VS induction from PLBs occurs at a higher intensity and frequency, and it is necessary to establish a number of subcultures to keep the VS frequencies low in clonal propagation. Ref. [92] reported increases in SV after the third subcultures of PLBs in proliferation medium (NDM + 0.1 mg L^{-1} TDZ and 10 mg L^{-1} chitosan) with same ISSR profile until third subculture, 95% at fourth and 80% at fifth subculture of PLBs.

The use of RAPD molecular markers (total of 1116 bands) did not allow the identification of these somaclonal variants in these plants, but isozyme pattern analysis demonstrates the difficulty of observing mutations in materials obtained from PLBs using RAPD molecular markers and the occurrence of conclusion errors or even underestimated data of somaclonal variants in the confirmation of clonal origin in other studies conducted with these markers [159].

Ref. [82] also used RAPD markers to analyze the clonal origin of PLBs and induced seedlings in in vitro leaf segments of *Phalaenopsis bellina* in ½MS medium with 3.0 mg L^{-1} TDZ. They observed that most somaclonal variants are obtained at the proliferation/multiplication phase, with no VS observed in the origin phase of the PLBs of the mother plant.

Analyses of SCoT and Target Region Amplification Polymorphism (TRAP) markers also showed the presence of somaclonal variants in *Dendrobium* Bobby Messina PLBs cryopreserved or not [163].

These differences in the frequencies of VS observed in different orchid species and genotypes are probably associated with higher sensitivity of different genotypes to the occurrence of mutations. Ref. [164] observed that the frequency of VS at the vegetative and reproductive stages in *Phalaenopsis* PLBs was dependent on the genotype used. These authors observed that there was a reduction in DNA methyltransferase (Dnmt)-related gene expression in *Phalaenopsis* 'Little Mary' VS.

Current advances in molecular marker techniques allow increasing the number of tools and the accuracy of these analyses and the greater possibility of identifying possible VS. There is little information about wide molecular genome characterization in *Oncidium*, and [133] used specific-locus amplified fragment sequencing (SLAF-seq) to analyze possible variations in single-nucleotide polymorphisms (SNPs) in *Oncidium* 'Milliongolds' obtained by PLBs grown for 10 years and observed high rates of variation and that adjacent SNPs adenine and thymine were more frequent than those related to guanine and cytosine, with prominence of mononucleotideInDels.

Ref. [157] isolated two most expressed transposable elements and identified a new Instability Factor (PIF)-like, one of which, called PePIF1 was identified by similarity to the *Phalaenopsis equestris* genome sequence, and which was transposed in the somaclonal variants of cultivars of *Phalaenopsis* from micropropagation, which resulted in the insertion of new genes identified and sequenced by the authors.

6. Conclusions

Induction, proliferation, and regeneration of PLBs (IPR–PLBs) in orchids is one of the most promising techniques to replace current conventional micropropagation techniques, in particular because it has wide application in clonal conservation, propagation, breeding, and phytossanitary-cleaning of elite plants used in the flower market. Although many authors used somatic embryogenesis to describe IPR–PLBs technique or their origin, recent molecular studies about the origin route of PLBs, at least in *Phalaenopsis* orchids, showed that IPR–PLBs routes are not the same of somatic embryonic origin. Some limitations of IPR–PLBs in orchids such as low repeatability of responses due to high genotype dependence and the presence of somaclonal variations (SV) still limit their large-scale use in the production of clone plantlets. Although the main causes of SV described in papers were the genotype-sensibility, the use of cytokinin thidiazuron and subsequent PLBs proliferation, only genotype sensibility looks conclusive, because SV was also observed in protocols using other cytokinins, such as BA and Kin. Nevertheless, the new findings associated with the identified instability factors, associated with the recent sequencing of the *Phalaenopsis equestris* genome, and the use of new molecular tools that increase the accuracy of quantitative identification analyses and the causes of somaclonal variation, are in agreement with the evolution of this technique, which represents the tool of greatest potential today to replace other less efficient micropropagation techniques in the production of plantlets in orchids.

Author Contributions: J.C.C. and C.A.Z. designed and wrote the manuscript. J.C.C. and J.-T.C. comprehensively revised and improved the quality of manuscript. All authors have read and agreed to the published version of the manuscript.

Acknowledgments: J.C.C. thanks to São Paulo Research Foundation for the project number 2018/20673-3 and to Conselho Nacional de Desenvolvimento Científico e Tecnológico for the project number 311083/2018-8.

References

1. Chase, M.W.; Cameron, K.M.; Freudestein, J.V.; Pridgeon, A.M.; Salazar, G.; Van den Berg, C.; Schuiteman, A. An update classification of Orchidaceae. *Bot. J. Linn. Soc.* **2015**, *177*, 151–174. [CrossRef]
2. The Plant List. 2019. Orchidaceae Family. Available online: http://www.theplantlist.org/1.1/browse/A/Orchidaceae/#statistics (accessed on 31 December 2019).
3. RHS 2019. The International Orchid Register. Available online: https://apps.rhs.org.uk/horticulturaldatabase/orchidregister/orchidregister.asp (accessed on 31 December 2019).
4. Bulpitt, C.J.; Li, Y.; Bulpitt, P.F.; Wang, J. The use of orchids in Chinese medicine. *J. R. Soc. Med.* **2007**, *100*, 558–563. [CrossRef] [PubMed]
5. Zuraida, A.R.; Izzati, K.H.F.L.; Nazreena, O.A.; Zaliha, W.S.W.; Radziah, C.M.Z.C.; Zamri, Z.; Sreeramanan, S. A simple and efficient protocol for the mass propagation of *Vanilla planifolia*. *Am. J. Plant Sci.* **2013**, *4*, 1685–1692. [CrossRef]
6. Chen, C. The Fundamental Issue in the Phalaenopsis Industry. 2018. Available online: http://amebse.nchu.edu.tw/orchids_cultivation21.htm (accessed on 30 December 2019).

7. Cardoso, J.C. *Dendrobium* 'Brazilian Fire 101'-New option of color of flowers for the orchid market. *Hortic. Bras.* **2012**, *30*, 561–564. [CrossRef]
8. Cardoso, J.C.; Martinelli, A.P.; Teixeira da Silva, J.A. A novel approach for the selection of Cattleya hybrids for precocious and season-independent flowering. *Euphytica* **2016**, *210*, 143–150. [CrossRef]
9. Cardoso, J.C. *Ionocidium* 'Cerrado101': Intergeneric orchid hybrid with high quality of blooming. *Ornam. Hortic.* **2017**, *23*, 351–356. [CrossRef]
10. Ho, T.-T.; Kwon, A.-R.; Yoon, Y.-J.; Paek, K.-Y.; Park, S.-Y. Endoreduplication level affects flower size and development by increasing cell size in *Phalaenopsis* and *Doritaenopsis*. *Acta Physiol. Plant* **2016**, *38*, 190. [CrossRef]
11. Lakshman, C.; Pathak, P.; Rao, A.N.; Rajeevan, P.K. *Commercial Orchids*; De Gruyter Open: Beijing, China, 2014; p. 322.
12. Van den Berg, C. Reaching a compromise between conflicting nuclear and plastid phylogenetic trees: A new classification for the genus *Cattleya* (Epidendreae; Epidendroideae; Orchidaceae). *Phytotaxa* **2014**, *186*, 75–86. [CrossRef]
13. Peraza-Flores, L.N.; Carnevali, G.; Van den Berg, C. A molecular phylogeny of the *Laelia* Alliance (Orchidaceae) and reassessment of *Laelia* and *Schomburgkia*. *Taxon* **2017**, *65*, 6. [CrossRef]
14. Dalström, S.; Higgins, W.E. New combinations and transfers to *Odontoglossum* Oncidiinae (Orchidaceae): Avoid creating new names. *Harv. Pap. Bot.* **2016**, *21*, 97–104. [CrossRef]
15. Yeung, E.C. A perspective on orchid seed and protocorm development. *Bot. Stud.* **2017**, *58*, 33. [CrossRef] [PubMed]
16. Oneal, E.; Willis, J.H.; Franks, R. Disruption of endosperm development is a major cause of hybrid seed inviability between *Mimulus guttatus* and *M. nudatus*. *New Phytol.* **2010**, *210*, 029223. [CrossRef] [PubMed]
17. Teixeira da Silva, J.A.; Cardoso, J.C.; Dobránszki, J.; Zeng, S. *Dendrobium* micropropagation: A review. *Plant Cell Rep.* **2015**, *34*, 671–704. [CrossRef] [PubMed]
18. Teixeira da Silva, J.A.; Tsavkelova, E.A.; Ng, T.B.; Parthibhan, S.; Dobránszki, J.; Cardoso, J.C.; Rao, M.V.; Zeng, S. Asymbiotic in vitro seed propagation of *Dendrobium*. *Plant Cell Rep.* **2015**, *34*, 1685–1706. [CrossRef] [PubMed]
19. Li, Y.-Y.; Chen, X.-M.; Zhang, Y.; Cho, Y.-H.; Wang, A.-R.; Yeung, E.C.; Zeng, X.; Guo, S.-X.; Lee, Y.-I. Immunolocalization and changes of hydroxyproline-rich glycoproteins during symbiotic germination of *Dendrobium officinale*. *Front. Plant Sci.* **2018**, *9*, 552. [CrossRef] [PubMed]
20. Mala, B.; Kuegkong, K.; Sa-ngiaemsri, N.; Nontachaiyapoom, S. Effect of germination media on in vitro symbiotic seed germination of three *Dendrobium* orchids. *S. Afr. J. Bot.* **2017**, *112*, 521–526. [CrossRef]
21. Arditti, J. Factors affecting the germination of orchid seeds. *Bot. Rev.* **1967**, *33*, 1–97. [CrossRef]
22. Santos, S.A.; Smidt, E.C.; Padial, A.A.; Ribas, L.L.F. Asymbiotic seed germination and in vitro propagation of *Brasiliorchis picta*. *Afr. J. Biotech.* **2016**, *15*, 134–144.
23. Kunakhonnuruk, B.; Inthima, P.; Kongbangkerd, A. In vitro propagation of *Epipactis flava* Seidenf, an endangered rheophytic orchid: A first study on factors affecting asymbiotic seed germination, seedling development and greenhouse acclimatization. *Plant Cell Tissue Organ Cult.* **2018**, *135*, 419–432. [CrossRef]
24. Rao, A.N. Tissue culture in orchid industry. In *Applied and Fundamental Aspects of Plant Cell Tissue and Organ Culture*; Reinert, J., Bajaj, Y.P.S., Eds.; Springer: Berlin, Germany, 1977; pp. 44–49.
25. Parmar, G.; Pant, B. In vitro seed germination and seedling development of the orchid *Coelogyne stricta* (D. Don) Schltr. *Afr. J. Biotechnol.* **2016**, *15*, 105–109.
26. Huang, H.; Zi, X.-M.; Lin, H.; Gao, J.-Y. Host-specificity of symbiotic mycorrhizal fungi for enhancing seed germination, protocorm formation and seedling development of over-collected medicinal orchid, *Dendrobium devonianum*. *J. Microb.* **2018**, *56*, 42–48. [CrossRef] [PubMed]
27. Fochi, V.; Chitarra, W.; Kohler, A.; Voyron, S.; Singan, V.R.; Lindquist, E.A.; Barry, K.W.; Girlanda, M.; Grigoriev, I.V.; Martin, F.; et al. Fungal and plant gene expression in the *Tulasnella calospora-Serapias vomeracea* symbiosis provides clues about nitrogen pathways in orchid mycorrizas. *New Phytol.* **2016**, *213*, 10–12.
28. Lee, Y.; Hsu, S.; Yeung, E.C. Orchid protocorm-like bodies are somatic embryos. *Am. J. Bot.* **2013**, *100*, 2121–2131. [CrossRef] [PubMed]
29. Fang, S.; Chen, J.C.; WEI, M.J. Protocorms and protocorm-like bodies are molecularly distinct from zygotic embryonic tissues. *Plant Physiol.* **2016**, *171*, 2682–2700. [CrossRef] [PubMed]

30. Zanello, C.A.; Cardoso, J.C. PLBs induction and clonal plantlet regeneration from leaf segment of commercial hybrids of *Phalaenopsis*. *J. Hortic. Sci. Biotech.* **2019**, *94*, 627–631. [CrossRef]
31. Chookoh, N.; Chiu, Y.; Chang, C.; Hu, W.; Dai, T. Micropropagation of *Tolumnia* orchids through induction of protocorm-like bodies from leaf segments. *Hortscience* **2019**, *54*, 1230–1236. [CrossRef]
32. Li, S.H.; Kuoh, C.S.; Chen, Y.H.; Chen, H.H.; Chen, W.H. Osmotic sucrose enhancement of single-cell embryogenesis and transformation efficiency in *Oncidium*. *Plant Cell Tissue Organ Cult.* **2005**, *81*, 183–192. [CrossRef]
33. Naing, A.H.; Chung, J.D.; Park, I.S.; Lim, K.B. Efficient plant regeneration of the endangered medicinal orchid, *Coelogyne cristata* using protocorm-like bodies. *Acta Physiol. Plant* **2011**, *33*, 659–666. [CrossRef]
34. Kalyan, K.; Sil, S. Protocorm-like bodies and plant regeneration from foliar explants of *Coelogyne flaccida*, a horticulturally and medicinally important endangered orchid of eastern himalaya. *Lanke* **2015**, *15*, 151–158.
35. Picolotto, D.R.N.; Paiva Neto, V.B.; Barros, F.; Padilha, D.R.C.; Cruz, A.C.F.; Otoni, W.C. Micropropagation of *Cyrtopodium paludicolum* (Orchidaceae) from root tip explants. *Crop Breed. App. Biotech.* **2017**, *17*, 191–197. [CrossRef]
36. Samala, S.; Te-chato, S.; Yenchon, S.; Thammasiri, K. Protocorm-like body of *Grammatophyllum speciosum* through asymbiotic seed germination. *ScienceAsia* **2014**, *40*, 379–383. [CrossRef]
37. Chen, C. Cost analysis of plant micropropagation of *Phalaenopsis*. *Plant Cell Tissue Organ Cult.* **2016**, *126*, 167–175. [CrossRef]
38. Tanaka, M.; Sakanishi, Y. Clonal propagation of *Phalaenopsis* by leaf culture. *Am. Orc. Soc. Bull.* **1977**, *46*, 733–737.
39. Tanaka, M.; Sakanishi, Y. Clonal propagation of *Phalaenopsis* through tissue culture. In *Proc. 9th World Orchid Conference*; Kashemsanta, M.R.S., Ed.; Amarin Press: Bangkok, Thailand, 1980; pp. 215–221.
40. Tanaka, M.; Sakanishi, Y. Regenerative capacity of in vitro cultured leaf segments excised from mature *Phalaenopsis* plants. *Bull. Univ. Osaka Ser. B* **1985**, *37*, 1–4.
41. Tokuhara, K.; Mii, M. Micropropagation of *Phalaenopsis* and *Doritaenopsis* by culturing shoot tips of flower stalk buds. *Plant Cell Rep.* **1993**, *13*, 7–11. [CrossRef]
42. Ishii, Y.; Takamura, T.; Goi, M.; Tanaka, M. Callus induction and somatic embryogenesis of *Phalaenopsis*. *Plant Cell Rep.* **1998**, *17*, 446–450. [CrossRef]
43. Vacin, E.F.; Went, F.W. Some pH in nutrient solutions. *Bot. Gaz.* **1949**, *110*, 605–617. [CrossRef]
44. Huan, L.V.T.; Takamura, T.; Tanaka, M. Callus formation and plant regeneration from callus through somatic embryo structures in *Cymbidium* orchid. *Plant Sci.* **2004**, *166*, 1443–1449. [CrossRef]
45. Ulisses, C.; Pereira, J.A.F.; Silva, S.S.; Arruda, E.; Morais, M. Indução e histologia de embriões somáticos primários e secundários do híbrido *Phalaenopsis* Classic Spotted Pink (Orchidaceae). *Acta Biol. Col.* **2016**, *21*, 571–580.
46. Goussard, P.G.; Wiid, J.; Kasdor, G.G.F. The effectiveness of in vitro somatic embryogenesis in eliminating fanleaf virus and leafroll associated viruses from grapevines. *S. Afr. J. Enol. Vitic.* **1991**, *12*, 77–81. [CrossRef]
47. Quainoo, A.K.; Wetten, A.C.; Allainguillaume, J. The effectiveness of somatic embryogenesis in eliminating the cocoa swollen shoot virus from infected cocoa trees. *J. Virol. Met.* **2008**, *149*, 91–96. [CrossRef] [PubMed]
48. Gambino, G.; Di Matteo, D.; Gribaudo, I. Elimination of *Grapevine fanleaf virus* from three *Vitis vinifera* cultivars by somatic embryogenesis. *Eur. J. Plant Pathol.* **2009**, *123*, 57–60. [CrossRef]
49. Nkaa, F.A.; Ene-Obong, E.E.; Taylor, N.; Fauquet, C.; Mbanaso, E.N.A. Elimination of African Cassava Mosaic Virus (ACMV) and East African Cassava Mosaic Virus (EACMV) from cassava (*Manihot esculenta* Crantz) cv. 'Nwugo' via somatic embryogenesis. *Am. J. Biotech. Molec. Sci.* **2013**, *3*, 33–40.
50. Chai, M.L.; Xu, C.-J.; Senthil, K.; Kim, J.Y. Stable transformation of protocorm-like bodies in *Phalaenopsis* orchid mediated by *Agrobacterium tumefasciens*. *Sci. Hort.* **2002**, *96*, 213–224. [CrossRef]
51. Mishiba, K.; Chin, D.P.; Mii, M. *Agrobacterium*-mediated transformation of *Phalaenopsis* by targeting protocorms at an early stage after germination. *Plant Cell Rep.* **2005**, *24*, 297–303. [CrossRef]
52. Huang, Y.W.; Tsai, Y.J.; Chen, F.C. Characterization and expression analysis of somatic embryogenesis receptor-like kinase genes from *Phalaenopsis*. *Genet. Mol. Res.* **2014**, *13*, 10690–10703. [CrossRef]
53. Hecht, V.; Vielle-Calzada, J.P.; Hartog, M.V.; Schmidt, E.D.; Boutilier, K.; Grossniklaus, U.; De Vries, S.C. The Arabidopsis *SOMATIC EMBRYOGENESIS RECEPTOR KINASE 1* gene is expressed in developing ovules and embryos and enhances embryogenic competence in culture. *Plant Physiol.* **2001**, *127*, 803–816. [CrossRef]

54. Chardin, C.; Girin, T.; Roudier, F.; Meyer, C.; Krapp, A. The plant RWP-RK transcription factors: Key regulators of nitrogen responses and of gametophyte development. *J. Exp. Bot.* **2014**, *65*, 5577–5587. [CrossRef]
55. Mursyanti, E.; Purwantoro, A.; Moeljopawiro, S.; Semiarti, E. Induction of somatic embryogenesis through overexpression of ATRKD4 genes in *Phalaenopsis* "Sogo Vivien". *Ind. J. Biotech.* **2015**, *20*, 42–53. [CrossRef]
56. Setiari, N.; Purwantoro, A.; Moeljopawiro, S.; Semiarti, E. Micropropagation of *Dendrobium phalaenopsis* orchid through overexpression of embryo gene *AtRKD4*. *Agriv. J. Agric. Sci.* **2018**, *40*, 284–294. [CrossRef]
57. Cai, J.; Liu, X.; Vanneste, K.; Proost, S.; Tsai, W.-C.; Liu, K.-W.; Chen, L.-J.; He, Y.; Xu, Q.; Bian, C.; et al. The genome sequence of the orchid *Phalaenopsis equestris*. *Nat. Genet.* **2015**, *47*, 65–72. [CrossRef] [PubMed]
58. Huang, J.-Z.; Lin, C.-P.; Cheng, T.-C.; Huang, Y.-W.; Tsai, Y.-J.; Cheng, S.-Y.; Chen, Y.-W.; Lee, C.-P.; Chung, W.-C.; Chang, B.C.-H.; et al. The genome and transcriptome of *Phalaenopsis* yield insights into floral organ development and flowering regulation. *PeerJ* **2016**, *4*, e2017. [CrossRef] [PubMed]
59. Gow, W.; Chen, J.; Chang, W. Effects of genotype, light regime, explant position, and orientation on direct somatic embryogenesis from leaf explants of *Phalaenopsis* orchid. *Acta Physiol. Plant* **2009**, *31*, 363–369. [CrossRef]
60. Mehraj, H.; Alam, M.M.; Habiba, S.U.; Mehbub, H. LEDs combined with CHO sources and CCC priming PLB regeneration of *Phalaenopsis*. *Horticulture* **2019**, *5*, 34. [CrossRef]
61. Teixeira da Silva, J.A.; Winarto, B. Somatic embryogenesis in two orchid genera (*Cymbidium, Dendrobium*). *Meth. Mol. Biol.* **2016**, *1359*, 371–386.
62. Reuter, E. The importance of propagating *Phalaenopsis* by tissue culture. *Orchid Rev.* **1983**, *91*, 199–201.
63. Meilasari, D.; Prayogo, I. Regeneration of plantlets through PLB (protocorm-like body) formation in *Phalaenopsis* 'Join Angle × Sogo Musadian'. *J. Math. Fund. Sci.* **2016**, *48*, 204–212. [CrossRef]
64. Chen, Y.-C.; Chang, C.; Chang, W.C. A reliable protocol for plant regeneration from callus culture of *Phalaenopsis*. *Vitr. Cell. Dev. Biol. Plant* **2000**, *36*, 420–423. [CrossRef]
65. Park, S.Y.; Murthy, H.N.; Paek, K.Y. Mass multiplication of protocorm-like bodies using bioreactor system and subsequent plant regeneration in *Phalaenopsis*. *Plant Cell Tissue Organ Cult.* **2000**, *63*, 67–72.
66. Tokuhara, K.; Mii, M. Induction of embryogenic callus and cell suspension culture from shoot tips excised from flower stalk buds in *Phalaenopsis* (Orchidaceae). *Vitr. Cell. Dev. Biol. Plant* **2001**, *37*, 457–461. [CrossRef]
67. Park, S.-Y.; Murthy, H.N.; Paek, K.Y. Rapid propagation of *Phalaenopsis* from flower stalk-derived leaves. *Vitr. Cell. Dev. Biol. Plant* **2002**, *38*, 168–172. [CrossRef]
68. Park, S.Y.; Yeung, E.C.; Chakrabarty, D.; Paek, K.Y. An efficient direct induction of protocorm-like bodies from leaf subepidermal cells of *Doritaenopsis* hybrid using thin-section culture. *Plant Cell Rep.* **2002**, *21*, 46–51.
69. Park, S.-Y.; Hosakatte, N.M.; Paek, K.Y. Protocorm-like body induction and subsequent plant regeneration from root tip cultures of *Doritaenopsis*. *Plant Sci.* **2003**, *164*, 919–923. [CrossRef]
70. Tokuhara, K.; Mii, M. Highly-efficient somatic embryogenesis from cell suspension cultures of *Phalaenopsis* orchids by adjusting carbohydrate sources. *Vitr. Cell. Dev. Biol. Plant* **2003**, *39*, 635. [CrossRef]
71. Kuo, H.; Chen, J.; Chang, W. Efficient plant regeneration through direct somatic embryogenesis from leaf explants of Phalaenopsis 'Little Steve'. *Vitr. Cell. Dev. Biol. Plant* **2005**, *41*, 453–456. [CrossRef]
72. Chen, J.T.; Chang, W.C. Direct somatic embryogenesis and plant regeneration from leaf explants of *Phalaenopsis amabilis*. *Biol. Plant* **2006**, *50*, 169–173. [CrossRef]
73. Murdad, R.; Hwa, K.S.; Seng, C.K.; Latip, M.A.; Aziz, Z.A.; Ripin, R. High frequency multiplication of *Phalaenopsis gigantea* using trimmed bases protocorms technique. *Sci. Hortic.* **2006**, *111*, 73–79. [CrossRef]
74. Minamiguchi, J.; Machado Neto, N.B. Embriogênese somática direta em folhas de *Phalaenopsis*: Orchidaceae. *Colloq. Agrar.* **2007**, *3*, 7–13. [CrossRef]
75. Sinha, P.; Hakim, M.; Alam, M. Efficient micropropagation of *Phalaenopsis amabilis* (L.) BL. cv. 'Cool Breeze' using inflorescence axis thin sections as explants. *Propag. Ornam. Plants* **2007**, *7*, 9–15.
76. Ling, A.C.K.; Yap, C.P.; Shaib, J.M.; Vilasini, P. Induction and morphogenesis of *Phalaenopsis* callus. *J. Trop. Agric. Food Sci.* **2007**, *35*, 147–152.
77. Gow, W.; Chen, J.; Chang, W. Influence of growth regulators on direct embryo formation from leaf explants of *Phalaenopsis* orchid. *Acta Physiol. Plant.* **2008**, *30*, 507–512. [CrossRef]
78. Gow, W.; Chen, J.; Chang, W. Enhancement of direct somatic embryogenesis and plantlet growth from leaf explants of *Phalaenopsis* by adjusting culture period and explant length. *Acta Physiol. Plant.* **2010**, *32*, 621–627. [CrossRef]

79. Chen, W.H.; Tang, C.Y.; Kao, Y.L. Ploidy doubling by in vitro culture of excised protocorms or protocorm-like bodies in *Phalaenopsis* species. *Plant Cell Tissue Organ Cult.* **2009**, *98*, 229–238. [CrossRef]
80. Subramaniam, S.; Balasubramaniam, V.R.M.T.; Poobathy, R.; Sasidharan, S. Chemotaxis Movement and Attachment of *Agrobacterium tumefaciens* to *Phalaenopsis violacea* Orchid Tissues an Assessment of Early Factors Influencing the Efficiency of Gene Transfer. *Trop. Life Sci. Res.* **2009**, *20*, 39–49.
81. Sinha, P.; Jahan, M.A.A.; Munshi, J.L.; Khatun, R. High frequency regeneration of *Phalaenopsis amabilis* (L.) Bl. cv. Lovely through in vitro culture. *Plant Tissue Cult. Biotech.* **2010**, *20*, 185–193. [CrossRef]
82. Khoddamzadeh, A.A.; Sinniah, U.R.; Kadir, M.A.; Kadzimin, S.B.; Mahmood, M.; Sreeramanan, S. Detection of somaclonal variation by random amplified polymorphic DNA analysis during micropropagation of *Phalaenopsis bellina* (Rchb.f.) Christenson. *Afr. J. Biotech.* **2010**, *9*, 6632–6639.
83. Khoddamzadeh, A.A.; Sinniah, U.R.; Kadir, M.A.; Kadzimin, S.B.; Mahmood, M.; Sreeramanan, S. In vitro induction and proliferation of protocorm-like bodies (PLBs) from leaf segments of *Phalaenopsis bellina* (Rchb.f.) Christenson. *Plant Grow. Reg.* **2011**, *65*, 381–387. [CrossRef]
84. Niknejad, A.; Kadir, M.A.; Kadzimin, S.B. In vitro plant regeneration from protocorms-like bodies (PLBs) and callus of *Phalaenopsis gigantea* (Epidendroideae: Orchidaceae). *Afr. J. Biotech.* **2001**, *10*, 11808–11816.
85. Sinha, P.; Jahan, M.A.A. Clonal propagation of *Phalaenopsis amabilis* (L.) BL. Cv. 'Golden Horizon' through in vitro culture of leaf segments. *BangladeshJ. Sci. Ind. Res.* **2011**, *46*, 163–168. [CrossRef]
86. Van Thanh, P.; Teixeira Da Silva, J.A.; Huy, H.E.; Tanaka, M. The effects of permanent magnetic fields on in vitro growth of *Phalaenopsis* plantlets. *J. Hortic. Sci. Biotech.* **2011**, *86*, 473–478. [CrossRef]
87. Rittirat, S.; Kongruk, S.; Te-Chato, S. Induction of protocorm-like bodies (PLBs) and plantlet regeneration from wounded protocorms of *Phalaenopsis cornu-cervi* (Breda) Blume & Rchb. f. *J. Agric. Tech.* **2012**, *8*, 2397–2407.
88. Samarfard, S.; Kadir, M.A.; Kadzimin, S.B.; Ravanfar, S.; Saud, H.M. Genetic stability of in vitro multiplied *Phalaenopsis gigantea* protocorm-like bodies as affected by chitosan. *Not. Bot. Horti Agrobot.* **2013**, *41*, 177–183. [CrossRef]
89. Antensari, F.; Mariani, T.S.; Wicaksono, A. Micropropagation of *Phalaenopsis* 'R11 × R10' Through Somatic Embryogenesis Method. *Asian J. Appl. Sci.* **2014**, *2*, 145–150.
90. Huang, Y.-W.; Tsai, Y.-J.; Cheng, T.-C.; Chen, J.-J.; Chen, F.C. Physical wounding and ethylene stimulated embryogenic stem cell proliferation and plantlet regeneration in protocorm-like bodies of *Phalaenopsis* orchids. *Genet. Mol. Res.* **2014**, *13*, 9543–9557. [CrossRef] [PubMed]
91. Rittirat, S.; Klaocheed, S.; Thammasiri, K. Enhanced efficiency for propagation of *Phalaenopsis cornu-cervi* (Breda) Blume & Rachb. F. using trimmed leaf technique. *Int. J. Agric. Biosyst. Eng.* **2014**, *8*, 336–339.
92. Samarfard, S.; Kadir, M.A.; Kadzimin, S.B.; Saud, H.M.; Ravanfar, S.A.; Danaee, M. In vitro propagation and detection of somaclonal variation in *Phalaenopsis gigantea* as affected by chitosan and thidiazuron combinations. *Hortscience* **2014**, *49*, 82–88. [CrossRef]
93. Soe, K.W.; Myint, K.T.; Naing, A.H.; Kim, C.K. Optimization of efficient protocorm-like body (PLB) formation of *Phalaenopsis* and *Dendrobium* hybrids. *Curr. Res. Agric. Life Sci.* **2014**, *32*, 179–183. [CrossRef]
94. Feng, J.; Chen, J. A novel in vitro protocol for inducing direct somatic embryogenesis in *Phalaenopsis aphrodite* without taking explants. *Sci. World J.* **2014**, *2014*, 263642. [CrossRef]
95. Balilashaki, K.; Vahedi, M.; Karimi, R. In vitro direct regeneration from node and leaf explants of *Phalaenopsis* cv. 'Surabaya'. *Plant Tissue Cult. Biotech.* **2015**, *25*, 193–205. [CrossRef]
96. Sultana, K.S.; Hasan, K.M.; Hasan, K.M.; Sultana, S.; Mehraj, H.; Ahasan, M.; Shimasaki, K.; Habiba, S.U. Effect of two elicitors on organogenesis in protocorm-like bodies (PLBs) of *Phalaenopsis* 'Fmk02010' cultured *in vitro*. *World Appl. Sci. J.* **2015**, *33*, 1528–1532.
97. Balilashaki, K.; Ghehsareh, M.G. Micropropagation of *Phalaenopsis amabilis* var. 'Manila' by leaves obtained from in vitro culturing the nodes of flower stalks. *Not. Sci. Biol.* **2016**, *8*, 164–169. [CrossRef]
98. Mose, W.; Indrianto, A.; Purwantoro, A.; Semiarti, E. The influence of Thidiazuron on direct somatic embryo formation from various types of explant in *Phalaenopsis amabilis* Blume Orchid. *Hayati J. Biosci.* **2017**, *24*, 201–205. [CrossRef]
99. Murashige, T.; Skoog, F. A revised medium for rapid growth and bio assays with tobacco tissue cultures. *Physiol. Plant.* **1962**, *15*, 473–497. [CrossRef]
100. Kano, K. Studies on the media for orchid seed germination. *Mem. Fac. Agri. Kagawa Univ.* **1965**, *20*, 1–68.

101. Ernst, R. Effects of thidiazuron on in vitro propagation of *Phalaenopsis* and *Doritaenopsis* (Orchidaceae). *Plant Cell Tissue Organ Cult.* **1994**, *39*, 273–275. [CrossRef]
102. Semiarti, E.; Indrianto, A.; Purwantoro, Y.H.; Martiwi, I.N.A.; Feroniasanti, Y.M.A.; Nadifah, F.; Mercuriana, I.S.; Dwiyani, R.; Iwakawa, H.; Yoshioka, Y.; et al. High-frequency genetic transformation of *Phalaenopsis amabilis* orchid using tomato extract-enriched medium for the pre-culture of protocorms. *J. Hortic. Sci. Biotech.* **2010**, *85*, 205–210. [CrossRef]
103. Chuanjun, X.; Zhiwei, R.; Ling, L.; Biyu, Z.; Junmei, H.; Wen, H.; Ou, H. The effects of polyphenol oxidase and cycloheximide on the early stage of Browning in *Phalaenopsis* explants. *Hortic. Plant J.* **2015**, *1*, 172–180.
104. Novak, S.D.; Luna, L.J.; Gamage, R.N. Role of auxin in orchid development. *Plant Sign. Behav.* **2014**, *9*, e972277. [CrossRef]
105. Bairu, M.W.; Aremu, A.O.; Van Staden, J. Somaclonal variation in plants: Causes and detection methods. *Plant Grow. Reg.* **2011**, *63*, 147–173. [CrossRef]
106. Raynalta, E.; Elina, J.; Sudarsono, S.; Sukma, D. Clonal fidelity of micropropagated *Phalaenopsis* plantlets based on assessment using eighteen Ph-Pto SNAP marker loci. *J. Agric. Sci.* **2018**, *40*, 390–402.
107. Park, S.I.; Yeung, E.C.; Paek, K.Y. Endoreduplication in *Phalaenopsis* is affected by light quality from light-emitting diodes during somatic embryogenesis. *Plant Biotec. Rep.* **2010**, *4*, 303–309. [CrossRef]
108. Young, P.S.; Murthy, H.N. Clonal fidelity of micropropagated *Phalaenopsis* plantlets based on assessment using eighteen Ph-Pto SNAP marker loci. Yoeup, P.K. Mass multiplication of protocorm-like bodies using bioreactor system Clonal fidelity of micropropagated *Phalaenopsis* plantlets based on assessment using eighteen Ph-Pto SNAP marker loci. and subsequent plant regeneration in *Phalaenopsis*. *Plant Cell Tissue Organ Cult.* **2000**, *63*, 67–72.
109. Liu, M.; Xu, Z.; Yang, Y.; Feng, Y. Effects of different spectral lights on *Oncidium* PLBs induction, proliferation, and plant regeneration. *Plant Cell Tissue Organ Cult.* **2011**, *106*, 1–10.
110. Wei, C.H. Optimization of PLB induction conditions for *Oncidium*. *Fuj. J. Agr. Sci.* **2007**, *22*, 332–335.
111. Li, W.-L.; Zhai, L.-S.; Ziu, Y.-P. Study on induction and culture of *Oncidium* protocorm-like body (PLB). *Hen. Sci.* **2004**, *3*.
112. Tran, M.V.; Nguyen, K.V.; Hoa, B.T. Rapid micropropagation of Vu Nu Orchid (*Oncidium* sp.) by using tissue culture technique. In Proceedings of the CBU International Conference, Prague, Czech Republic, 22–24 March 2017.
113. Yang, J.F.; Piao, X.C.; Sun, D.; Lian, M.L. Production of protocorm-like bodies with bioreactor and regeneration in vitro of *Oncidium* 'Sugar Sweet'. *Sci. Hortic.* **2010**, *125*, 712–717. [CrossRef]
114. Chen, J.T.; Chang, W.C. Effects of auxins and cytokinins on direct somatic embryogenesis from leaf explants of *Oncidium* 'Gower Ramsey'. *Plant Growth Regul.* **2001**, *34*, 229–232. [CrossRef]
115. Chen, J.-T.; Chang, C.; Chang, W.C. Direct somatic embryogenesis from leaf explants of *Oncidium* 'Gower Ramsey' and subsequent plant regeneration. *Plant Cell Rep.* **1999**, *19*, 143–149. [CrossRef]
116. Chen, J.-T.; Chang, W.C. Plant regeneration via embryo and shoot bud formation from flower-stalk explants of *Oncidium* Sweet Sugar. *Plant Cell. Tiss. Organ Cult.* **2000**, *62*, 95–100. [CrossRef]
117. Chen, J.; Chang, W. Efficient plant regeneration through somatic embryogenesis from callus cultures of *Oncidium* (Orchidaceae). *Plant Sci.* **2000**, *160*, 87–93. [CrossRef]
118. Chen, J.-T.; Chang, W.-C. Effects of GA_3, ancymidol, cycocel and paclobutrazol on direct somatic embryogenesis of *Oncidium* in vitro. *Plant Cell Tissue Organ Cult.* **2003**, *72*, 105–108. [CrossRef]
119. Kerbauy, G.B. Plant regeneration of *Oncidium varicosum* (Orchidaceae) by means of root tip culture. *Plant Cell Rep.* **1984**, *3*, 27–29. [CrossRef] [PubMed]
120. Flachsland, E.A.; Graciela-Mroginski, L.A. Regeneración de Protocormos y Yemas de *Oncidium bifolium* Sims. Por cultivo in vitro de láminas foliares. 2001. Available online: researchgate.net/publication/267300737_Regeneracion_de_protocormos_y_yemas_de_Oncidium_bifolium_Sims_por_cultivo_in_vitro_de_laminas_foliares (accessed on 30 January 2020).
121. Chen, J.T.; Chang, W.C. Effects of tissue culture conditions and explant characteristics on direct somatic embryogenesis in *Oncidium* 'Gower Ramsey'. *Plant Cell Tissue Organ Cult.* **2002**, *69*, 41–44. [CrossRef]
122. Rahman, S.M.M.; Islam, M.S.; Sen, P.K.; Begum, F. In vitro propagation of *Oncidium taka*. *Biotechnology* **2005**, *4*, 225–229.

123. Jheng, F.Y.; Do, Y.Y.; Liauh, Y.W.; Chung, J.P.; Huang, P.L. Enhancement of growth and regeneration efficiency from embryogenic callus cutures of *Oncidium* "Gower Ramsey" by adjusting carbohydrate sources. *Plant Sci.* **2006** *170*, 1133–1140. [CrossRef]
124. Hong, P.I.; Chen, J.T.; Chang, W.C. Promotion of direct somatic embryogenesis of *Oncidium* by adjusting carbon sources. *Biol. Plant.* **2008**, *52*, 597–600. [CrossRef]
125. Wang, A.-S.; Lin, M.-G.; Liu, F.-X. Rapid propagation of cut flower varieties of *Oncidium* by tissue culture. *Guangxi Agric. Sci.* **2009**, *40*, 801–806, (In Chinese with abstract in English).
126. Chung, J.P.; Huang, C.Y.; Dai, T.E. Spectral effects on embryogenesis and plantlet growth of *Oncidium* Gower Ramsey. *Sci. Hortic.* **2010**, *124*, 511–516. [CrossRef]
127. Mayer, J.L.S.; Stancato, G.C.; Appezzato-da-Glória, B. Direct regeneration of Protocorm-like bodies (PLBs) from leaf apices of *Oncidium flexuosum* Sims (Orchidaceae). *Plant Cell Tissue Organ Cult.* **2010**, *103*, 411–416. [CrossRef]
128. Chen, J.-T. Induction of petal-bearing embryos from root-derived callus of *Oncidium* 'Gower Ramsey'. *Acta Physiol. Plant* **2012**, *34*, 1337–1343. [CrossRef]
129. Gomes, L.R.P.; Franceschi, C.R.B.; Ribas, L.L.F. Micropropagation of *Brasilidium forbesii* (Orchidaceae) through transverse and longitudinal thin cell layer culture. *Acta Sci. Biol. Sci.* **2015**, *37*, 143–149. [CrossRef]
130. Mahesh, R.; Kumar, H.G.A.; Satyanarayana, S. Synthesis and characterization of 2-mercapto-N methyl imidazole substituted benzimidazole derivatives and investigation of their effect on production of plantlets in *Oncidium* Gower Ramsey. *Mater. Today Proc.* **2018**, *5*, 21505–21511.
131. Lloyd, G.; McCown, B. Commercially-feasible micropropagation of mountain laurel, *Kalmia latifolia*, by use of shoot-tip culture. *Int. Plant Propag. Soc. Proc.* **1980**, *30*, 421–427.
132. Chen, Y.H.; Chang, Y.S.; Chen, W.H. Tissue culture advances for mass propagation of *Oncidium* mericlones. *Rep. Taiwan Sugar Res. Inst.* **2001**, *173*, 67–76.
133. Wang, T.C.; Zhang, M.; Tong, Y.O. Molecular specstrum of somaclonal variation in PLB-regenerated *Oncidium* revealed by SLAF-seq. *Plant Cell Tissue Organ Cult.* **2019**, *137*, 541–552. [CrossRef]
134. Mohanty, P.; Paul, S.; Das, M.C.; Kumaria, S.; Tandon, P. A simple and efficient protocol for the mass propagation of *Cymbidium mastersii*: An ornamental orchid from Northeast India. *Aob Plants.* **2012**, *2012*, pls023. [CrossRef]
135. Teixeira da Silva, J.A.; Tanaka, M. Multiple regeneration pathways via Thin Cell Layers in hybrid *Cymbidium* (Orchidaceae). *J. Plant Growth Regul.* **2006**, *25*, 203. [CrossRef]
136. Malabadi, R.B.; Mulgund, G.S.; Kallappa, N. Micropropagation of *Dendrobium nobile* from shoot tip sections. *J. Plant Physiol.* **2005**, *162*, 473–478. [CrossRef]
137. Parthibhan, S.; Venkateswara Rao, M.; Teixeira da Silva, J.A.; Senthil Kumar, T. Somatic embryogenesis from stem thin cell layers of *Dendrobium aqueum*. *Biol. Plant.* **2018**, *62*, 439–450. [CrossRef]
138. Teixeira da Silva, J.A.; Jin, X.; Dobránszki, J.; Lu, J.; Wang, H.; Zotz, G.; Cardoso, J.C.; Zeng, S. Advances in *Dendrobium* molecular research: Applications in genetic variations, identification and breeding. *Mol. Phylogen. Evol.* **2016**, *95*, 196–216. [CrossRef]
139. Bhattacharyya, P.; Kumaria, S.; Tandon, P. High frequency regeneration protocol for *Dendrobium nobile*: A model tissue culture approach for propagation of medicinally important orchid species. *S. Afr. J. Bot.* **2016**, *104*, 232–243. [CrossRef]
140. Chien, K.W.; Agrawal, D.C.; Tsay, H.S.; Chang, C.A. Elimination of mixed '*Odontoglossum ringspot*' and '*Cymbidium mosaic*' *viruses* from *Phalaenopsis* hybrid 'V3' through shoot-tip culture and protocorm-like body selection. *Crop Protect.* **2015**, *67*, 1–6. [CrossRef]
141. Chen, F.C. *Phalaenopsis in Vitro Cloning: Strategy for PLB or Shoots?* Taiwan International Orchid Symposium: Taipei, Taiwan, 2009.
142. Miguel, T.P.; Leonhardt, K.W. In vitro polyploid induction of orchids using oryzalin. *Sci. Hortic.* **2011**, *130*, 314–319. [CrossRef]
143. Sarathum, S.; Hegele, M.; Tantiviwat, S.; Nanakorn, M. Effect of concentration and duration of colchicine treatment on polyploid induction in *Dendrobium scabrilingue* L. *Eur. J. Hort. Sci.* **2010**, *75*, 123 127.
144. Wannajindaporn, A.; Kativat, C.; Tantasawat, P.A. Mutation induction of *Dendrobium* 'Earsakul' using sodium azide. *Hortscience* **2016**, *51*, 1363–1370. [CrossRef]

145. Billore, V.; Mirajkar, S.J.; Suprasanna, P.; Jain, M. Gamma irradiation induced effects on in vitro shoot cultures and influence of monochromatic light regimes on irradiated shoot cultures of *Dendrobium sonia* orchid. *Biotech. Rep.* **2019**, *22*, e00343. [CrossRef]
146. Chew, Y.-C.; Abdullah, W.; Kok, D.A.; Ong-Abdullah, J.; Mahmood, M.; Lai, K.-S. Development of an efficient particle bombardment transformation system for de endemic orchid *Phalaenopsis bellina*. *Sains Malays.* **2019**, *48*, 1867–1877. [CrossRef]
147. Mii, M.; Chin, D.P. Genetic transformation of orchid species: An overview of approaches and methodologies. In *Orchid Propagation: From Laboratories to Greenhouses – Methods and Protocol*; Lee, Y.I., Yeung, E.T., Eds.; Humana Press: New York, NY, USA, 2018; pp. 347–365.
148. Liau, C.-H.; You, S.-J.; Prasad, V.; Hsiao, H.-H.; Lu, J.-C.; Yang, N.-S.; Chan, M.-T. *Agrobacterium tumefasciens*-mediated transformation of *Oncidium* orchid. *Plant Cell Rep.* **2003**, *21*, 993–998. [CrossRef]
149. Thiruvengadam, M.; Hsu, W.-H.; Yang, C.-H. Phosphomannose-isomerase as a selectable marker to recover transgenic orchid plants (*Oncidium* Gower Ramsey). *Plant Cell Tissue Organ Cult.* **2011**, *104*, 239–246. [CrossRef]
150. Morel, G.M. Producing virus free Cymbidiums. *Am. Orchid Soc. Bull.* **1960**, *29*, 495–497.
151. Pradhan, S.; Regmi, T.; Ranjit, M.; Pant, B. Production of virus-free orchid *Cymbidium aloifolium* (L.) Sw. by various tissue culture techniques. *Heliyon* **2016**, *2*, e00176. [CrossRef] [PubMed]
152. Khentry, Y.; Paradornuwat, A.; Tantiwiwat, S.; Phansiri, S.; Thaveechail, N. Incidence of *Cymbidium mosaic virus* and *Odontoglossum Ringspot Virus* on in vitro Thai native orchid seedlings and cultivated orchid Mericlones. *Kasertsat J. Nat. Sci.* **2006**, *40*, 49–57.
153. Shen, R.-S.; Hsu, S.-T. Virus elimination through meristem culture and rapid clonal propagation using a temporary immersion system. In *Orchid Propagation: From Laboratories to Greenhouses–Methods and Protocols*; Lee, Y.I., Yeung, E.T., Eds.; Humana Press: New York, NY, USA, 2018; pp. 267–282.
154. Saiprasad, G.V.S.; Polisetty, R. Propagation of three orchid genera using encapsulated protocorm-like bodies. *Vitr. Cell. Dev. Biol. Plant* **2003**, *39*, 42–48. [CrossRef]
155. Wang, H.-Q.; Jin, M.-Y.; Paek, K.-Y.; Piao, X.-C.; Lian, M.-L. An efficient strategy for enhancement of bioactive compounds by protocorm-like body culture of *Dendrobium candidum*. *Ind. Crop. Prod.* **2016**, *84*, 121–130. [CrossRef]
156. Paek, K.Y.; Hahn, E.J.; Park, S.Y. Micropropagation of *Phalaenopsis* orchids via protocorms and protocorm-like bodies. *Methods Mol. Biol.* **2011**, *710*, 293–306. [PubMed]
157. Hsu, C.-C.; Lai, P.-H.; Chen, T.-C.; Tsai, W.-C.; Hsu, J.-L.; Hsiao, Y.-Y.; Wu, W.-L.; Tsai, C.-H.; Chen, W.-H.; Chen, H.-H. PePIF1, a P-lineage of PIF-like transposable element identified in protocorm-like bodies of Phalaenopsis orchids. *Bmc Genom.* **2019**, *20*, 25. [CrossRef] [PubMed]
158. Chin, C.K.; Lee, Z.H.; Mubbarakh, S.A.; Antony, J.J.J.; Chew, B.L.; Subramanian, S. Effects of plant growth regulators and activated charcoal on somaclonal variations of protocorm-like bodies (PLBs) of *Dendrobium* Sabin Blue orchid. *Biocatal. Agric. Biotech.* **2019**, *22*, 101426. [CrossRef]
159. Chen, W.H.; Chen, T.M.; Fu, Y.M.; Hsieh, R.M. Studies on somaclonal variation in *Phalaenopsis*. *Plant Cell Rep.* **1998**, *18*, 7–13. [CrossRef]
160. Bairu, M.W.; Stirk, W.A.; Dolezal, K.; van Staden, J. The role of topolins in micropropagation and somaclonal variation of banana cultivars 'Williams' and 'Grand Naine' (*Musa* spp. AAA). *Plant Cell Tissue Organ Cult.* **2008**, *95*, 373–379. [CrossRef]
161. Tokuhara, K.; Mii, M. Somaclonal variation in flower and inflorescence axis in micropropagated plants through flower stalk bud culture of *Phalaenopsis* and *Doritaenopsis*. *Plant Biotech.* **1998**, *15*, 23–28. [CrossRef]
162. Chen, Y.H.; Tsai, Y.J.; Huang, J.Z.; Chen, F.C. Transcription analysis of peloric mutants of *Phalaenopsis* orchids derived from tissue culture. *Cell Res.* **2005**, *15*, 639–657. [CrossRef] [PubMed]
163. Antony, J.J.J.; Shanshir, R.A.; Poobathy, R.; Chew, B.L.; Subramanian, S. Somaclonal variations were not induced by cryopreservation: Levels of somaclonal variation of in vitro and thawed protocorms of *Dendrobium* Bobby Messina analysed by SCoT and TRAP DNA markers. *South Afr. J. Bot.* **2015**, *100*, 148–157. [CrossRef]
164. Chen, F.-C.; Yu, J.-Y.; Chen, P.-Y.; Huang, Y.-W. Somaclonal variation in orchids and the application of biotechnology. *Acta Hortic.* **2008**, *766*, 315–322. [CrossRef]

12

Combined Metabolome and Transcriptome Analyses Reveal the Effects of Mycorrhizal Fungus *Ceratobasidium* sp. AR2 on the Flavonoid Accumulation in *Anoectochilus roxburghii* during Different Growth Stages

Ying Zhang, Yuanyuan Li, Xiaomei Chen, Zhixia Meng * and Shunxing Guo *

Institute of Medicinal Plant Development, Chinese Academy of Medical Sciences & Peking Union Medical College, Beijing 100193, China; zhangying908@163.com (Y.Z.); liyuanyuan184@163.com (Y.L.); cxm_implad@163.com (X.C.)

* Correspondence: mengzhixianj@163.com (Z.M.); sxguo1986@163.com (S.G.)

Abstract: *Anoectochilus roxburghii* is a traditional Chinese herb with high medicinal value, with main bioactive constituents which are flavonoids. It commonly associates with mycorrhizal fungi for its growth and development. Moreover, mycorrhizal fungi can induce changes in the internal metabolism of host plants. However, its role in the flavonoid accumulation in *A. roxburghii* at different growth stages is not well studied. In this study, combined metabolome and transcriptome analyses were performed to investigate the metabolic and transcriptional profiling in mycorrhizal *A. roxburghii* (M) and non-mycorrhizal *A. roxburghii* (NM) growth for six months. An association analysis revealed that flavonoid biosynthetic pathway presented significant differences between the M and NM. Additionally, the structural genes related to flavonoid synthesis and different flavonoid metabolites in both groups over a period of six months were validated using quantitative real-time polymerase chain reaction (qRT-PCR) and high-performance liquid chromatography coupled with tandem mass spectrometry (HPLC-MS/MS). The results showed that *Ceratobasidium* sp. AR2 could increase the accumulation of five flavonol-glycosides (i.e., narcissin, rutin, isorhamnetin-3-O-beta-D-glucoside, quercetin-7-O-glucoside, and kaempferol-3-O-glucoside), two flavonols (i.e., quercetin and isorhamnetin), and two flavones (i.e., nobiletin and tangeretin) to some degrees. The qRT-PCR showed that the flavonoid biosynthetic genes (*PAL*, *4CL*, *CHS*, *GT*, and *RT*) were significantly differentially expressed between the M and NM. Overall, our findings indicate that AR2 induces flavonoid metabolism in *A. roxburghii* during different growth stages, especially in the third month. This shows great potential of *Ceratobasidium* sp. AR2 for the quality improvement of *A. roxburghii*.

Keywords: *Anoectochilus roxburghii*; *Ceratobasidium* sp.; metabolome and transcriptome analyses; flavonoid; HPLC-MS/MS; qRT-PCR

1. Introduction

Anoectochilus roxburghii (Wall.) Lindl., also called "Jin Xianlian" and "Jin Xianlan" is a perennial herb of the genus *Anoectochilus* of the Orchidaceae family. It is widespread in southern China and considered a famous drug in the provinces of Jiangxi, Taiwan, Guangdong, Guizhou, Zhejiang, and Fujian which is its main markets [1]. The herb is a valuable Chinese medicinal material that is known as the "King Medicine", the "Golden Grass", and the "Bird Ginseng" by countryfolk [2]. The whole plant is used as a medicine and has the efficacy to clear the heat, cool the blood, eliminate dampness, and detoxification. Many pharmacological studies have demonstrated its wide range of pharmacological effects including its antioxidant [3,4], hepatoprotective [3], anti-diabetic [5,6], anti-rheumatoid arthritis [7], anti-hyperglycemic [4,8], anti-inflammatory [9] and many other properties. Due to the fact of its high medical values, *A. roxburghii* is not only widely used in medicine and health care but also in beauty and drinking products with an increasing market demand. The average annual demand in South Korea and Japan is more than 1000 t, 70% of which depend on imports [10]. The vigorous market demand results in excessive harvesting and exploitation of the plant, leading to resource depletion. Thus, tissue culture has become the main source of commercial *A. roxburghii*.

Most orchids form mycorrhizae with mycorrhizal fungi [11]. The fungal hyphae form coiled structures termed "peloton" inside the cells of the plant roots which play a key role in the nutrients' exchange and absorption between the orchid and its mycorrhizal fungi [12]. The process is unique and complex involving various processes related to growth and development, such as colonization, increasing the survival rate and morphological growth. Moreover, the mycorrhizal fungal elicitor can rapidly induce the expression of relative genes that are related to secondary metabolic pathways which result in a significant accumulation of active ingredients in the host plant [13]; this was demonstrated in orchid by a few studies. The symbiosis between *Dendrobium nobile* and *Mycena* sp. MF23 caused the accumulation of dendrobine and polysaccharide [14,15]. *Mycena* sp. MF23 could also stimulate the accumulation of flavonoid and kinsenoside in *A. formosanus* [16]. For *A. roxburghii*, the major bioactive components include polysaccharides, kinsenosides, steroids, triterpenes, amino acids, alkaloids, and flavonoids that have been regarded as the quality standard of *A. roxburghii* [10]. However, *A. roxburghii* contains a very limited number of flavonoids, which limits the development and utilization of *A. roxburghii*-based medicines. In order to improve the content of flavonoid in *A. roxburghii* and to avoid excessive exploitation, many methods have recently been put forward by researchers [17–19]; the role of mycorrhizal fungi in the accumulation of flavonoid in *A. roxburghii* has been gradually recognized. Wang et al. reported that flavonoid accumulated significantly in *A. roxburghii* growth for 8 weeks treated with different fungi such as *Rhizoctonia* sp. cw-6 and cw-13, *Exophila pisciphila* (cw-8), *Nemania* sp. (cw-10), and *Umbelopsis* sp. (cw-1) [20]. However, few studies investigated the effects of mycorrhizal fungi on flavonoid accumulation in *A. roxburghii* during different growth stages.

The flavonoid biosynthetic pathway has been well characterized in some medicinal plants such as *Gnetum parvifolium* [21], *Chrysanthemum morifolium* [22], *Lotus japonicus* [23]. Its biosynthesis can be divided into two stages: phenylpropanoid and flavonoid pathways. Phenylalanine ammonia-lyase (PAL) is the first enzyme of the phenylpropanoid pathway which can convert phenylalanine into cinnamic acid [24]. Cinnamic acid is then converted into p-coumaric acid by trans-cinnamate 4-hydroxylase (C4H). Next, 4-coumarate CoA ligase (4CL) converts coumaric acid into its CoA ester. 4CL is one of the key branch point enzymes in the phenylpropanoid pathway and its products are subsequently used by various oxygenases, reductases, and transferases for the biosynthesis of lignin, flavonoids, anthocyanins, aurones, stilbenes, coumarins, suberin, cutin, and sporopollenin [25]. Chalcone synthase (CHS) and chalcone isomerase (CHI) are involved in two step condensation reaction, producing naringenin chalcone and naringenin, respectively. Then, flavanone is catalyzed by flavonoid 3′-hydroxylase (F3′H) and other enzymes. Subsequently, flavanone produces the branches of flavone and dihydroflavonol under the catalysis of flavone synthase (FNS) and flavanone 3-hydroxylase (F3H), respectively. Next, flavonol synthase (FLS) catalyzes C-3 hydroxylation in the structure of dihydroflavonols to form various flavonols, and flavonol-glycosides are formed by flavonoid 3-O-glucosyltransferase (GT) and rhamnosyltransferase (RT) or GT.

With the rapid development of high-throughput sequencing technology and systems biology, multi-omics technology has become an indispensable research method in the field of life science [26,27]. It can provide the dynamic changes of the plant's growth and development from the cell to the individual level. The metabolome is a powerful approach to qualitatively and quantitatively analysis all the small-molecule metabolites (mass $\leq$ 1000 Da) in the cells or tissues of an organism at any physiological period using different analysis technologies including nuclear magnetic resonance (NMR) spectroscopy, liquid chromatograph-mass spectrometer (LC-MS), and gas chromatography-mass spectrometer (GC-MS) [28]. In other words, this method could provide the global metabolic changes. Similarly, the transcriptome means the detection of all RNA transcripts in a sample and reflects gene expression differences between different treatments [29]. Integrated transcriptome and metabolome analyses have been successfully applied to study the metabolic pathways of some substances [30,31], the color formation of vegetables, fruits, and flowers [32,33], the stress resistance mechanisms [34,35], and the growth and development mechanisms of the plants [36,37]. The combination can not only elucidate changes in the content of a series of metabolites, but it can also analyze the corresponding differentially expressed genes.

To investigate the changes of metabolites in *A. roxburghii* that is infected with mycorrhizal fungus, we performed metabolome and transcriptome analyses on six-month growth data of the mycorrhizal *A. roxburghii* (M) and non-mycorrhizal ones (NM). The results indicated that AR2 significantly promoted flavonoid biosynthesis in the plant. During the growth stage, the flavonoid content (two flavones: nobiletin and tangeretin; two flavonols: quercetin and isorhamnetin; five flavonol-glycosides: narcissin, rutin, isorhamnetin-3-O-beta-D-glucoside, quercetin 7-o-glucoside, and kaempferol 3-O-glucoside) in different metabolites and the expression of the genes that were related to the biosynthesis of flavonoid were further tested over a period of six months using high-performance liquid chromatography coupled with tandem mass spectrometry (HPLC-MS/MS) then quantitative real-time polymerase chain reaction (qRT-PCR). These will provide valuable information to reveal the effects of *Ceratobasidium* sp. AR2 on the flavonoid accumulation in *A. roxburghii*.

2. Results

2.1. Detecting Mycorrhizal Fungus Colonization in A. roxburghii

Pelotons (hyphae coils) formation is the important characteristic of orchid mycorrhizal association [38]. In the study, *Ceratobasidium* sp. AR2 would soon grow over the whole substrate for about two weeks. At this time, both morphological observation and toluidine blue staining, which showed no hyphae in the control group (Figure 1A) and that the intracellular pelotons existed in the treatment (Figure 1B), indicated that a symbiotic relationship had been established between *A. roxburghii* and mycorrhizal fungus *Ceratobasidium* sp. AR2.

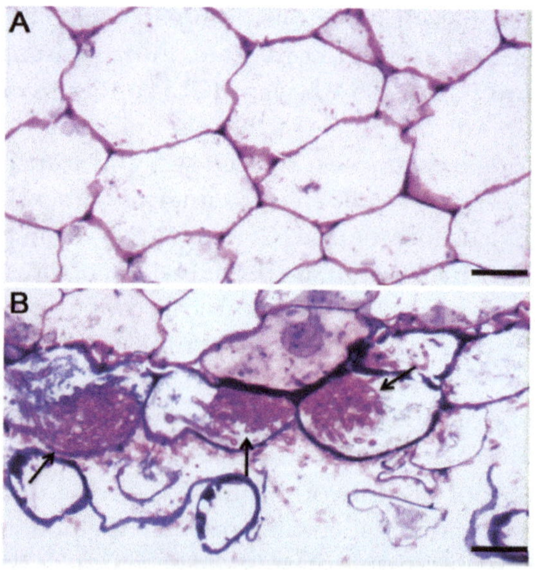

Figure 1. The semi-thin sections of root of *A. roxburghii* after two weeks of symbiotic cultivation. (**A**) the control, scale bar = 50 mm; (**B**) the treatment; arrows represent pelotons, scale bar = 50 mm.

2.2. Identification of Metabolites

The M and NM growth for six months were used for metabolome analysis. Overlapping analysis of the total ion chromatography (TIC) in different quality control (QC) samples showed that the retention time and peak intensities were consistent (Figure S1), which indicated that the instrument had a good stability and could thus be used for subsequent analysis. Principal component analysis (PCA) was performed on the NM, M, and mixed samples. Principal component 1 (PC1) and principal component 2 (PC2) were 54.7% and 26.3%, respectively (Figure 2A). The metabolite profiles of *A. roxburghii* were then subjected to orthogonal partial least squares discriminant analysis (OPLS-DA). The result showed that the R2X, R2Y, and Q2 were 0.862, 1.000, and 0.995, respectively (Figure 2B), which indicated the model of OPLS-DA was stable and reliable. The score plots of PCA and OPLS-DA exhibited an obvious separation between the M and NM, and each formed a cluster. These suggested that mycorrhizal fungus AR2 affected the metabolism in *A. roxburghii*.

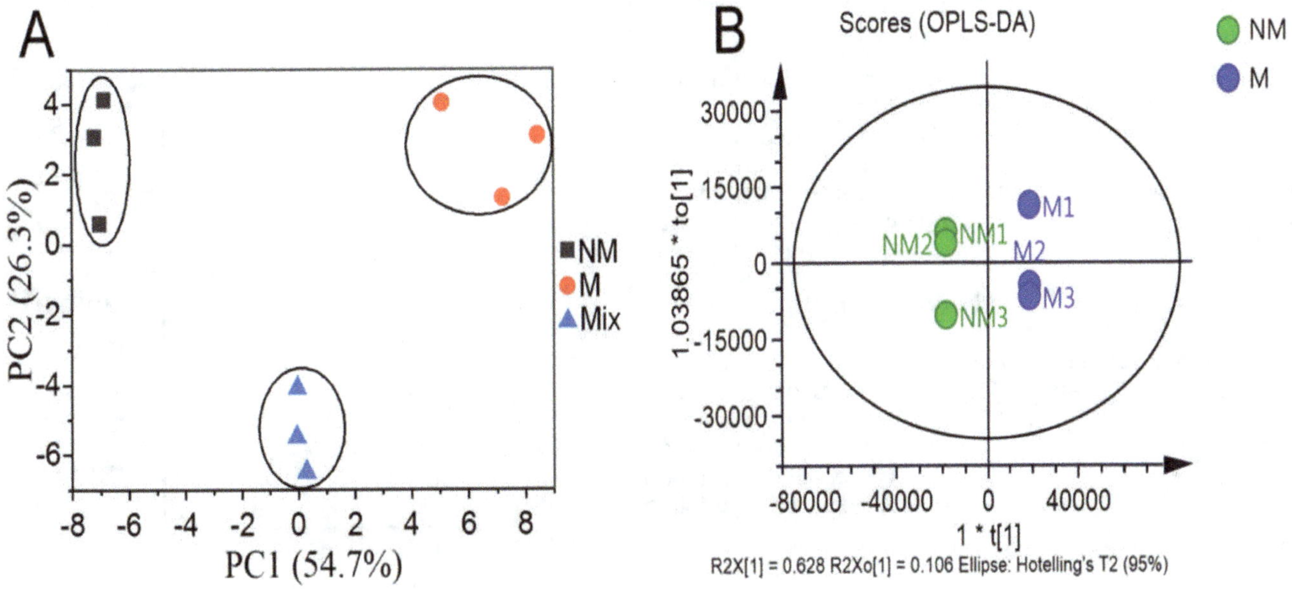

Figure 2. PCA and OPLS-DA scores plots derived from ultra-performance liquid chromatography-electrospray ionization-tandem mass spectrometry (UPLC-ESI-MS/MS) profiling of non-mycorrhizal *A. roxburghii* (NM) and mycorrhizal *A. roxburghii* (M) growth for six months. (**A**) PCA scores plot of the two samples (NM and M growth for six months) and the quality control sample (mix, the same volume of sample extract from the NM and M growth for six months was prepared by mixing); the x-axis represents the PC1 and the y-axis represents PC2. (**B**) OPLS-DA scores plot of the putatively annotated metabolites from NM and M growth for six months. The x-axis represents the score value of main components in the orthogonal signal correction process and the differences between the groups can be seen from the direction of the x-axis; the y-axis represents the scores of orthogonal components in the orthogonal signal correction process and the differences within the groups can be seen from the direction of the y-axis.

A total of 709 metabolites with known structures were identified in M and NM under quality validation, each of which was analyzed using three biological replicates. Detailed information about the identified metabolites, including the compounds, classes, molecular weights, ionization models, Kyoto encyclopedia of genes and genomes (KEGG) pathways, and quantities for each of the three periods is shown in Table S1. Flavonoid (20.9%), organic acids, and derivatives (15.4%), amino acids and derivatives (12.8%), lipid (9.6%), and phenylpropanoid (8.7%) accounted for a large proportion of these 709 metabolites (Figure 3A).

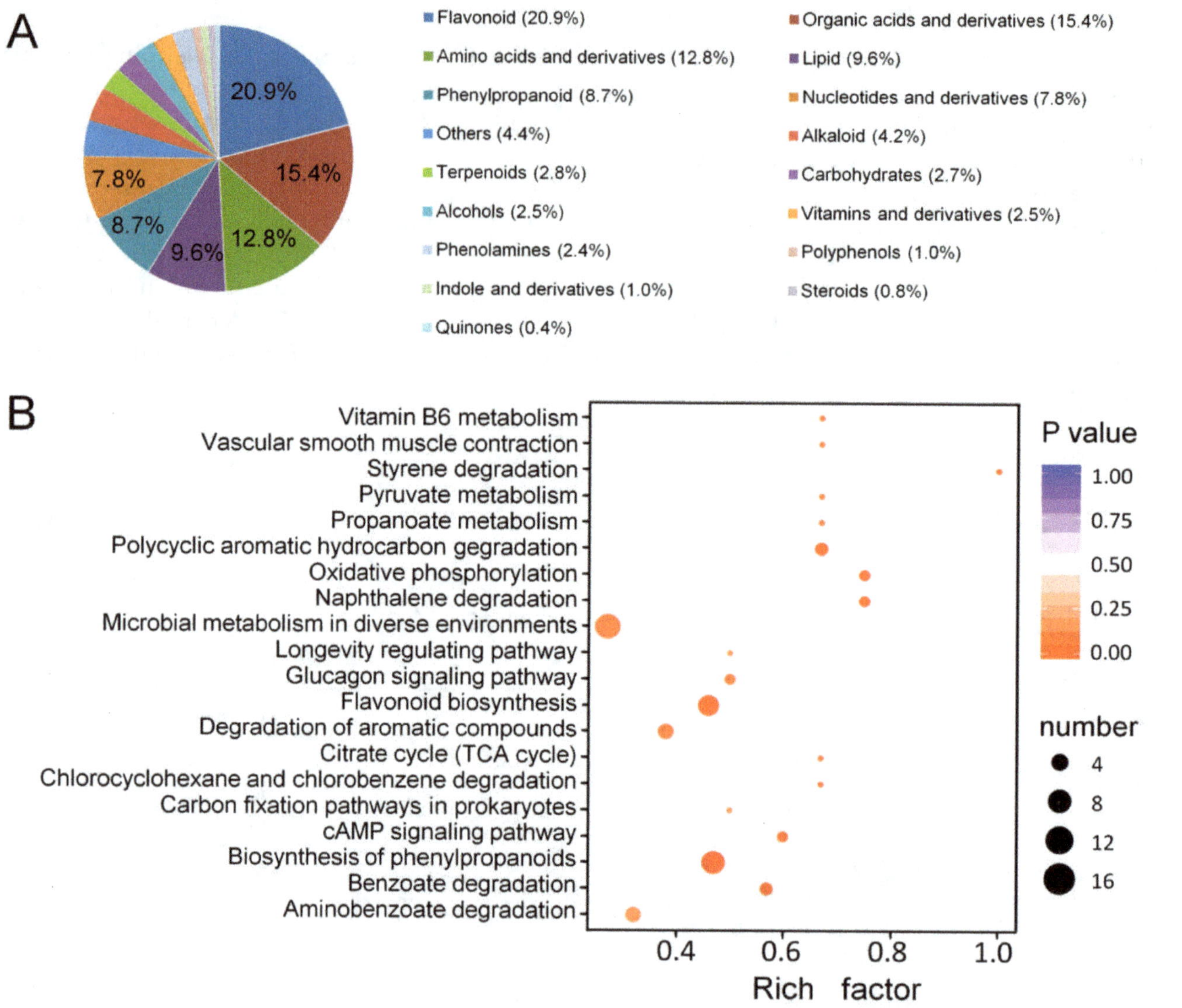

Figure 3. Component analysis of the putatively annotated metabolites and pathway enrichment analysis of the DAMs. (**A**) Component analysis of the putatively annotated metabolites from non- mycorrhizal *A. roxburghii* (NM) and mycorrhizal *A. roxburghii* (M) growth for six months. The percentage of the top six metabolites are shown in the graph. The percentage after each compound represents the percentage of the number of DAMs of a certain class of compounds in the total DAMs. (**B**) Pathway (top 20) enrichment analysis of the DAMs between the NM and M growth for six months. The x-axis represents the corresponding rich factor of each pathway. The y-axis represents the name of pathway. The color of the dot is *p*-value, and the closer it is to 0, the more significant the enrichment is. The size of the point represents the number of DAMs enriched in the corresponding pathway. The rich factor is the ratio of the number of metabolites in the corresponding pathway to the total number of metabolites detected and annotated in the pathway. The higher the value of rich factor is, the higher the enrichment degree is.

2.3. Identification of Differentially Accumulated Metabolites and Differentially Accumulated Flavonoids

Differentially accumulated metabolites (DAMs) were defined as those exhibiting a fold change ≥ 2 or ≤ 0.5 and a variable importance of projection (VIP) ≥ 1. In total, 135 DAMs were identified, among which 63 and 72 metabolites were upregulated and downregulated, respectively. To further understand the DAMs function and the related biological processes they participated in, pathway enrichment analysis of the DAMs was conducted using KEGG. The results showed that the terms "Biosynthesis of phenylpropanoids", "Flavonoid biosynthesis", "Polycyclic aromatic hydrocarbon degradation", "Oxidative phosphorylation", and "Naphthalene degradation" were significantly enriched (Figure 3B). Among them, the flavonoid biosynthesis began with the biosynthesis of phenylpropanoids. To some extent, these data further indicated that the flavonoid metabolism in *A. roxburghii* was significantly affected by mycorrhizal fungus AR2.

Furthermore, 148 metabolites involved in the flavonoid metabolism were identified, among which nine metabolites that belonged to two flavones (i.e., nobiletin and tangeretin), two flavonols (i.e., quercetin and isorhamnetin) and five flavonol-glycosides (i.e., narcissin, rutin, isorhamnetin-3-O-beta-D-glucoside, quercetin-7-O-glucoside, and kaempferol-3-O-glucoside) were obviously different between M and NM.

2.4. *Transcriptome Profiles of Mycorrhizal and Non-Mycorrhizal A. roxburghii*

The expression profiles of M and NM grown for six months were also analyzed using RNA-sequencing (RNA-seq). The Q20 and Q30 base percentages were greater than or equal to 97.79% and 93.61%, respectively (Table 1). The GC contents of the M and NM were in the range of 48.10–49.06% and 47.93–48.43%, respectively (Table 1). Compared with the control group, 4341 genes were shown to be differentially expressed, including 2915 upregulated genes and 1426 downregulated genes.

Table 1. Summary of the analysis of transcriptome sequences from the NM and M growth for six months.

Sample	Raw Reads	Clean Reads	Clean Base (G)	Error Rate (%)	Q20 (%)	Q30 (%)	GC Content (%)
NM1	61,226,728	61,071,914	9.09	0.017	97.79	93.61	48.19
NM2	60,542,772	60,425,910	8.99	0.016	98.11	94.42	47.93
NM3	67,559,786	67,410,292	10.05	0.016	98.07	94.28	48.43
M1	55,632,192	55,492,010	8.28	0.016	98.02	94.20	49.06
M2	65,007,376	64,859,884	9.68	0.016	98.19	94.58	48.10
M3	67,125,158	66,965,132	9.99	0.016	98.09	94.34	48.23

NM: represents non-mycorrhizal *A. roxburghii*; M: represents mycorrhizal *A. roxburghii*.

Genes with $|\log_2(\text{fold change})| > 1$ and $q < 0.001$ were defined as differentially expressed genes (DEGs). To further understand the functions of DEGs and the related biological processes they have a role in, gene ontology (GO) and KEGG analyses were conducted. The GO analysis classified DEGs into three categories: "molecular function", "cellular component", and "biological process" with a total of 45 GO terms (Figure 4A). The enriched GO terms were "binding" and "catalytic activity" within molecular function, "cell part", "cell", and "organelle" within cellular component and "cellular process", "metabolic process", and "response to stimulus" within biological process. The pathway enrichment analysis of the DEGs using KEGG identified significantly enriched "metabolic pathways" and "biosynthesis of secondary metabolites" (Figure 4B). As a result, the transcription analysis also showed that mycorrhizal fungus AR2 significantly affected the metabolic pathways in *A. roxburghii*.

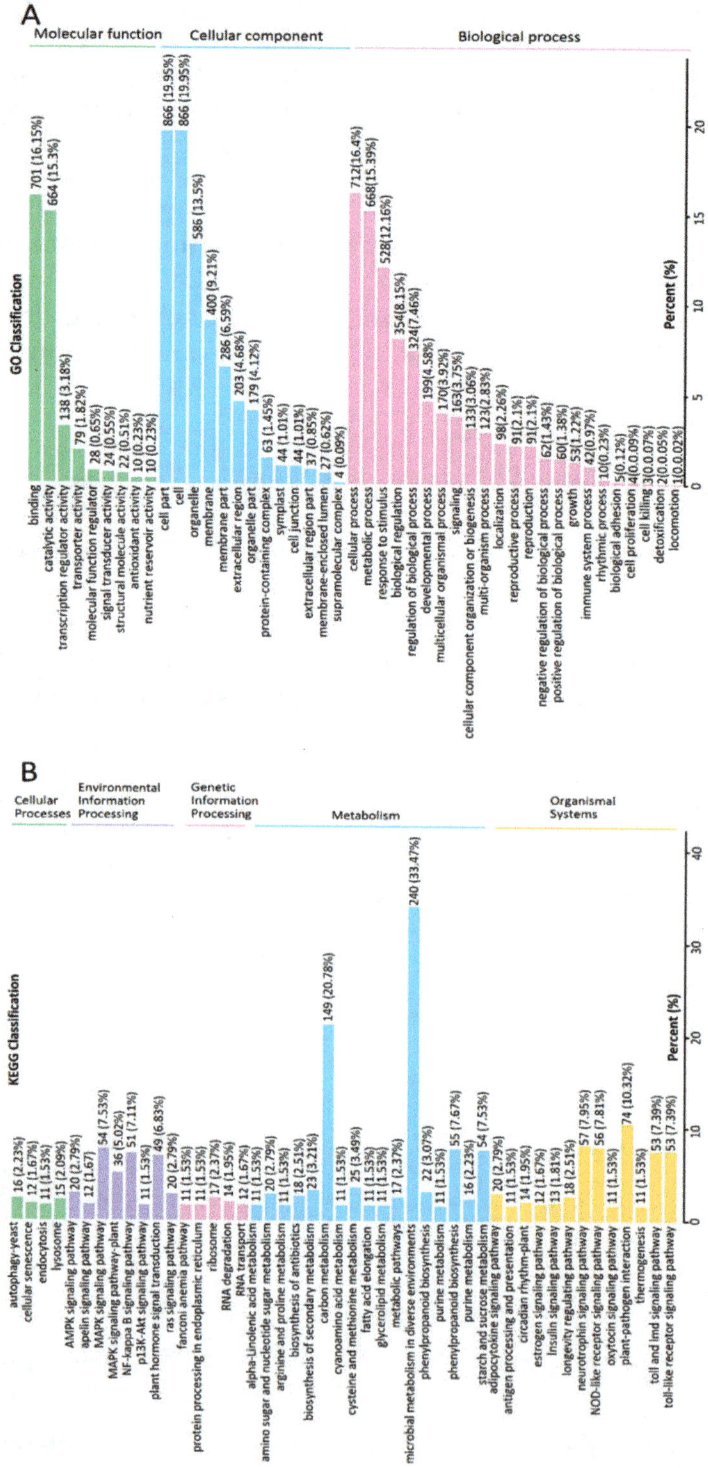

Figure 4. The classification column of GO and KEGG from the DEGs between the NM and M growth for six months. (**A**) GO classification of DEGs. The x-axis represents the secondary GO item. The y-axis represents the proportion of the DEGs in the total number of DEGs. The labels above the columns is the number and proportion of DEGs of this GO item. (**B**) KEGG classification of DEGs. The x-axis represents the name of KEGG pathway. The y-axis represents the proportion of genes annotated to the pathway in the total of annotated genes. The labels above the columns represent the classification of KEGG pathway.

2.5. *Association Analysis of the DAMs and DEGs*

Correlation analysis was carried out on the DAMs and DEGs. The variations in the metabolites and their corresponding genes with the Pearson correlation coefficient over 0.8 were selected to draw nine quadrant diagrams and the correlation coefficient cluster heat map. As shown in Figure 5, the higher number of DAMs and DEGs was in the seventh and ninth quadrants; and they were positively correlated in the seventh quadrant and negatively correlated in the ninth quadrant.

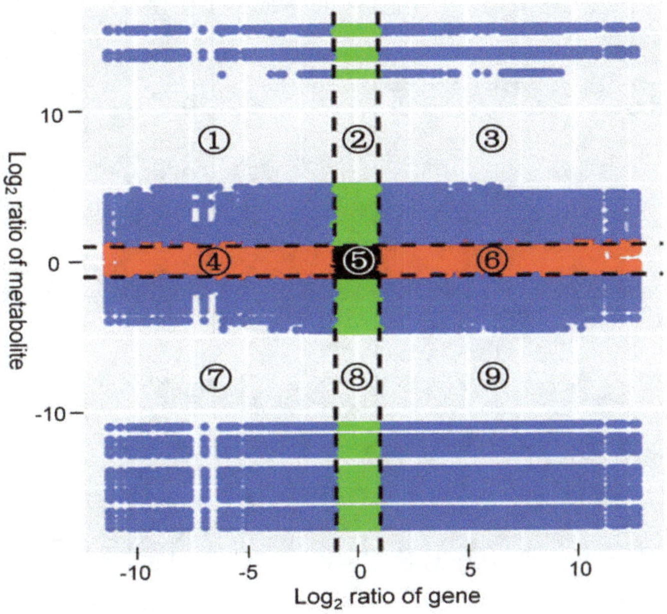

Figure 5. Quadrant diagrams representing the association of the DAMs and DEGs between the non-mycorrhizal and mycorrhizal *A. roxburghii* growth for six months. The x-axis represents that the $\log_2$ ratio of gene and the y-axis represents the $\log_2$ ratio of metabolite; black dotted lines represent the different threshold; each point represents a gene or metabolite; black dots represent the unchanged genes or metabolites; green dots represent differentially accumulated metabolites with unchanged genes; red dots represent differentially expressed genes with unchanged metabolites; blue dots represent both differentially expressed genes and differentially accumulated metabolites; it is divided into ①–⑨ quadrants from left to right and from top to bottom with black dotted lines; the ①, ② and ④ quadrants indicate that the expression abundance of metabolites is higher than that of genes; the ③ and ⑦ quadrants indicate that the expression patterns of genes are consistent with the metabolites; the ⑤ quadrant indicates that the genes and metabolites are not differentially expressed; the ⑥, ⑧ and ⑨ quadrants indicate that the expression abundance of metabolites is lower than that of genes.

2.6. *Dynamic Variations in Flavonoid DAMs*

In order to measure the dynamic variations of nine flavonoid DAMs in M and NM at different growth stages, the extracts were detected using HPLC-MS/MS. The optimized parameters for each analyte are shown in Table 2. Figure 6 showed the TIC diagrams of nine standards in the positive and negative ion mode. The method validation including linearity, limit of detection (LOD), limit of quantification (LOQ), stability, precision, and repeatability is shown in Table 3 with an R^2 higher than 0.9901, LOD ranging from 0.488 ng/mL to 15.63 ng/mL, LOQ ranging from 0.977 ng/mL to 31.25 ng/mL, the relative standard deviations (RSDs) of the stability ranging from 1.94% to 4.22%, the RSDs of the precision ranging from 1.43% to 4.49%, and the RSDs of the repeatability in the range of 1.74% to 4.93%. Additionally, the recovery rate was determined by adding 50%, 100% and 150% of the standard content of the sample. The average recovery rate of nine standards was from 86.03% to 99.54% with RSD between 1.09% and 4.55% (Table S2). These data indicated the reliability of the method.

Table 2. The optimized parameters for nine flavonoid standards by HPLC-MS/MS.

No.	Name	Precursor Ion (m/z)	Product Ions (m/z) PI [q]	PI [i]	DP (V)	CE (V)	EP (V)	CXP (V)	IS (V)	Ionization Mode	Retention Time (min)
1	nobiletin	402.9	373.1	388.1	20	33	10	13	5500	ESI^+	4.56
2	narcissin	624.9	316.7	479.6	50	25	10	13	5500	ESI^+	3.51
3	isorhamnetin-3-O-beta-D-glucoside	479.3	317	253	150	34	10	13	5500	ESI^+	3.53
4	tangeretin	372.8	343.1	358.1	10	32	10	13	5500	ESI^+	4.67
5	rutin	609	299.9	279.5	−50	−43	−10	−15	−4500	ESI^-	3.44
6	quercetin	301	150.8	178.9	−100	−35	−10	−15	−4500	ESI^-	3.69
7	isorhamnetin	314.8	300	150.8	−150	−30	−10	−15	−4500	ESI^-	3.84
8	quercetin-7-O-glucoside	462.9	300.9	342.9	−50	−28	−10	−15	−4500	ESI^-	3.43
9	kaempferol-3-O-glucoside	592.9	284.9	255	−50	−40	−10	−15	−4500	ESI^-	3.50

DP de-clustering potential; EP entrance potential; CXP collision cell exit potential; CE collision energy; PI product ions; [q] for quantification; [i] for identification; CUR curtain gas; IS ion spray voltage; TEM temperature; GS1 ion source gas 1; GS2 ion source gas2; CUR = 35 psi; TEM = 500 °C; both GS1 and GS2 = 55 psi.

Table 3. Method validation results including linearity, LOD, LOQ, stability, precision and repeatability.

No.	Name	Linearity: Regression Equations	R^2	Ranges (ng/mL)	LOD (ng/mL)	LOQ (ng/mL)	Stability (RSD, %)	Precision (RSD, %)	Repeatability (RSD, %)
1	nobiletin	y = 62,293,800x − 1,576,600	0.9968	7.81–1000	0.488	0.977	3.05	1.43	4.44
2	narcissin	y = 3,029,650x − 314,384	0.9942	31.25–4000	7.81	15.63	3.26	3.22	2.26
3	isorhamnetin-3-O-beta-D-glucoside	y = 279,387x − 11,500	0.9973	62.5–4000	15.63	31.25	4.22	4.37	4.93
4	tangeretin	y = 4,034,570x − 18,086	0.9902	3.91–250	0.488	0.977	1.94	1.76	4.87
5	rutin	y = 3,089,740x − 61,316	0.9978	15.63–1000	7.81	31.25	2.99	3.71	1.74
6	quercetin	y = 17,556,400x − 169,184	0.9921	15.63–500	7.81	15.63	3.82	4.03	4.9
7	isorhamnetin	y = 116,434,300x − 149,754	0.9903	7.81–250	3.91	7.81	2.53	2.94	2.82
8	quercetin-7-O-glucoside	y = 5,857,970x − 144,514	0.9989	31.25–1000	7.81	15.63	3.26	3.31	2.39
9	kaempferol-3-O-glucoside	y = 4,499,260x − 117,172	0.9901	31.25–1000	7.81	15.63	2.06	4.49	4.9

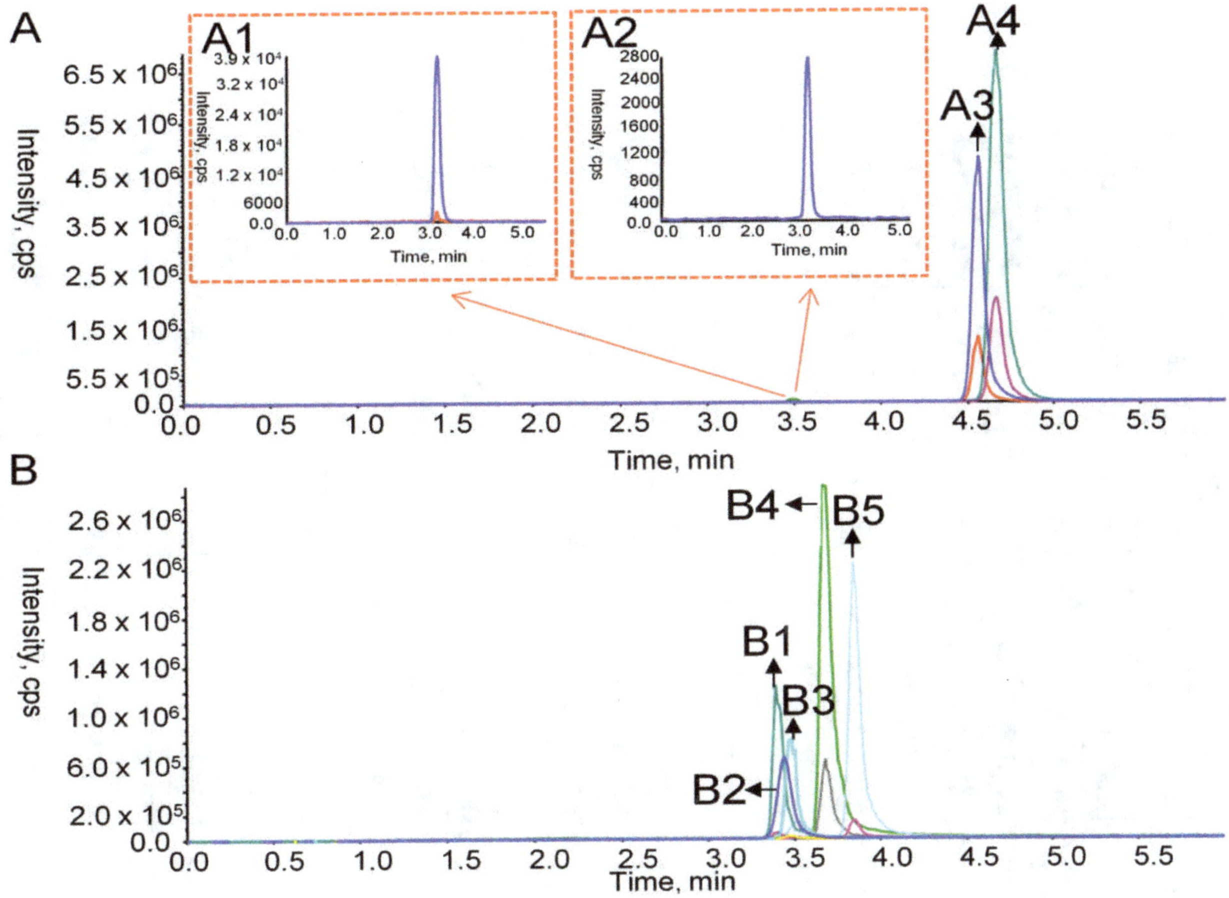

Figure 6. TIC diagrams of nine flavonoid standards in the positive and negative ion mode. (**A**) the positive ion mode; A1 represents the chromatograms of narcissin (retention time = 3.51 min); A2 is the ones of isorhamnetin-3-O-beta-D-glucoside (retention time = 3.53 min); A3 is the ones of nobiletin (retention time = 4.56 min); A4 is the ones of tangeretin (retention time = 4.67 min); (**B**) the negative ion mode; B1 represents the chromatograms of quercetin-7-O-glucoside (retention time = 3.43 min); B2 is the ones of rutin (retention time = 3.44 min); B3 is the ones of kaempferol-3-O-glucoside (retention time = 3.50 min); B4 is the ones of quercetin (retention time = 3.69 min); B5 is the ones of isorhamnetin (retention time = 3.84 min).

For M and NM growth for six months, the tangeretin content was inhibited, while the content of the other 8 flavonoid DAMs was significantly upregulated which was consistent with the results of the metabolome and showed that the metabolome data were reliable. The levels of four metabolites (i.e., isorhamnetin-3-o-beta-D-glucoside, rutin, isorhamnetin, and kaempferol-3-O-glucoside) gradually increased across the growth time. For narcissin, quercetin, and quercetin-7-O-glucoside, the content tended to increase at first, then decrease and then increase again. While the tangeretin content reached the peak of accumulation at day 0 and the fourth month. Flavonoid accumulation in the plantlets showed significant difference between the fungal and no-fungal inoculations. The AR2 had a positive effect on narcissin, rutin, quercetin, and quercetin-7-O-glucoside content in *A. roxburghii* growth for one month, while it inhibited the accumulation of isorhamnetin-3-O-beta-D-glucoside, isorhamnetin, and kaempferol-3-O-glucoside and had no significant effect on nobiletin and tangeretin. Compared with the control group, significantly higher amount of narcissin, rutin and quercetin-7-O-glucoside accumulated in mycorrhizal plantlets from the first month to the sixth month. Up until the sixth month, AR2 could significantly promote the accumulation of nobiletin, but there was no significant difference at other times. The detailed results are shown in Figure 7. In conclusion, AR2 could significantly affect the accumulation of different flavonoids in *A. roxburghii*.

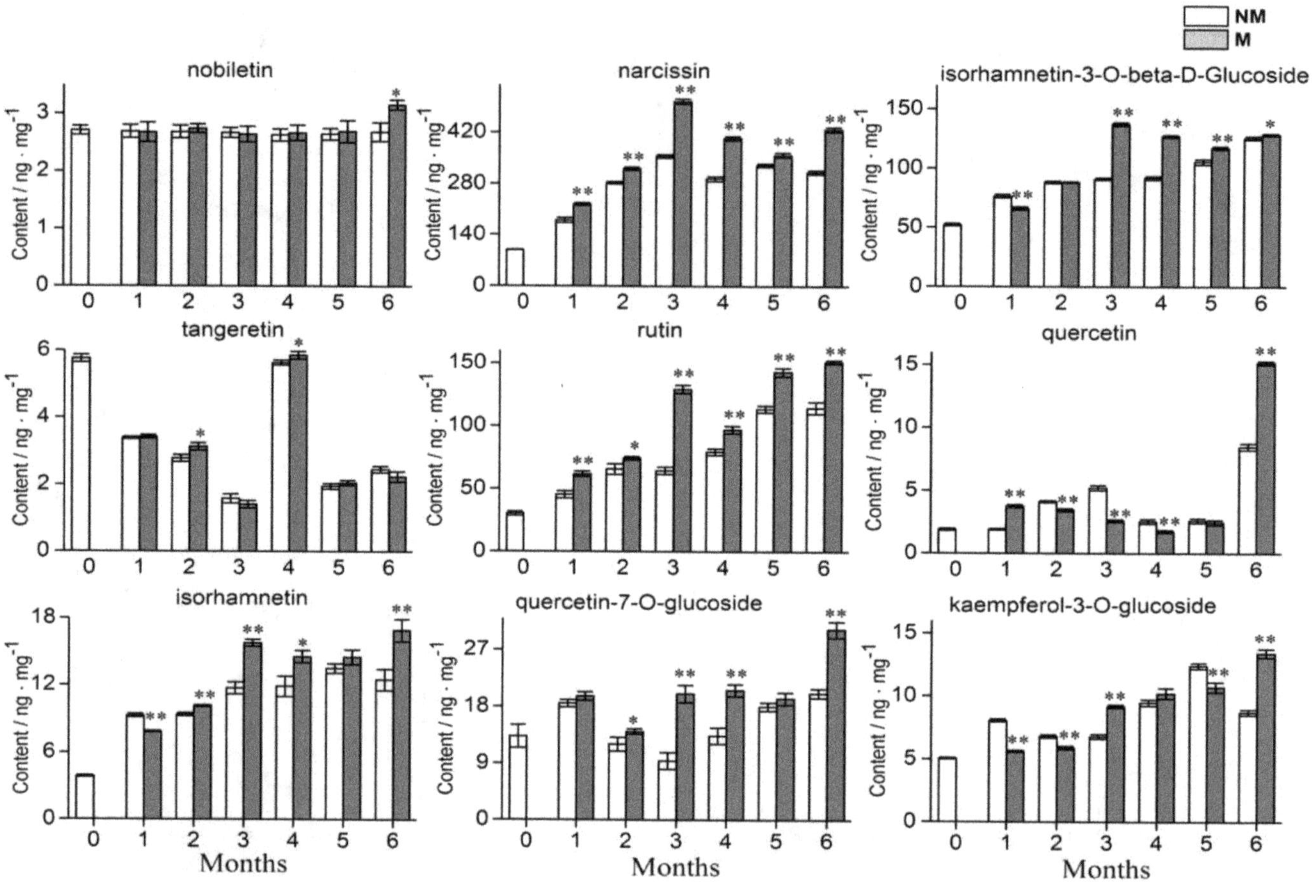

Figure 7. Dynamic variations of 9 flavonoids in the mycorrhizal and non-mycorrhizal *A. roxburghii* growth 0 month to 6 months. NM and M represents non-mycorrhizal *A. roxburghii* and mycorrhizal *A. roxburghii*, respectively. Each value is the mean of three replicates, and error bars indicate standard deviations. Statistical analysis of the data was performed by independent samples *t*-test using the SPSS 22.0 software (IBM, Chicago, IL, USA). * and ** above the columns are significantly different at $p \leq 0.05$ and $p \leq 0.01$, respectively.

2.7. Dynamic Variations of Expression Levels of Flavonoid Biosynthetic Genes

The metabolic pathways of three common flavonoids in *A. roxburghii* are shown in Figure 8. qRT-PCR was used to assess the relative expression levels of nine key enzyme genes in the conserved flavonoid biosynthesis pathway in every treatment group. The results showed that the expression of the *4CL* gene in the treatment group was significantly upregulated during the whole growth process, compared with the control group (Figure 9). The expression of the *PAL* and *CHS* genes in the M was significantly upregulated at the same growth stage (2 months, 3 months, and 4 months), while it was almost the same at other growth stages. The AR2 showed an inhibitory effect on the expression of the *C4H* and *F3'H* genes in the 1 month old and 2 month old *A. roxburghii*, while it showed an inhibitory effect on the expression of the *CHI* gene in the 1 month old, 5 month old and 6 month old *A. roxburghii*. The *F3H/FLS* gene had an upregulated expression in the M grown for 1 month and 2 months and the *GT* gene also had an upregulated expression in the M grown for 1 month, 2 months, 3 months, 5 months, and 6 months. Finally, the expression of the *RT* gene was upregulated in the M grown for 1 month and 6 months. In conclusion, AR2 could induce the expression of flavonoid synthesis related genes in varying degrees and at different growth stages.

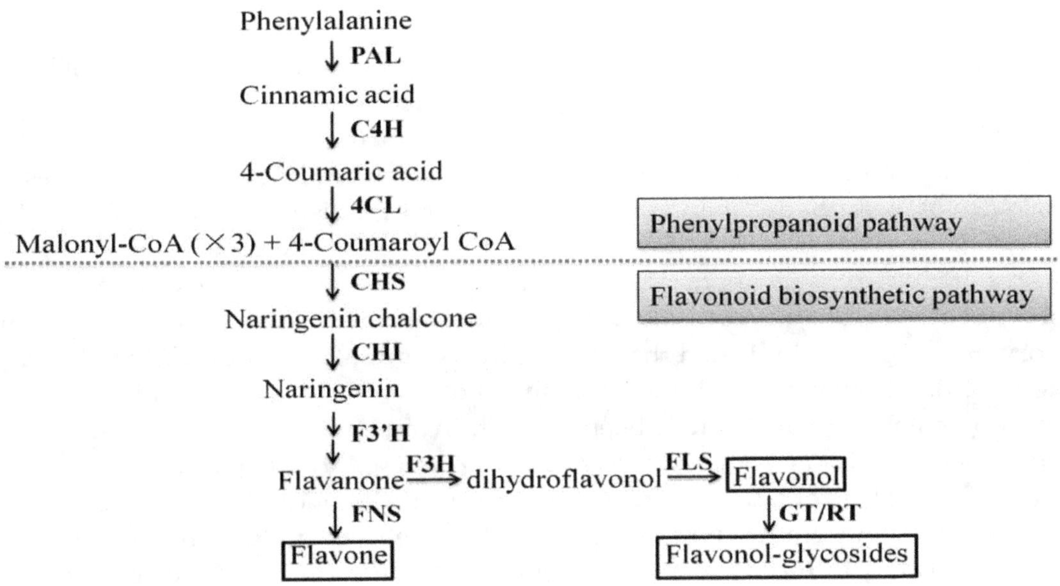

Figure 8. The flavonoid biosynthetic pathway in *A. roxburghii*. Bold words indicate the key enzymes in flavonoid biosynthesis. Compounds in the box show flavones, flavonols and flavonol-glycosides studied in this study.

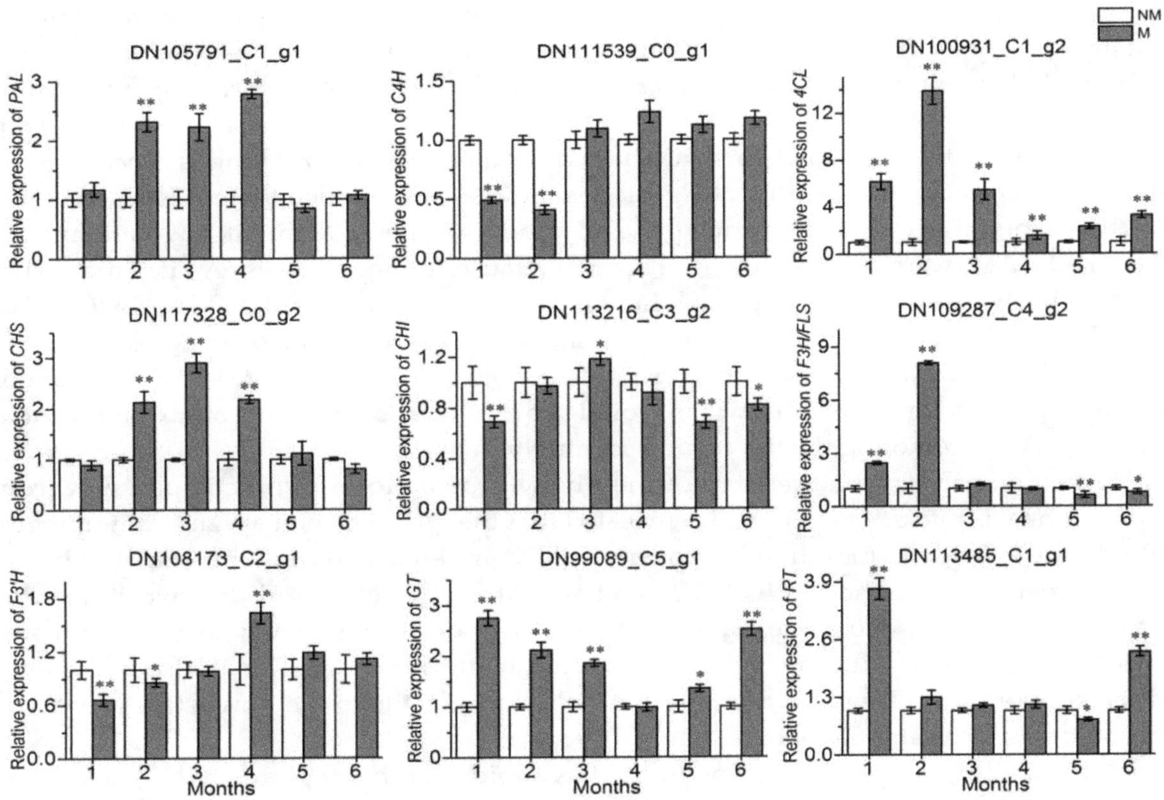

Figure 9. Expression levels of flavonoid biosynthetic genes in the M and NM growth 0 month to 6 months. M and NM represent the mycorrhizal *A. roxburghii* and non-mycorrhizal *A. roxburghii*, respectively. DNxx_Cx_g1 represents the ID of gene. The x-axis indicates the relative expression level of the genes. Each value is the mean of three replicates, and error bars indicate standard deviations. Statistical analysis of the data was performed by independent samples *t*-test using the SPSS 22.0 software (IBM, Chicago, IL, USA). * and ** above the columns are significantly different at $p \leq 0.05$ and $p \leq 0.01$, respectively.

3. Discussion

Symbiotic association of mycorrhizal fungi with plants has been shown to affect flavonoid content. The application of co-cultivation of plants with orchid mycorrhizal (OM) fungi and arbuscular mycorrhizal (AM) fungi are progressing gradually. For AM fungi, they can penetrate and colonize the root of the host to form intracellular haustoria-like structures known as arbuscules, which are the principal sites of metabolic exchange between the two organisms [39,40]. Flavonoid content in *Medicago truncatula* was increased by AM inoculation [41]. Xie et al. [42] reported that AM colonization in the soybeans attributed to the increase of certain flavonoids in the root exudates. For OM fungi, they can form pleloton inside the root cells. And the association of OM with the orchid is the focus of our laboratory. Some studies indicated that flavonoid accumulated significantly in the mycorrhizal orchidaceae [16,43]. However, the dynamic changes of flavonoid content in mycorrhizal host at different developmental stages have rarely been studied. In our study, AR2 was belonging to a member of the orchid mycorrhizal fungi, and the co-cultivation between *A. roxburghii* and AR2 was performed. The dynamic changes of several flavonoids showed that the flavonoid had its special accumulation content at a defined growth time and that AR2 had different effects on different flavonoids at different growth stages. Also, AR2 induced the narcissin, rutin and quercetin-7-o-glucoside accumulations in mycorrhizal plantlets across the growth stage. For narcissin and isorhamnetin-3-o-beta-D-glucoside, their content in mycorrhizal *A. roxburghii* growth at three months reached the highest and was more than 420 ng/g and 120 ng/g, respectively. This was the first report regarding the changes of flavonoid content induced by AR2 in *A. roxburghii* at different growth stages.

To investigate the effects of mycorrhizal fungi on metabolites in its host, transcriptome and metabolome analyses were performed. Zhao et al. [29] reported that the secondary biosynthesis and hormone balance in the *Cymbidium hybridum* were induced by mycorrhizal fungus through transcriptome analysis. Schliemann et al. [44] reported that the biosynthesis of some constitutive isoflavonoids and plastidial metabolism could be activated by mycorrhizal fungus *Glomus intraradices* through metabolome analyses. In our study, metabolome analysis revealed that all 709 metabolites and 135 DAMs were putatively annotated among the NM and M. Among them, 148 flavonoid metabolites and 9 flavonoid DAMs were investigated. Furthermore, transcriptome analysis revealed that 4341 DEGs were identified between the two groups, of which 2915 DEGs were up-regulated and 1426 DEGs were down-regulated; KEGG pathways of the more DEGs were involved in the biosynthesis of secondary metabolites including flavonoid. These results implied that AR2 might change internal metabolism in *A. roxburghii*, especially for flavonoids, which would provide a basis for further study on the molecular mechanisms of AR2 promoting the flavonoid accumulation in *A. roxburghii*.

PAL, a key enzyme in the first step of the phenylpropanoid biosynthetic pathway, could be activated by fungal elicitors. Our study also revealed that the *PAL* gene had a significant upregulation, especially in the 2, 3, and 4 month mycorrhizal herbs, compared with the uninoculated ones. This result is in agreement with the results of Zhou et al. [45] and Xu et al. [46]. It is worth mentioning, the expression of the *4CL* gene in the plantlets inoculated AR2 during the whole growth process was significantly upregulated, with the highest expression being 13.3 fold. This is also consistent with Wang et al.'s report [20]. These data imply that AR2 might activate the downstream pathways of phenylpropanoids including flavonoids.

In addition, CHS, is the key enzyme in the flavonoids synthesis pathway [47]. Harrison and Dixon [48] reported that the expression level of the gene *CHS* in the roots of *Medicago truncatula* was enhanced by mycorrhizal fungus *Glomus versiforme*. Xie et al. [49] reported that mycorrhizal symbiosis induced the expression of the *CHS* gene of *Glycyrrhiza uralensis*, and the liquiritin accumulation and the expression of *CHS* gene showed a positive correlation. In our study, the expression level of the *CHS* gene was also upregulated in the mycorrhizal herbs growth for 2, 3, and 4 months; Meanwhile, the corresponding flavonoids (narcissin, rutin, isorhamnetin and quercetin-7-O-glucoside) accumulated in different degrees. Our data added new evidence to support mycorrhizal symbiosis induced the expression of the *CHS* gene and promoted the flavonoids accumulation. Additionally, our study showed that the *GT* gene expression was significantly upregulated in the 1–4 month mycorrhizal herbs, while the *RT* gene was induced in the 1–6 month ones. The corresponding flavonol-glycoside (narcissin, rutin, isorhamnetin-3-O-beta-D-glucoside, quercetin-7-O-glucoside and kaempferol-3-O-glucoside) showed basically the same induction trend in mycorrhizal *A. roxburghii*. These data again indicated that AR2 might activate the metabolic pathway of flavonoids.

In summary, this study provides much information about the changes that occur in the main active ingredient flavonoids and its related genes during different growth stages in M and NM. AR2 has different induction effects on flavonoid content and gene expression in *A. roxburghii* at different growth stages. These will provide a theoretical basis for reasonable harvest time of *A. roxburghii* and a new insight into improving the quality of the *A. roxburghii*.

4. Materials and Methods

4.1. Plant and Mycorrhizal Fungus Materials

Tissue culture plantlets of *A. roxburghii* were about 3 month-old and 2–3 cm in height from Yongan city, Fujian province, China. The mycorrhizal fungus AR2 (NCBI accession No: MN068847) was previously isolated in our laboratory and deposited at the microbiological center of the Institute of Medicinal Plant Development, Chinese Academy of Medical Sciences and Peking Union Medical College, Beijing, China. Before symbiotic cultures, the fungal strain was inoculated on the potato dextrose agar medium (PDA: potato 200 $g{\cdot}L^{-1}$, glucose 20 $g{\cdot}L^{-1}$, agar 12 $g{\cdot}L^{-1}$, pH 5.2) in darkness at a temperature of 25 ± 1 °C for 5 days. Two pieces (0.5 cm^3) of the fungal inoculum were inoculated on solid matrix medium (sawdust: wheat bran: water = 3:1:1.5, *v/v/v*; 101.33 kPa and 121 °C for 180 min) in darkness at a temperature of 25 ± 1 °C for 15 days. After that, the fungal solid substrates were obtained for subsequent symbiotic culture.

4.2. Symbiotic Cultures of A. roxburghii Plantlets

In symbiotic cultures of plantlets, the substrate (humus soil: vermiculite: water = 3:1:1) was used. Each culture bottle (9 cm in diameter, 12.5 cm in height) containing 70 g substrate was sterilized at 101.33 kPa and a temperature of 121 °C for 180 min. Six plantlets, derived from asexual reproduction, were transferred into each culture bottle which was inoculated with 0.5 g of fungal solid substrates, while the culture bottles without fungal inoculum served as the control. The symbiotic cultures were placed in the growth room under a 12/12 h photoperiod at a temperature of 25 ± 1 °C and an illumination intensity of 1500 Lx. The samples (the whole plantlet) were collected once a month from the beginning of co-culture until the sixth month. Samples were immediately frozen in liquid nitrogen and stored at −80 °C for later analyses using LC-MS/MS, RNA-seq and qRT-PCR. Among them, the collected samples grown for six months were used for the metabolome and transcriptome analyses. All data were obtained based on three independent biological replicates.

4.3. Histological Study

Two weeks after inoculation, fresh root segments were fixed in 50 mM phosphate buffer (pH 6.8) containing 2.5% glutaraldehyde and 1.6% paraformaldehyde for 4 h at room temperature. After fixation, the samples were rinsed three times with phosphoric acid buffer (pH 6.8, 0.1 M) for about 15 min each time and then dehydrated with a graded ethanol series (15% ethanol for 30 min; 30% ethanol for 30 min; 50% ethanol for 30 min; 70% ethanol for 1 h; 85% ethanol for 1 h; 95% ethanol for 1 h; absolute ethanol for 1 h). After dehydration, the samples were embedded in LRwhite gradient mixture (25% LRwhite for 24 h; 50% LRwhite for 24 h; 75% LRwhite for 24 h; 100% LRwhite for 24 h) and polymerized for 48 h at 60 °C. Next, three-µm-thick sections were cut using glass knives with a rotary microtome (Autocut 2040; Reichert-Jung; Germany). The sections were collected on slides and stained with 0.05% (*w/v*) toluidine blue O in benzoate buffer for general histology examinations. The sections were examined, and images were captured digitally using a digital camera attached to the microscope (Axio Imager A1; Carl Zeiss, Oberkochen, Germany).

4.4. *Sample Extraction and Metabolome Analysis*

4.4.1. Sample Extraction

The cryopreserved samples were freeze-dried and then ground for 1 min at 30 Hz using a MixerMill MM400 (Retsch Technology, Haan, Germany). Subsequently, 100 mg powder was weighed and extracted for 24 h at 4 °C in 1.0 mL of 70% methanol. During this period, the samples were vortexed (10 s, 40 Hz) once per 10 min for a total of three times. After extraction, the pellets were centrifuged at 10,000× *g* for 10 min at 4 °C and the extracts were then filtered through a 0.22 μm microporous membrane and stored in a sample vial. The QC was prepared by mixing all the samples. In order to examine the repeatability of the analysis process, a QC sample was injected after every five test samples during the instrumental analysis.

4.4.2. Liquid Chromatographic Mass Spectrometry Analysis

The sample extracts were analyzed using a UPLC-ESI-MS/MS system, which mainly includes UPLC (Shim-pack UFLC SHIMADZU CBM30A, SHIMADZU, Kyoto, Japan) and MS/MS (Applied Biosystems 6500 QTRAP, AB SCIEX, Foster City, CA, USA). The UPLC separation was completed on a Waters Acquity UPLC HSS T3 C18 column (100 × 2.1 mm, 1.8 μm) (Waters Corp., Milford, MA, USA). The mobile phase A solvent was water containing 0.04% acetic acid, and the mobile phase B solvent was acetonitrile containing 0.04% acetic acid. The elution gradient was shown as follows: 0–11 min 95–5% A, 11–12 min 5–5% A, 12–12.1 min 5–95% A and 12–15 min 95–95% A. The flow was 0.4 mL/min and the injection volume was 2 μL. The column temperature was set to 40 °C.

The effluents were alternatively connected to an electrospray ionization-triple quadrupole-linear ion trap MS/MS (ESI-QTRAP-MS/MS). Linear ion trap (LIT) and triple quadrupole (QQQ) scans were carried out using QTRAP. The mass spectrometry conditions were as follows: the ESI temperature was set to 500 °C, the mass spectrometry voltage was 5500 V, the GS 1 and GS 2 were set to 55 psi and 60 psi, respectively, and the CUR was set to 25 psi. QQQ scans were obtained as MRM experiments with collision gas (nitrogen) set to 5 psi. The DP and CE for individual MRM transitions were optimized in the QQQ. The data were analyzed by the mass spectrometry software (Version 1.6.1 Applied Biosystems Company, Framingham, MA, USA).

4.4.3. Metabolite Identification

Based on the public database of metabolite information and the MVDB V2.0 database of Wuhan Metware Biotechnology Co., Ltd. (Wuhan, China), qualitative analysis of the primary and secondary mass spectrometry data was performed by referencing the existing mass spectrometry databases such as MassBank, KNAPSAcK, HMDB and METLIN; thus, the structural analysis of metabolites was determined. Regarding the quantitative analysis of the metabolites, MRM was used, and PCA and OPLS-DA were then carried out to identify differential metabolites. A VIP ≥ 1 and a fold change ≥ 2 or ≤ 0.5 were set for metabolites with significant difference.

4.5. *Illumina Sequencing*

4.5.1. RNA Extraction, cDNA Library Construction and Sequencing

The total RNA was extracted from NM and M using the RNeasy Plant Mini Kit (Qiagen, Hilden, Germany) following the manufacturer's instructions, and treated with an RNase-free DNase I digestion kit (Aidlab, Beijing, China) in order to remove contaminated genomic DNA. RNA degradation was measured using 1% agarose gel, RNA concentration was measured using NanoDrop 2000 spectrophotometer (Thermo Scientific, Wilmington, DE, USA) and RNA integrity was assessed on an Agilent 2100 Bioanalyzer (Agilent, Palo Alto, CA, USA).

mRNA was enriched with oligo (dT)-attached magnetic beads and then the fragmentation buffer was exploited to randomly fragment the mRNA into short fragments. Using these cleaved RNA fragments as a template, the first cDNA strand was synthesized by random hexamers, then the second cDNA strand was synthesized by adding buffer, dNTPs, RNaseH and DNA polymerase I and purification of the double-stranded cDNA was done by NEBNext Ultra RNA Library Prep Kit for Illumina (NEB, Ipswich, MA, USA). The purified double-stranded cDNA was repaired by end-to-end, added poly A tail and connected to the sequencing connector. cDNA of about 200 bp in length was selected by AMPure XP beads, and PCR amplification was performed to enrich the purified cDNA template. Finally, the libraries were sequenced using an Illumina Hiseq 2000. Sequence data were deposited in the NCBI SRA database (accession number: PRJNA579778).

4.5.2. De Novo Transcriptome Assembly and Annotation

In order to achieve high-quality and clean data, the raw data was filtered by removing the reads with adapter sequence and low-quality reads using the fastp software [50], then the clean reads were assembled into transcriptome by the Trinity v.2.0.6 software (Broad Institute, Cambridge, MA, USA). Unigene function was annotated based on these databases: NCBI non-redundant protein sequences (Nr, https://ftp.ncbi.nlm.nih.gov/blast/db/FASTA/), KEGG (https://www.genome.jp/kegg), Clusters of Orthologous Groups of proteins (COG, https://www.ncbi.nlm.nih.gov/COG/), GO (https://www.geneontology.org), the Swiss-Prot (http://www.ebi.ac.uk/uniprot/) and the translation of EMBL (TrEMBL).

4.6. Determination of the Flavonoid Contents During Different Growth Stages

4.6.1. Preparation of Standard Solutions

In order to validate the changes of flavonoids during different growth stages after symbiotic culture, several monomer flavonoids from DAMs were tested. The standards of nobiletin (478-01-3), narcissin (604-80-8), isorhamnetin-3-O-beta-D-Glucoside (5041-82-7), tangeretin (481-53-8), rutin (153-18-4), quercetin (117-39-5), isorhamnetin (480-19-3), quercetin-7-O-glucoside (491-50-9), and kaempferol-3-O-glucoside (480-10-4) were purchased from the National Institutes for Food and Drug Control (Beijing, China) and the purity of all of them was over 98%. Each standard sample was dissolved in methanol to obtain 1 μg/mL of standard working solutions and used within 1 month at −20 °C.

4.6.2. Preparation of Sample Solutions

The preparation of the sample (approximately 20 mg) followed the same process as the above mentioned extract. In order to enrich the flavonoids, the extract was processed through C18-SPE solid phase extraction column (Waters Technology Co., Ltd., USA) and the column was rinsed with 10%, 20%, 30%, 40%, 50%, 60%, 70%, 80%, and 90% absolute alcohol. The rinse solutions of 70%, 80%, 90%, and absolute alcohol, that were verified to contain flavonoids, were collected and freeze dried using a vacuum freeze-drying machine LGJ-18 (Beijing Songyuanhuaxing Technology Develop Co., Ltd, Beijing, China). The extract was re-dissolved with 2 mL methanol, then the sample solution was passed through a 0.22 μm syringe nylon filter and stored at −20 °C prior to analysis.

4.6.3. Apparatus and Analytical Conditions

The HPLC separation was performed by an Agilent 1260 Infinity II series HPLC System that was equipped with Waters Atlantis C18 column (150 × 3.9 mm, 5 μm). The mobile phase consisted of solvent A (water containing 0.1% formic acid) and solvent B (methanol). The gradient procedure was as follows: 0–3 min 60–5% A, 3–5 min 5–5% A, 5–5.02 min 5–60% A, and 5.02–6 min 60–60% A. The flow rate was 0.5 mL/min and the injection volume was 5 μL.

An Applied Biosystems Sciex QTRAP®4500 MS/MS spectrometer equipped with a version of 1.6 Analyst software (AB SCIEX, Massachusetts, USA) was used for the analysis. The instrument was equipped with an ESI source, and the targeted analytes were performed in positive and negative ion modes for all the targeted analytes. Compressed air was used as GS1 and GS2, and high-purity (99.99%) nitrogen was used as CUR and CAD. The operation conditions were as follows: the EP was 10.0/−10.0 V, the TEM at 500 °C, the IS 5500/–4500 V, GS1 set to 55 psi, GS2 set to 55 psi, CUR set to 35 psi, and the CXP 13.0/−15.0 V for ESI^+/ESI^- mode, respectively. The dwell time for each MRM transition was 10 ms.

4.7. Expression of the Flavonoid Biosynthesis Related Genes During Different Growth Stages

Nine unigenes that were related to flavonoid biosynthesis following Zhang et al. [51] and Park et al. [52] were selected for validation using qRT-PCR. The RNA was reverse-transcribed to cDNA using PrimeScriptTM RT reagent Kit (TaKaRa, Dalian, China). The qRT-PCR was performed using the SYBR®Premix ExTaq TM (TaKaRa, Dalian, China) on the LightCycler®480 II Real-Time PCR System (Roche, Carlsbad, CA, USA). Three biological replicates and three technical replicates were performed, and all the primer names and corresponding sequences are listed in Table S3. The qRT-PCR was performed in a 20 μL reaction volume containing SYBR Premix Ex Taq II (10 μL), forward prime (10 μM, 0.8 μL), reverse primer (10 μM, 0.8 μL), cDNA template (5 ng/μL, 2 μL), and ddH_2O (6.4 μL). The PCR conditions were as follows: denaturation at 95 °C for 30 s, followed by 40 cycles of amplification (95 °C for 5 s, 60 °C for 30 s). The melting curves were measured at 95 °C for 5 s and 60 °C for 1 min. The elongation factor 1 alpha (*EF-1α*) gene of *A. roxburghii* was used as the internal control reference gene. Finally, gene expression was calculated using the $2^{-\Delta\Delta Ct}$ method [53].

4.8. Statistical Analysis

All the transcriptome and metabolome samples were designed for three biological replicates. Statistical analyses were conducted using the SPSS 22.0 software (IBM, Chicago, IL, USA). An independent samples *t*-test was used for statistical evaluations between the control and treatment groups.

Author Contributions: Conceptualization, X.C. and Z.M.; Methodology, Y.Z. and Y.L.; Software, Y.Z.; Validation, Y.Z.; Resources, Z.M. and S.G.; Data curation, Y.Z., Y.L. and X.C.; Writing—Original draft preparation, Y.Z.; Writing—Review and editing, Z.M. and S.G.; Supervision, X.C.; Funding acquisition, S.G. and X.C. All authors have read and agreed to the published version of the manuscript.

Acknowledgments: We thank Ruibin Hu from the Institute of Chemistry, Chinese Academy of Sciences for his help for data analysis and article logic structure; We thank Yung-I Lee from Biology Department, National Museum of Natural Science (Taiwan) for providing guidance for semi-thin section observation experiment.

Abbreviations

CE	collision energy
CUR	curtain gas
CXP	collision cell exit potential
DAMs	differentially accumulated metabolites

DEGs	differentially expressed genes
DP	de-clustering potential
EP	entrance potential
ESI-QTRAP-MS/MS	electrospray ionization-triple quadrupole-linear ion trap MS/MS
GC-MS	gas chromatography-mass spectrometer
GO	gene ontology
GS1	ion source gas 1
GS2	ion source gas 2
HPLC-MS/MS	high-performance liquid chromatography coupled with tandem mass spectrometry
IS	ion spray voltage
KEGG	Kyoto encyclopedia of genes and genomes
LC-MS	liquid chromatograph-mass spectrometer
LIT	linear ion trap
LOD	limit of detection
LOQ	limit of quantification
NMR	nuclear magnetic resonance
OPLS-DA	orthogonal partial least squares discriminant
PCA	principal component analysis
PC1	principal component 1
PC2	principal component 2
PI	product ions
QC	quality control
QQQ	triple quadrupole
qRT-PCR	quantitative real-time polymerase chain reaction
RNA-seq	RNA-sequencing
RSDs	relative standard deviations
TEM	temperature
TIC	total ion chromatography
UPLC-ESI-MS/MS	ultra-performance liquid chromatography-electrospray ionization-tandem mass spectrometry
VIP	variable importance in project
cps	count per second
MRM	multiple reaction monitoring
psi	pounds per square inch
NM	non-mycorrhizal *A. roxburghii*
M	mycorrhizal *A. roxburghii*
PAL	phenylalanine ammonia-lyase
C4H	cinnamate 4-hydroxylase
4CL	4-coumarate CoA ligase
CHS	chalcone synthase
CHI	chalcone isomerase
F3′H	flavonoid 3′-hydroxylase
FNS	flavone synthase
F3H	flavanone 3-hydroxylase
FLS	flavonol synthase
GT	flavonoid 3-*O*-glucosyltransferase
RT	rhamnosyltransferase

References

1. Li, S.; Wang, Z.; Shao, Q.; Fang, H.; Zhu, J.; Wu, X.; Zheng, B. Rapid detection of adulteration in *Anoectochilus roxburghii* by near-infrared spectroscopy coupled with chemometric methods. *J. Food Sci. Technol.* **2018**, *55*, 3518–3525. [CrossRef]

2. Chen, Y.; Huang, J.; Yeap, Z.Q.; Zhang, X.; Wu, S.; Ng, C.H.; Yam, M.F. Rapid authentication and identification of different types of *A. roxburghii* by Tri-step FT-IR spectroscopy. *Spectrochim. Acta A Mol. Biomol. Spectrosc.* **2018**, *199*, 271–282. [CrossRef] [PubMed]
3. Zeng, B.Y.; Su, M.H.; Chen, Q.X.; Chang, Q.; Wang, W.; Li, H.H. Antioxidant and hepatoprotective activities of polysaccharides from *Anoectochilus roxburghii*. *Carbohydr. Polym.* **2016**, *153*, 391–398. [CrossRef] [PubMed]
4. Cui, S.C.; Yu, J.; Zhang, X.H.; Cheng, M.Z.; Yang, L.W.; Xu, J.Y. Antihyperglycemic and antioxidant activity of water extract from *Anoectochilus roxburghii* in experimental diabetes. *Exp. Toxicol. Pathol.* **2013**, *65*, 485–488. [CrossRef] [PubMed]
5. Tang, T.T.; Duan, X.Y.; Zhang, L.; Shen, Y.B.; Hu, B.; Liu, A.P.; Chen, H.; Li, C.; Wu, W.J.; Shen, L.; et al. Antidiabetic activities of polysaccharides from *Anoectochilus roxburghii* and *Anoectochilus formosanus* in STZ-induced diabetic mice. *Int. J. Biol. Macromol.* **2018**, *112*, 882–888. [CrossRef] [PubMed]
6. Li, L.; Li, Y.M.; Liu, Z.L.; Zhang, J.G.; Liu, Q.; Yi, L.T. The renal protective effects of *Anoectochilus roxburghii* polysaccharose on diabetic mice induced by high-fat diet and streptozotocin. *J. Ethnopharmacol.* **2016**, *178*, 58–65. [CrossRef] [PubMed]
7. Guo, Y.L.; Ye, Q.; Yang, S.L. Therapeutic effects of polysaccharides from *Anoectochilus roxburghii* on type II collagen-induced arthritis in rats. *Int. J. Biol. Macromol.* **2019**, *122*, 882–892. [CrossRef]
8. Zhang, Y.; Cai, J.; Ruan, H.; Pi, H.; Wu, J. Antihyperglycemic activity of kinsenoside, a high yielding constituent from *Anoectochilus roxburghii* in streptozotocin diabetic rats. *J. Ethnopharmacol.* **2007**, *114*, 141–145. [CrossRef]
9. Hsiao, H.B.; Wu, J.B.; Lin, H.; Lin, W.C. Kinsenoside isolated from *Anoectochilus formosanus* suppresses LPS-stimulated inflammatory reactions in macrophages and endotoxin shock in mice. *Shock* **2011**, *35*, 184–190. [CrossRef]
10. Ye, S.Y.; Shao, Q.S.; Zhang, A.L. *Anoectochilus roxburghii*: A review of its phytochemistry, pharmacology, and clinical applications. *J. Ethnopharmacol.* **2017**, *209*, 184–202. [CrossRef]
11. Dearnaley, J.D.W.; Martos, F.; Selosse, M.A. Orchid mycorrhizas: Molecular ecology, physiology, evolution and conservation aspects. In *Fungal Association*, 1st ed.; Hock, B., Ed.; Springer: Berlin/Heidelberg, Germany, 2012; Volume 9, pp. 207–230.
12. Dearnaley, J.D.; Cameron, D.D. Nitrogen transport in the orchid mycorrhizal symbiosis-further evidence for a mutualistic association. *New Phytol.* **2017**, *213*, 10–12. [CrossRef] [PubMed]
13. Zhai, X.; Jia, M.; Chen, L.; Zheng, C.J.; Rahman, K.; Han, T.; Qin, L.P. The regulatory mechanism of fungal elicitor-induced secondary metabolite biosynthesis in medical plants. *Crit. Rev. Microbiol.* **2017**, *43*, 238–261. [CrossRef] [PubMed]
14. Li, Q.; Ding, G.; Li, B.; Guo, S.X. Transcriptome analysis of genes involved in dendrobine biosynthesis in *Dendrobium nobile* Lindl. infected with mycorrhizal fungus MF23 (*Mycena* sp.). *Sci. Rep.* **2017**, *7*, 316. [CrossRef] [PubMed]
15. Li, Q.; Li, B.; Zhou, L.S.; Ding, G.; Li, B.; Guo, S.X. Molecular analysis of polysaccharide accumulation in *Dendrobium nobile* infected with the mycorrhizal fungus *Mycena* sp. *RSC Adv.* **2017**, *7*, 25872–25884. [CrossRef]
16. Zhang, F.S.; Lv, Y.L.; Zhao, Y.; Guo, S.X. Promoting role of an endophyte on the growth and contents of kinsenosides and flavonoids of *Anoectochilus formosanus* hayata, a rare and threatened medicinal orchidaceae plant. *J. Zhejiang Univ. Sci. B* **2013**, *14*, 785–792. [CrossRef]
17. Zhu, J.J.; Huang, Y.J.; Jin, J.H.; Shen, J.Y. Effect of cultivation substrate on growth and active component contents of *Anoectochilus roxburghii* from three different origins. *China J. Chin. Mater. Med.* **2019**, *44*, 2467–2471.
18. Niu, H.; Xie, Z.M.; Gu, L.; Liang, Y.; Wei, K.H.; Wang, J.M.; Li, M.J.; Zhang, Z.Y. Effects of planting density and harvesting stages for *Anoectochilus roxburghii* planted under forest on its yield and quality. *Mod. Chin. Med.* **2018**, *20*, 837–865.
19. Gan, J.J.; Mao, L.L.; Huang, R.L.; Jiang, H.Y.; Huang, X.Y.; Li, H. Effects of different cultivation methods on the growth and quality of *Anoectochilus roxburghii*. *J. Agric. Sci. Technol.* **2018**, *20*, 130–136.
20. Wang, H.; Lin, Q.Q.; Hu, X.J.; Wu, G.H.; Xu, Q. Screening of *Anoectochilus roxburghii* growth-promoting endophytic fungi and their promoting mechanism. *J. Fujian Norm. Univ. (Nat. Sci. Ed.)* **2019**, *35*, 72–79.
21. Deng, N.; Chang, E.; Li, M.H.; Ji, J.; Yao, X.M.; Bartish, I.V.; Liu, J.F.; Ma, J.; Chen, L.Z.; Jiang, Z.P.; et al. Transcriptome characterization of *Gnetum parvifolium* reveals candidate genes involved in important secondary metabolic pathways of flavonoids and stilbenoids. *Front. Plant Sci.* **2016** *7*, 174. [CrossRef]

22. Yue, J.Y.; Zhu, C.X.; Zhou, Y.; Niu, X.L.; Miao, M.; Tang, X.F.; Chen, F.D.; Zhao, W.P.; Liu, Y.S. Transcriptome analysis of differentially expressed unigenes involved in flavonoid biosynthesis during flower development of *Chrysanthemum morifolium* 'Chu ju'. *Sci. Rep.* **2018**, *8*, 13414. [CrossRef] [PubMed]
23. Shimada, N.; Aoki, T.; Sato, S.; Nakamura, Y.; Tabata, S.; Ayabe, S. A cluster of genes encodes the two types of chalcone is omerase involved in the biosynthesis of general falavonoids and legume specific 5-deoxy (iso) flavonoids in *Lotus japonicus*. *Plant Physiol.* **2003**, *131*, 941–951. [CrossRef] [PubMed]
24. Nag, S.; Kumaria, S. In silico characterization and transcriptional modulation of phenylalanine ammonia lyase (PAL) by abiotic stresses in the medicinal orchid *Vanda coerulea* Griff. ex Lindl. *Phytochemistry* **2018**, *156*, 176–183. [CrossRef] [PubMed]
25. Wang, C.H.; Yu, J.; Cai, Y.X.; Zhu, P.P.; Liu, C.Y.; Zhao, A.C.; Lü, R.H.; Li, M.J.; Xu, F.X.; Yu, M.D. Characterization and functional analysis of 4-coumarate: CoA ligase genes in mulberry. *PLoS ONE* **2016**, *11*, e0155814.
26. Kim, J.; Woo, H.R.; Nam, H.G. Toward systems understanding of leaf senescence: An integrated multi-omics perspective on leaf senescence research. *Mol. Plant* **2016**, *9*, 813–825. [CrossRef]
27. Zhang, T.Z.; Hu, Y.; Jiang, W.K.; Fang, L.; Guan, X.Y.; Chen, J.D.; Zhang, J.B.; Saski, C.A.; Scheffler, B.E.; Stelly, D.M.; et al. Sequencing of allotetraploid cotton (*Gossypium hirsutum* L. acc. TM-1) provides a resource for fiber improvement. *Nat. Biotechnol.* **2015**, *33*, 531–537. [CrossRef]
28. Qin, Z.X.; Wang, W.; Liao, D.Q.; Wu, X.Y.; Li, X.E. UPLC-Q/TOF-MS-based serum metabolomics reveals hypoglycemic effects of *Rehmannia glutinosa*, *Coptis chinensis* and their combination on high-fat-diet-induced diabetes in KK-Ay Mice. *Int. J. Mol. Sci.* **2018**, *19*, 3984. [CrossRef]
29. Zhao, X.L.; Zhang, J.X.; Chen, C.L.; Yang, J.Z.; Zhu, H.Y.; Liu, M.; Lv, F.B. Deep sequencing-based comparative transcriptional profiles of *Cymbidium hybridum* roots in response to mycorrhizal and non-mycorrhizal beneficial fungi. *BMC Genom.* **2014**, *15*, 747. [CrossRef]
30. Sadre, R.; Magallanes-Lundback, M.; Pradhan, S.; Salim, V.; Mesberg, A.; Jones, A.D.; DellaPenna, D. Metabolite diversity in alkaloid biosynthesis: A multi-lane (diastereomer) highway for camptothecin synthesis in *Camptotheca acuminate*. *Plant Cell* **2016**, *28*, 1926–1944. [CrossRef]
31. Liu, Y.; Luo, S.H.; Schmidt, A.; Wang, G.D.; Sun, G.L.; Grant, M.; Kuang, C.; Yang, M.J.; Jing, S.X.; Li, C.H.; et al. A geranylfarnesyl diphosphate synthase provides the precursor for sesterterpenoid (C25) formation in the glandular trichomes of the mint species *Leucosceptrum canum*. *Plant Cell* **2016**, *28*, 804–822. [CrossRef]
32. Lou, Q.; Liu, Y.L.; Qi, Y.Y.; Jiao, S.Z.; Tian, F.F.; Jiang, L.; Wang, Y.J. Transcriptome sequencing and metabolite analysis reveals the role of delphinidin metabolism in flower colour in grape hyacinth. *J. Exp. Bot.* **2014**, *65*, 3157–3164. [CrossRef] [PubMed]
33. Cho, K.; Cho, K.S.; Sohn, H.B.; Ha, I.J.; Hong, S.Y.; Lee, H.; Kim, Y.M.; Nam, M.H. Network analysis of the metabolome and transcriptome reveals novel regulation of potato pigmentation. *J. Exp. Bot.* **2016**, *67*, 1519–1533. [CrossRef] [PubMed]
34. Zhang, Y.; Li, D.; Zhou, R.; Wang, X.; Dossa, K.; Wang, L.; Zhang, Y.; Yu, J.; Gong, H.; Zhang, X.; et al. Transcriptome and metabolome analyses of two contrasting sesame genotypes reveal the crucial biological pathways involved in rapid adaptive response to salt stress. *BMC Plant Biol.* **2019**, *19*, 66. [CrossRef] [PubMed]
35. Shi, H.; Jiang, C.; Ye, T.; Tan, D.X.; Reiter, R.J.; Zhang, H.; Liu, R.; Chan, Z. Comparative physiological, metabolomic, and transcriptomic analyses reveal mechanisms of improved abiotic stress resistance in bermudagrass [*Cynodon dactylon* (L). Pers.] by exogenous melatonin. *J. Exp. Bot.* **2015**, *66*, 681–694. [CrossRef] [PubMed]
36. Broekgaarden, C.; Pelgrom, K.T.B.; Bucher, J.; van Dam, N.M.; Grosser, K.; Pieterse, C.M.J.; van Kaauwen, M.; Steenhuis, G.; Voorrips, R.E.; de Vos, M.; et al. Combining QTL mapping with transcriptome and metabolome profiling reveals a possible role for ABA signaling in resistance against the cabbage whitefly in cabbage. *PLoS ONE* **2018**, *13*, e0206103. [CrossRef] [PubMed]
37. Wang, J.; Sun, L.; Xie, L.; He, Y.; Luo, T.; Sheng, L.; Luo, Y.; Zeng, Y.; Xu, J.; Deng, X.; et al. Regulation of cuticle formation during fruit development and ripening in 'Newhall' navel orange (*Citrus sinensis* Osbeck) revealed by transcriptomic and metabolomic profiling. *Plant Sci.* **2016**, *243*, 131–144. [CrossRef]
38. Smith, S.E.; Read, D.J. *Mycorrhizal Symbiosis*, 3rd ed.; Academic Press: San Diego, CA, USA, 2008; pp. 419–506.
39. Harrison, M.J. Signaling in the arbuscular mycorrhizal symbiosis. *Annu. Rev. Microbiol.* **2005**, *59*, 19–42. [CrossRef]

40. Genre, A.; Chabaud, M.; Timmers, T.; Bonfante, P.; Barker, D.G. Arbuscular mycorrhizal fungi elicit a novel intracellular apparatus in *Medicago truncatula* root epidermal cells before infection. *Plant Cell* **2005**, *17*, 3489–3499. [CrossRef]
41. Adolfsson, L.; Nziengui, H.; Abreu, I.N.; Šimura, J.; Beebo, A.; Herdean, A.; Aboalizadeh, J.; Široká, J.; Moritz, T.; Novák, O.; et al. Enhanced secondary-and hormone metabolism in leaves of arbuscular mycorrhizal *Medicago truncatula*. *Plant Physiol.* **2017**, *175*, 392–411. [CrossRef]
42. Xie, Z.P.; Staehelin, C.; Vierheilig, H.; Wiemken, A.; Jabbouri, S.; Broughton, W.J.; Vogeli-Lange, R.; Boller, T. Rhizobial nodulation factors stimulate mycorrhizal colonization of nodulating and nonnodulating soybeans. *Plant Physiol.* **1995**, *108*, 1519–1525. [CrossRef]
43. Zhang, Y.; Li, Y.Y.; Guo, S.X. Effects of the mycorrhizal fungus *Ceratobasidium* sp. AR2 on growth and flavonoid accumulation in *Anoectochilus roxburghii*. *PeerJ* **2019**, *7*, e8346.
44. Schliemann, W.; Ammer, C.; Strack, D. Metabolite profiling of mycorrhizal roots of *Medicago truncatula*. *Phytochemistry* **2008**, *69*, 112–146. [CrossRef]
45. Zhou, Y.J.; Hong, W.J.; Huang, J.X.; Tang, G.D. Impact of inoculation with mycorrhizal fungi in vitro on growth and resistant enzymes of *Rhododendron moulmainense*. *Southwest China J. Argric. Sci.* **2017**, *30*, 2687–2692.
46. Xu, C.; Zhang, H.Y.; Liu, G.H.; Yang, H.T.; Xi, G.J. The effect and disease-resistant mechanism of mycorrhizal fungus on *Dendrobium officinale* Kimura et Migo. *J. West China Sci.* **2017**, *46*, 1–5.
47. Han, Y.Y.; Ming, F.; Wang, W. Molecular evolution and functional specialization of chalcone synthase superfamily from *Phalaenopsis* orchid. *Genetica* **2006**, *128*, 429–438. [CrossRef] [PubMed]
48. Harrison, M.J.; Dixon, R.A. Spatial patterns of expression of flavonoid/isoflavonoid pathway genes during interactions between roots of *Medicago truncatula* and the mycorrhizal fungus *Glomus versiforme*. *Plant J.* **1994**, *6*, 9–20. [CrossRef]
49. Xie, W.; Hao, Z.P.; Zhou, X.F.; Jiang, X.L.; Xu, L.J.; Wu, S.L.; Zhao, A.H.; Zhang, X.; Chen, B.D. Arbuscular mycorrhiza facilitates the accumulation of glycyrrhizin and liquiritin in *Glycyrrhiza uralensis* under drought stress. *Mycorrhiza* **2018**, *28*, 285–300. [CrossRef] [PubMed]
50. Chen, S.F.; Zhou, Y.Q.; Chen, Y.R.; Gu, J. Fastp: an ultra-fast all-in-one FASTQ preprocessor. *Bioinformatics* **2018**, *34*, i884–i890. [CrossRef] [PubMed]
51. Zhang, Y.Z.; Wei, K.; Li, H.L.; Wang, L.Y.; Li, R.; Pang, D.D.; Cheng, H. Identification of key genes involved in catechin metabolism in tea seedlings based on transcriptomic and HPLC analysis. *Plant Physiol. Biochem.* **2018**, *133*, 107–115. [CrossRef]
52. Park, S.; Kim, D.H.; Lee, J.Y.; Ha, S.H.; Lim, S.H. Comparative analysis of two flavonol synthases from different-colored onions provides insights into flavonoid biosynthesis. *J. Agric. Food Chem.* **2017**, *65*, 5287–5298. [CrossRef]
53. Livak, K.J.; Schmittgen, T.D. Analysis of relative gene expression data using real-time quantitative PCR and the $2^{-\Delta\Delta Ct}$ method. *Methods* **2001**, *25*, 402–408. [CrossRef] [PubMed]

13

Comparative Analysis and Expression Patterns of the *PLP_deC* Genes in *Dendrobium officinale*

Lei Zhang [1,2], Chunyan Jiao [1,2], Yunpeng Cao [3,4], Xi Cheng [1,2], Jian Wang [1,2], Qing Jin [1,2,*] and Yongping Cai [1,2,*]

1 School of Life Sciences, Anhui Agricultural University, Hefei 230036, China; zhanglei123@ahau.edu.cn (L.Z.); 15212426671@163.com (C.J.); chengxi90@ahau.edu.cn (X.C.); jian@ahau.edu.cn (J.W.)
2 Anhui Provincial Engineering Technology Reserach Center for Development & Utilization of Regional Characteristic Plants, Anhui Agricultural University, No. 130, Changjiang West Road, Hefei 230036, China
3 Key Laboratory of Cultivation and Protection for Non-Wood Forest Trees, Ministry of Education, Central South University of Forestry and Technology, Changsha 410004, China; xfcypeng@126.com
4 Key Lab of Non-wood Forest Products of State Forestry Administration, College of Forestry, Central South University of Forestry and Technology, Changsha 410004, China
* Correspondence: qingjin@ahau.edu.cn (Q.J.); ypcaiah@163.com (Y.C.)

Abstract: Studies have shown that the type II pyridoxal phosphate-dependent decarboxylase (*PLP_deC*) genes produce secondary metabolites and flavor volatiles in plants, and TDC (tryptophan decarboxylase), a member of the *PLP_deC* family, plays an important role in the biosynthesis of terpenoid indole alkaloids (TIAs). In this study, we identified eight *PLP_deC* genes in *Dendrobium officinale* (*D. officinale*) and six in *Phalaenopsis equestris* (*P. equestris*), and their structures, physicochemical properties, response elements, evolutionary relationships, and expression patterns were preliminarily predicted and analyzed. The results showed that *PLP_deC* genes play important roles in *D. officinale* and respond to different exogenous hormone treatments; additionally, the results support the selection of appropriate candidates for further functional characterization of *PLP_deC* genes in *D. officinale*.

Keywords: *Dendrobium officinale*; *PLP_deC*; bioinformatics; expression pattern analysis; evolution

1. Introduction

Dendrobium officinale Kimura et Migo (also known as *D. catenatum*) is a perennial herb that is commonly used as a valuable Chinese herbal medicine and has a long evolutionary history among orchids. *D. officinale* is rich in alkaloids [1,2], and its genome, transcriptome, and metabolome indicate that *D. officinale* may also contain terpenoid indole alkaloids (TIAs) [3–5]. The common precursor of TIAs is strictosidine, which is formed by the combination of tryptamine and secologanin [6,7]. Tryptophan decarboxylase (TDC), which catalyses the formation of tryptamine, belongs to the type II pyridoxal phosphate-dependent decarboxylase (PLP_deC) family [8]. To date, the actual roles of many PLP_deCs in plants are still unknown due to a lack of relevant protein sequences and information about the biochemical properties. In particular, the role of PLP_deC in the alkaloid synthesis pathway of *D. officinale* has not been reported.

Pyridoxal 5′-phosphate (PLP) is the active form of vitamin B6 and is used by a variety of enzymes in all organisms [8]. Previously, we classified all PLP-dependent enzymes into at least five structural groups based on their protein structures [8,9]. Among them, the type I group is the most common and contains aminotransferases, decarboxylases, and an enzyme that catalyzes α-, β-, or γ-eliminations. Type II encodes the enzymes involved in β-elimination reactions. Type III is primarily alanine-racemase specific, while type IV enzymes typically include D-alanine aminotransferases. Type V enzymes are the most diverse, including glycogen and starch phosphorylases. One important group of PLP-dependent enzymes belongs to the PLP_deC family, which includes aromatic-L-amino acid decarboxylase (AAD), glutamic acid decarboxylase (GAD), and histidine decarboxylase (HDC) [10]. The biological functions of plant and animal AADs are closely related to their corresponding substrate selectivity and catalytic reactions; thus, some AADs, such as tyrosine decarboxylase (TYDC) and tryptophan decarboxylase (TDC), are further annotated based on their principal substrates [8,11]. These enzymes catalyze the decarboxylation of aromatic L-amino acids and are primarily involved in the synthesis of secondary metabolites in plants [12,13].

Numerous data indicate that *PLP_deC* exhibits tissue-specific and inducible transcript accumulation during plant development. In addition, several roles of *PLP_deC* in plant development have been identified. For example, TDC is a key enzyme that links primary and secondary metabolism with high substrate specificity [14,15]. In addition, the transcript levels of *PLP_deC* genes are affected by abscisic acid (ABA), methyl jasmonate (MeJA), salicylic acid (SA), and abiotic stress [16,17].

In 2015, Chinese scientists announced that they had completed the genome sequence of the orchid *Phalaenopsis equestris* [18]. *D. officinale* and *P. equestris* (Schauer) Rchb.f. are epiphytes in the family Orchidaceae. The draft of the *D. officinale* genome sequence was reported recently [4,19]. To further understand the *PLP_deC* gene family in orchids, we identified 8 and 6 *PLP_deC* genes from the genomes of *D. officinale* and *P. equestris*, respectively, and analyzed their phylogenetic relationships, gene structures, cis-regulatory elements, tissue expression patterns, and expression profiles under MeJA, ABA, and SA treatments. Our results may not only improve the current understanding of the evolutionary expansion, sequence conservation, and functional differentiation of *PLP_deC* genes but also provide in-depth basic biological information for further studies of the evolution of these genes in Orchidaceae.

2. Results

2.1. *Sequence Analysis of* PLP_deC *Family Members*

HMMER software was used to identify candidate genes in the genomes of *D. officinale* and *P. equestris*. All candidate genes were submitted to Pfam and SMART for verifying the presence of the PLP_deC domain. The sequences without the conserved domain and the redundant sequences were removed. Finally, a total of 14 *PLP_deC*-family sequences were obtained in *D. officinale* and *P. equestris*. The basic information of each *PLP_deC* was listed in Table 1. We found that in *D. officinale*, the molecular weights ranged from 34.48 kDa (DoGAD4) to 79.29 kDa (DoAAD2), with an average molecular weight of 57.64 kDa, and the theoretical isoelectric points ranged from 5.27 (DoHDC1) to 7.52 (DoAAD1), with an average value of 6.12. In *P. equestris*, the molecular weights ranged from 53.44 kDa (PeHDC1) to 57.26 kDa (PeGAD3), with an average molecular weight of 55.53 kDa, while the theoretical isoelectric points ranged from 5.53 (PeGAD1) to 8.55 (PeGAD3), with an average of 6.52. These results showed that most genes have an acidic pI. We analyzed the subcellular localization of 8 and 6 PLP_deC-family protein sequences in *D. officinale* and *P. equestris* by Target P 1.1. and the results showed the presence of a signal peptide in DoAAD1 and DoAAD3, suggesting their cellular localization in extracellular with a probability of 0.735 and 0.561, The proteins of DoHDC1, PeAAD1, PeAAD2 and PeHDC1 are possibly located in other cellular compartments, e.g. chloroplast transit peptide and mitochondrial, whereas the location of the remaining eight proteins is unknown (Table 2).

Table 1. Sequence analysis of pyridoxal phosphate-dependent decarboxylase (*PLP_deC*) family members.

Gene Name	Gene ID	Location	pI	MW (kDa)	Pyridoxal_deC Domain
PeGAD1	PAXXG012610	scaffold3	5.53	55.66	√
PeAAD1	PAXXG064070	scaffold28	6.43	54.31	√
PeAAD2	PAXXG110580	scaffold67	6.68	55.69	√
PeGAD2	PAXXG162190	scaffold130	5.96	56.84	√
PeGAD3	PAXXG162200	scaffold130	8.55	57.26	√
PeHDC1	PAXXG233810	scaffold291	5.96	53.44	√
DoAAD2	*Dendrobium*_GLEAN_10136748	scaffold166	6.32	79.29	√
DoGAD3	*Dendrobium*_GLEAN_10126118	scaffold506	5.88	68.10	√
DoAAD3	*Dendrobium*_GLEAN_10085660	scaffold2906	5.83	62.7	√
DoGAD4	*Dendrobium*_GLEAN_10051783	scaffold6549	6.28	34.48	√
DoGAD2	*Dendrobium*_GLEAN_10051784	scaffold6549	6.2	51.98	√
DoAAD1	*Dendrobium*_GLEAN_10046548	scaffold7413	7.52	42.84	√
DoHDC1	*Dendrobium*_GLEAN_10045094	scaffold7529	5.67	73.82	√
DoGAD1	*Dendrobium*_GLEAN_10044649	scaffold7738	5.27	47.84	√

Note: √ shows conserved domain of Pyridoxal_deC.

Table 2. Subcellular localization of PLP_deC proteins.

Gene Name	Chloroplast Transit Peptide	Mitochondrial Targeting Peptide	Signal Peptide	Other	Location	Reliability Class
DoAAD1	0.040	0.026	0.735	0.346	S	4
DoAAD2	0.093	0.239	0.029	0.806	*	3
DoAAD3	0.039	0.109	0.561	0.525	S	5
DoGAD1	0.218	0.279	0.203	0.148	*	5
DoGAD2	0.135	0.068	0.180	0.727	*	3
DoGAD3	0.570	0.190	0.106	0.161	*	4
DoGAD4	0.015	0.261	0.229	0.061	*	5
DoHDC1	0.096	0.194	0.134	0.846	_	2
PeAAD1	0.038	0.201	0.091	0.893	_	2
PeAAD2	0.128	0.098	0.088	0.900	_	2
PeGAD1	0.324	0.135	0.110	0.383	*	5
PeGAD2	0.460	0.188	0.134	0.243	*	4
PeGAD3	0.283	0.595	0.056	0.072	*	4
PeHDC1	0.064	0.116	0.219	0.845	_	2

Note: Confidence levels range from 1 to 5, and 1 shows the highest predicted reliability with difference > 0.8; 2: 0.8 > difference > 0.6; 3: 0.6 > difference > 0.4; 4: 0.4 > difference > 0.2; 5: 0.2 > difference. Location: C, M, and S, represent chloroplast, mitochondrial and extracellular, respectively. _ other location, * unknown location.

2.2. Analysis of the Gene Structures, Conserved Motifs, and Phylogenetic Relationships of PLP_deC *Genes*

To clarify the evolutionary relationships among the *PLP_deC* genes, we compared the PLP_deC proteins from *Arabidopsis thaliana (A. thaliana)*, *Oryza sativa (O. sativa)*, *D. officinale* and *P. equestris*. We used the maximum likelihood (ML) method to construct a phylogenetic tree using IQ-TREE software. As shown in Figure 1, the 42 *PLP_deC* genes could be divided into three subfamilies: GAD, HDC, and aromatic-L-AAD. The *PLP_deC* genes in *D. officinale* and *P. equestris* were named according to their relative homology with *A. thaliana* and *O. sativa* genes. Among them, the GAD subfamily was the largest, with 18 *PLP_deC* genes, and the HDC subfamily was the smallest, with 6 members.

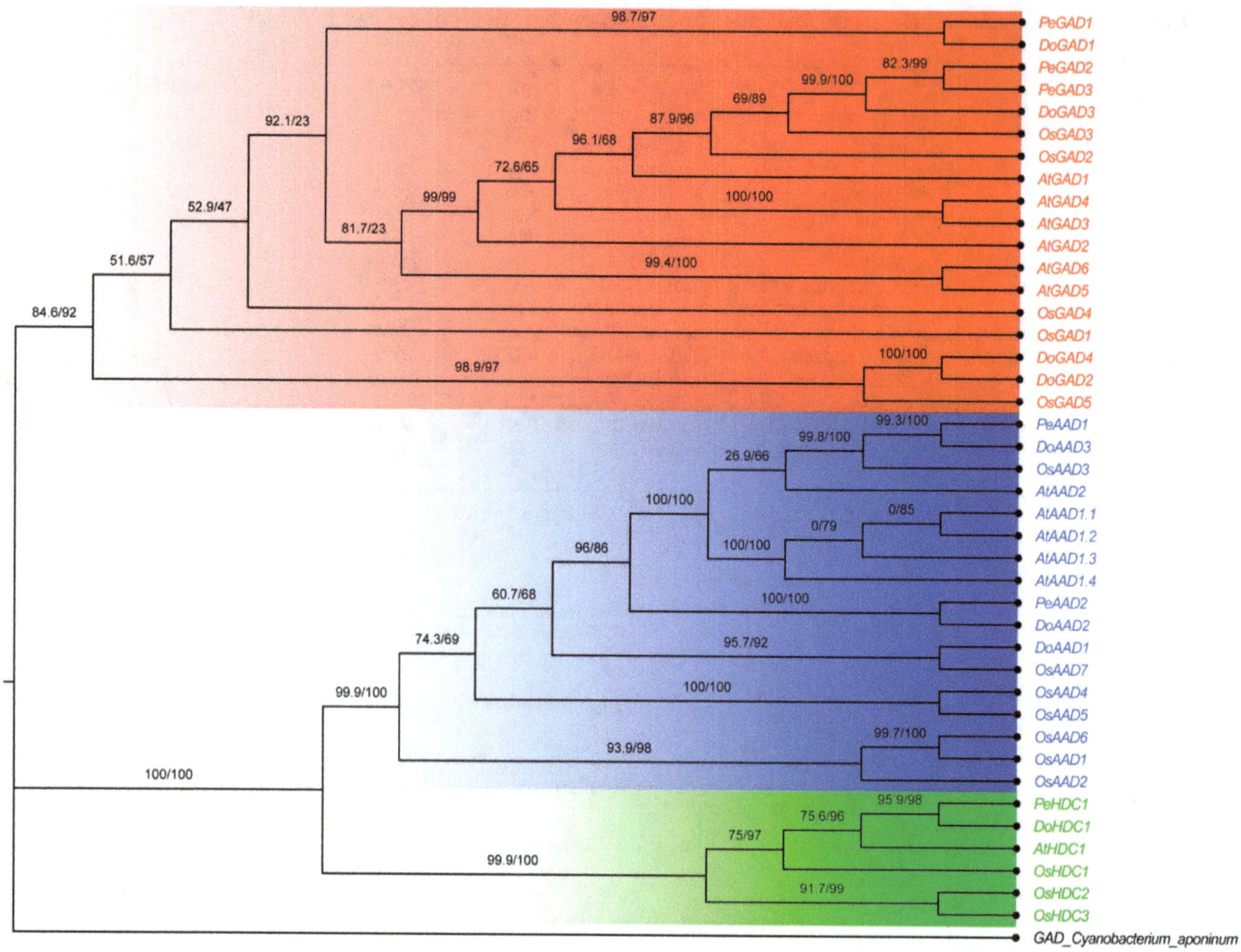

Figure 1. Phylogenetic analysis of *PLP_deC* genes from *Dendrobium officinale*, *Phalaenopsis equestris*, *Oryza sativa*, and *Arabidopsis thaliana*. The maximum likelihood (ML) tree was created using IQ-TREE with 8 *D. officinale* (Do), 6 *P. equestris* (Pe), 15 *O. sativa* (Os) and 12 *A. thaliana* (At) PLP_deC protein sequences. The red, green, and blue colors indicate GADs, HDCs, and AADs, respectively. Bootstrap supports are indicated at each branch.

To further analyze the gene structures and conserved motifs of the *PLP_deC* family members in *D. officinale* and *P. equestris*, a total of 10 motifs were identified from the amino acid sequences of the *PLP_deC* family members using the Multiple EM for Motif Elicitation (MEME) software (Figure 2). We checked these motifs to verify they are known domains by pfam and the motif logos were generated using online MEME program (Figure S1). Group I, which includes GAD sequences, harbors all the motifs; group II, which includes AAD sequences, harbors motifs 5, 6, 9, and 10; and group III, which includes HDC sequences, harbors motifs 4, 5, and 10. These results show that most *PLP_deC* genes in the same subfamily have highly similar motifs, which supports their close evolutionary relationships and the reliability of the constructed phylogenetic tree. Remarkably, motif 10 and motif 5 are present in all subfamilies, but they are not part of the PLP_deC domain. Thus, we speculate that they may perform other specific functions. In addition, we used the online Gene Structure Display Server to analyze the gene structures. The results showed that the number of exons in *PLP_deC* genes ranged from 2 to 14. For example, there are 3–8 exons in subfamily GAD, 2–14 exons in subfamily AAD, and 5 or 7 exons in subfamily HDC. These results indicate that the number of exons in the *PLP_deC* gene family has increased or decreased during evolution, providing a basis for functional differences among the homologous *PLP_deC* genes (Figure S2).

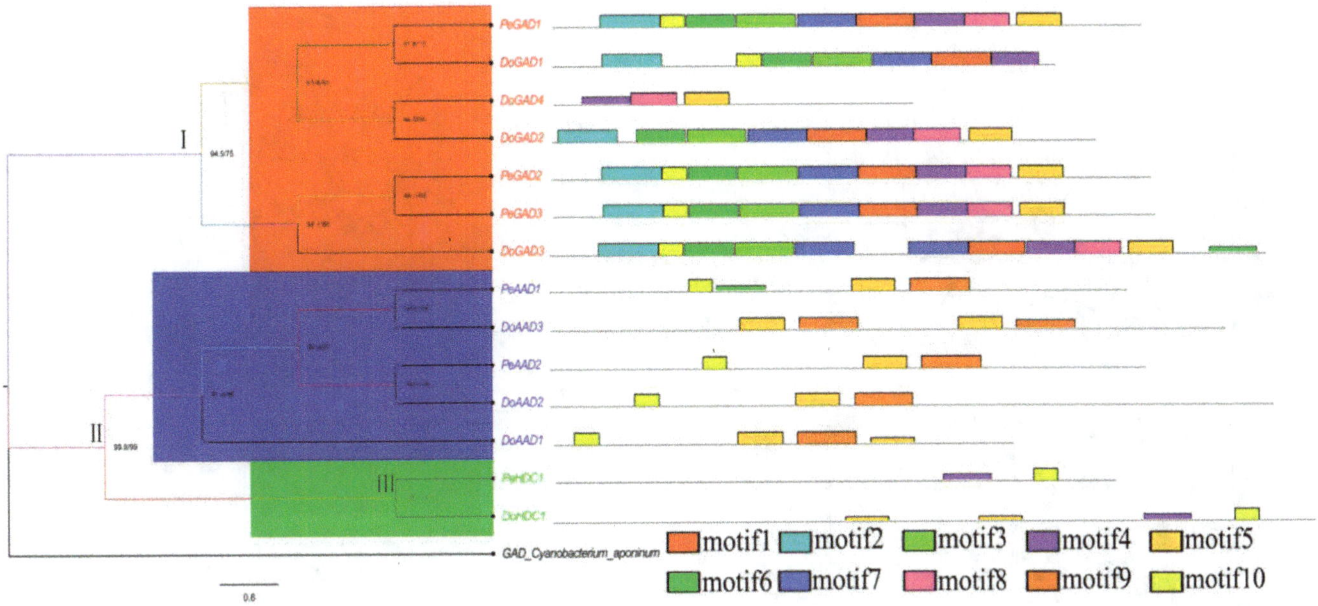

Figure 2. Distribution of 10 putative conserved motifs in *PLP_deC* proteins.

2.3. *Analysis of Evolutionary Patterns of* PLP_deC *Genes*

We analyzed the relationships of *D. officinale* and *P. equestris* homologous gene pairs to further analyze the evolutionary patterns of PLP_deC genes. We obtained six homologous gene pairs (*PeGAD1–DoGAD1, DoGAD2–DoGAD4, PeAAD1–DoAAD3, DoAAD2–PeAAD2, PeHDC1–DoHDC1,* and *PeGAD2–PeGAD3*) and calculated their Ka, Ks, and Ka/Ks values. The estimation of the synonymous (Ks) and nonsynonymous (Ka) nucleotide substitution rates is one of the important parameters for molecular evolutionary analyses, which are, respectively, defined as the number of synonymous substitutions per synonymous site and the number of nonsynonymous substitutions per nonsynonymous site per year or per generation. It is generally recognized that Ka/Ks > 1, Ka/Ks = 1, and Ka/Ks < 1 indicate positive, neutral, and purifying selection, respectively [20]. The experimental results showed that the Ka/Ks values of two homologous pairs (*DoAAD2–PeAAD2, PeHDC1–DoHDC1*) were lower than 0.3, the Ka/Ks values of the three homologous pairs (*PeAAD1–DoAAD3, DoGAD2–DoGAD4,* and *PeGAD1–DoGAD1*) were between 0.3 and 1, and the Ka/Ks value of one of the gene pairs (*PeGAD2–PeGAD3*) was greater than 1 (Table 3). These data indicated that most homologous *PLP_deC* gene pairs are subjected to purifying selection. To further understand the influence of selection on the homologous gene pairs, we performed a sliding window analysis. The grey regions shown in Figure 3 represent the conserved domains. The Ka/Ks values for the conserved domains were typically less than 1, indicating that these pairs have purifying selection. These findings suggest that purifying selection may have played a key role in the evolution of this gene family.

Table 3. The Ka, Ks, and Ka/Ks values of gene pairs.

Gene Pair		Ka	Ks	Ka/Ks
PeGAD1	*DoGAD1*	0.6314	0.8256	0.7645
DoGAD2	*DoGAD4*	0.194	0.37	0.5243
PeAAD1	*DoAAD3*	0.3143	0.5474	0.5742
DoAAD2	*PeAAD2*	0.1117	0.76	0.147
PeHDC1	*DoHDC1*	0.1642	0.5726	0.2868
PeGAD2	*PeGAD3*	0.2015	0.1384	1.4559

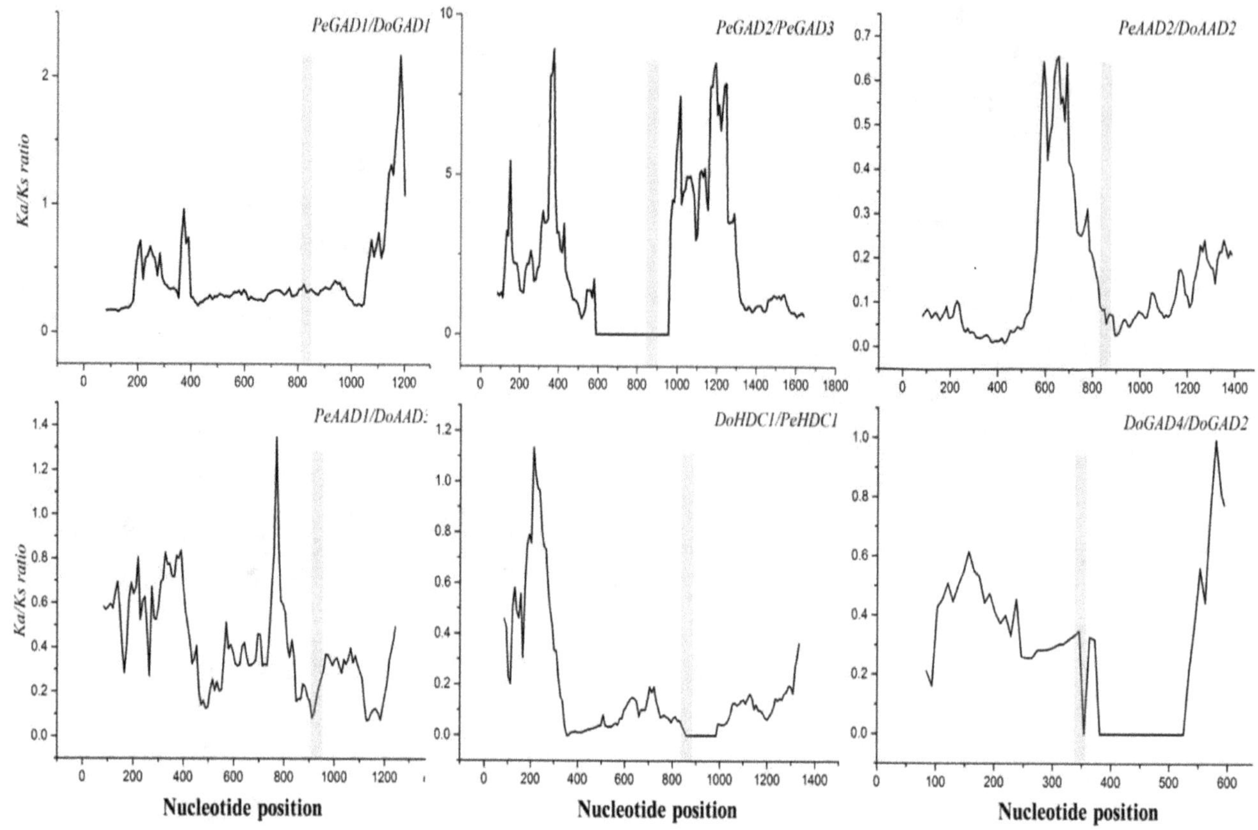

Figure 3. Sliding window analysis of Ka/Ks for each gene pair. The window size is 150 bp, and the step size is 9 bp. The grey region represents the conserved domain.

2.4. D. officinale PLP_deC *Gene Expression in Different Tissues*

To understand the gene expression pattern of the *PLP_deC* genes in *D. officinale*, we performed an overall in silico analysis of gene expression profiles in eight tissues (root, root tip, stem, leaf, sepal, column, lip, and flower bud). These *PLP_deC* genes exhibited distinct organ-specific expression and could be divided into three groups. As shown in Figure 4, in group A, two genes (*DoAAD1* and *DoAAD2*) were highly expressed in the flower bud and lip columns, indicating that these *PLP_deC* genes may be involved in the development of these tissues. In addition, *DoAAD2* had a high expression level in the sepal, indicating that *DoAAD2* may be involved in sepal development. The four *PLP_deC* genes in group B were generally expressed in low amounts in all eight tissues. Remarkably, a homologous gene pair (*DoGAD2–DoGAD4*) exhibited similar patterns of expression and had relatively low transcript abundance in different tissues. In group C, *DoAAD3* had low expression in all tissues. *DoGAD3* had higher expression levels in the column and sepal. Overall, the *PLP_deC* genes with high expression levels may be involved in tissue growth and differentiation in *D. officinale*.

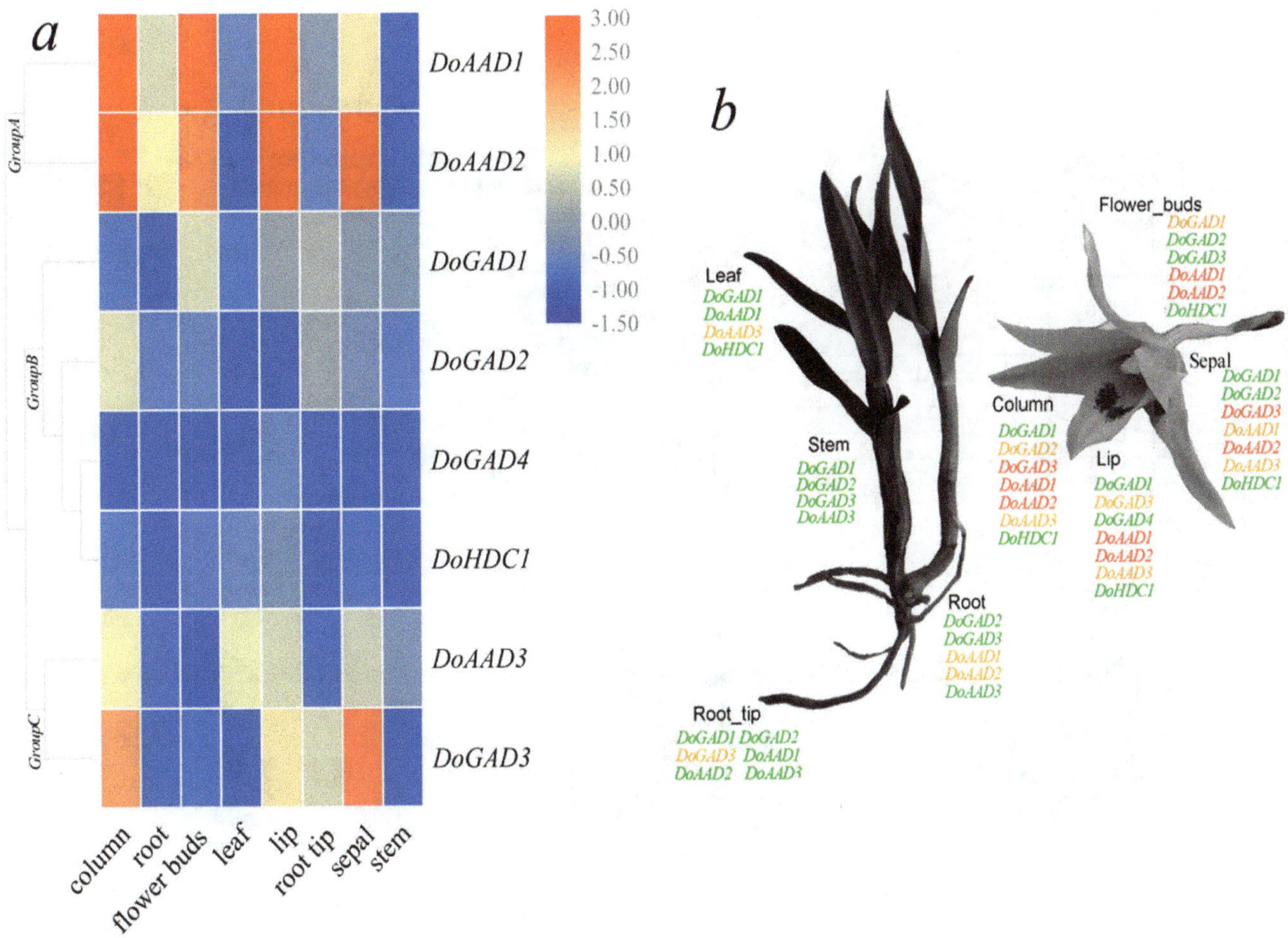

Figure 4. Expression profiles of *PLP_deC* genes in different tissues. (**a**) Heatmap of the in silico expression analysis in different tissues and organs. Blue and red indicate lower and higher transcript abundance, respectively. (**b**) *PLP_deC* genes expressed in different tissues and organs. Green, yellow, and red indicate low (0.01–0.71 FPKM), medium (1–1.71 FPKM), and high (3.42–28.4 FPKM) expression, respectively.

2.5. *Identification of Cis-Acting Elements of* PLP_deC *Genes*

Plant growth and development are regulated by different cis-elements in genes. Therefore, we used the PlantCARE database to identify and analyze cis-elements in the *PLP_deC* genes and identified three categories of cis-elements: plant growth and development, phytohormone response, and biotic and abiotic stress response. The cis-acting elements in the growth and development category included the CAT-box for meristem expression, O_2-site for zein metabolic regulation, MRE and Box 4 for light response, and others. We identified 35 Box-4 motifs, which compose the largest portion of the growth and development category. All the *PLP_deC* genes contained Box-4 elements (Figure 5a), indicating that the expression of all these genes is closely related to light. In the phytohormone response category, TCA elements for SA response and P-boxes and TATC-boxes for gibberellic acid response were identified. Notably, TGACG motifs for MeJA response accounted for 47% of the phytohormone response category (Figure 5c), while ABRE elements for ABA response accounted for 33%. The last category was biotic and abiotic-stress-response elements, including AREs and GC motifs for anerobic response, TC-rich repeats for defence and stress response, and MBSs and LTRs for low temperature response. Our data suggested that the *PLP_deC* genes may respond to abiotic stresses in *D. officinale*.

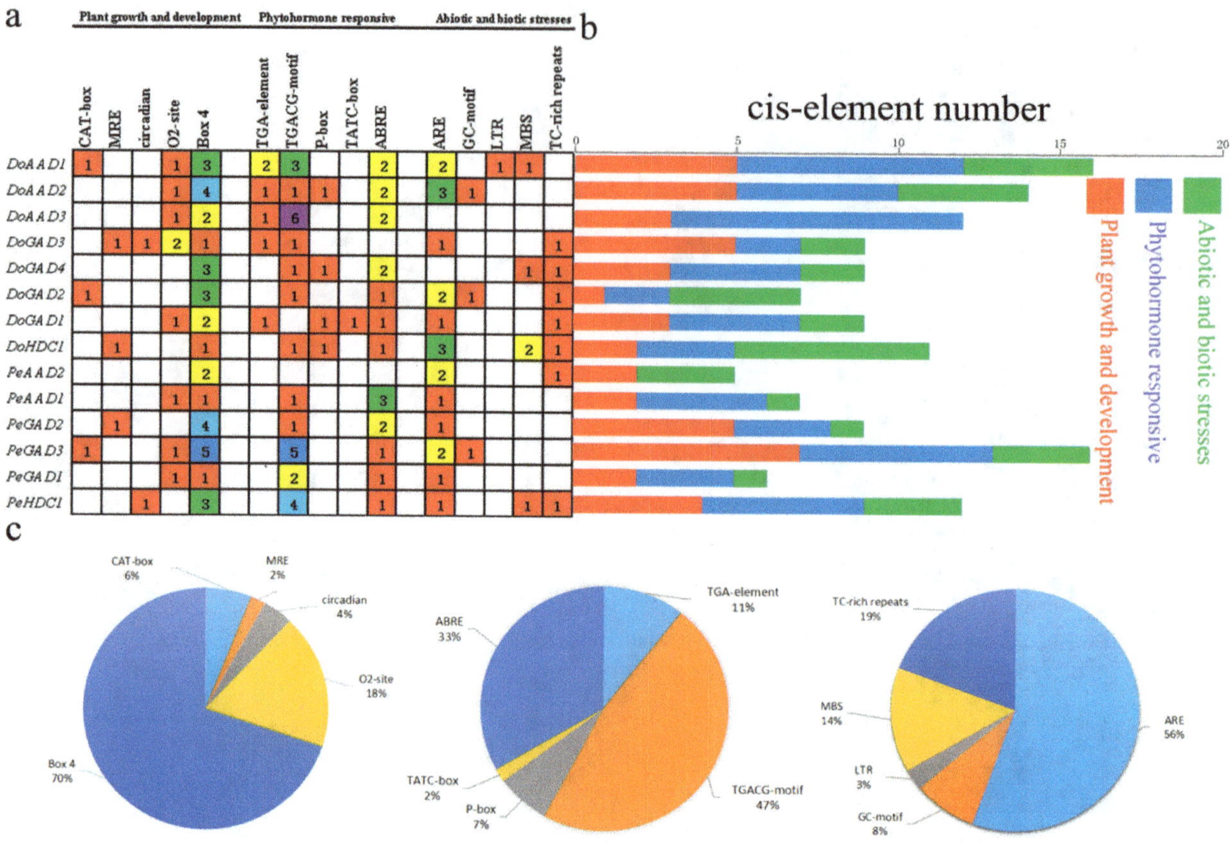

Figure 5. Numbers of cis-acting elements in all the *PLP_deC* genes of *D. officinale* and *P. equestris*. (**a**) The different colors and numbers in the grid indicate the numbers of different promoter elements in each *PLP_deC* gene. (**b**) The different colored histogram represents the numbers of cis-acting elements in the different categories. The red, blue, and green indicate plant growth and devolepment, phytohormone response, and biotic and abiotic stress, respectively. (**c**) The pie charts indicate the percentages of different promoter elements in the different categories.

2.6. *Analysis of the Expression Patterns of* PLP_deC *Genes in* D. officinale

In plants, many stress responses are modulated or mediated by various signaling pathways that are inseparable from gene expression and regulation [21]. To investigate the responses of the *PLP_deC* genes to different hormone treatments, we used qRT-PCR to analyze their expression under MeJA, ABA, and SA treatments.

In the ABA treatment, we found that *DoAAD1*, *DoAAD2*, *DoAAD3*, *DoGAD1*, *DoGAD3*, and *DoGAD4* reached their highest expression levels after 72 hours, and *DoGAD1* and *DoGAD3* were strongly upregulated (by more than 110-fold and 50-fold, respectively). The expression of two *PLP_deC* genes (*DoGAD2* and *DoHDC1*) peaked at 96 h, and *DoGAD2* was strongly upregulated (more than 250-fold) (Figure 6). In the MeJA treatment, three *PLP_deC* genes (*DoAAD3*, *DoGAD3*, and *DoHDC1*) showed strong upregulation at 2 h (Figure 7), while five PLP_deC genes (*DoAAD1*, *DoAAD2*, *DoGAD1*, *DoGAD2*, and *DoGAD4*) were strongly upregulated after 4 h of treatment. In the SA treatment, the expression levels of *DoAAD1*, *DoAAD3*, *DoGAD1*, *DoGAD2*, and *DoHDC1* were strongly upregulated at 2 h (Figure 8). However, *DoAAD2* and *DoGAD3* were strongly upregulated after treated with SA for 72 h of treatment (over 645- and 508-fold, respectively).

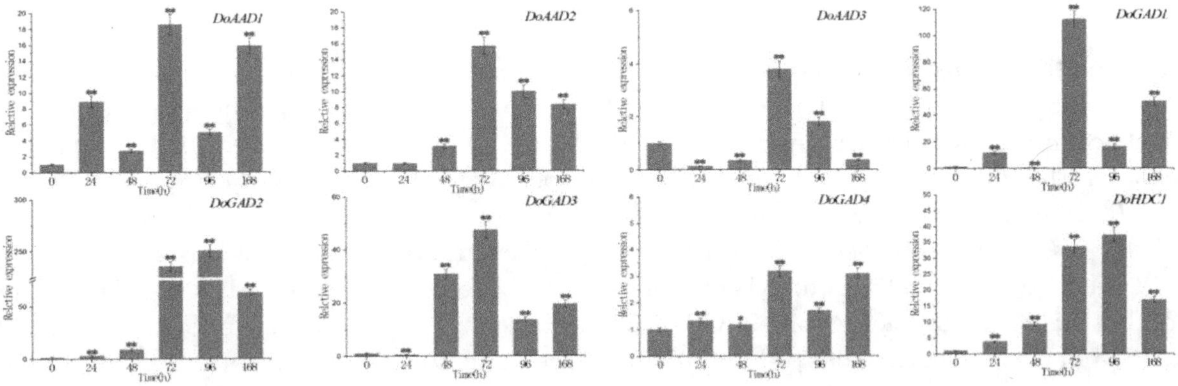

Figure 6. The expression levels of *PLP_deC* genes in *D. officinale* under abscisic acid (ABA) treatment. The x-axis represents the treatment time, and the y-axis represents the gene expression level. Error bars indicate the mean and standard deviation (SD). The asterisks indicate significant difference relative to the time 0. ** significant difference ($p < 0.01$), * significant difference ($p < 0.05$).

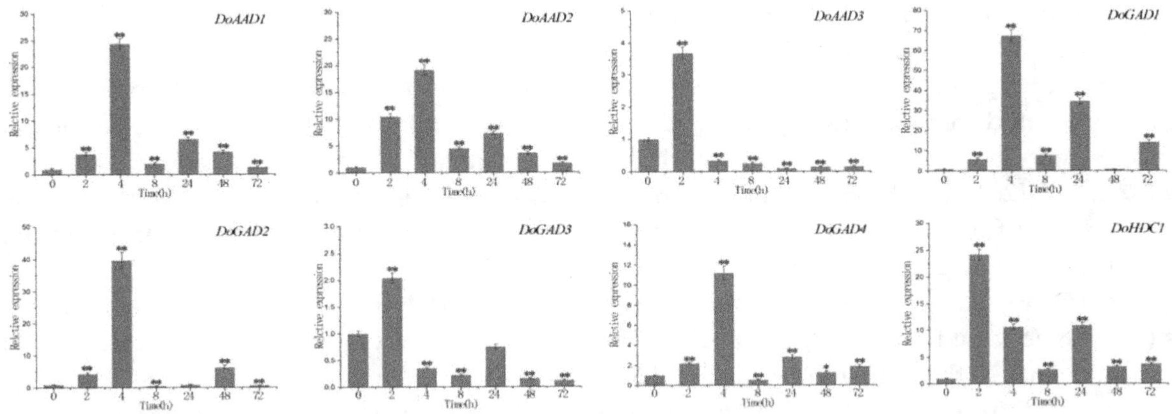

Figure 7. The expression level of *PLP_deC* genes in *D. officinale* under methyl jasmonate (MeJA) treatment stress. The x-axis represents the treatment time, and the y-axis represents the gene expression level. Error bars indicate the mean and standard deviation (SD). The asterisks indicate significant difference relative to the time 0. ** significant difference ($p < 0.01$), * significant difference ($p < 0.05$).

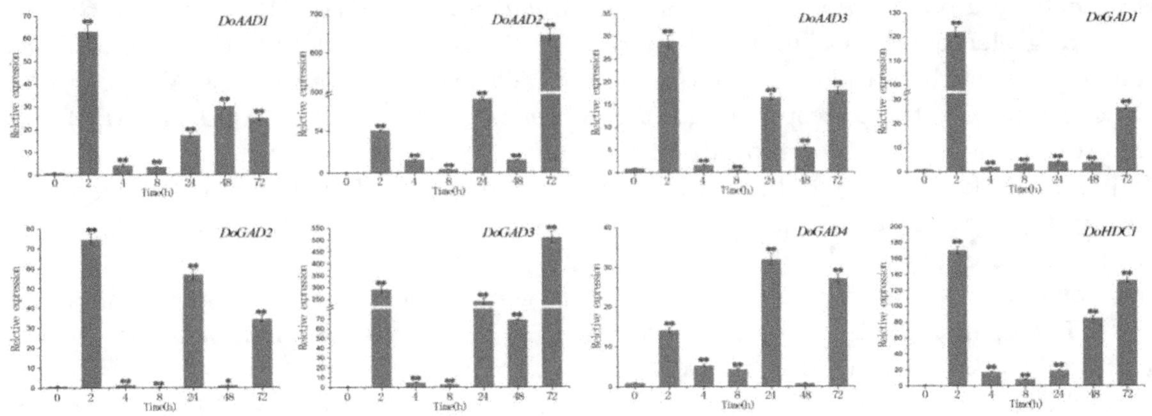

Figure 8. The expression levels of *PLP_deC* genes in *D. officinale* under salicylic acid (SA) treatment. The x-axis represents the treatment time, and the y-axis represents the gene expression level. Error bars indicate the mean and standard deviation (SD). The asterisks indicate significant difference relative to the time 0. ** significant difference ($p < 0.01$), * significant difference ($p < 0.05$).

3. Discussion

The type II PLP_deC enzymes are an important group of carboxylases among the PLP-dependent enzymes. Many data indicate that PLP_deCs show developmental, tissue-specific, and inducible transcript accumulation during plant development [19,22]. In this paper, we identified 8 and 6 *PLP_deC* genes from the *D. officinale* and *P. equestris* genomes, respectively. According to the phylogenetic analysis, all the *PLP_deC* genes from *A. thaliana*, *O. sativa*, *D. officinale*, and *P. equestris* were clustered into GAD, AAD, and HDC subclasses based on their high sequence similarity, which is consistent with the ML tree of *PLP_deC* genes from the genomes of 18 species and previously published articles [17]. However, some of these genes might have evolved with different functions. For example, in tomato, *SlHDC19* and *SlHDC6* do not act on histidine but prefer tyrosine as their substrate [17]. Furthermore, many HDCs are biased toward serine rather than histidine based on biochemical analysis [18]. Therefore, although their sequences have high similarity, *PLP_deC* genes have individual substrate specificities; we should perform an in-depth biochemical characterization to understand their precise functions [23].

Gene duplication is a common phenomenon in species and contributes to the generation of biodiversity during evolution [24]. To date, the chromosome assemblies of the *D. officinale* and *P. equestris* genomes have not yet been finished [25], and thus, the homologous genes of *D. officinale* and *P. equestris* cannot yet be clearly shown on the chromosomes. Therefore, we are unable to determine the type of replication events that have occurred between these species. To further understand the evolutionary patterns of the *PLP_deC* genes, we calculated the Ka and Ks values of homologous gene pairs. We predicted that two gene pairs (*PeGAD1–DoGAD1* and *PeAAD2–AAD2*) are evolved from the genome-wide duplication events shared by *D. officinale* and *P. equestris*, because their values of Ks are 0.7 to 1.1 [26]. The Ka/Ks values in this experiment were less than 1 for all the homologous gene pairs except for *PeGAD2–PeGAD3*, implying that these gene pairs have undergone purifying selection during evolution. In addition, we noticed four homologous gene pairs (*PeGAD1–DoGAD1*, *PeAAD2–AAD2*, *PeAAD1–DoAAD3*, and *PeHDC1–DoHDC1*) had the comparatively high Ka/Ks values (>0.5), showing that these gene pairs have undergone rapid evolutionary diversification after duplication events in the course of evolution [24].

Analysis of *D. officinale PLP_deC* gene expression in different tissues can help us better understand the tissue specificity of the *PLP_deC* genes. Therefore, expression profiles for all the *PLP_deC* genes were established using published RNA-sequence data. Among them, *DoAAD1*, *DoAAD2*, and *DoGAD3* were highly expressed in different tissues, indicating that these *PLP_deC* genes play important roles during *D. officinale* growth and development. For example, GADs are involved in many cellular processes, including pollen-tube development in *Arabidopsis* and *Picea wilsonii* [25,27]. In this study, some cis-acting elements associated with particular tissues were identified in the *PLP_deC* gene promoter regions, such as the O_2-site required for seed expression and the CAT-box required for meristem organization. The corresponding *PLP_deC* genes (such as *DoAAD1* and *DoAAD2*) might play important role in the formation of reproductive organs.

Many studies have suggested that the expression levels of *PLP_deC* genes are also influenced by abiotic and biotic stresses [28–30]. Furthermore, plant hormones such as ABA, SA, and ethylene also modulate the expression of these genes [17,29,30]. In this study, we identified a number of cis-acting elements in the promoter regions of *PLP_deC* genes in both *D. officinale* and *P. equestris*, such as MBS, MRB, Box 4, and ABRE. We found that these *PLP_deC* genes contain at least one abiotic stress cis-element, which showed that they may contribute to biotic and abiotic stress responses. To further investigate the responses of the *PLP_deC* genes to different hormones, we analyzed their expression with the treatments of MeJA, ABA, and SA by qRT-PCR. We observed that the *PLP_deC* genes had significantly differential expression patterns under different treatments. Some of the *PLP_deC* genes showed strong upregulation under the treatments, indicating that these genes play key roles in the abiotic stress responses of *D. officinale*. For example, *DoAAD2* was strongly upregulated (645-fold) after 72 h of SA treatment. Overall, we found that the *PLP_deC* genes of *D. officinale* responded to abiotic stress, such as MeJA, ABA, and SA stresses. These results provide strong evidence that the *PLP_deC* genes in plants are involved in abiotic stress responses.

4. Materials and Methods

4.1. Materials and Treatments

Seedlings of *D. officinale* were planted on Murashige and Skoog (MS) medium and placed in a tissue culture chamber at a constant temperature of 25 °C (16 h light/8 h dark) for 1 month. The tissue culture seedlings were then transferred to MS medium containing 1.0 mg/L 6-BA (biosharp, Shanghai, China), 0.1 mg/L NAA (Aladdin, Shanghai, China), and 30 g/L sucrose (Aladdin, Shanghai, China). The induced protocorms (PLBs) were supplemented with 1/2 MS liquid medium containing 0.1 mg/L α-naphthylacetic acid (NAA), 0.1 g/L whey protein hydrolysate, and 30 g/L sucrose (pH 5.8), and then cultured in darkness at 25 °C for 2 months. The PLBs were cut into 0.5 cm × 0.5 cm pellets, and 7 g of the pellet was inoculated into an Erlenmeyer flask containing 40 mL of MS medium. MeJA (100 μM methyl jasmonate; Aladdin, Shanghai, China), SA (100 μM salicylic acid; Aladdin, Shanghai, China), and ABA (100 μM abscisic acid; Aladdin, Shanghai, China) were filtered through 0.22 μm filter membrane and added to the MS medium, based on previously published articles [31]. After induction with SA and MeJA, samples were harvested at 0, 2, 4, 8, 24, 48, and 72 h. The original bulbs under ABA induction were harvested at 0, 24, 48, 72, 96, and 168 h. All samples were immediately stored at −80°C after harvesting for RNA extraction. All results were based on three biological repeats and each biological repeat had three technical replicates. The extraction of total RNA from PLBs was carried out with Plant Total RNA Isolation Kit (Sangon Biotech, Shanghai, China) using 300 mg tissue homogenized in liquid nitrogen according to the manufacturer's protocol, which was subsequently reverse transcribed into the first DNA strand using a One Step RT-qPCR Kit (BBI Life Science, Shanghai, China).

4.2. Screening and Identification of the PLP_deC Genes

We obtained the HMM (Hidden Markov Model) configuration file for the PLP_deC domain (Pfam00282) from Pfam (http://pfam.xfam.org/). All *PLP_deC* genes were then identified in the *D. officinale* and *P. equestris* genomes using HMMER software (*E*-value = 0.001) [32]. All candidate genes were submitted to Pfam and SMART for verifying the presence of the PLP_deC domain. The sequences that did not contain the conserved domain and the redundant sequence of the repeat were removed, and finally, the *PLP_deC* genes were obtained. In Table S1 are listed the sequences of the *PLP_deC* genes of *O. sativa* and *A. thaliana* described in previous studies (add citations) and used in the present work [17].

4.3. Sequence Attribute Analysis and Phylogenetic Tree Construction of PLP_deC Genes

To analyze the sequence attributes and characteristics of the amino acids of the *PLP_deC* family members, the isoelectric points (pIs) of the obtained PLP_deC amino acid sequences were determined using online analysis with the ProtParam tool (https://web.expasy.org/protparam/) [33] and subcellular localization of 8 and 6 PLP_deC-family protein sequences in *D. officinale* and *P. equestris* by Target P 1.1 (http://www.cbs.dtu.dk/ services / Target P). Properties such as molecular weight (MW) were predicted. The obtained PLP_deC protein sequences were aligned using Clustal X [34], implemented in MEGA 5.0 [35], and a maximum likelihood (ML) phylogenetic tree was generated using IQ-TREE software [36] with 1000 bootstrap replicates. The PLP_deC genes were classified according to the phylogenetic relationships. If two different species of genes are located in the phylogenetic tree at the same node and the sequence similarity is more than 80%, we consider two of these are homologous genes [37]. The conserved motifs on the orchid sequences of PLP_deC were defined by MEME (http://meme-suite.org/) using the following parameters: maximum number of motifs = 10, number of repetitions—any, and only motifs with an *E*-value < 0.01 were retained for further analysis. The motif logos of the PLP_deC domains were generated using online MEME program (Figure S2) [38], and GSDS was used to determine the exon–intron structure (http://gsds.cbi.pku.edu.cn /) [39].

4.4. *Calculation of Ks and Ka Values of the* PLP_deC Genes

The protein sequences of the gene pairs were first aligned using Clustal X 2.0, and then the multiple sequence alignments of proteins and the corresponding cDNA sequences were converted to codon alignments using PAL2NAL (http://www.bork.embl.de/pal2nal/) [40]. Finally, the resulting codon alignment was used to calculate Ks and Ka using DnaSP 5.0 (http://www.ub.edu/dnasp/) [41].

4.5. *Analysis of Cis-Acting Elements of the* PLP_deC *Genes*

A 2000 bp sequence upstream of the translation start site (ATG) of each *PLP_deC* gene was obtained, and an analysis of the upstream regulatory promoter elements was performed using the online tool PlantCARE (http://bioinformatics.psb.ugent.be/webtools/plantcare/html/) [42].

4.6. *Expression of the* D. officinale PLP_deC *Genes in Different Tissues*

To obtain expression data for *D. officinale*, we searched the NCBI SRA database (PRJNA348403) for RNA-sequence data from different tissues [27] (Additional file: Table S2). The raw data were stripped of adapters and low-quality reads (and bases) and then rRNA and virus reads were filtered out by using Trimmomatic software with the default parameters. The clean reads were aligned to the *D. officinale* genome using Hisat2 [43] with the options -dta and -no-unal. The aligned outputs were converted from SAM to BAM format by using SAMtools [44]. Then, the Stringtie software was used to estimate the transcript abundances with FPKM method. The heat map of the *PLP_deC* genes expression profiles was obtained by TBtools software (https://github.com/CJ-Chen/TBtools/releases).

4.7. *Quantitative Fluorescence Analysis of the* PLP_deC *Genes*

We used the CFX96 Touch^{TM} Real-Time PCR Detection System (Singapore) to perform a quantitative fluorescence analysis of the *PLP_deC* genes in the cDNA samples from *D. officinale* protocorms with three replicates. β-actin [45–47] was used as an internal reference, the relative expression levels of genes were calculated using the $2^{-\Delta\Delta CT}$ method [48], and the primers were designed using Primer Premier 5.0 software (Table S3). Each reaction contained the following: 8 μL of SYBR Premix Ex Taq II (2x), 2 μL of template cDNA, 1 μL of forward and reverse primers, and addition of ddH_2O to a final volume of 20 μL. The reaction conditions were as follows: 95 °C for 3 min followed by 40 cycles of 95 °C for 10 s, 52 °C for 15 s, and 72 °C for 30 s.

To determine whether the differential expression was significant, the difference in the relative expression of each target gene in the different treatment groups was analyzed by Student's *t*-test in SPSS 25.0 software [49].

5. Conclusions

Overall, we conducted a comprehensive analysis of *PLP_deC* genes in both *D. officinale* and *P. equestris*. Comparative analysis has shown that eight *PLP_deC* genes from *D. officinale* could be divided into three subfamilies: GAD, HDC, and AAD. Most genes have an acidic pI. Purifying selection may have played a key role in the evolution of this *PLP_deC* genes in *D. officinale* based on the Ka/Ks value. Among them, both *DoAAD1* and *DoAAD2* were highly expressed in the column, flower bud, and lip. Under three hormone treatments, MeJA, ABA, and SA, the *PLP_deC* genes responded to abiotic stresses. These results provide preliminary biological information for further studies of the evolution of *PLP_deC* genes in Orchidaceae.

Author Contributions: Y.C. (Yongping Cai) and Q.J. conceived and designed the experiments; L.Z., C.J., Y.C. (Yunpeng Cao), X.C., and J.W. performed the experiments; L.Z. and C.J. analyzed the data; L.Z. wrote the paper. All authors have read and agreed to the published version of the manuscript.

Acknowledgments: This work was supported by the Universities Natural Science Research Project of Anhui, China (Project number KJ2016A224), the University Synergy Innovation Program of Anhui Province (Project number GXXT-2019-043), and the Major Science and Technology Project in Anhui Province (Project number 17030701031).

References

1. Ng, T.B.; Liu, J.; Wong, J.H.; Ye, X.; Sze, S.C.W.; Tong, Y.; Zhang, K.Y. Review of research on Dendrobium, a prized folk medicine. *Appl. Microbiol. Biotechnol.* **2012**, *93*, 1795–1803. [CrossRef]
2. Zhang, J.; He, C.; Wu, K.; Teixeira da Silva, J.A.; Zeng, S.; Zhang, X.; Yu, Z.; Xia, H.; Duan, J. Transcriptome Analysis of Dendrobium officinale and its Application to the Identification of Genes Associated with Polysaccharide Synthesis. *Front. Plant Sci.* **2016**, *7*, 5. [CrossRef]
3. Guo, X.; Li, Y.; Li, C.; Luo, H.; Wang, L.; Qian, J.; Luo, X.; Xiang, L.; Song, J.; Sun, C.; et al. Analysis of the Dendrobium officinale transcriptome reveals putative alkaloid biosynthetic genes and genetic markers. *Gene* **2013**, *527*, 131–138. [CrossRef]
4. Yan, L.; Wang, X.; Liu, H.; Tian, Y.; Lian, J.; Yang, R.; Hao, S.; Wang, X.; Yang, S.; Li, Q.; et al. The Genome of Dendrobium officinale Illuminates the Biology of the Important Traditional Chinese Orchid Herb. *Mol. Plant* **2015**, *8*, 922–934. [CrossRef]
5. Shen, C.; Guo, H.; Chen, H.; Shi, Y.; Meng, Y.; Lu, J.; Feng, S.; Wang, H. Identification and analysis of genes associated with the synthesis of bioactive constituents in Dendrobium officinale using RNA-Seq. *Sci. Rep.* **2017**, *7*, 187. [CrossRef] [PubMed]
6. Zhu, X.; Zeng, X.; Sun, C.; Chen, S. Biosynthetic pathway of terpenoid indole alkaloids in Catharanthus roseus. *Front. Med.* **2014**, *8*, 285–293. [CrossRef] [PubMed]
7. Rischer, H.; Orešič, M.; Seppänen-Laakso, T.; Katajamaa, M.; Lammertyn, F.; Ardiles-Diaz, W.; Van Montagu, M.C.; Inzé, D.; Oksman-Caldentey, K.M.; Goossens, A. Gene-to-metabolite networks for terpenoid indole alkaloid biosynthesis in Catharanthus roseus cells. *Proc. Natl. Acad. Sci. USA* **2006**, *103*, 5614–5619. [CrossRef] [PubMed]
8. Torrens-Spence, M.P.; Liu, P.; Ding, H.; Harich, K.; Gillaspy, G.; Li, J. Biochemical Evaluation of the Decarboxylation and Decarboxylation-Deamination Activities of Plant Aromatic Amino Acid Decarboxylases. *J. Biol. Chem.* **2013**, *288*, 2376–2387. [CrossRef]
9. Milano, T.; Paiardini, A.; Grgurina, I.; Pascarella, S. Type I pyridoxal 5′-phosphate dependent enzymatic domains embedded within multimodular nonribosomal peptide synthetase and polyketide synthase assembly lines. *BMC Struct. Biol.* **2013**, *13*, 26. [CrossRef]
10. Percudani, R.; Peracchi, A. A genomic overview of pyridoxal-phosphate-dependent enzymes. *Embo Rep.* **2003**, *4*, 850–854. [CrossRef]
11. Facchini, P.J.; Huber-Allanach, K.L.; Tari, L.W. ChemInform Abstract: Plant Aromatic L-AminoAcid Decarboxylases: Evolution, Biochemistry, Regulation, and Metabolic Engineering Applications. *Phytochemistry* **2010**, *31*, 121–138. [CrossRef]
12. Lehmann, T.; Pollmann, S. Gene expression and characterization of a stress-induced tyrosine decarboxylase from Arabidopsis thaliana. *FEBS Lett.* **2009**, *583*, 1895–1900. [CrossRef] [PubMed]
13. Noé, W.; Mollenschott, C.; Berlin, J. Tryptophan decarboxylase from Catharanthus roseus cell suspension cultures: Purification, molecular and kinetic data of the homogenous protein. *Plant Mol. Biol.* **1984**, *3*, 281–288. [CrossRef] [PubMed]
14. Jadaun, J.S.; Sangwan, N.S.; Narnoliya, L.K.; Tripathi, S.; Sangwan, R.S. Withania coagulans tryptophan decarboxylase gene cloning, heterologous expression, and catalytic characteristics of the recombinant enzyme. *Protoplasma* **2016**, *254*, 181–192. [CrossRef] [PubMed]

15. Goddijn, O.J.; Lohman, F.P.; de Kam, R.J.; Hoge, J.H.C. Nucleotide sequence of the tryptophan decarboxylase gene of Catharanthus roseus, and expression of tdc-gus a gene fusions in Nicotiana tabacum. *Mol. Gen. Genet. MGG* **1994**, *242*, 217–225. [CrossRef]
16. De Luca, V.; Fernandez, J.A.; Campbell, D.; Kurz, W.G. Developmental Regulation of Enzymes of Indole Alkaloid Biosynthesis in Catharanthus roseus. *Plant Physiol.* **1988**, *86*, 447–450. [CrossRef]
17. Kumar, R.; Jiwani, G.; Pareek, A.; Sravan Kumar, T.; Khurana, A.; Sharma, A.K. Evolutionary Profiling of Group II Pyridoxal-Phosphate-Dependent Decarboxylases Suggests Expansion and Functional Diversification of Histidine Decarboxylases in Tomato. *Plant Genome* **2016**, *9*. [CrossRef]
18. Cai, J.; Liu, X.; Vanneste, K.; Proost, S.; Tsai, W.C.; Liu, K.W.; Chen, L.J.; He, Y.; Xu, Q.; Bian, C.; et al. The genome sequence of the orchid Phalaenopsis equestris. *Nat. Genet.* **2015**, *47*, 65–72. [CrossRef]
19. Zhang, G.-Q.; Xu, Q.; Bian, C.; Tsai, W.-C.; Yeh, C.-M.; Liu, K.-W.; Yoshida, K.; Zhang, L.-S.; Chang, S.-B.; Chen, F. The Dendrobium catenatum Lindl. Genome sequence provides insights into polysaccharide synthase, floral development and adaptive evolution. *Sci. Rep.* **2016**, *6*, 19029. [CrossRef]
20. Wang, D.; Zhang, S.; He, F.; Zhu, J.; Hu, S.; Yu, J. How Do Variable Substitution Rates Influence Ka and Ks Calculations? *Genom. Proteom. Bioinform.* **2009**, *7*, 42–53. [CrossRef]
21. Akihiro, T.; Koike, S.; Tani, R.; Tominaga, T.; Watanabe, S.; Iijima, Y.; Aoki, K.; Shibata, D.; Ashihara, H.; Matsukura, C.; et al. Biochemical Mechanism on GABA Accumulation During Fruit Development in Tomato. *Plant Cell Physiol.* **2008**, *49*, 1378–1389. [CrossRef] [PubMed]
22. Gutensohn, M.; Klempien, A.; Kaminaga, Y.; Nagegowda, D.A.; Negre-Zakharov, F.; Huh, J.H.; Luo, H.; Weizbauer, R.; Mengiste, T.; Tholl, D.; et al. Role of aromatic aldehyde synthase in wounding/herbivory response and flower scent production in different Arabidopsis ecotypes. *Plant J.* **2011**, *66*, 591–602. [CrossRef] [PubMed]
23. Moore, R.C.; Purugganan, M.D. The early stages of duplicate gene evolution. *Proc. Natl. Acad. Sci. USA* **2003**, *100*, 15682–15687. [CrossRef] [PubMed]
24. Cao, Y.; Meng, D.; Han, Y.; Chen, T.; Jiao, C.; Chen, Y.; Jin, Q.; Cai, Y. Comparative analysis of B-BOX genes and their expression pattern analysis under various treatments in Dendrobium officinale. *BMC Plant Biol.* **2019**, *19*, 245. [CrossRef]
25. Palanivelu, R.; Brass, L.; Edlund, A.F.; Preuss, D. Pollen Tube Growth and Guidance Is Regulated by POP2, an Arabidopsis Gene that Controls GABA Levels. *Cell* **2003**, *114*, 47–59. [CrossRef]
26. Zhang, G.-Q.; Liu, K.-W.; Li, Z.; Lohaus, R.; Hsiao, Y.-Y.; Niu, S.-C.; Wang, J.-Y.; Lin, Y.-C.; Xu, Q.; Chen, L.-J. The Apostasia genome and the evolution of orchids. *Nature* **2017**, *549*, 379. [CrossRef]
27. Ling, Y.; Chen, T.; Jing, Y.; Fan, L.; Wan, Y.; Lin, J. γ-Aminobutyric acid (GABA) homeostasis regulates pollen germination and polarized growth inPicea wilsonii. *Planta* **2013**, *238*, 831–843. [CrossRef]
28. Walther, D.; Brunnemann, R.; Selbig, J. The Regulatory Code for Transcriptional Response Diversity and Its Relation to Genome Structural Properties in A. thaliana. *PLoS Genet.* **2007**, *3*, e11. [CrossRef]
29. Pan, Y.J.; Lin, Y.C.; Yu, B.F.; Zu, Y.G.; Yu, F.; Tang, Z.H. Transcriptomics comparison reveals the diversity of ethylene and methyl-jasmonate in roles of TIA metabolism in Catharanthus roseus. *BMC Genom.* **2018**, *19*, 508. [CrossRef]
30. Shen, E.M.; Singh, S.K.; Ghosh, J.S.; Patra, B.; Paul, P.; Yuan, L.; Pattanaik, S. The miRNAome of Catharanthus roseus: Identification, expression analysis, and potential roles of microRNAs in regulation of terpenoid indole alkaloid biosynthesis. *Sci. Rep.* **2017**, *7*, 43027. [CrossRef]
31. Wang, Y.; Hammes, F.; Duggelin, M.; Egli, T. Influence of size, shape, and flexibility on bacterial passage through micropore membrane filters. *Environ. Sci. Technol.* **2008**, *42*, 6749–6754. [CrossRef] [PubMed]
32. Mistry, J.; Finn, R.D.; Eddy, S.R.; Bateman, A.; Punta, M. Challenges in homology search: HMMER3 and convergent evolution of coiled-coil regions. *Nucleic Acids Res.* **2013**, *41*, e121. [CrossRef]
33. Gasteiger, E.; Hoogland, C.; Gattiker, A.; Duvaud, S.E.; Wilkins, M.R.; Appel, R.D.; Bairoch, A. Protein identification and analysis tools on the ExPASy server. *Proteom. Protoc. Handb.* **2005**, *112*, 571–607.
34. Thompson, J.D.; Gibson, T.J.; Higgins, D.G. Multiple Sequence Alignment Using ClustalW and ClustalX. *Curr. Protoc. Bioinform.* **2003**, *1*, 2–3. [CrossRef] [PubMed]
35. Tamura, K.; Peterson, D.; Peterson, N.; Stecher, G.; Nei, M.; Kumar, S. MEGA5: Molecular evolutionary genetics analysis using maximum likelihood, evolutionary distance, and maximum parsimony methods. *Mol. Biol. Evol.* **2011**, *28*, 2731–2739. [CrossRef]

36. Cao, Y.; Liu, W.; Zhao, Q.; Long, H.; Li, Z.; Liu, M.; Zhou, X.; Zhang, L. Integrative analysis reveals evolutionary patterns and potential functions of SWEET transporters in Euphorbiaceae. *Int. J. Biol. Macromol.* **2019**, *139*, 1–11. [CrossRef] [PubMed]
37. van der Heijden, R.T.; Snel, B.; van Noort, V.; Huynen, M.A. Orthology prediction at scalable resolution by phylogenetic tree analysis. *BMC Bioinform.* **2007**, *8*, 83. [CrossRef] [PubMed]
38. Bailey, T.L.; Johnson, J.; Grant, C.E.; Noble, W.S. The MEME suite. *Nucleic Acids Res.* **2015**, *43*, W39–W49. [CrossRef]
39. Hu, B.; Jin, J.; Guo, Y.A.; Zhang, H.; Luo, J.; Gao, G. GSDS 2.0: An upgraded gene feature visualization server. *Bioinformatics* **2014**, *31*, 1296. [CrossRef]
40. Mikita, S.; David, T.; Peer, B. PAL2NAL: Robust conversion of protein sequence alignments into the corresponding codon alignments. *Nucleic Acids Res.* **2002**, *34*, W609–W612.
41. Rozas, J.; Rozas, R. DnaSP, DNA sequence polymorphism: An interactive program for estimating population genetics parameters from DNA sequence data. *Bioinformatics* **1995**, *11*, 621–625. [CrossRef] [PubMed]
42. Lescot, M. PlantCARE, a database of plant cis-acting regulatory elements and a portal to tools for in silico analysis of promoter sequences. *Nucleic Acids Res.* **2002**, *30*, 325–327. [CrossRef] [PubMed]
43. Kim, D.; Langmead, B.; Salzberg, S.L. HISAT: A fast spliced aligner with low memory requirements. *Nat. Methods* **2015**, *12*, 357–360. [CrossRef] [PubMed]
44. Li, H.; Handsaker, B.; Wysoker, A.; Fennell, T.; Ruan, J.; Homer, N.; Marth, G.; Abecasis, G.; Durbin, R.; 1000 genome project data processing subgroup. The sequence alignment/map (SAM) format and SAMtools. *Bioinformatics* **2009**, *25*, 2078–2079. [CrossRef] [PubMed]
45. Pertea, M.; Kim, D.; Pertea, G.M.; Leek, J.T.; Salzberg, S.L. Transcript-level expression analysis of RNA-seq experiments with HISAT, StringTie and Ballgown. *Nat. Protoc.* **2016**, *11*, 1650–1667. [CrossRef]
46. Wang, T.; Song, Z.; Meng, W.L.; Li, L.B. Identification, characterization, and expression of the SWEET gene family in Phalaenopsis equestris and Dendrobium officinale. *Biol. Plant.* **2017**, *62*, 24–32. [CrossRef]
47. Fan, H.; Wu, Q.; Wang, X.; Wu, L.; Cai, Y.; Lin, Y. Molecular cloning and expression of 1-deoxy-d-xylulose-5-phosphate synthase and 1-deoxy-d-xylulose-5-phosphate reductoisomerase inDendrobium officinale. *Plant Cell Tissue Organ Cult. (PCTOC)* **2016**, *125*, 381–385. [CrossRef]
48. Livak, K.J.; Schmittgen, T.D. Analysis of relative gene expression data using real-time quantitative PCR and the $2-\Delta\Delta CT$ Method. *Methods* **2001**, *25*, 402–408. [CrossRef]
49. Verma, J.P. *Data Analysis in Management with SPSS Software*; Springer: Berlin, Germany, 2013. [CrossRef]

Permissions

All chapters in this book were first published in MDPI; hereby published with permission under the Creative Commons Attribution License or equivalent. Every chapter published in this book has been scrutinized by our experts. Their significance has been extensively debated. The topics covered herein carry significant findings which will fuel the growth of the discipline. They may even be implemented as practical applications or may be referred to as a beginning point for another development.

The contributors of this book come from diverse backgrounds, making this book a truly international effort. This book will bring forth new frontiers with its revolutionizing research information and detailed analysis of the nascent developments around the world.

We would like to thank all the contributing authors for lending their expertise to make the book truly unique. They have played a crucial role in the development of this book. Without their invaluable contributions this book wouldn't have been possible. They have made vital efforts to compile up to date information on the varied aspects of this subject to make this book a valuable addition to the collection of many professionals and students.

This book was conceptualized with the vision of imparting up-to-date information and advanced data in this field. To ensure the same, a matchless editorial board was set up. Every individual on the board went through rigorous rounds of assessment to prove their worth. After which they invested a large part of their time researching and compiling the most relevant data for our readers.

The editorial board has been involved in producing this book since its inception. They have spent rigorous hours researching and exploring the diverse topics which have resulted in the successful publishing of this book. They have passed on their knowledge of decades through this book. To expedite this challenging task, the publisher supported the team at every step. A small team of assistant editors was also appointed to further simplify the editing procedure and attain best results for the readers.

Apart from the editorial board, the designing team has also invested a significant amount of their time in understanding the subject and creating the most relevant covers. They scrutinized every image to scout for the most suitable representation of the subject and create an appropriate cover for the book.

The publishing team has been an ardent support to the editorial, designing and production team. Their endless efforts to recruit the best for this project, has resulted in the accomplishment of this book. They are a veteran in the field of academics and their pool of knowledge is as vast as their experience in printing. Their expertise and guidance has proved useful at every step. Their uncompromising quality standards have made this book an exceptional effort. Their encouragement from time to time has been an inspiration for everyone.

The publisher and the editorial board hope that this book will prove to be a valuable piece of knowledge for researchers, students, practitioners and scholars across the globe.

List of Contributors

Tao Wang
State Key Laboratory of Forest Genetics and Tree Breeding, Key Laboratory of Silviculture of the State Forestry Administration, Research Institute of Forestry, Chinese Academy of Forestry, Beijing 100091, China
Beijing Botanical Garden, Beijing 100093, China

Zheng Song, Xiaojing Wang, Lijun Xu, Qiwu Sun and Lubin Li
State Key Laboratory of Forest Genetics and Tree Breeding, Key Laboratory of Silviculture of the State Forestry Administration, Research Institute of Forestry, Chinese Academy of Forestry, Beijing 100091, China

Yue Wang
Key Laboratory of Economic Plants and Biotechnology, Yunnan Key Laboratory for Wild Plant Resources, Kunming Institute of Botany, Chinese Academy of Sciences, Kunming, Yunnan 650201, China
Bio-Innovation Center of DR PLANT, Kunming Institute of Botany, Chinese Academy of Sciences, Kunming, Yunnan 650201, China

Aizhong Liu
Key Laboratory for Forest Resources Conservation and Utilization in the Southwest Mountains of China, Ministry of Education, Southwest Forestry University, Kunming, Yunnan 650224, China

Yingdan Yuan, Xinggang Tang, Jinchi Zhang and Jie Lin
Co-Innovation Center for Sustainable Forestry in Southern China, Nanjing Forestry University, Nanjing 210037, China
Jiangsu Province Key Laboratory of Soil and Water Conservation and Ecological Restoration, Nanjing Forestry University, Nanjing 210037, China

Bo Zhang
Co-Innovation Center for Sustainable Forestry in Southern China, Nanjing Forestry University, Nanjing 210037, China
Jiangsu Province Key Laboratory of Soil and Water Conservation and Ecological Restoration, Nanjing Forestry University, Nanjing 210037, China
Department of Environmental Science and Policy, University of California, Davis, CA 95616, USA

Surendra Sarsaiya
Key Laboratory of Basic Pharmacology and Joint International Research Laboratory of Ethnomedicine of Ministry of Education, Zunyi Medical University, Zunyi 563003, China
Bioresource Institute for Healthy Utilization, Zunyi Medical University, Zunyi 563003, China

Archana Jain and Jingshan Shi
Key Laboratory of Basic Pharmacology and Joint International Research Laboratory of Ethnomedicine of Ministry of Education, Zunyi Medical University, Zunyi 563003, China

Qi Jia, Fuxing Shu, Zhongwen Chen and Qinian Zhou
Bioresource Institute for Healthy Utilization, Zunyi Medical University, Zunyi 563003, China

Xiaokuan Fan
College of Biotechnology and Pharmaceutical Engineering, Nanjing Tech University, Nanjing 211800, China

Jishuang Chen
Key Laboratory of Basic Pharmacology and Joint International Research Laboratory of Ethnomedicine of Ministry of Education, Zunyi Medical University, Zunyi 563003, China
Bioresource Institute for Healthy Utilization, Zunyi Medical University, Zunyi 563003, China
College of Biotechnology and Pharmaceutical Engineering, Nanjing Tech University, Nanjing 211800, China

Fusheng Jiang and Bin Ding
College of Life Science, Zhejiang Chinese Medical University, Hangzhou 310053, China

Meiya Li
Academy of Chinese Medical Sciences, Zhejiang Chinese Medical University, Hangzhou 310053, China

Hongye Wang and Xiaobo Yu
State Key Laboratory of Proteomics, Beijing Proteome Research Center, National Center for Protein Sciences (PHOENIX Center, Beijing), Beijing Institute of Lifeomics, Beijing 102206, China

Chunchun Zhang and Guiyuan Lv
College of Pharmaceutical Science, Zhejiang Chinese Medical University, Hangzhou 310053, China

Zhishan Ding
College of Medical Technology, Zhejiang Chinese Medical University, Hangzhou 310053, China

Jie Gao, Rui Ren, Yonglu Wei, Jianpeng Jin, Sagheer Ahmad, Chuqiao Lu, Jieqiu Wu, Chuanyuan Zheng, Fengxi Yang and Genfa Zhu
Guangdong Key Laboratory of Ornamental Plant Germplasm Innovation and Utilization, Environmental Horticulture Research Institute, Guangdong Academy of Agricultural Sciences, Guangzhou 510640, China

Yiyi Zhang, Zhongwu Dai, Xiaoyu Dai, Wei Li, Mengxia Cao, Chengru Li, Xiaoqian Wu, Junwen Zhai, Zhongjian Liu and Shasha Wu
Key Laboratory of National Forestry and Grassland Administration for Orchid Conservation and Utilization, College of Landscape Architecture, Fujian Agriculture and Forestry University, Fuzhou 350002, China

Tinghong Zhou
Huanglong National Scenic Reserve, Songpan 623300, China

Wen-Chieh Tsai
Key Laboratory of National Forestry and Grassland Administration for Orchid Conservation and Utilization, College of Landscape Architecture, Fujian Agriculture and Forestry University, Fuzhou 350002, China
Institute of Tropical Plant Sciences and Microbiology, National Cheng Kung University, Tainan City 701, China

Anjin Zhao, Zheng Cui, Tingge Li, Huiqin Pei, Yuhui Sheng and Xueqing Li
Key Laboratory of Ministry of Education for Genetics and Germplasm Innovation of Tropical Special Trees and Ornamental Plants, Hainan Key Laboratory for Biology of Tropical Ornamental Plants Germplasm, College of Forestry, Hainan University, Haikou 570228, China

Ying Zhao, Yang Zhou and Xiqiang Song
Key Laboratory of Ministry of Education for Genetics and Germplasm Innovation of Tropical Special Trees and Ornamental Plants, Hainan Key Laboratory for Biology of Tropical Ornamental Plants Germplasm, College of Forestry, Hainan University, Haikou 570228, China
Research Center for Terrestrial Biodiversity of the South China Sea, College of Forestry, Hainan University, Haikou 570228, China

Wenjun Huang
Department of Development and Design, Hainan University, Haikou 570228, China

Ting Peng
Key Laboratory of Germplasm Innovation on Protection and Conservation of Mountain Plant Resources, Ministry of Education/College of Agriculture, Guizhou University, Guiyang 550025, China

Mummadireddy Ramya, Hye-Ryun An, Su-Young Lee and Pil-Man Park
Floriculture Research Division, National Institute of Horticultural and Herbal Science, RDA, Wanju-gun, Jellabuk-do 55365, Korea

Seonghoe Jang
World Vegetable Center Korea Office (WKO), Wanju-gun, Jellabuk-do 55365, Korea

Pue Hee Park
Floriculture Research Division, National Institute of Horticultural and Herbal Science, RDA, Wanju-gun, Jellabuk-do 55365, Korea
Department of Horticultural Science and Biotechnology, Seoul National University (SNU), Seoul 08826, Korea

Björn Bohman
School of Molecular Sciences, the University of Western Australia, Crawley, WA 6009, Australia
Department of Plant Protection Biology, Swedish University of Agricultural Sciences, 23053 Alnarp, Sweden

Alyssa M. Weinstein
Research School of Biology, the Australian National University, Acton, ACT 2601, Australia

Raimondas Mozuraitis
Department of Zoology, Stockholm University, 106 91, Stockholm, Sweden

Gavin R. Flematti
School of Molecular Sciences, the University of Western Australia, Crawley, WA 6009, Australia

Anna-Karin Borg-Karlson
Department of Chemical Engineering, Mid Sweden University, 85170 Sundsvall, Sweden

Jean Carlos Cardoso
Laboratory of Plant Physiology and Tissue Culture, Department of Biotechnology, Plant and Animal Production, Centro de Ciências Agrárias, Universidade Federal de São Carlos, Rodovia Anhanguera, km 174, CEP 13600-970 Araras, SP, Brazil

Cesar Augusto Zanello
Masterscience degree by Programa de Pós Graduação em Produção Vegetal e Bioprocessos Associados, Centro de Ciências Agrárias, Universidade Federal de São Carlos, CEP 13600-970 Araras, SP, Brazil

Jen-Tsung Chen
Department of Life Sciences, National University of Kaohsiung, Kaohsiung 811, Taiwan

Ying Zhang, Yuanyuan Li, Xiaomei Chen, Zhixia Meng and Shunxing Guo
Institute of Medicinal Plant Development, Chinese Academy of Medical Sciences & Peking Union Medical College, Beijing 100193, China

Lei Zhang, Chunyan Jiao, Xi Cheng, Jian Wang, Qing Jin and Yongping Cai
School of Life Sciences, Anhui Agricultural University, Hefei 230036, China
Anhui Provincial Engineering Technology Reserach Center for Development & Utilization of Regional Characteristic Plants, Anhui Agricultural University, No. 130, Changjiang West Road, Hefei 230036, China

Yunpeng Cao
Key Laboratory of Cultivation and Protection for Non-Wood Forest Trees, Ministry of Education, Central South University of Forestry and Technology, Changsha 410004, China
Key Lab of Non-wood Forest Products of State Forestry Administration, College of Forestry, Central South University of Forestry and Technology, Changsha 410004, China

Index

www.ingramcontent.com/pod-product-compliance
Lightning Source LLC
LaVergne TN
LVHW081020210926
848485LV00004B/908